Quantum Mechanics, Mathematics, Cognition and Action

Fundamental Theories of Physics

An International Book Series on The Fundamental Theories of Physics: Their Clarification, Development and Application

Volume 129

Quantum Mechanics, Mathematics, Cognition and Action

Proposals for a Formalized Epistemology

Edited by

Mioara Mugur-Schächter

Centre pour la Synthèse d'une Épistemologie Formalisée, Paris, France

and

Alwyn van der Merwe

University of Denver, Denver, Colorado, U.S.A.

KLUWER ACADEMIC PUBLISHERS

DORDRECHT / BOSTON / LONDON

A C.I.P. Catalogue record for this book is available from the Library of Congress.

ISBN 1-4020-1120-2

Published by Kluwer Academic Publishers,
P.O. Box 17, 3300 AA Dordrecht, The Netherlands.

Sold and distributed in North, Central and South America
by Kluwer Academic Publishers,
101 Philip Drive, Norwell, MA 02061, U.S.A.

In all other countries, sold and distributed
by Kluwer Academic Publishers,
P.O. Box 322, 3300 AH Dordrecht, The Netherlands.

Printed on acid-free paper

Printed in the Netherlands.

CONTENTS

PREFACE

The *Centre pour la Synthèse d'une Épistemologie Formalisée*, henceforth briefly named CeSEF, was founded in June 1994 by a small group of scientists working in various disciplines, with the definite aim to synthesize a "formalized epistemology" founded on the methods identifiable within the foremost modern scientific disciplines. Most of the founders were already authors of well-known works displaying a particular sensitivity to epistemological questions. But the aim that united us was new. This aim along with the peculiar choice of its verbal expression are thoroughly discussed in the Introduction.

In the present volume, we publish the first harvest of explorations and constructive proposals advanced in pursuit of our goal. The contributions are expressive also of the views of those who shared only our beginnings and then left us[1]; they equally reflect input from those who participated in our workshops but did not contribute to this volume.

We are indebted to the Association *Naturalia et Biologica* for having supported with a donation the publication of this volume.

The camera-ready form of this book we owe to the patient and meticulous labor of Ms. Jackie Gratrix. The superb job she has done is herewith gratefully acknowledged.

Mioara Mugur-Schächter and Alwyn van der Merwe

[1]Paul Bourgine and, quite specially, Bernard Walliser.

INTRODUCTION

The purpose of this book is to initiate a new discipline, namely *a formalized epistemological method* drawn both from the cognitive strategies practised in the main modern sciences and from general philosophical thinking. Our progress in this direction will be attempted by general discussions concerning the concept itself, by constructive attempts, and by informative-critical explorations. Our goal has been triggered by the following considerations.

Everywhere at the present frontiers of scientific thought one can watch how absolute assertions and absolute separations that formerly seemed unshakeable are fading away.

So, for instance, in logic and mathematics the belief in the possibility of an uninterrupted progression of unlimited purely formal developments, which dominated the beginning of the last century, has collapsed. It has become clear that any definite domain of exclusively formal action is confined, even if in principle it can always be extended, while the process of extension itself escapes formalization, as also, quite radically, the process of creation of a domain of formal operationality does.

For living systems, the definition of what is called the system raises nontrivial problems. Biologists have been led to introduce notions like "self-organization" and "organizational closure" in order to point the way in which a living system constantly re-constructs its own matter, forms, and functions by processes where the feedback upon the system, of its interactions with the environment, are as important as the characters of the system itself.

As soon as life is involved, the concept of cause resists any attempt to clearly distinguish it from the concept of aim. For living beings as well as for those meta-living beings called social organizations, the importance of pragmatic models conforming to aims located in the future but shaping the actions accomplished now in order to reach the aims, becomes decisive. The aims—tied to belief and anticipation—operate backwards upon the action that furthers the aims, whereas the action, while it develops, changes the aims. This entails a dynamic that depends upon its history and its context, and of which the characterization requires a cognitivistic and evolutionary approach.

The theory of (the communication of) information deals with the transmission of messages by making use of a probabilistic representation of a peculiar sort, according to which any received message unavoidably depends not exclusively on the message sent but also on the "channel" used in the process. Thus the message received is quite fundamentally dependent on the way in which it becomes perceptible to the receiver. As a consequence, the possiblity of reconstructing the sent message out of the received one has to be studied explicitly as a function of the modalities of transmission; and the conditions required for such a reconstruction are highly nontrivial.

The investigators of "chaos" have resolved a millenary confusion by elaborating abstract mathematical examples, on the one hand, and simulations, on the other hand, which prove that determinism does not entail predictability: Deterministic modelings, and the full recognition of the randomness of the facts such as they are directly perceived by us exist side by side in mutual independence. Thus the fictitious belief that a choice has to be made evaporates, and a world of new questions arises concerning a pertinent representation of the relations between perceptual randomness and deterministic models of physical processes.

In the approaches concerning the treatment of "complex" systems or processes, the "agents", their "environment" and "actions", and the feedbacks from these, constitute inextricably entwined hierarchies of matter, situations, conscious aims and behaviours, knowledges, social organizations, and devices. What is named how, what is treated how, becomes a matter of *method* much more than a matter of fact. The boundaries between categories with fixed inner content fade away, and *roles* take their place.

And so on. We could continue the list. Everywhere the contours of separations that seemed obvious, clear-cut and absolute become shaky and full of gaps. And these superficial symptoms make us feel that we are witnessing changes which, though superficially appearing to be unrelated, are connected beneath the level of the directly perceptible. We also feel that the implications of these changes go down very deep, that they touch and modify the slopes of the first layer of our conceptualization, the very place where the general structure of our modern way of thinking and speaking has been forged. But the nature of changes of this sort—precisely because they concern established manners of thinking and speaking—is very difficult to grasp by use of the established manner of thinking and speaking. So the existence of these changes is revealed by their effects long before we become able to discern and express their precise content.

The very existence of these changes as such, before any attempt to define their contents, already raises questions. The conceptualization by man, of what he calls "reality", is itself an element of "reality". Is it then not subjected to some *laws*, to some *in*variances? This should be the case in some sense; but in which sense exactly? What changes and what stays the same? How could one delve deep enough, and how should we proceed in order to be sure that we capture and fully seize the essence of the developing transmutation as well as the stable structure that meanwhile persists? Without permitting decades to pass while the process is accomplished implicitely by osmotic assimilation of random, disparate bits of knowledge and interactions among them, without generating any perceivable contour?

It would be of crucial utility to succeed. Only what is explicitly known acquires a definite form, perceptible from the "outside". And only once this happens does it become possible to then detach what has been formed, optimize it with respect to definite purposes, and shape it into a genuine instrument that can be deliberately employed and indefinitely improved.

At the beginning of the last century, the theory of special relativity reduced the structure of the concept of spacetime that underlies the descriptions of physical phenomena, in the sense that the fracture of a bone is reduced by a surgeon. And later, starting in 1924, quantum mechanics crafted conceptual-operational-formal channels that have enabled the human mind to apply itself directly to the unobservable and to construct concerning it observable predictions that are realized with impressive precision. Of course, these are arcane revolutions which so far have penetrated the thinking of only a very few people. Moreover, they are as yet unfinished revolutions. But some philosophers, helped by a small number of physicists, have generated a process of communication by which, osmotically, the essence of some views of modern physics has more or less infused many minds. The germs of new approaches that are developing in various areas of scientific investigation have sprouted in this modified earth, which has nourished their further growth.

I now make the following possibly surprising assertion, which I hold to be crucial:

> *Quantum mechanics, like a diver, can take us down to the level of the* `very first` *actions of our conceptualization of reality. And starting from there, it can induce an* `explicit` *understanding of certain fundamental features of the new scientific thinking.*

The following remarks can give a first idea of the content of this assertion. Our way of conceiving the "object", which is what we separate from the "rest" in order to enable us to definitely examine and reason about it, marks our whole way of thinking as well as all our actions. Now, intuitively, the word "object" is still quasi-unanimously felt to be essentially tied to invariance, material, morphological, and functional, and thus to what could be called an "intrinsic objectivity", independent of observation, pre-existing such as it is perceived. More or less implicitly, all of current language and the entire classical logical and probabilistic thinking are founded on this presupposition. But quantum mechanics opposes a direct, radical and definitive *veto* of this presupposition. If its cognitive strategy is fully decoded and conveniently generalized, the formalism of quantum mechanics acts like a strong magnifying lens under which the static contour of the classical concept of object dissolves into a complex *process* inextricably tied to human cognitive actions, most usually reflex actions, but often also deliberate ones; and, in any case, the result of this process is indelibly marked by *relativities* to all the cognitive actions involved. In essence this conclusion has been known well for a long time. But the specific way in which quantum mechanics conveys this old conclusion is new, and it amounts potentially to an overt seizure by *physics* of the basic metaphysical question of realism. Physics thereby merges with philosophy in a basic, massive way, and it injects into philosophy a stream of innovation that leads directly into epistemology :

> *Quantum mechanics has captured and represented—for the particular case of microstates and in an implicit, cryptic way, but for the first time in the history of human thought and directly in mathematical terms—certain* **universal** *features belonging to the* **very first stage** *of the processes by which man extracts chains of communicable knowledge from the physical reality in which he is immersed and of which he partakes.*

This is what the epistemological universality of quantum mechanics consists of. By no means does it consist, as is often asserted, of the fact that any material system is made of microsystems—which is a physical circumstance, not an epistemological one. The feeling of essentiality conveyed by the quantum mechanical formalism to those who can read it, does not stem from this physical circumstance; it stems exclusively from the universal character of the *peculiar cognitive situation* dealt with in quantum mechanics. And, while reflections of it are encrypted in the general features of the formalism considered as a whole, this cognitive situation marks also directly the specific formal features that are pointed toward by the expressions "quantum probabilities" and "quantum logic". These simply are not intelligible in terms of what is classically called probabilities and logic. This manifests strikingly that the *general* epistemological consequences of the quantum mechanical formalism, if elaborated, modify the structure of our classical representations of probabilities and of logic, the two most basic and worked out representations of domains of our everyday thinking and acting. Indeed, when the universal representation of the very first stage of our conceptualization processes, drawn by generalization from quantum mechanics, is injected into classical probabilities and classical logic, they undergo a sort of spectral decomposition; and this places into evidence that, far down beneath language, probabilistic and logical conceptualization merge into one unified probabilistic-logical structure. This circumstance entails deep conceptual clarifications as well as corresponding formal modifications. No other theory of a domain of reality, not even Einstein's relativity, has ever triggered an outflow of a comparable scope, so deep-set and so powerfully innovating.

This, however, though variously felt and much discussed and analyzed for more than 70 years, often with remarkable penetration, nevertheless is still very far from being fully known and understood. The general epistemological implications of quantum mechanics are still cryptic, even for most physicists and even for many who currently manipulate the formalism, often in a masterly manner. *A fortiori*, quantum mechanics is very superficially and feebly connected to the development of other new scientific approaches. This is a *huge* lacuna. It hinders a free, rapid, and maximal development of the revolution of the basic concept of object, implicitly started by quantum mechanics, but the pressure of which manifests itself also in biology, systemics, information theory, etc. Thus it also inhibits the perception and full elaboration of the consequences of this revolution upon logic and probabilities that guide our everyday thinking. Thereby it obstructs the now-*possible* radical progress in our knowledge of our manner of producing knowledge. Which furthermore delays a now-possible dramatic improvement of an explicit and deliberate domination of our epistemological behaviour, and thus also of our actions.

One of the main aims of this book is to fill this lacuna.

This aim joins with a still larger one, which stems from the postulate that *any* big theory of a domain of reality fixes in the concepts and the structures defined by it, certain essential features of the epistemological processes by which the human mind generates representations of what we call reality. But, as happens in the special case of quantum mechanics, these features tend always to remain more or less implicit in the descriptional substance that has incorporated them, which entails that their universal value remains unused. *A fortiori*, the different epistemological innovations that accompany different scientific approaches, in general remain unreferred to one another, which blocks the emergence of an integration.

So, for instance, the theory of information obviously involves a certain epistemological universality. Any "transmission of knowledge"—even if it is a natural, non-intentional process of just the *acquisition* of knowledge, or a scientifically normed process of measurement, i.e., of deliberately organized transmission of data from an object of study to the mind of an investigator, etc.—can be cast in the canonical mould of the theory of information, according to which there always exists a source of "information" that issues "messages", a "channel" for the transmission of information which can alter in various ways the messages sent through it, and a "receiver" that attempts to restore the original message out of the received one. This remarkable generality entails a tendency to apply the informational representation (initially conceived for the engineering of communication devices) to the most diverse domains, in biology, in the theory of physical measurements, in linguistics, and so on. It would therefore certainly be fruitful to explicate thoroughly the general epistemological presuppositions of the information-theoretical formalism and to confront them systematically with those involved in other approaches. The theory of quantum mechanical measurements clearly offers an opportunity for a particularly interesting confrontation. Indeed, this theory distills the essence of fundamental quantum mechanics *and* quite essentially addresses an informational problem. Nevertheless, the *formalism of the quantum mechanical measurement theory possesses certain formal features that are essentially* `different` *from those of the informational formalism.* It would by interesting to explore what facts, assumptions, and methodological choices underlie this unexpected difference. While it might produce a deeper understanding of the, so central, general concept of "information", such an investgation could perhaps furthermore lead to a reformulation of the theory of information in terms of Hilbert mathematics,[2] which probably would be a formulation much deeper, more precise and general than the present one. In turn, a re-expression in terms of Hilbert mathematics of the theorems from information theory (especially the second theorem of Shannon) could draw the famous question of hidden parameters into an organized and mathematical framework; additionally it should foster important clarifications concerning the concept of physical superposition as well as throw further light on the concept of "object".

Considerations of a similar nature could be advanced for several other modern disciplines, in particular for the various computational approaches, for molecular and genetic biology and, quite specially, the modern cognitivistic approaches.

But the preceding considerations suffice already to convey the following conclusion:

> What is lacking in order to improve our knowledge and control of the modes available for the generation and communication of knowledge, thoroughly and rapidly and with precision and detail, is a systematic research within the mutually isolated special languages belonging to all the major modern scientific disciplines, of the epistemological essence inherent in every one of them, and a systematic cross-referencing of the explicated results.

[2] I do not write Hilbert *vectors*, because evidently a principle of superposition permitting a pertinent use of vector spaces does *not* hold for any transmission of information

Part One

Preliminary Explorations: What? Why? How?

1

REMARKS ABOUT THE PROGRAM FOR A FORMALIZED EPISTEMOLOGY*

Francis Bailly

Laboratoire de Physique du Solide de Bellevue
CNRS, 1 Place Aristide Briand
92195 Meudon Cedex, France

The question of relationships between mathematical structures and language analysis in epistemology is considered briefly in the framework of a program for a formalized epistemology.

Key words: language, formalization, mathematical structures.

1. PRESENTATION

This short paper does not pretend to analyze either the full importance or the stakes of a formalized epistemology such as the one proposed by the CeSEF. We shall limit ourselves to pointing out a few tracks likely to prove interesting to follow and to show the long-range aim and relevance of such a project. My own position is determined, of course, both by personal attitudes about general commitments (in philosophical, ethical, political domains) and by a professional practice in research in physics, i.e., in a discipline where mathematics have proved to be both deeply explanatory and fruitful in building new concepts and producing counter-intuitive notions. It appears particularly that natural language and every day conceptualization remain unable to account directly for physical features and properties, while mathematical formalisms made them easily understood. On the basis of this experience it is tempting to look at what, in epistemology and reflections about scientific knowledge, can be defined and formalized in order to free this specific domain from the over-determinations of natural language and commonplace representations.

*See "Important Note" on p. xviii.

2. WHY A FORMALIZED EPISTEMOLOGY?

Mathematical structures have their own developments, mutations and mutual transformations. Strongly formalized physical theories remain "self-sufficient" insofar as they have to describe and explain the material world. Biological sciences are continuously forging their concepts and constructing their objects and seem on the way to explain living organisms and their intrinsic complexity. Even social sciences have developed to some extent their own meta-languages about the problematic reality they are in charge to treat. For its part, epistemology has succeeded in elaborating discursive and conceptual methods that enable it to characterize and analyze the specific ways of scientific knowledge. What needs, then, to be questioned are the interest and the possibility for a "formalized epistemology" to exist. However, being inspired by the earlier movement of axiomatization in mathematics and its consequences, we could retain at least three types of considerations in order to justify pursuing such a program:

(i) It provides the possibility, through the requirement of some formalization, to elicit many presuppositions and implicit postulates involved in scientific theories as well as in the epistemological analyses linked to them, thus helping to clarify the involved contents and procedures.

(ii) It makes it possible to bring into evidence the incompatibilities or even contradictions contained in certain analyzes, which are difficult to express through the pure discursive argumentation. It may thus be used as a tool for criticism of interpretations and representations.

(iii) Thanks to the internal generativity of the formalisms themselves, it might make possible the discovery of new ways of research, in the same manner as the mathematical modelization of the phenomena do. It may thus play a heuristic and fruitful role for analyzes. Beyond the opening up of these possibilities, it is tempting to formalize epistemology in a way that could lead to make more explicit and even to redefine the role and the use of language in a theory of scientific knowledge.[1] This point will be briefly considered in Sec. 3

3. NEITHER A COGNITIVE SCIENCE NOR A REDUCTIONIST PROJECT

Indeed, before discussing this last point, we have to stress the fact that despite the appearances and even if some similarities may be found, such a

[1] In the same spirit but in a different manner as that which has been attempted by some recent researches [1–3].

project is not equivalent to the development of a research program in cognitive sciences, nor, conversely, to an attempt for a renewed logical reduction, as did in their time the philosophers of the *Wiener Kreis*. Let us point out the differences between these two perspectives.

(i) On the one hand, a formalized epistemology cannot simply be a part of cognitive sciences to the extent that its investigation range does not identify with constructing a "scientific object" as has to be the case for cognition: as emphasized by G.-G. Granger [4], philosophical knowledge is a "knowledge without object," and epistemology in its philosophical version does not aim at constructing an object, but rather at elucidating the processes of such a constitution in sciences. To this assumption it could be objected that elaborating a formalism, as epistemological as it would be, determines *ipso facto* some objectivity as a correlate and a referent for this formalism if it appears to be adequate. Answering this objection requires the notion of "formal content" (as introduced by Granger [5] in his epistemological analysis of the mathematical science), extending its relevance according to two points of view: first, a formal epistemology might be considered as a formalized epistemology of such formal contents and, second, it might be considered as the research of the mutual articulations, in a given scientific theory, of the formal contents this theory produces. Thus, if a formalized epistemology leads to the rising of some "pseudo-object", the latter refers in fact to a mathematical universe of concept construction dealing with the interpretation and mutual coupling of the implied theoretical concepts. It follows that this "pseudo-object" remains determined less by the formalization of the epistemology itself, than by the scientific disciplines which have generated it.

(ii) On the other hand, it is known that the abstract logicism of the *Wiener Kreis*, has to do with an empiricism as regards phenomena. It leads to a quasi-ontological disjunction between two worlds: the one of logic and the one of phenomena (considered as sets of data). Conversely, a formalized epistemology would develop the aim to explore a unique world: the one of the "scientific object" as such [6], through the analysis of its effective construction in the discipline where it is produced. Formalizing this analysis would offer a double advantage: the first one, already mentioned, is to detect through their traces the cognitive operations making possible this process; the second one is to permit the formation of a new meta-language regarding simultaneously both these operations and the concepts they treat. With the hope to make more evident the conditions of possibility for such a construction of objectivity, reviving thus, but only to this extent, a transcendental approach [8, 9].

4. FORMALISMS AND LANGUAGE

This point leads us to a quite general feature, which seems to be linked with every formalization of knowledge. Indeed, as we stressed elsewhere [7], when we have to deal with more or less formalized sciences, natural language acquires two distinct functions, that formalism enables us to distinguish and separate: a referring function and a referred function. In its referring function, language provides the means to express and establish the axiomatic of the formalisms or the main theoretical principles underlying the discipline. Somehow, it governs the objectifying activity. In its referred function, the language uses more (technical) terms than (usual) words, more conceptual relations than signification. It appears as submitted to the proper determinations of the abstract structures it contributed earlier to construct. Until new scientific theorization leads to use this referred state of language in order to confer it a new referring function in view of new formalisms or new principles, more abstract or more general. And so on, from paradigm to paradigm, from themata to themata, from epistemological cut to epistemological cut. In this continually acting process, the formalism as such keeps the space open as well as the splitting - which remain fundamental for constructing an objectivity - between these two functions of the language, thus enabling the mediation between them. More and more assured and in evolution, thanks to the first, it modifies continually the second through its proper internal dynamics, as is well shown through the intrinsic generativity of mathematics in modelizations. Meanwhile the formalism contributes in generating the language through both the functions the latter has alternatively to fulfill and between which the former assures a ruled communication.

In the usual practice of epistemological analysis, these two functions are very feebly mutually individualized. Their relationships are deciphered in the light of the conceptual analysis of scientific theories themselves and the referring function is made use of in considering the new relations induced by the formalisms in the referred function, while this last one is made use of for putting into evidence the theoretical concepts involved in the formalisms.

In order to achieve this, epistemology calls for a philosophy of knowledge, at the same time that it uses the disciplinary language with its own concepts. Thus, to aim at a formalized epistemology amounts to aim at reiterating the proper device of sciences, on the interpretative and comprehensive level, and at renewing its power of explanation. It also raises hopes that such a reiteration would lead us to build some real formal hermeneutics (in the same sense that mathematics can be regarded as such a formal hermeneutics [10].)

5. CONCLUSION

Finally, and despite the fact that it is not the first time it has been tried to build an approach deserving being called a formalized epistemology (cf., e.g., the above references to the Leibnizian "universal characteristic", to the logical research of the *Wiener Kreis*, to the set-theoretical approaches of Stegmüller and Sneed), the program proposed by the CeSEF may appear as a new possible progress in understanding the elaboration and the status of scientific knowledge, as well as the role played by linguistic representations and by language itself in their interpretation. Of course, it is out of question to reduce the interpretative process to the benefit of some excessive "formalicism", or to reduce epistemology itself to a purely cognitive approach. But, according to the concern of Wittgenstein about philosophy, it could be seen as a concern of freeing epistemology from some spontaneous determinations induced by natural language, by taking into account scientific results: it could play the role of some "therapy" (following Wittgenstein's provocative terminology) for a reflection about science and its cultural appropriation, as well as a revival in the research for the deep cognitive invariant structures underlying science.

REFERENCES

1. J. D. Sneed, "Philosophical problems in the empirical science of science: A formal approach," *Erkenntnis* **10**, 115 (1976).
2. T. S. Kuhn, "Theory change as structure-change: Comments on the Sneed formalism," *Erkenntnis* **10**, 179 (1976).
3. W. Stegmüller, "A combined approach to the dynamics of theories. How to improve historical intepretations of theory change by applying set theoretical structures," in *The Structure and Development of Science*, G. Radnitsky and G. Anderson, eds. (Reidel, Dordrecht, 1979).
4. G.-G. Granger, *Pour la connaissance philosophique* (Odile Jacob, 1988).
5. G.-G. Granger, *Formes, operations, objets* (Vrin, 1994).
6. F. Bailly, "Sur le statut contemporain du concept d''objet-scientifique'," *Rev. Int. Syst.* 9 (4), 385 (1995).
7. F. Bailly and J. Petitot, "Les mathématiques, de la diversité a l'unification," *Encyclopedia Universalis, Symposium* (1990), p. 700.

8. J. Petitot, "Logique transcendantale: Synthétique a priori et herméneutique mathématique des objectivités," *Fundamenta Scientiae* **10**, 1 (1990).
9. J. Petitot, "Idéalités mathématiques et Réalité objective. Approche transcendantale," in *Hommage à Jean-Toussaint Desanti*, G. Granel, ed. (TER, 1991).
10. J.-M. Salanskis, *L'herméneutique formelle* (Editions du CNRS, 1999).

2

FORMALIZED EPISTEMOLOGY IN A PHILOSOPHICAL PERSPECTIVE

Hervé Barreau

23 rue Goethe
67000 Strasbourg, France

The need for a formalized epistemology is recognized by all scholars who think that the relativity of all sciences must not be referred to a social relativism. In XXth century, Husserl was the protagonist of such an epistemic philosophy. But this philosophy was more successful in social and human sciences than in natural sciences. In this latter domain, quantum mechanics obeys the requirements of a Kantien perspective in a more precise sense that was the case with Newtonian mechanics.

Key words: relativity, relativism, phenomenology, epistemology, quantum mechanics.

1. INTRODUCTION

To my mind, what motivates thinkers of diverse schools and tendencies to adhere to a project of a *formalized* epistemology is the fact that there is no other plausible alternative, even if each one among us conceives such a project in an original manner.

And why is there no other plausible alternative? Simply, I think, because, should we not take heed, the task of epistemology would be more and more taken up by authors who, although they have impeccable communication skills, possess a far less solid, not to say dubious grasp of scientific knowledge. Over the last thirty years we have witnessed the publication of works characterized by such loose accounts of the principles and results of the theories of mathematical physics, that the very essence of these is dissolved. The authors of such works distort and misrepresent the scientific discourse, on the basis of the misguided conception that scientific theories are a mere object of blind faith agreed upon amongst specialists, nothing more. In their mind, the remarkable relativizing *methods* by which modern physics succeeds to include into rigorous descriptional structures the subjective elements which, in a non-removable way, mark any piece of knowledge, are identified with a wholly arbitrary 'relativism'. The existence, also, of constraints stemming from a source distinct from human minds, is entirely overlooked.

On the other hand – besides a well known and widely-experienced need to gain some well-constructed insight into the results of the mathematical sciences – there emerges the new goal to furthermore extract useful *general* epistemological methods from such an insight.

Under these circumstances, an epistemology that aims to incorporate the essence of the methods of modern-day scientific theories into precise, formalized, but nevertheless intelligible, general representations of the processes of generation of knowledge seems to be the only genuinely acceptable perspective. This essence cannot be left unexploited. And the slope which, from the observational principles of relativity that found any intersubjective consensus concerning reproducible physical phenomena, leads to mere relativism, *must* somehow be suppressed.

This slope is a very slippery one indeed. Nietzsche glided down to its very bottom. He considered scientific adventure as only an avatar of the search for power, which needs to be closely kept under supervision because it is continually deluding itself and others into the belief that what is attained was truth. Heidegger, in his own way, went down the same slope.

Husserl, on the other hand, continuing Kant, developed a view that strongly opposed relativism.

In what follows, I would like to make some remarks concerning the developments that withstood relativism and the *limits* of these. For, beyond these limits lie the main conceptual rooms where a formalized epistemology could now build useful new contributions.

2. THE HERMENEUTIC ROLE OF PHENOMENOLOGY

Edmund Husserl fully understood what systematic belittlement of science relativism was to cause. The forms which it took in the thinking of his former assistant, Martin Heidegger, yielded already a striking illustration. In *The Crisis of European Sciences and Transcendental Philosophy* [1], which is his philosophical testament, Husserl wrote (pp. 21-22):

> "Today we are aware that the rationalism of the XVIII century – its way of wanting to guarantee the solidity and the proper behavior

> required for the European humanity – was a naïveté. Yet, together with this naive rationalism, which is even contradictory if followed up to its final consequences, are we obliged to abandon also the *authentic* significance of rationalism? And what about a serious explanation of such a naïveté and contradiction? Where is the rationality of this irrationalism that is being vaunted and towards which some want to compel us?"

In his manuscript, written after a lecture given in Prague in 1935 under a slightly different title, Husserl shows that the naïveté and the contradictory character of the rationalism of the period of Enlightenment consisted in its objectivism (or its positivism, as one might prefer to say), i.e., in the fact that it remained unaware of the subjective contributions to scientific objectivities that these objectivities were presented as purely *factual data* to be accepted as entirely 'real'. As early as the XVIII century, trends have been manifested toward *relativizing* scientific knowledge by taking into explicit account also the unavoidable subjective features that mark it, and in this respect Husserl pays homage to Kant who initiated the transcendental philosophy, of which Husserl's transcendental phenomenology can be regarded as both a critical review and a development.

From the point of view of a formalized epistemology, which is ours, Husserl's critical assessment of Kant is particularly interesting, for it shows well the trap where any type of formalism may become ensnared. Of course, for Husserl, Kant had the merit of having proclaimed the insufficiency of the rationalism of the period of Enlightenment, notwithstanding that he shared the essential ideals of rationalism. However, according to Husserl, Kant did so in a manner that remained dogmatic and even "mythical". As it is well known, Kant, while showing scant interest for common knowledge, imposed very elaborate conditions upon scientific knowledge, but without really analyzing the intuitive elements that he posited at the basis of any knowledge. Whereas according to Husserl, transcendental life – endowed with an *a priori* character, and generative of 'objects' – is so rich and cumulative that it possesses a depth into which it is possible to penetrate by an analytic effort. There each subjective phenomenon appears to act, in turn, as form and as substance for other such phenomena, by a process that has no end. It is mostly in this sense, I think, that Husserl's approach is so original. The philosophical ideals nourished by Husserl, the aim that he wished to assign to the field of philosophy, are particularly well expressed in the following text (*op. cit.* pp. 127-128):

> "As soon as, philosophizing with Kant, instead of starting from the same point and continuing with him along his own path, we turn

back upon his assumptions to question them (Kantian thought also, as any thought, makes use of certain assumptions considered to be obvious and beyond questioning), so as soon as we become aware of these as of mere 'presuppositions' and we consider them as endowed with an own universal and theoretical interest, from that moment on, to our greatest astonishment, there opens up for us an infinity of ever-new phenomena belonging to a new dimension and that come to light only if one realizes coherently the implications of significance and validity of these assumed obviousness: an infinity, I say, for, while we penetrate further and further, it appears that each one of them, such as we find it by unfolding its meaning, and also such as it is initially lived and given as just obviously being, carries already in itself implications of meaning and validity of which the interpretation, in its turn, leads to new phenomena, etc. These – wholly – are purely subjective phenomena, but thereby one should not conceive them as mere factualities, mere psycho-physical processes concerning sensual data, they are on the contrary mind processes, which, by an essential necessity, perform the function of constituting forms of meaning. But this, they always realize starting from a definite 'material' of the mind which always reveals itself anew, by an essential necessity, to be in its turn some figure of the mind, but called to become a 'material' in its turn, that is, to function as the constituent of a further figure, just like what lastly appeared as a figure, then became a material.

No objective science, no psychology, despite the desire to set itself up as a universal science of the subjective, nor any philosophy either, has ever posited as a theme this realm of the subjective, and consequently, never really discovered it. Not even the Kantian philosophy, even though it wanted to come back upon the conditions of subjective possibility of the experimentable and knowable world. This is a subjective realm totally closed upon itself, which, in its own manner, exists, which – in an ever-inseparable manner – functions in every experience, in every thought, in every life, and nevertheless has never been perceived, never been grasped nor conceived of."

So, according to Husserl himself, his specificity with respect to Kant lies much in the structure and functioning assigned to individual subjectivities, and in the consequences of these assignments upon knowledge.

The richness of the new field thus introduced by Husserl in the domain of philosophy is attested by the impressive number of authors who have cultivated it and have therefrom reaped a notable harvest. However it is astonishing that all the various methods that have been tried out on this field – which often intersect with one another – have *failed* to generate a *science*, the science of which Husserl dreamt, that would lie at the bases of all the sciences and act as a common foundation for them.

This situation, of course, might be provisional. On the basis of a phenomenologically-induced psychology are now developed 'cognitive sciences' for which the links with phenomenology have not been severed. And, to my mind, there is no contradiction between a phenomenology that is devoted to the study of the relationships between the strata of living consciousness and, on the other hand, an analysis of these strata carried out *via* the methods of 'objective psychology' and of neuro-biology. These last methods suggest and finally will impose certain formalisms, just as in the case of the other sciences. Phenomenology, then, will have to take care that its own adductions, associated with the formal elements, be unified into a coherent *corpus* where the contributions from each source shall stay inside the limits of their own validity, as confirmed by specific experiments. But such a role for phenomenology confronts a real difficulty: it possesses no other weapon than that passed down by Husserl, namely a striving to develop a direct introspective knowledge of the, anonymous, essential processes from the human mind. These, in order to create concerning them some sort of 'normalized' inter-subjective consensus, should somehow become *communicable* in a clear and standardized way. Phenomenology in this sense is quite different from introspection in the sense of just consciousness of one's own psychological operations. This, however, does not entail that it became a current reference in the principles of some established science. Piaget, when he made use of the clinical method, practiced something that lay between introspection and what Husserl meant, but he refused to acknowledge any link with phenomenology, as if any connection between his own genetic psychology, and psychology in some other, new sense, were suspect.

I personally think that, in the long run, phenomenology will settle into a role of the type indicated above, a hermeneutic role, of interpretation of the principles and results stemming from the cognitive sciences. And in this sort of role it will indeed some day accompany *all* the sciences of man and of society, not only the cognitive sciences, since all these sciences can acquire validity only under the condition of an intimate and clearly established connection with human psychological functionings. This condition should constantly be called to mind as a criterion of trial among the incessant production of 'objective' studies which in fact *lose their objectivity* in so far that the subjectivity from which they stem becomes a subjectivity of scientists where there subsists an only loose relation with the initial, founding, individual subjectivity.

3. PHENOMENOLOGY AND THE NATURAL SCIENCES

For the moment, however, and whatever the future of phenomenology will be, it is a fact that the Husserlian vision of phenomenology, in spite of its undeniable impact upon humanistics, has practically no influence at all upon the natural sciences. In particular, it does not explicitly mark mathematical physics (I put the purely logico-mathematical sciences in another category). Strongly centered upon the *individual* subjectivities, Husserl's view concerning the relations of these with scientific constructs are weakly worked out: in any case insufficiently for radically damming up the tendencies toward relativism. This boundary of the domain where nowadays Husserlian phenomenology does notably act, brings into evidence – beyond it – a vast vacant ground on which could very usefully be constructed a definite pattern of formal representations of the processes of conceptualization mainly drawn from the methods practiced inside the modern mathematical sciences.

Among the numerous difficulties that confront such a project, I would like to mention a particular one that is totally different from that encountered in the field of human sciences. There, the difficulty was to make us share an individual intuition. Here the difficulty resides in the fact that intuitions are almost always misleading, as Bachelard liked to point out, and that they constitute obstacles that have to be dissolved. The formalisms of natural sciences are not simple extensions of the lived symbolism. They involve purifications and abrupt innovations commanded from *outside* the *individual* psychological sources.

Husserl is quite right when he notes that in Galileo-Newtonian mechanics a 'blanket of ideas' is thrown over a world that was pre-geometric and already structured. Yet he fails to point out also that the reason why this Galileo-Newtonian science managed to develop roots is that it had founded itself upon postulates and conventions that were efficient but far from being unquestionable, and which – as any postulate or convention – can be refuted if reasons emerge for doing so. Poincaré brought this into evidence forcefully during Husserl's youth. But, while Einstein understood the message fully, Husserl was not moved by Poincaré's views. According to him, the Galileo-Newtonian revolution consisted mainly in the mathematization of the representation of nature, which, he moreover thought, was not in the least inconsistent with the debate on determinism which in 1935 animated the field of atomic physics (*ibid.* pp. 61-62):

> "Galileo, who discovered – or, to be equitable toward his predecessors, who completed the discovery of – physics, so of nature in the sense of physics, is a genius, both a dis-covering and a re-covering one. He

> discovers mathematical nature, the notion of method, he paves the way for the infinite number of discoverers and discoveries in the realm of physics. He discovers, as opposed to the universal causality of the sensible world (regarded as an invariant form of the latter), which has since been termed simply the law of causality, the " a priori form of the 'real' world" (idealized and mathematized), 'the exact law of legality,' according to which every event of 'nature' (of that which is idealized) has to obey to exact laws. All this is a dis-covery and a re-covery which we have taken, up to this day, to be just the pure and simple truth. For, in the field of principles, nothing has been changed by the so-called philosophical revolution consisting of the criticism, by the new atomic physics of 'the classical law of causality'. Indeed, in my opinion, despite all this novelty, all which is capital with regard to principles, remains, namely: the mathematical character itself, the nature expressible by formulae and to be interpreted only on the basis of formulae."

One cannot but agree with the assertion that classical physics mathematized the representation of nature and that this was an essential step. Yet it seems clear that the deep methodological significances involved in the mathematical formulae whereby our *human* knowledge of nature is expressed, should not be overseen. The very core of scientific revolutions consists of what the mathematical formulae express concerning the *way* in which 'natural phenomena' are actively *determined* by the scientist. Since Husserl makes no explicit remarks on this point, he seems to have underestimated the importance of the methodological content of the mathematical formulae from a physical theory. These – apart from their obvious predictive-operational-observational powers – also involve peculiar features of a procedural kind, wherefrom their performative capacities stem. The elements from the physical world and those from our minds have to be combined in certain definite ways in order to reach the type of efficiency that characterizes mathematical sciences:

> These ways *impose* r e l a t i v i t i e s and they *exclude* r e l a t i v i s m.

Heidegger's unfortunate formulation that "science does not think" is strongly misleading. Mathematical scientists are the deepest and most creative thinkers of present times.

4. OPERATIONALISM AND QUANTUM MECHANICS

If indeed science were nothing more than a collection of operational recipes for predicting, we would not even know how to make a creative, non automatic use of it. Nor would physicists be able to reconstruct a theory if this somehow went lost. They would be comparable to a man who, in case of need, is just able to follow step by step the indications from a cooking book. Whereas a real cook, even deprived of his book of recipes, can always prepare a wonderful meal because he *understands* cooking, and he even can write a new book of recipes. So a formal representation of the processes of conceptualization, drawn from the methods practiced inside the present-day mathematical sciences, certainly cannot be reduced to operationalistic aims. In what a sense, then, should it transgress such aims?

Concerning the positivistic requirement of a radical suppression of any 'ontological' search in connection with the quantum mechanical formalism, Mugur-Schächter [2] writes (pp. 179-180):

> "But I hold that such a purging is at the same time impossible and frustrating. Whether verbal reference is made to it, or not, ontological back-up has been infused throughout the whole action of construction of the quantum mechanical formalism, in particular by the fact and the way of generating the studied states, and above all, by the choice, for each observable X, of a name for its mathematical representation O(X), and of a method M(X) for 'measuring' it. Indeed, why has been chosen precisely that association name-mathematical-representation-method-for-measuring, rather than another one? Each stage has been based on non-declared models, that is why..... It is not the formalism that imposes the choices of M(X) and A(X) (A(X): Apparatus for measuring X (our specification)), it is the task of the physicist to make these choices, outside the theory, confined to the more or less explicit use of the intuitions and models he bears in his mind. And all these ontological adductions have become incorporated into the form and efficiency of the obtained algorithms. A certain ontological content is there, dissolved and assimilated in the very algorithms, inseparable and confirmed. So what is the point of juggling this ontological content away from our final manner of speaking, thereby breaking the bridges toward our own modes of mental action?"

To begin with, let us stress that in this context the term 'ontological' is used in the sense of 'methodological model', which is quite different, if not even *opposed*, to the classical philosophical sense of 'a-description-of-things-such-as-they-are-in-themselves'. Now:

> It is precisely in the nature of the possible 'ontological' contents – in the sense of a methodological model – associable with the quantum mechanical formalism, that consists the break between classical and quantum mechanics.

Husserl failed to notice the schism because he was too absorbed with reconstructing 'a world of life' from behind a world of formulae in which he only saw a 'clothing'. Whereas, if correctly interpreted, the formulae of quantum mechanics modify the 'body' itself, the object constructed by the physicists for qualification. According to many physicists the quantum object called 'state of a microsystem' is conceived of as involving an 'essential' probabilism. This – in so far that one has indeed to accept it – is the most fundamental revolution that has ever taken place in the realm of physics since Galileo. But even if the 'essential' character of the quantum mechanical probabilities turned out to be avoidable, still

> the way in which actuality and potentiality intertwine in quantum mechanics – so the status of contextuality there involved – mark a radical departure from classical physics.

Thus, even though quantum mechanics by no means abandons the mathematical style of physics initiated by Galileo, its writings point toward contents that are so different from those of classical physical theories, that the terms of these, and *a fortiori* those of usual language, cannot be used any more with respect to the quantum mechanical writings without thereby introducing a flood of misunderstandings. So quantum mechanics brings in truly new types of significance which, in certain respects that will have to be specified closely, *resist* the scheme constructed by Kant (and by Husserl) for the development of a physical theory.

Nevertheless – in a certain sense that is defined by Mugur-Schächter – the Kantian category of causality can be maintained: it is possible to build a certain 'causalizing' modelization, namely a 'minimal' one (that has nothing in common, neither with the currently quoted deterministic character of the Schrödinger evolution law, nor with the de Broglie-Bohm attempt, but concerns exclusively the connection between operations, potentialities, and processes of actualization of these, which characterizes quantum mechanics (V. Fock has also expressed a very similar view)). This, she thinks, is of fundamental importance, for the following reason (*ibid.*, p. 184):

> "Ontological models do not 'exist' outside ourselves, in space, like a star, and they cannot be 'discovered'. We forge them ourselves. And

our manner of forging, starting from our perceptions, an ontology that shall please us, shall appease us, that shall place us in a position of psycho-intellectual equilibrium – is causalizing. This feature probably corresponds to certain optimalities of adaptation (possibly maximisations of the rapidity and adequacy of our reactions to the environment) in the absence of which the species would perhaps have foundered. Which irrepressibly suggests some harmony with the unknowable reality-as-it-is-in-itself."

By mentioning such a 'harmony' with 'the unknowable reality-as-it-is-in-itself', Mugur-Schächter, as she admits herself, departs from both Kant and Wittgenstein.

5. KANT AND WITTGENSTEIN COMPARED TO THE NEW FORMALIZED EPISTEMOLOGY

Though, as I have said already, I fully agree that quantum mechanics does indeed impose a distance with respect to Kant, I cannot completely agree with the *reason* for which Mugur-Schächter holds that such a distance emerges. I think that it is important to stress this point, because it concerns a misinterpretation of Kant that is wide-spread among physicists. I quote again (*ibid.* p. 184):

> "According to Kant the properties of physical entities cannot be known 'such-as-they-are-in-themselves'. Nevertheless, he left us free to conceive that our perceptions, so our knowledge, are produced by these very properties, which the entities possess intrinsically in an actual way. But, as we have seen, the formalism of quantum mechanics interdicts even this concerning the state of a microsystem.[1] Our knowledge concerning the state of a microsystem, in so far that it is possible to make it emerge and to 'explain' it, can only be conceived in terms of properties that are actualized starting from the properties of the studied state, but that are different from these."

The interpretation of Kant's view asserted in the quotation is abusive, I

[1] "Intrinsically", in Mugur-Schächter's explanation of quantum mechanics, is used in a sense quite different from 'such-as-it-is-in-itself' and furthermore, different also from certain more curent but vague acceptation. It refers to the author's concept of "intrinsic metaconceptualization [3] (pp. 260-264, pp. 270-273), a *second* stage in the processes of 'relativized conceptualization' of which the first stage – always – is a 'transferred description'.

think. For Kant there exist no physical entities, nor space and time, independently of the view of the knowing subject. Kant did *not* leave us free to conceive that the entities which we distinguish correspond to 'physical entities'. He rather forebade this. For him such entities would be *noumena* accessible only by 'intellectual intuition', a kind of intuition of which, in fact, we are radically refused possession. For Kant the only intuitions we are capable of are in space and time, the forms of our sensitivity. Which means that we experience intuitions of – exclusively – phenomena. Out of these, by using categories projected into the pure intuitions of space and time by way of what he calls schemata, we create 'objects' endowed with properties and linked by causal connections. All this is our own work. What does not come from us is solely the 'matter' of our sensitive intuitions. But this 'matter' comes from the anonymous 'thing-in-itself', not from 'physical entities'.

So it can be said that quantum mechanics *concurs* with the Kantian interdiction to posit our knowledge as being produced by real entities. But there is *disagreement* with the way in which Kant permits the constitution and mutual connection of physical objects within his 'analogies of experience' and 'postulates of general empirical thinking'. For nothing, in Kantian epistemology, authorizes the conception of a physical explanation in terms of 'relative potentialities and actualizations', such as quantum mechanics *does* authorize. While an alternative ontology, founded on the acceptance of an essential randomness, would be even further removed from Kantian epistemology: in this sense – contrary to what Husserl asserted – the criticism of 'the classical law of causality', in so far that it is accepted, clearly is a philosophical revolution induced by physics. In this respect Husserl appears to have been insensitive to the novelties brought forth by the science of his time, in consequence of which the mathematical style introduced by Galileo is now endowed with a fundamentally new sort of bearing.

As to the distance with respect to Wittgenstein, it concerns the famous remark that the limit of what can be said consists of what can only be 'shown': Mugur-Schächter remarks that quantum mechanical objects – states of microsystems – cannot be shown, and nevertheless they are (retroactively) said, 'dicted', and then even 'predicted'. It seems that the 'grammar' of quantum mechanics has not drawn upon it Wittgenstein's attention, nor that of his students, as far as I know.

So it can be concluded, I think, that quantum mechanics indeed reveals to us new forms of objectivity, totally unknown to classical science and epistemology. Thereby it contains precious guiding lines for a formalized epistemology. And I quite agree with Mugur-Schächter that the theory of information should also be used as another most important guide. Given its remarkable generality, its quintessential extract from the phenomena of communication, its successful applications in genetics and molecular biology, it would certainly be extremely enriching to carefully incorporate its epistemic contents. I would like now to draw attention on a very important point. In a certain sense each great philosopher, in his own time, could be said to have perceived science by way of some epistemology, that is, by insisting on the fact that science never is founded exclusively on experience, that it also involves choices of types of explanation or 'causalities' (one might have said 'paradigms', had not this word, under the pen of Thomas Kuhn, taken a too complex and ambiguous significance). Now, this circumstance might lead to a confusion with what is here called a 'formalized epistemology'. It is quite essential, I think, to hinder such a confusion from the start on:

> A 'formalized epistemology' in our sense is not an epistemology of science in general or of some particular sciences. It is a formalization of general epistemological methods, drawn from the most performing nowadays sciences.

To close now, I ask: Would a formalized epistemology make the theories converge? Personally, I have doubts, since, starting from its origins, the scientific undertaking manifests much more a tendency toward divergence than toward convergence, and it seems unreasonable to believe that an epistemology drawn from it shall head in another direction. Yet if we managed to explain the divergent tendencies of the multiple forms of our representations of empirical reality, their resistance to converge, then the epistemological development undertaken here would acquire not only scientific value but also a philosophical one.

REFERENCES

1. Edmund Husserl, *The Crisis of European Sciences and Transcendental Philosophy*, translated into French from German by Gérard Granel (Editions Gallimard, 1976).
2. Mioara Mugur-Schächter, "Les leçons de la mécanique quantique (vers une épistémologie formelle)," *Le débat* **94**, March-April, 1997 (Editions Gallimard).
3. Mioara Mugur-Schächter, "Spacetime quantum probabilities II: Relativized descriptions and Popperian propensities," *Found. Phys.* **22** (2) (1992).

3

FORMALIZED EPISTEMOLOGY, LOGIC, AND GRAMMAR*

Michel Bitbol

Centre de Recherche en Epistemologie Appliquee du CNRS
1 rue Descartes
75005 Paris, France

The task of a formal epistemology is defined. It appears that a formal epistemology must be a generalization of "logic" in the sense of Wittgenstein's *Tractatus*. The generalization is required because, whereas logic presupposes a strict relation between activity and language, this relation may be broken in some domains of experimental enquiry (e.g., in microscopic physics). However, a formal epistemology should also retain a major feature of Wittgenstein's "logic": It must not be a discourse *about* scientific knowledge, but rather a way of making *manifest* the structures usually implicit in knowledge-gaining activity. This strategy is applied to the formalism of quantum mechanics.

Key words: quantum mechanics, logical structure, Wittgenstein, philosophy.

1. INTRODUCTION

What makes possible the background against which is set our knowledge of the order of nature? "There," said Kant, "solutions and answers are brought to a halt; because we must always go back to (this background) for all answers and all thought of objects" [1].

At least this setting of limits, typical of transcendental philosophy, points towards that which a formal epistemology cannot be. It *cannot be* the formalism of an objectified theory of knowledge which would take the subject-object relation as a second-order natural object, and would then leave unquestioned the grounding of normative presuppositions on which all science, including epistemology itself, depends. Further, a formal episte-

*See "Important Note" on p. xvii.

mology cannot conform to the definition "of a clearly too foolish ambition" which H. Putnam depicts as "[...] a superb theory of the normative grasped in its own terms" [2] ; a sort of redoubling of the realm of norms of thought, by which the theory would try to explain itself in objectifying the system of its own principles, without being able, except by an infinite regression, to question itself in return on its use of those very principles. Of course, these remarks are not to deny the current attempts at naturalising epistemology any interest. They are only aimed at pointing out that naturalisation of epistemology can only be a *process* with *no foreseeable completion*; and that at the provisional end of each step of this process there is a set of non-explicit norms of investigation which we can but call the "pragmatico-transcendental background" of the current state of research.

It is of no apparent advantage either to cast formal epistemology in the role of a mathematicised or logicised variant of epistemology in the modest sense traditionally intended in France: that of a multiplicity of critical analyses of the premisses and results of particular sciences. Because in epistemology, as in the sciences, formalisation consists in the abstraction of particular contents in order to reach universal rules. A formal epistemology must therefore be of value to any science, even if it is especially profitable (as we shall see below) when elaborating on the knowledge acquired by certain methodologically advanced sciences.

Having discarded some tentative definitions of formal epistemology it remains to give it a plausible positive characterisation. To that end a comparative rather than a directly constructive strategy will be used. A parallel will be established with the case of logic; the remarkable isomorphisms between logic and what we would expect of a formal epistemology will be underlined; then, at the end of the discussion, the bringing into consideration of some major differences between the two disciplines will allow the formulation of the specific project of formal epistemology. It will thus appear that formal epistemology can be understood as a generalisation of logic; a generalisation of considerable range because it principally consists in recognising the expansion of the form of the sciences beyond the closed domain delimited by the Logos, taken in its narrow sense of explaining by means of *discourse consisting in predicative judgments.*

2. LOGIC AND FORMAL EPISTEMOLOGY

Let us start from the dualistic prejudice of the theory of knowledge; because it is by way of criticising it that we will most quickly arrive at the point of neutral equilibrium where both logic and formal epistemology stand. Knowledge, according to Piaget [3] , consists in a certain relation between a subject and an object. It manifests itself by way of *judgment* (which consists of ascribing a predicate to an object) or more generally by way of thought shaped by the structure of judgment. Each science can be said to attain *knowledge*, in the local sphere of objectivity to which it is assigned, if it expresses itself *via* a certain network of judgments whose interdependence and coherence fence off the temptation to systematically resort to *ad hoc* explanations. But in that case, logic, which we traditionally present as a general doctrine of judgment and of relations between judgments, is at once "science's doctrine"; logic, as Husserl points out, "[...] aims at bringing to light the essential forms of knowledge [...] as well as the essential presuppositions with which its forms are linked" [4] . In that, at least, the programme of logic covers exactly the programme of a formal epistemology. In the dualistic framework provisionally adopted, however, logic (and the formal epistemology which matches it) takes on a sort of constitutive ambivalence. Logic and formal epistemology are what Husserl calls "double-sided" disciplines [5] ; disciplines having at once a subjective and an objective side. They have a subjective side because they seek to extract the necessary states and regulative principles of a "rational subjectivity in general" [6] . And they have an objective side for two reasons. Firstly, because they engender ideal objective formations, as the product of their work of abstraction and deductive generation; and secondly because, although they situate themselves below the level of the concrete determination of objects and of classes of objects of the particular sciences, they relate to the form of judgments, of which the prime function is to characterise *objects*. It is this last thought which led Husserl to characterise formal logic as "*a priori* formal doctrine of the object" [7].

The two orientations of logic (subjective and objective) were given privileged roles in turn by the actors of the history of philosophy. But this process only led one to show the inadequacy of each single orientation as a paradoxical result of the attempts to assure it the exclusivity. Let us consider a first example. In the framework of the critical philosophy of Kant the distinction between the reflexive and objective orientations of a discipline does not rest on an exterior account of the face-to-face subject-object, but rather on an internal analysis of the conditions for the possibility of experience. The "fundamental proposition" [8] of the critical philosophy effectively announces that "The conditions of the *possibility of experience* in general are at the same time conditions of the *possibility of objects of experience*" [9]; so there can be no question of a confrontation between a pre-constituted subject and object, but rather a *co-constitution* of experience and its objects. That being allowed, we notice that Kant's internal analysis gives him two motives for insisting on the *reflexive* orientation of logic. Firstly, logic situates itself entirely on the side of the formal aspect of our knowledge, without any reference to its material and "objective" aspect [10] . Thus, according to Kant, logic somehow situates on the "subjective" side of knowledge. Secondly, logic is freed from any link with the form of the *sensible intuition*, by which "we can perceive objects *a priori*" [11] ; it proceeds without consideration of perceived objects, turns on the understanding alone, and consists in announcing the formal laws necessary to the *thought* of any object [12]. It is certainly not subjective according to the *psychological* conception of subjectivity, because it is not content to describe in empirical terms the intellectual mode of functioning of particular subjects; but we can call it "reflexive" in so far as it is linked to the principles which order the thought of the subject in general. Logic relates not to particular subjective facts but to the norms which bear on the intelligent activity of any subject.

In that way, Kant does not just oppose logic to psychology, but also to transcendental philosophy in its entirety. Because if transcendental philosophy *also* deals, like logic, with the *a priori* formal conditions of thought of objects, it does it *via* a very broad perspective in order to elucidate the connection between knowledge and the faculty of knowing in general [13]; on the contrary, logic is supposed to content itself with furnishing a "canon" of agreement of *one* of the constitutive elements of the faculty of knowing (understanding) with itself. "In logic" writes Kant, "the question is uniquely: *how does the understanding know itself?*" [14]. In the evolution of the Kantian project after Kant, the joining of form with the subjective side of knowledge, the stratification of the sensible and intellectual and the limitation of logic to a task of self-validation of the operations of the understanding, have been regarded as the weakest and least indispensable aspects of the project's completion. With the impetus of Cassirer and the many protagonists of the "linguistic turn", the integrated forms of symbolic expression have replaced the hierarchical forms of the faculty of knowing in the role of preconditions of objectivity. Since then, as G. G. Granger points out, the opposition between logic and transcendental philosophy has had no *raison d'être* : "logic can [...] appear to be the most elementary and the most radical aspect of the transcendental" [15]. "Logic is transcendental" [16], writes the early Wittgenstein, and, in the intention which it shares with logic, formal epistemology is too.

In opposition to this process of abstraction and identification of the Kantian *a priori* forms with the symbolic, another current of thought has tended to put them into relation with the concrete turning points of the phylogenesis and ontogenesis of the human subject. A psychogenetic reinterpretation of the Kantian hierarchy of the constituents of the faculty of knowing has been proposed by Piaget, for example. According to Piaget, the underlying structures of natural thought issue from the stepwise coordination of the diverse operational activities of the subject in the world. But contrary to physics, which partially takes its information from the manipulated objects, by way of perceptual or experimental phenomena which are supposed to open access to objects, logic proceeds from the exclusive cordination of the *actions* which impinge *on* these objects and transform them [17]. To the Kantian duplex of sensibility and understanding there corresponds here a duplex of sensible receptivity and structured motor activity of which only the second term concerns logic. According to Piaget, "That which is axiomatised by formal logic is certainly an *activity of the subject*" [18]. More precisely, it is a systematic activity of the subject whose psychogenetic evolution has passed an essential stage: the conquest of the *reversibility* of operations, which allows their formalised outcome to constitute an ensemble of timeless and necessary connections [19].

But doesn't exclusive concern with an ideal and isolated *subject* in evolution keep us within a framework too narrow to yield reasons for the emergence of the norms of thought? Does it not mask other genetic components which are indispensable to the shaping of a logic? These additional components are not denied by Piaget, but they do not constitute, in his work, the material of systematic development. They concern just as much the social interaction *between subjects* as that which is presupposed by them *regarding objects*. On the one hand, although it is true that the construction of logic is in the first place, according to Piaget, the work of a subject in activity, its operational structures require "the collective contribution" of other communicating subjects in order to be "reinforced and multiplied" . "Reinforcement" ends in the stabilisation of norms by means of the symbols used to communicate them. And "multiplication" could well refer to the construction of non-classical logics which, not contenting themselves with stating the normed forms of the *effective* operational activity of subjects, formalise many *possible* operational activities by exploiting the supplementary free space which is offered by symbolism. The orientation towards a theory of communicational intersubjectivity, favoured by contemporary German philosophers such as Apel and Habermas [21], is thus able to complete and enrich the focus on this work of "inquiry" of a generic subject which, before the work of Piaget, already formed the principal theme of many currents of American pragmatism of the beginning of the 20th century [22].

On the other hand, the throwing back of the Piagetian problematic of the normed activity onto the *subject(s)*, its liberation in regard to the object, calls at least to be nuanced. Activity is certainly, in the first instance, that of the subject, but what about its regulatory forms which alone concern logic? Piaget admits that it is "[...] impossible to know in advance if (these forms) belong to the subject, to the object, to both, or solely to their relation" [23]. It must not be forgotten that activity consists in operations-of-the-subject-on-objects. Even if it is indispensable to remove from the description of this activity any mention of the particular features of the handled objects, it must be recognised that the activity and its formal sediment rest on two suppositions which bear on the objects in general: the supposition of the *permanence* of objects and that of a *minimal degree of stability* in their properties.

Let us dwell a moment on these two elementary suppositions of operational activity, because they will have particular importance in the rest of this paper. What must be noted about them from now on is that they correspond term for term with those which the very use of the proposition implies (by way, respectively, of the two fundamental acts of *reference* and *predication*). For that reason, the formal kernel of the coördination of the operations of the subject in the world corresponds closely to the formal kernel of *language*. And so we understand that the axiomatisation of the motor activities of the subject, on which Piaget focuses, converges at once towards an axiomatisation of discursive activities to constitute that which can be called a *Logic*. Nevertheless, one must bear in mind that the circumstances which allow this remarkable convergence between the norm of the activity and the norm of linguistic activity, are very peculiar. They are linked to everyday life and speech. This urges us to introduce a reservation from now on: nothing guarantees the durability of the relation obtained between the domain of activity and the domain of discourse when we pass from a gestural activity exercised at the heart of the everyday environment to an experimental activity aimed at exploring its limits.

3. LOGIC, GRAMMAR, AND FORMAL EPISTEMOLOGY

On the other side of the dualistic demarcation, to wit according to the philosophers who have privileged the *objective* side of logic, symmetrical difficulties have provoked a swing of the pendulum back towards the same point of equilibrium.

For the Wittgenstein of the *Tractatus*, the status of logic is that of a representational framework. Logic, he writes, is a picture which reflects the world [24], its propositions represent the scaffolding of the world [25]; "logical pictures can depict the world [26]" . Accordingly, logic merges with the limit of the world. Indeed, the form of representation cannot for its part be represented in the logical picture; it can only be *shown* by it [27]. This remark, directed against the possibility of an *authentic* meta-representation, justifies in its turn the crucial distinction which Wittgenstein makes between *concepts* and *formal concepts* [28]. We can say (with the help of a proposition) that something falls under a concept, but we can only *show* that something falls under a formal concept. "Object" and "property" are such formal concepts. That something is subsumed under them is not claimed but shows itself by way of the type of sign which is used to denote it, or by means of the position of that sign in the propositional network. The formal concepts of object and property are, so to speak, *structurally presupposed* by language.

But as it is well known, that statics of mimesis typical of the *Tractatus* is exactly one of the principal targets of the Wittgenstein of the *Philosophical Investigations* [29]. The meaning of a proposition no longer establishes itself in a projective relation to the world, but in a lateral relation to its use in a "language game" or to its being put to work as a moment of a "form of life". The dynamic of this use appear to be constrained to a body of rules which we call *grammar* in a wide sense; but it must not be believed, Wittgenstein insists, that those who speak and act in conformity with those rules are *guided* by them. Grammar is only the formalised residue of the practice of language-games. In order to idenfy this formalised residue, one can rely on the so-called "hinge" propositions of language. These propositions are "[...] devoid of content because they do not admit of a negation endowed with meaning" [30]. In other terms, they are devoid of content because they constitute the minimal basis of tacitly accepted affirmations in relation to which the affirmation or negation of all other propositions makes sense.

At this stage, if we would situate the thesis of the *Philosophical Investigations* in terms of the dualism in the theory of knowledge, we would need to ask ourselves some second-order questions: from what emerges the symbolic practice with which that thesis deals? Is it imposed by the subject or the world? The later Wittgenstein turned at length around these questions, but it was more to denounce their double lack of relevance than to answer them. Even though practices are proper to the subject, they do not reduce to series of arbitrary gesticulations and vocalisations. And even though the grammar of practice is constrained by some "reality", it does not constitute a *copy* of this reality [31]. To paraphrase a remark of J. Bouveresse's [32] concerning arithmetic, we should say that*the connection which exists between grammar and reality is something which can only be shown in the application of grammar, and so it must not be described in terms of correspondence with the facts accessible from a point of view exterior to the practice of language-game.* Just like logic in the *Tractatus*, or formal epistemology according to the sketch which we have traced, grammar in the *Investigations* is thus transcendental. It simply is so in a quite particular way; not in the rigid style of Kantian *a priori* or the pictorial skeleton of the *Tractatus*, but in the mobile manner of the functional *a priori* of Dewey, qualified as *quasi- a priori* by Putnam [33].

Furthermore, grammar retains a feature which we have so far reckoned as characteristic of *logic* : the merging of presuppositions of discourse and of action. "I act with complete certainty" [34] in accordance with the norms which underpin the forms of life in which I participate; I speak in complete confidence within the framework of rules of language of which I make use; and I hold as unquestionable that background of propositions "against which I distinguish between true and false" [35]. Forms of life, background of "beliefs", and rules of use of language, constitute for the later Wittgenstein a coherent and undivided communal basis.

It is now possible to clarify the project of a formal epistemology by means of a detailed play of similarities and differences between it and logic and "grammar" in the sense of the later Wittgenstein.

To begin with, we have said that logic, "grammar" and formal epistemology are all transcendental. That is to say, they tend to reveal the formal conditions of possibility for a state of knowledge (or a practical and verbal orientation in the world). From that standpoint, they constitute second-order disciplines, as against the first-order disciplines which are the sciences. But they are not, for all that, meta-sciences or meta-theories relative to the theories of particular domains of objects. They take as their object of investigation neither the sciences nor knowledge-gaining activity as a whole. They content themselves with codifying a symbolising procedure, and thus are able to *make manifest* the structures usually implicit in knowledge-gaining activity. They say nothing; they show. They do not represent; they present. They are typical examples of what G. G. Granger very appropriately calls *non-meta-theoretic meta-disciplines* [36]; examples of disciplines which in coming *after* a discipline do not establish between them and the discipline a distancing relation as between a science and its objects.

Further, we have underlined a considerable difference between logic in the sense of the *Tractatus* and "grammar" in the sense of the *Investigations.* The former has the rigid and hierarchical character of a structure which presents itself as *grounded* in the giving of a world of which it exhibits the "scaffolding". "Grammar" has the mobility of a system of rules following the lines of force of an interlacing of operational and linguistic practices which is certainly constrained in some way by the real "other", but which has the elasticity to modify both the mode of expression of this constraint and its mode of response. If we want formal epistemology to be able to meet the challenge of scientific revolutions with a sure hand it must resemble the "grammar" of the later Wittgenstein in its elasticity. It must, like "grammar", belong to the class of evolutionary and non-foundational meta-disciplines.

Finally, there is a common feature of logic and "grammar" which we are inclined *not* to ascribe to formal epistemology; it is the presupposed certainty of a concordance between the form of discursive practice and the form of operational practices. This additional degree of freedom should allow formal epistemology to take care of a situation like that which confronts quantum mechanics, in which there is no obvious agreement between the formal coördination of operational activities and the structure of language. It justifies in every way our calling formal epistemology *an evolutionary meta-discipline leaving in suspense the linguistico-operational concordance.* And it also justifies the expression "formal epistemology" when one attempts to display the formal structure of a physical theory, as opposed to Y. Gauthier's expression "internal logic". We can sum up these thoughts in the following table:

Meta-disciplines	Logic	"Grammar"	Formal Epistemology
evolutionary		"Grammar"	Formal Epistemology
lacking linguistico-operational concordance			Formal Epistemology

4. TO ACT BEFORE PREDICATING

The idea of a meta-discipline leaving in suspense the linguistico-operational concordance is not entirely new. It is brought out very well, albeit in negative relief, in a critique which Husserl addresses at formal logic. Formal logic, Husserl explains, is of value for "[...] a real world thought of as already given beforehand". In traditional logic, the predicative structure of judgment, together with the presupposition of the permanent existence of that of which something is predicated, "[...] was self-evident and was never examined" [37]. This constitutive pre-judged is equally brought to light by M. Mugur-Schächter when she emphasises that language, logic and the classical theory of probability rest on the common postulate of an "[...] *intrinsic* 'objectivity' which would prexist all acts of observation and conceptualisation" [38].

In contrast, Husserl proposes to go below the categorical structures of language, below the form of judgment and below the formal concepts of object, property or relation. In *Formal and Transcendental Logic*, and more systematically in *Experience and Judgment*, he undertakes to put "[...] in question their innate *production* and their *springing up* in the lower stage of knowledge" [39]. Husserl calls this lower stage of knowledge "ante-predicative experience"; and he shows page after page, with all the refinement of his specific analyses, how from this can emerge the characteristic moments of predicative judgment. The emergence of the substratum of predication and that of the predicate arises respectively by way of two modes of ante-predicative experience: the "identificatory aiming" and the "explanatory experience" [40]. The identificatory aiming synthetically unites the multiplicity of perspectives, of profiles or of aspects presented by perception in an open experience of *same* and constitutes the precondition of the act of *reference* to an *identified* object. As for the explanatory experience, with its anticipatory tension, with its way of projecting interest towards the aspects which we *expect* to find if we modify our point of view on one and the same object, it is at the basis of *predication*. An anticipation attested and confirmed by the reproduction of a phenomenon when a certain perspectival situation is reiterated can, in effect, translate into a *predicate* assigned to the object aimed at.

But what would be the result if the phenomenological circumstances of this twin stabilisation, of predicate as well as substratum of predication, were not realised? What would happen in circumstances of the total disorder of "ante- predicative experience"? Nothing less than this would result: the disappearance of the conditions of an objectifying discourse making use of predicative judgment; and consequently the equivocation of the means of *saying* what this would amount to.

This *aporia* of the inexpressible can nevertheless be defused on two conditions (which are not mutually exclusive):

(1) If the disorder of the experience is only *partial* and not total; because in that case the failure of the anticipations associated with aiming at an object could simply point towards the opportunity to change the type of object or quite profoundly modify the mode of aiming towards it.
(2) If, in response to the disorder, we can limit ourselves to a calling into question of the lower levels of a logic which be *less universal* than that which Husserl proposes.

The restricted calling into question which we propose is certainly superficial when compared to the programme of genesis and foundation which the creator of phenomenology formulated; but it is sufficient to deal with the difficulties of contemporary physics. Instead of opposing, like Husserl, the pre-logical circumscription of ante-predicative experience to the domain where (predicative) logic is to be established, we would oppose the combined domain of everyday life and instrumental operations, where the validity of logic remains unquestioned, to the domain of the putative objects of experimental investigation, in which the relevance of logical structures remains an open question [41]. If we proceed in this way, the loss of the conditions for an objective mode of expression using predicative judgments within the particular domain aimed at by experimental investigation does not have as a consequence a global slide into the inexpressible, but only the restriction of the sphere of relevance of the forms of discourse to the description of instrumental operations.

5. QUANTUM MECHANICS AND FORMAL EPISTEMOLOGY

In quantum mechanics we are exactly at this point. On the one hand the traditional forms of discourse using predicative judgments remain valid in the domain of instrumental operations; better, they *must* remain so in as much as they are the bearers of the preconditions of an intersubjectively shared experimental knowledge [42]. But on the other hand, the *expectations* which are induced by the aiming at a traditional type of object (corpuscular bearers of properties) *beyond* the experimental apparatus, are generally confounded. The expectation of being able to re-identify a corpuscular object founders on the impossibility of experimentally following its trajectory continuously, and on the indirect consequences of this impossibility (i.e., combinatorial and statistical consequences). The expectation of seeing a phenomenon reproduce itself is for its part systematically confounded in certain well-documented cases: a value of a variable is not reproducible if, between two occurrences of its measurement, we insert a measurement of a variable called "incompatible" or "conjugate" (e.g. position and momentum).

So none of the phenomenological criteria for reference to a corpuscular type of object, and for the predication of determinations to that type of object, are satisfied in the experimental domain of microscopic physics. We are left in the presence of something which *prima facie* resembles an isolated flux of singular experimental results, indissolubly dependent on the experimental conditions which have given rise to them. In effect, these results do not have a sufficient degree of invariance with changing experimental sequences for us to be able to detach them from the instrumental context of their occurrence and to treat them *as if* they were evidence of a determination which an object would possess. In short, the events of microscopic physics are essentially *contextual*, or again, as M. Mugur-Schächter says, they arise from a *"descriptional relativity"*.

What is to be done in facing this critical situation, in which the corroborated theoretical anticipations of the results of operational activity do *not* satisfy the presuppositions of discourse reflected by traditional logic? The first strategy, urgent and clarificatory, consists in *showing*, in *manifest-*

ing, the coördinated structure of these anticipations as it is extracted by the formalism of quantum mechanics in a rigorous but not very explicit way. It consists in capturing the *meta-contextual* structure which P. Heelan spoke of [43], or the *algebra of observation* which S. Watanabe developed [44], or the ordered system of relativising glances in the sense of M. Mugur-Schächter [45]. To summarize, the strategy amounts to extensively utilising the resources of a meta-discipline freed of the constraint of an isomorphism between language and operations. A meta-discipline which corresponds exactly to the definition which we have given to *formal epistemology*.

As a second strategy, we could always ask ourselves if it is not possible to go back to the golden age of the linguistico-operational concordance in changing logic ("quantum" logics), in choosing a new partitioning of the world into objects having nothing in common with the material bodies which bear localised properties (e.g., the referents of state vectors, as according to Schrödinger [46]), or in assuming (as in hidden variable theories) that the properties of corpuscular objects are instantly influenced by the instrumental or environmental conditions of their manifestation [47].

But none of these second-level endeavours will be able to ignore the lesson to be drawn from the first-level analysis brought to fruition by formal epistemology. Quantum logicians face considerable difficulties in defining what they mean by "property of an object" without conceding too much to contextuality; the new ways of partitioning the world (i.e. the "new ontologies") remain reliant on a level of discourse where a tacit "natural ontology" operates; and hidden variable theories must have recourse to the artifice consisting in substituting "contextualism" for contextuality: that is to say, invoking a holistic influence of experimental circumstances on the underlying intrinsic processes, rather than drawing directly the consequences of the co-definition of the phenomenon and the conditions of its manifestation.

6. EPILOGUE

In the manner of the Euclidean geometers of Michel Serres, the physicist of the classical epoch "[...] hopped on the moving train, at a moment when everything was already worked out, when the concepts were a thousand times over- determined" [48]. Then, without clearly understanding what he was doing or why he was doing it the physicist of the 20th century adopted the path of a radical rexamination of the previously unquestioned articulation between the operational and discursive domains. To that extent he puts himself in the predicament of the modern mathematician who, in a paradoxical development, " [...] steers himself towards his unforseeable horizon and his starting point" [49]; a mathematician who, to put it differently, ap-

4

EPISTEMIC OPERATIONS AND FORMALIZED EPISTEMOLOGY: CONTRIBUTION TO THE STUDY OF THE ROLE OF EPISTEMIC OPERATIONS IN SCIENTIFIC THEORIES

Michel Paty

Equipe REHSEIS
CNRS and Université Paris 7-Denis Diderot
Centre Javelot, 2 Place Jussieu
F-75251-Paris-Cedex 05, France

We ponder the kind of problems and perspectives of a "formalized epistemology", by considering the advantages than one get from a concern with the "formal", with its structural orientation, that would favour comprehensive, unifying and synthetic, intelligibility. We confront this perspective with that of the changes in knowledge, considering the relation between form and meaning for knowledge contents, and examine the notion of "epistemic operation" as instrumental for creating new forms, at the theoretical and meta-theoretical levels. Actually, the notions of form, of formal and of object are not independent of the problem of a subject that decides on conventions and choices. "Epistemic operations" might suggest a link with "algorithmic functions" for knowledge statements, that themselves entail the risk of reductionism in a naturalistic conception of representation.

Key words: changes, contents, epistemology, form, meaning, object, operation.

1. INTRODUCTION

The expression "formalized epistemology" admits of a whole variety of possible definitions, thus staying widely open to interpretations.[1] Taken intu-

[1]The following reflections have benefited from exchanges and discussions inside the working group entitled "Centre pour la Synthese d'une Epistémologie Formalisée (CeSEF, Paris), animated by Mioara Mugur-Schächter. I acknowledge the friends and colleagues

itively, it is perceived to point toward a particular interest in formal representations, in any given area of knowledge, and in the connection of such representations with scope and meaning, and with the relationships between different fields of knowledge. In what follows, we much less intend to reach a precise definition of what a formalized epistemology should or could be, than to focus on some problems characteristic of an investigation of the type suggested by such a name and to stress both its interest and its limits. Therefore, from the start, we tolerate vagueness in the definition of the concept, in order to avoid an *a priori* confinement inside a too-narrow and artificial set of boundaries: we mainly aim at exploring what the label "formalized epistemology" could reasonably point toward.

Let us begin by specifying our project still more. Three quarters of a century ago, Ernst Cassirer tried to establish a "morphology of the mind". First, in his beautiful work *Substance and Function* [13] he exposed an inquiry into "the structure of thought" as manifested in mathematics and natural sciences. Then, under the global title *The Philosophy of Symbolic Forms* [15], he dedicated three new volumes to an analysis of language, of mythical thought, and of the phenomenology of knowledge.[2] In what follows, Cassirer's basic idea that any area of human thought—from its "infrastructure" to the "architectonic organization" of superstructures that constitute the sciences [3]—is expressed *via* symbolic forms, while keeping oriented towards the real world, will remain for us a fundamental source of inspiration.[4] Now, questioning form and formal problems when one deals with exact sciences, in general involves emphasizing some current aspects of the present scientific theories, with an endeavour to bring forth their characteristics in this respect. We do not want, however, to confine ourselves to static structural features. Our interest, especially from a comparative viewpoint, is also directed toward the *movement*, the streams by which the formal aspects are brought about, and is strongly impulsed by questions as to how in future other formal aspects could possibly settle in. If we restricted ourselves to examination of only the formalization of recent knowledges, we could "lose the prey for the shade" retaining merely a schematic, a "logically" recon-

---

from this group for the rich debates we had together. A first version in French of the present work has already been published: "Opérations épistémiques et épistémologie formelle. Contribution à l'étude des opérations épistémiques dans les théories scientifiques", *Principia* (Florianopolis, Br.) **3** (2, December), 257-306 (1999).

[2]To each one of these subjects, he dedicated a whole volume of *The Philosophy of Symbolic Forms*.

[3] [15], Vol. 3, pp. 13-14.

[4]This idea has already oriented us in other studies, in particular in our *La matière dérobée* (*The Stealing of Matter*), dealing with the conceptions of contemporary physics [64].

rating certain contingent and conventional features; that these contingent and conventional features affect not only the objects of our representations but also the modalities of representations, *up to the very manner to conceive them*; and that *all* these changing characters at *all* these various levels, affect equally—and possibly even more—*all* the other areas of knowledge and of human experience which are not reducible to the scientific knowledge, such as aesthetics or morals, these pillars of any culture which, albeit differently, are also linked to the use of reason, while, like any knowledge, they bring into play also the other functions of thought, like imagination and memory.

So, coming back to our nowadays knowledge, it should be clear that in many areas the way in which we now conceive of what this or that phenomenon is, to what sort of object it refers or what it admits of, differs appreciably from preceding views in these respects, including some of those which we ourselves have initially learned. However, although we might be conscious of many among these changes, since we experience them directly, we do not yet know clearly *what* they modify in our current manner of thinking and speaking, nor how [53-55]. This knowledge is indeed most difficult to be gained while still staying inside the previous global traditions of thought. We are living these changes before knowing how to think them. However, in spite of all, it is unavoidable for science, and it is essential for philosophy, to somehow undertake an effort for thinking them in a coherent way. But is it possible, right now, to know something more about these ongoing changes, and to formulate explicitly the new rules of thought required by them implicitly?

A preliminary step would be to explore what can be hoped in this direction. Since the present situation is not *entirely* new, since it has precedents from certain points of view, we can try to draw lessons from the past.

Also, focusing attention to just the changes which are now occurring, can be instructive even *without* necessarily understanding them, nor -*a fortiori*—being immediately able to forge a new global view able to encompass them. Our ambition, even if it is kept modest, might nevertheless be fertile; while if it is too big at the start it might generate merely illusion.

Another, more constructive step, would consist in explicating *directions* inside the already known evolutions of thought. Here we shall evoke those which took place inside physics, but similar examples can be found in biology, geology, cosmology, mathematics, as well as in the social and human sciences. For instance, the theory of relativity (special and general) has obliged to rebuild the notions of space and time which were considered to be the most obvious and stable ones and were the deepest rooted in our cognitive structures. It has also brought forth the necessity to re-examine the concept of physical theory itself, namely to regard it more and more as a symbolic, conceptual and formal construction which nevertheless pertinently represents the real *world* [64,68]. Thereby it has enabled us to acquire a more appropriate perception of the very peculiar role played by mathematics in this construction. We can thus conclude that *mathematics manifest an abruptlyincreasing importance in our representations of "reality"*. This conclusion yields a dimension where past evolutions can be pertinently embedded, thus clarifying certain historical facts. Indeed, since already a long time it became obvious that "the space-time continuum" is just an abstract entity elaborated by thought, of which the justification cannot be regarded to consist exclusively of evidence provided by intuition, but also—mainly perhaps—of the operational power of this construct: the use of differential equations as well as the basic concept of field of forces are based on this construct, and this triggered accurate descriptions and explanations of a huge amount of phenomena. But the operational power of the concept of space-time continuum is fundamentally tied with its mathematical representation, which strongly guided our most basic schemes of intelligibility throughout long, progressive elaborations [71].

Similar remarks hold concerning also other conceptual, symbolic constructs, like material points, forces, etc., which, strictly speaking, are devoid of a genuinely "real" counterpart. Newton's mechanic of material points, while it introduced the space-time continuum, also brought into evidence the *general* necessity and outstanding usefulness of idea-like ("idéels") mathematical constructs founded on abstractions drawn from—supposed—realities. Indeed most among the constructs of this sort were first introduced inside the framework of newtonian mechanics, with its general principles and laws, its own inner relationships and basic concepts. But later they have constantly and progressively been reformulated, completed, and generalized (to complex solid bodies, to fluids, to gravitational attractions between more than two bodies, etc.). And, whatever be their still unclear relations with what we call physical reality, such constructs proved so extremely fertile in the hands of physicists and mathematicians that one feels strongly inclined to conclude that, though nowadays they have become so very remote from the real in the sense of our intuitive perceptions, nevertheless they somehow exceed the schematic character of a mere mathematically convenient "ideality", that they somehow have definitively incorporated something drawn from the "real". So the direction of increasing mathematization seems to belong organically to the past-and-future history of scientific knowledge.

But it is quite noteworthy that this pertinence and viability with respect to the physical reality, of certain abstract constructs from mathematical physics, *owes also much to a deep epistemological concern of major scientists* ,[6] who throughout centuries kept discussing the conditions of validity and the limits of applicability of the mathematical representations of the basic concepts involved by physics. Which, again, illustrates the intimate relation between change and epistemology.

As for the changes which quantum physics introduced in the conceptualization of physical reality, these, from both the viewpoint of the physical meaning of the theory, and the viewpoint of its epistemological impact, are still far from having been fully evaluated. Very soon, the quantum physicists themselves have explicitly tried to extract from this so peculiar discipline, a general methodology for physical theories, and a philosophy of knowledge. But precisely the concern to insure for this new and remarkably fertile scientific theory, a legitimacy and an impact that seemed in danger of being denied, fixed—prematurely, and in a rigid way—fictitious *limits* of interpretability. The conceptual difficulties, pertaining to the physical argumentation as well to the theoretical problematic and the epistemological analyses, were buried under the automatic answers of an *ad hoc* "positivistic" philosophy. It is quite true that, in the light of the knowledge produced by the exploration of the new phenomenal area, it was necessary to reconsider attentively basic categories—like causality or determinism, observation, object, objectivity—of which the function had seemed before to be definitively understood. But at the same time many other concepts—like those of state of a microsystem, of magnitudes or descriptional quantities characterizing such a state, or the significance of the probability linked with a state—concepts which possibly could come out to hold, in the questions of interpretation, a role no less fundamental than the classical categories mentioned above, were *totally omitted* in the epistemological and philosophical discussion. The quantum formalism introduced mathematical definitions for concepts of which the relationship with a physical reference was left obscure, while the very relevance of specifying such relationships was denied and banished from the start on, on the basis of alleged philosophical reasons [83,81,86]. In this way the intelligibility of quantum physics has been artificially immobilized in an incipient stage. This is illustrated by the recent debates on the meaning of what is called local non-separability (of genetically tied quantum microstates) [6,63,27,110,18] and by the subsisting doubts concerning the theoretical status and the physical meaning of the "principle of reduction of the wave packet" (the problem of quantum measurements) [119]): these debates point clearly toward important zones of obscurity that persist in the question of the significance to be assigned to the quantum mechanical formalism.

This situation is clearly unsatisfactory. Indeed the quantum the-

[6] From d'Alembert to Mach, Boltzmann, Einstein, etc.

ory is not restricted to fundamental quantum mechanics, it extends to the quantum field theories which today possess a considerable importance, and which, in the long range, will manifest their own conceptual and physical implications, certainly related to those of fundamental quantum mechanics.[7] Therefore, when we speak of the general lessons to be drawn from quantum physics, we do not mean what is currently called "the philosophy of quantum mechanics", which refers mainly to Bohr's concept of "complementarity" or to its variants according to others physicists, insofar that they put theoretical and conceptual criticism under the dependence of a philosophy of the observation. Much more generally, we mean a global understanding of all the theoretical, epistemological and philosophical questions raised by the new knowledge concerning *quantum phenomena*, which, with the relativistic approaches, constitutes one of the two most important corpuses of knowledge from modern physics [82,83,81,86,85]. So the representations of quantum phenomena have to be thoroughly studied from an epistemological point of view: this is a main task of a nowadays formalized epistemology.

Various other contemporary physical theories have also surprising implications concerning relationships that were usually considered to be unquestionable, but which now require a rigorous critical reconsideration. Consider for instance the one-to-one link posited between determinism and prediction, which lasted unchallenged such a long time and now is found to be inconsistent with the modern theory of nonlinear dynamical systems and of the "chaotic" phenomena manifested by these: the dogma of predictibility in terms of space-time coordinates, of the individual trajectory of a moving body, looses importance, while other global descriptors, like the *strange attractors*, are pertinently defined [108,109]. Many other such epistemological re-evaluations are strongly needed today if one wants to grasp the whole of our present knowledges accordingly to a wider, global view endowed with inner consistency and offering a deeper intelligibility. Which possibly—probably—would also entail *reformulations.* Obviously, an integration of the kind alluded to above, could be achieved only by precise, "differential" analyses of each of the already acquired knowledges. But these, if achieved and then confronted with *history*, might surprisingly bring forth *general structural lines concerning some definite characters*: in this sense, history and new knowledge can be brought to work together for the elaboration of a more integrated and deeper epistemology. Of course, regularities of the mentioned kind cannot be perceived at a first glance, nor with full generality from the start on. Passage of time, with the sedimentation entailed by it, is necessary for the regularities to draw attention upon them. Any regularity, when it first becomes perceptible, is perceived as a mere coincidence, while

[7] [64], Chap. 8.

coincidences and analogies are—rightly—felt to be too weak to be assigned thorough consideration.[8]

One may wonder, however, whether it would not be possible to imagine a point of view or an axis for research able to constantly yield a global perspective on the facts and problems with which the regional, specific epistemologies of the various theories are confronted: a perspective that would keep encompassing their wealth and variety while also giving at the same time an insight into the main structural currents that organize them into one whole according to some common, unifying, synthetic intelligibility. Getting at such a point of view might allow to better control the unavoidable changes which occur inside any sort of knowledge, to discern the main directions generated by them, to set up a reasoned inventory of these, and perhaps to anticipate other ones.

Such an anticipation, founded on a systematic search for regularities or trends, is not unthinkable. Indeed, one can investigate concerning the reason of similarities, analogies and convergences observed in the transformations occurred inside totally different regions of knowledge, and to strive connecting them to morphological or functional features which, like a pre-set conditioner, enveloped them all from the start on and was doomed to eventually manifest itself. If this were true, previous systematic examinations would possibly have been able to become aware of such acting pre-set conditioners, without having had to wait for the whole variety of cognitive events to occur in a contingent way. Such considerations suggest a kind of meta-epistemology, to be invented, if only it is possible. This, with respect to the regional epistemologies of the various definite theoretical representations, would yield a kind of an analogue of what, in Hermann Minkowski's eyes, would have been an *a priori* mathematical theory of space-time if it had been developed before the physical theory of special relativity [51,52]. Let us note that hypotheses of this kind seem to involve a wholly rational conception on the emergence and evolution of the theories and knowledges, and of their links with the meta-theories liable to frame them. Which, however, would require specification of what is to be understood by a "rational" conception of the mentioned kind, compatible with some allowance left for also invention. For, if not, could we conceive of invention to be merely the unpragmatic apprehension of our accessing to progresses in knowledge which "in fact" were determined with an absolute necessity?

Anyhow, without prejudging on the answers to all these questions, the preceding considerations incite to bestow particular attention upon that which, in the stage of mutual readjustments of previously established knowledges and even while the elaboration of new knowledges, comes up in the

[8] On analogy, see Paty [88].

kinds of the "operative" and the "formal". What is to be understood by these terms, however, has to be closer examined before trying to set landmarks for a reflection on what can possibly be expected from an approach founded on the concepts they refer to.

3. EPISTEMIC OPERATIONS

We shall call *epistemic operation* an act of thought (or a series of such acts) by which a body of knowledge is constituted, no matter whether yes or not the nature of this act is consciously perceived while the process of generation of knowledge goes on: this nature can be recognized in only a further stage, when the contents and procedures from a given area of knowledge are examined.

A simple example of epistemic operation can be found among the current methods of contemporary physics, namely in the search for *invariants* while building a definite physical theory: Lorentz-invariants are selected inside the framework of special relativity; what stays invariant under a given "gauge" symmetry posited to be fundamental for the representation of some given kind of dynamic interactions between "quantum" particles, is selected as a relevant quantity in the theory of these interactions; etc. This practice has become usual in physics since the advent of general relativity and of quantum physics, in the years 1920-1930. But its origin can be found in the mmoire composed in 1905 by Henri Poincaré on the dynamic of the electron [92] (*La dynamique de l'electron*), at the time when this author was planning to build a theory of gravitation modified with respect to that of Newton by imposing a condition of "covariance", or invariance under the "Lorentz transformations" of space and time coordinates, of the equations expressing laws, which means subjection to the principle of (special) relativity. This practice in theoretical physics is related with the importance gained since then by the connected notions of *group of transformation, symmetry and invariance*, as defined inside the corresponding mathematical theory. This importance has been fully understood with Einstein's theory of general relativity. It received a first formal systematization with the theorem of Emmy Noether [58,59]. It has later guided the elaboration of quantum mechanics and then of the quantum field theory, up to the recent developments concerning the fundamental interaction fields obeying gauge symmetry (invariance).

This sort of epistemic operation has considerably modified the concept of physical theory as well as the everyday practice of the theorists. Once one admits it to be justified, it is easy to formulate; but an a priori justification by use of some simple explanatory scheme would distort or ignore the "facts that resist" of historical reality. After its "invention", an invariant usually is so well understood that no justification seems to be needed. But this does not entail that we can regard it as a natural *a priori* evidence. If in our retrospective look the invariants we make use of nowadays appear to be endowed with such obviousness, this is so because we consider them from inside a conceptual universe where they already work as a reference, in consequence of a radical reorganization of our knowledges and of the methods of theoretical physics. But the concept of invariant itself clearly possesses a historical origin that can be traced back to structural and conceptual changes occurred in physics at the beginning of XXth century. This concept appeared—it has been *invented*—in a conceptual world still strongly marked by conceptions and practices radically different from the nowadays ones, and in circumstances the study of which pertains to history, and more specifically to historical epistemology.

This example, by its relative simplicity, permits to clearly perceive that, in our study of epistemic operations, we should distinguish two levels: a first level of study of the use of this or that epistemic operation, inside the scientific work (in our example, the use of invariants for formulating laws and physical theories); and a second level of study of this same operation but considered to a "second degree", as stemming from procedures involving historical aspects which have triggered its invention (invariants, in our case). This second level is that of the *constitution of epistemic operations*, of their elaboration, and it is not reducible to only descriptions of the operations themselves: this is connected with the question of the genesis of new conceptions, of creation of novelty in science, and more generally, with the question of the emergence of new forms in the cognitive thinking.

One can also consider as an epistemic operation the fact of reasoning inside the framework of some given logical structure and some category of thought that inform our "interpretations", our manner to assign meaning to the concepts and the theoretical statements we make use of. For instance, concerning the propositions from a physical theory, the notion of *causality*, some specific way of understanding *determinism*, the meaning ascribed to the concept of *probability*. Interpretations in this sense act on the manner in which a problem is processed, this manner being possibly common to researchers and specialists from a given period, or else, differing according to the individuals or the schools of thought. But they concern even more the way in which knowledges are understood and justified; this way, moreover, acts upon the dynamic of thinking and upon the decision to continue on a given direction of investigation, or not. For instance, the notion of "theoretical completeness", raised by general relativity and by quantum physics, suggests a program of research in view of modified theories, program that is accepted or not according to the position adopted with respect to the considered problem [65,84]. None of the two mentioned theories is complete in the "strong" sense of a "self-generation" of its own objects, which is indeed the sense involved by the present attempts at unifying the field theories, so the question which persists is to know whether they are complete in some weaker sense (are they sufficient for determining the properties that can be assigned to their objects?):[9] this, for Einstein, was a fundamental condition required for trying to accomplish a unified description of the physical object-entities.

New results obtained in a given science might challenge epistemic operations which otherwise would have been considered to be definitively obvious. Such has been the case for the concept of causality that has been modified by the theory of special relativity which, by obliging to distinguish between the space-like and the time-like regions of the light cone of an event,[10] has entailed changes in our conception of the relationship between cause and effect. All the regions of the space-time diagram are not equivalent: if the time-like region is physical, the space-like region is non physical (there is no possible causal relationship between its hyperpoints). Henri Poincaré himself, with regard to his own ideas, asserted this concerning Minkowski's spacetime.[11]

On the other hand, the quantum mechanical concept of "probability amplitude" entails a modification of the idea of probabilities as formed in the classical physical theory. Probability is a mathematical concept. When applied in physics, it is generally identified with the limit of convergent relative frequencies of events, according to the law of great numbers. The construction of quantum theory makes use of it in a way which is indirect. Namely, the theoretical probabilities postulated concerning the quantum mechanical states are calculated from the "probability amplitudes" or state functions or vectors belonging to the Hilbert space of the microsystem which represent the considered microstates of this system, with operators taken as the quantum mechanical variables. These theoretical probabilities are afterward put in relation with the experimental relative frequencies of physical observed events. But since a theory defines the meaning of the quantities it makes use of (via the relationships between these) independently of any experiment

[9] This was the essence of the "EPR argument" [26]; see Paty [72,89].

[10] The light cone, defined by the equation $x^2 + y^2 + z^2 - c^2t^2 = 0$, determines an inner time-like region, such as $x^2 + y^2 + z^2 < c^2t^2$, and an external space-like region, such as $x^2 + y^2 + z^2 > c^2t^2$. The former is the region of causal relationships between spacetime points, the latter is that of a-causal relationships ("nonphysical" region).

[11] Poincaré [95]. Cf. Paty [73,74]. This was shortly after Paul Langevin had discussed in philosophical meetings, in Poincaré's presence, the physical implications of the new relativist conceptions on causality (Langevin [46,47]; cf. Paty [73,84].

(the experiments providing only knowledge of particular values occurring for these quantities), the theoretical quantum mechanical probabilities possess a specific theoretical meaning, and *this is not reducible to the classical frequency meaning of the probability of an event* (think of the probability of events involved in the self-interference for one single photon); it posseses a meaning which differs from that of the probability of the result of a cast of a die, as Dirac remarked already in 1930 [22,81,86].[12] All the well-known controversies on this subject are nothing more, in fact, than intuitive substitutes for a crucial conceptual breakthrough, namely that which led from the classical concept of a descriptional quantity endowed with numerical values *that can be conceived to pre-exist*, and the quantum mechanical concept of a quantity, which is more complex, rejecting the mentioned conception [85,86,87].

The question whether the epistemic operations can be identified with certain *algorithms* will be discussed later. First we shall have to make clearer what we mean by the term "formalized" from the expression "formalized epistemology": We would like to show that, and how, this meaning largely transcends that usually assigned to the word algorithm. We shall also consider more thoroughly the concept of "object", which will permit us to specify more narrowly the purpose of a formalized epistemology, with respect to general features of the epistemic operations.

4. VARIATIONS ON THE THEME OF A FORMALIZED EPISTEMOLOGY

In the classical sense, *formal* —thus also *formalized*—somehow opposes to material, as for example in the case of Aristotle's formal cause (bearing on an idea or an essence), or as in the antagonistic couple *form-matter*, or else, in the most current meaning of the word "formalism" ("purely formal" is meant as devoid of a "real" content in the sense of a "material" one)[13]. This opposition, exploited by scholastics, refers to the *shape of the relations* existing between the elements of a cognitive operation, abstraction being made of the "matter" or *meaning* (reference) associated to these elements. With this meaning the terms form applies, for instance, to expressions such as *formal relationship* in algebra, which point toward connections that stay valid for any numbers so that literal symbols can be used instead of this or

[12]We leave aside here the various interpretations that have been proposed for probability in this context, from Werner Heisenberg's "potentialities" to Karl Popper's "propensities" [99,101]; these notions, as interesting as they might be for pointing toward certain problems, nevertheless are only "intuitive" and vague.

[13]Lalande [45]: articles Cause, Formalisme, Forme, Formel.

that definite number.[14] The concept of form is also part of the terminology concerning natural laws (the form of a law), while in philosophy, for example the kantian philosophy, the laws of thought involve pure forms of the sensible intuition, *a priori forms of the sensibility* (time and space), as well as *forms of intelligibility* (categories) or *forms of reason* (ideas). In a different, more recent sense introduced in the *Gestalttheorie*, the theory of structure in psychology, the word form points toward what obliges to consider an element as part of an organized totality, as participating of the structure and its structural laws.

Taking into account this whole variety of meanings, we can now get closer to a significance able to be directly useful to us in an attempt to reach the fundamental features of a gnoseological approach of the contents of knowledge.

Gilles-G. Granger speaks of "formal contents" in the case of *mathematics*, to be distinguished from the empirical contents involved in the natural sciences while at the same time bringing them nearer to these last ones: the mathematical forms are not *empty* forms, they also possess certain contents, namely those expressed by *relationships that are not reducible to tautologies* like the logical axioms, nor to mere well-formed expressions in the sense of symbolic logic [33,34]. Indeed the "formal" in this meaning, i.e., of *formalized contents*, cannot be identified to the "purely" logical, which is exclusively formal *by construction*, since, by definition, it is not opposed to content. In what follows we intent precisely to clarify the relationship between the formal and the corresponding content in the case of a formalization of a content (aspects of this relationship will appear in the discussion on the concept of object). Thereby the opposition between content and form will be transcended.

"Formalized" can be opposed to what is a *particular* empirical description at a phenomenal level, even if this description has been obtained inside a natural science, via a theoretical approach. Consider mathematical physics and theoretical physics, between which we distinguish. What mathematical physics is mainly interested in, are the formal relations between the mathematical quantities posited to qualify physical entities (objects or phenomena). This way of being formal is of the same kind as that which occurs in mathematics; while in theoretical physics one is mainly concerned with the physical contents pointed toward by the involved mathematical relations. Mathematical physics appears in various respects as a formalized approach if it is compared with experimental physics regarded as the source of contents; and the same assertion holds even if it is compared with theoretical

[14]In ancient writings (like the scholastic ones) "formal" calls forth "actual", while in the mathematical or logical sense it calls forth the idea of "general".

physics. However, rather often theoretical physics converges with mathematical physics, whereby the referents of these two denominations are brought to a momentary identification (like in the case of the analytical mechanics as exposed by Joseph Louis Lagrange and Rowland Hamilton [70,78]; or Hermann Minkowski's formulation of special relativity [51,52,116]; or Einstein's theory of general relativity, or certain presentations of quantum mechanics (in particular those of Hermann Weyl and John von Neumann [17,57]), or many developments of gauge field-theories (consider the works of Yang and Mills), up to the most recent researches on quantum gravitation).[15] But the above distinction between mathematical physics and theoretical physics is only relative, and the periodic identifications between a piece of mathematical physics and the corresponding piece of theoretical physics prove that, as highly formalized a piece of mathematical physics might be with respect to the representations of natural phenomena achieved in theoretical physics, the mathematical relationships involved by such a representation never become strictly alien to the physical contents of phenomena: only, at certain privileged limits, the *determination* of the form of the mathematical relations that express these physical contents is so *unambiguous*, so achieved, that it is exactly the same inside both theoretical physics and mathematical physics, and then this superposition, by suppression of any comparability, generates an illusion of absence of physical content, so of "purely" formal.

Coming back to the previously mentioned case of invariants, these, at a first sight, might be perceived by thought as purely formal relationships, but in fact they express general and fundamental properties of the physical systems and of the quantities by the help of which these are described in the considered physical theory. So, far from being external and superficial, the invariants express *physical contents*, they are bearers of *meaning*, privileged bearers of certain specified sorts of meaning, maybe the only possible bearers for those sorts of meaning. It is in this way that Poincaré considered the "mathematical analogies," [16] and Einstein considered the "formal analogies," [17] which amount to the same thing (for mathematics is the "formal" of the physicist). Poincaré and, some time later, Einstein did not hesitate to speak of a "heuristic of the mathematical formalism" that drives the "physical thinking", precisely because this formalism, in the cases considered by them, was impregnated with, and informed by the physical meanings which it served to express.[18]

Although the nature and role of the mathematical formalization in

---

[15]Notably, Ashtekhar [1]. Cf. Kouneiher [41].
[16]Poincaré [91]. Cf. Paty [88].
[17]Einstein [23]. Cf. Paty [68], Chap. 4, pp. 164-172.
[18]Paty [68], Chap. 5.

mathematical physics draws already attention to the fact that formalization by no means excludes content and meaning, this does not yet tell us what, exactly, should be conceived to be the purpose of a "formalized epistemology".

Does one mean, for example, a study of the formal aspects of various epistemological approaches, in the way in which certain art-critics study the formal aspects of some given set of works from some given area of artistic expression? As for instance in the case of the critical analyses of the literary forms occurring in Baudelaire's or in Mallarmé's poetry? Or otherwise, does one imagine a study of forms the aim of which be not exclusively to describe or to characterize existing "styles" of writing or thinking,[19] but which are furthermore intended to disclose new forms of expression, or even—operationally—manners to *generate new contents, new meanings* by concentrating attention upon the form, for example by imposing *formal constraints* while achieving a work of some given charcter. This tendency is frequent in contemporary art, in painting, in music, or in literature (consider the exercises of the *Oulipo* group[20]) where poetical or literary writing is submitted to formal constraints,[21] which generates aesthetic innovations and "effects of meaning".

Consider what might be called an *epistemological reflection*: it also deals in part with forms, and it also takes on forms itself, forms that depend on the modalities of the practised approach and that are connected to questions of meaning, in a somehow more direct and compelling manner than the aesthetic forms: a meaning which, even if it is conceived to pre-exist, is not necessarily also known for that, and which could be tried to be brought into light from under the facts and appearances. Such an epistemological reflection would certainly not be reducible to a *free* creation of form, nor to a creation of pure form. In any study, the object to be studied is in a certain sense, at least to a certain extent, given in advance from outside the study, and the study suffers constraints entailed by this object and tied with its externality. In particular, in the case of any given sort of scientific knowledge, mathematics included, the source of the constraints which restrict the representation can indeed be mainly assigned to the object of the knowledge to be elaborated. Now, will a similar situation manifest itself in the case of this meta-study which we want to call a formalized epistemology, devoted to scientific knowledge as a whole regarded as the "object" of study? (We let

[19] On the notion of style in science, see Granger [31], Paty [66], Chap. 4, [68], Chap. 1.

[20] *Ouvroir de littérature potentielle (Opening device of potential litterature)* created in 1960 and animated in particular by Italo Calvino, François Le Lionnais, Georges Pérec, Raymond Queneau, Jacques Roubaud (Oulipo [61]).

[21] See, e.g., *La disparition*, by Georges Pérec (Pérec [89]), a novel written without using the letter e, a vowel yet so ubiquitous).

aside for the moment the question of formal constraints possibly imposed by the gnoseological structures from the subject's mind).

A "formalized epistemology" should be, so to say, in close connivance with its object. This excludes any formalisms that would be more or less alien to this object, that is, to scientific descriptions of domains of real facts. It also excludes any identification of the desired formalized epistemology itself, with a formalism that would not be *directly* relevant to its object (like arbitrary algorithms or exercises dealing with arbitrary models). So a formalized epistemology might focus either on forms that characterize formalizations of pieces of scientific knowledge, or upon the operations by which these forms have been established (which we called epistemic operations). If this is agreed upon, the margin still remains large for defining now the manner, or manners, of realizing the aim just sketched out; which, incidentally, is an advantage: it thus remains possible to choose the manner which *a priori* seems to be the most adequate, and so the most fruitful one.

On the other hand, *formalized does not mean quantified.* The mathematical representation of physical phenomena illustrates this via the distinction entailed by it between the *qualitative* consequences of the mathematization (in the modern sense, not in that of Aristotle and of the scolastics) like the mutual disposition of the defined quantities and of the levels of *order* of magnitude that reflect the involved *conceptual* constructs, the ideas, and on the other hand the quantified, the measures, of which the expression is numerical.[22]

Furthermore, *formalized does not mean schematic*, in the same sense in which describing the form of an animal does not mean to reduce it to its skeleton: it concerns the fact and the manner of taking on a form. And the question of the form in which scientific knowledge in general and the particular sciences manifest themselves, either when already constituted or while coming into being, suggests that form has to do with *intelligibility.* Indeed what sort of intelligibility is offered by a given scientific knowledge via the form acquired by it?

Also, though between form and *structure* there exists a narrow link, it nevertheless is necessary to distinguish them from one another, form being more global than structure.[23] The form expresses the structure, but only partly; it also expresses non structural characters but which must be compatible with the structure. On the other hand, a same piece of knowledge may dress on various forms and one might wonder whether these correspond

---

[22]See Paty [69].

[23]On *structure* in science, see, e.g., Stegmuller [112,113]; and often less defined. On structure for history, cf. Foucault [29], Veyne [114,115]; and, for anthropological representations, cf. Levi-Strauss [49].

to different intelligibilities. This question, in fact, linked to that of "interpretations", appears as a quite fundamental one: one could see there one of the key-articulations of a "formalized epistemology".

"Epistemology of formalized knowledges", "epistemology of form and of formalization", "formalized epistemology", ... A scientific meta-approach concerning the function of form in knowledge can also be seen as a *methodology* on the basis of its heuristic-oriented preferences. It remains to know to what an extent it can pretend to generate norms for conceptualizing and for reasoning, norms which would enable to reproduce or to anticipate inventions (that are precisely inventions of forms): this is not obvious *a priori* given what is known up to now concerning understanding and creation of concepts. In certain sense reconsidering a concept amounts to recreating it, and this happens inside the unity of an isolated, unique, subjective mind (subjectivity being regarded to be the place where reasoning can occur). Reasoning, creation of new conceptual forms, does not operate exclusively with countable rules that lend themselves to classification according to some typology; it goes on inside a consciousness which, throughout the processes of understanding and constructing, mobilizes many other instances besides those consisting of the identified "elements" of the problem to be solved. By "other instances" we do not mean the psychological or sociological ones of which, in a first stage at least, abstraction has to be made in the present investigation, we mean factors which in general keep implicit, unconscious, but which nevertheless play their part in the global economy of our acts of conceptualization and reasoning.[24]

The preceding remarks lead us to try to situate the concepts of a formalized epistemology and of epistemic operations, with respect to *judgements* and to the *decisions, choices*, occurring inside the mind of the subject of knowledge regarded as an *epistemic subject.* Which leaves then to be specified what remains of this subject—the unique place for intelligibility—that can be assigned to "objective" knowledge which, as a matter of principle, is required to abstract away any subjective *singularity*, keeping exclusively the "general subjectivity" as expressed by *operation, process and content "in any subject"* or, as some have told, "without subject". However, insofar that knowledge is produced, evaluated, communicated, by means of acts of creation and of judgement, the existence of *subjects* as the *loci* of such acts is unavoidably necessary in an epistemology, even in a formalized one. Without them knowledge would bear on contents devoid of intelligibility, or at best endowed with an anonymous and abstract sort of intelligibility: but one is in right to wonder whether contents in this sense are genuinely

[24]On scientific invention, see Hadamard [35]; concerning the rationality of this invention as confessed by several scientists, see Paty [79].

conceivable: *can* a content of *knowledge* be *merely* schematic?

A subject is *always* present in filigree beneath certain elements which, upstream or downstream of any knowledge, qualify the conditions, the modalities and the effects of the epistemic actions that have produced that knowledge. Here we restrict to only mentioning the presence of such elements. Their enumeration and study belong to epistemology in the general sense, structural or historical: conditions of possibility,[25] fields of rationality, styles, programs, intelligibility, intuition ... .[26] In the present context, the difference between formalized epistemology and epistemology in the general sense is that, if the second one takes into account the above mentioned elements as objects for study, the first one takes them into account only as a *given* conditioning datum, which it wants to transcend or, more exactly, with respect to which it situates itself in order to bring forth "structural invariants".

5. OBJECT AND CONVENTION

As epistemology in general does, examining in particular epistemic operations leads to discern in the operations of knowledge a preliminary part, of a conceptual pre-organization by the mind, which "prepares" an object of knowledge, or rather, the conditions of its identification.[27]

All sciences, whether exact, natural or social ones, are nowadays conscious of the necessity of *a critical re-consideration of the concept of object* by taking into account that an object is defined by separation from the subject who introduces it, which entails a critique of also the concept of *objectivity*. On the one hand, no object can be designated in the absence of a *mental act tied with an intention of the acting subject*. On the other hand, any object is defined by its *distinction* or *separation* from—both—the subject, and a *background* against which it stands out. It is furthermore well-known that science cannot be restricted, as for the conceptualisation of its objects, to the characters assigned by common sense: in this respect again, quantum physics offers particularly precise lessons, to which we refer without being able to detail them here.[28]

There undoubtedly exists in science, especially since the belief in the

---

[25]Kant [40].

[26]Granger [31], Lakatos [44], Zahar [121], Chap. 1, Paty [64,65,66,68].

[27]The term "preparation", rightly stressed by M. Mugur-Schächter (*op. cit.*), comes from quantum mechanics. But awareness of this rather universal pre-organizing procedure (to which quantum mechanics has associated a peculiar flavour and an increased precision) was present already in previous epistemological reflections (see, in particular, Margenau [51]).

[28]Paty [64,65,63,83].

uniqueness of geometry has dissolved, a remarkable trend toward a *representational transformation of the objects, in relashionships.* This trend is most definite in mathematics, but in physics also it is quite clear, supported by the fact that physical theory takes on increasingly the form of a mathematical physics. Now, many have tried, or still try, to present this trend as a "de-ontologization" of the sciences. However, by this trend, the concept of object is by no means *abolished* in that of "mere" relationship. For in order to deal with relationships, there must be some *entities* which are related, say, some "elements", even if the nature of these is undefined or problematic. These—at least—are the "objects of the relationship". And it is indeed in this way that one tends to conceive of the concept object inside mathematics and inside modern physics as well. It is also true that, while the progress of increasingly abstract formalization of the physical theories proceeds, the related "elements" themselves, in their turn, go over into new relationships, partly at least. In a mathematical and in a physical theory these "elements" can, and actually are, defined in and by the relationships in which they partake (such is eminently the case in the quantum theory).

But, exactly insofar the system of such relationships is not entirely "transparent" or "tautological", it expresses a—structured—content that resists total dissolution into the "exclusively relational", whatever this might mean; or, to specify it, meaning relations without anything being related.

This content which resists total dissolution in the relational and which is what has to be known, possesses the attributes generally pointed toward by the term "reality". An example which at a first sight is simple is offered by the system of integer numbers as organized by arithmetic, for at a closer inspection this system brings forth the complexity of the involved relations, with the type of "reality" underlying it. Though it is possible to generate any integer by starting with the number 1—the unit—and by operating "additions" of this unit, first to 1 itself and then, progressively, to the number constructed in this way by preceding additions (and labelled 2,3, etc.), it is not possible to know in advance all the properties of the numbers constructed in this way, for instance whether some given such number is prime or not, what relationships with other numbers it does satisfy (think of Fermat's great theorem), etc. Note, however, that the generating relationship (repetition of the elementary additions of one unit) implies not only numbers, but also opeation and number of operations in numbers, so that the mode of generation looks simple in appearance. The related "elements" and the relations between these form together one organic whole of form-with-content wherefrom the content cannot be eliminated thereby loosing also the form.

Physics also provides good examples. Consider the concept of field, cleared by the special theory of relativity of the material support called ether: all the physical features of what is called a field are defined by the mathematical *equations of that field.* Or else, consider the non-distinguishability of identical quantum systems ("particles") which defines at the same time the relationship and the object of this relationship; and also the symmetries of elementary particles that determine these particles-in-their-mutual-relationships, via their interaction-fields (gauge fields).[29]

It might even be more adequate to say that the relations are what becomes *as concrete as something real.* For if the transformation, always possible, of given elements into others "more relational" ones makes the "objectal" nature of these more and more relative, it nevertheless remains that the relationship itself finally appears as endowed with all the characters of what calls an "object" endowed with an own consistency of a peculiar nature. A nucleus of structured relationships closely interwoven is devoid of none of the characteristic features of what an "object" means, once freed from the "substances" of the ancient metaphysical doctrines. So the concept of "object"—mathematical or physical (and, by extension, chemical, biological)—looses its traditional relation with a directly *ontological* perspective. With regard to it the category of "being" is not deleted, but is nevertheless removed at a certain distance by increasingly many relational mediations. *But the predicates of existence remain, because they are of a non removable epistemological nature.* So the ambiguities of the meaning assigned to the word *ontology* cannot justify a suppression of the concept of object. A representation or a theory is always representation or theory of *something*: by definition, this "something", the necessity of which is a logical one, is the "object" that this theory designates and characterizes.

As soon as the contours of the representation or of the theory at stake have been drawn, *ipso facto* its object has somehow been determined. Obviously it is by an *act of the thought* that we have designated it as the object to be described by that representation or theory. Such an act contains a *choice* to separate this object from the "rest", and it indicates a program involving certain *conventions* which the object, in this sense, carries with it. The conception on *objectivity* is correlatively affected by this state of affairs: it is not only *given* (because thought is drifted along by the object), but it is also *decided* (according to some norms or conditions) and *built* as well (in relation with our choices in positing the object). The chosen conventions depend upon the concepts and the theoretical system which weave them together in order to describe the object. They are relative to this system, and alternative conventions are thinkable, that do not bear on

[29]Paty [82,81,86].

only the theory, but also on the totality of the elements of meanings for the theory which insure for it a definite intelligibility, for which the criteria are themselves partly meta-theoretical. It important to study and to specify in precise terms the interplay of these two concepts, object *versus* convention, which calls forth requirements that vary with the considered type of theory.

We add only a few more words concerning this point. Quantum mechanics is often asserted to have eliminated the concept of object, at least insofar as this concept is considered to be independent of any act of observation and of conceptualization and to preexist to any such act. It is true— and commonly admitted, even beyond quantum physics, as already stated above—that the object has become object by an operation of the mind which has separated it from the rest (or has prepared this the conditions of this separation). But this "condition of possibility" of thinking the object does not exhaust the descriptions which one can perform of it: in fact, it only makes these possible by opening up the field for their realization.

In this respect one may consider that the specific problem of quantum mechanics is that of the nature of the acts of thought and of the operations required to get access to the description of the object to be studied, or, more exactly, of its state. I have tried elsewhere to show (here I cannot detail on this) that quantum systems and their states can be conceived in terms of objects endowed with properties, at the cost of an extension of the meaning assigned to the concept of physical quantity, beyond its usually accepted meaning of a directly numerically valued quantity, and of a correlative extension of the concept of state function:[30] in the practical understanding of the physicists these extensions are already performed but they are not yet explicitly admitted. Notice that already before the construction of quantum physics one could have find of "objects" represented by abstract magnitudes that had already lost the features usually assigned by common sense to any magnitude (such has been the the case of the light wave, spreading out in the whole space, or the field without the support of an ether, as evoked above). Let us also remark that the concept of state function, already involved in Hamilton's mechanics, was suggested to him by optics with its principle of minimum, which refers to another still more ancient origin that can be found in both mechanics and optics, namely the principle of least action.

Let us retain at least what follows: it has been possible to build representations that were said to correspond to a definite object, without being in possession of a deep justification other than the mere fact that these representations proved to be effective: this sufficed for asserting that a description of the object and of the phenomena related with it was available. And the encountered epistemological difficulties always concerned the tools

[30]Paty [81,86,87].

that were made use of, rather than the nature of the object of the considered theory. In fact the conceptual-theoretical tools from a theory have always been adjusted in view of the object they were intended to make us conceive of: not to make us perceive it directly, but just to make us conceive of it via the mediation of the theory constructed in order to represent it. In particular, the operational tool consisting of the quantum mechanical algorithms, has been elaborated by a process of adequation to the aim of representing coherently a certain system of objects, namely the world of quantum objects and of phenomena tied with them. By the logic of their fabrication, the tool and the elements of the quantum mechanical representation are made of the same stuff. And the formalism of state functions defined in a Hilbert space and of operators acting on these, which are aimed at representing states of quantum systems, involves *by construction* the rules for making use of it. However, inside their own physical domain of existence, the "objects" toward which the state functions point, namely the states of quantum systems, do not need any more, in order to be thought and specifically referred to as *physical* entities, to still be constantly referred to the abstract tools which, in a certain sense, have led to bring them into being, and which also permit to detect them (in terms of events coded in a language of eigenvalues of eigenfunctions of quantum mechanical operations and of probabilities of such events). But in fact they are systematically conceived of accordingly to the way in which they have been designated, so in full agreement with the quantum mechanical formalism.[31] In a way, the well known problems raised by the quantum mechanical formalism is not so much that of the represented quantum objects, as that of the relation between this quantum representation and the classical "mechanical" representation, adapted to the involved experimental devices.[32]

These remarks suggest a process which in a certain sense is opposite to that of a progressive "syntactization of semantics", in the sense of the philosophy of language, of the mathematization of physical contents, or of the transformation of the objects in relations.[33] If one considers the evolution of the question of the interpretation of quantum mechanics since the first debates on the subject, and the subsequent familiarization acquired in this area by physicists, up to the new knowledge made available during the last years and to the reinterpretations that may be formulated consequently, it is tempting to speak, on the contrary, of a *semantization of the syntax*. The description of quantum physical systems was conceived previously only by means of operations, while henceforth (by means of the transformations

[31]It is possible in such a way to "think quantum non-separability" (Paty [63]).
[32]Paty [85].
[33]The expression is from Ernst Cassirer, cf. Cassirer [13]. See Granger [33].

alluded to above in our definitions of the involved physical magnitudes) it might evolve toward an interpretation in terms of physical systems conceived as objects possessing properties: this being at the expense of modifying our definitions of what is a physical quantity for the description of such properties. The previous "syntax" remained exterior to the physical content it concerned, insofar that it considered itself to be strictly confined to purely formal means of the description, without declaring a definite position as to the physical feature of this physical content, and even its very existence. A full achievement of the program of the *semantization of the theory* would consist of formulating from the start on the quantum theory as the theory of a category of physical objects and of their physical properties. This would be a necessary task before entering again upon new syntaxizations. (One should be able to substitute to the axiomatic formulation *à la* von Neumann [34] an equivalent axiomatic formulated in terms of physical properties concerning physical entities at the quantum level).

6. ALGORITHM AND NATURALISTIC REDUCTION OF REPRESENTATION

The notion of epistemic operation quite naturally leads to ask to what extent such operations can be expressed by algorithms. In the case of simple operations, such as the search of invariants, it is perhaps possible to imagine equivalent algorithms. However, though an algorithm for the construction of an already known invariant might be found, this does not entail the possibility of also an algorithm that would have led to the discovery of that invariant. For complex operations bearing on the acquisition of new and more precise knowledge, the opinions concerning the possibility of algorithms diverge. The experts in artificial intelligence and its adepts will willingly reply affirmatively. According to them, in the long range, any cognitive operation, scientific inventions included, will be reconstituted by machines: for simple operations they are already proposing models, and they are proclaiming an obvious necessity of principle in any case, sending skeptics back to the archaic matter-mind dualism.[35] Why, do they claim, should we hold that brain is essentially different from a machine, a neuronal machine?

Of course, by certain aspects, this attitude has ancient precedents. The Cartesian research of a method to get a certain knowledge, Leibniz's search of a modality for formulating the totality of knowledge in a universal and perfectly logical language, and even the *parti-pris*, in Spinoza's *Ethics, of a more geometrico proof* of each statements, not to go back still farther

[34] Neumann [57].
[35] Changeux [16], Damasio [21].

until Aristotle, testify of the permanence of similar concerns the history of philosophy; stemming from a belief in the possibility of a powerful algorithm permitting to found, to grasp and to organize the totality of "true" knowledge. This quest of synthesis by means of formal unity does not necessarily possess a reductionist character: in Descartes' view it was protected by the dualism (acted matter)-(thinking mind), and the Spinozian monism left full room for the specificities of all the different sciences. But in the case of the modern conception of the brain as a machine, a character of a quite different nature is involved: what is imagined in this case is not only an algorithm for the *representation* of any knowledge, but furthermore also a naturalistic reduction of the processs of generation of knowledge,[36] a question which cannot be discussed here.

The claim for algorithms of production of the scientific knowledge has been strengthened in our time by the logicist views of the philosophers from the Vienna and the Berlin circles [37] and of their successors, dissidents or not, up to analytic philosophy.[38] According to the logical positivists and empiricists, science is doomed to generate a *compelling* philosophy of knowledge (the so called "scientific philosophy", exposed and supported by Hans Reichenbach[39]), a philosophy rooted in the experience data taken as the fundamental reference for knowledge. The search, with Rudolf Carnap, for an inductive logic, be it only a probabilistic one,[40] presupposed the idea that any scientific knowledge can be reduced to rules valid everywhere, perhaps in all times, which have to be discovered: which amounts to assigning to scientific knowledge an algorithmic essence. The assertion of confidence in methodology,[41] and of the normative legitimacy of a "rational reconstruction" of the scientific knowledge (Reichenbach, Popper,[42] among others) that would allow to collect the irrationalities due to the intervention of the subject in a knowledge of which the vocation is to be objective, so a "knowledge without subject", are other attitudes that go in the same direction.

The Popperian "third world" of the forms of objective knowledge [43] can also be regarded as an indication (or an effect) of this sort of view: such

[36]See the quite interesting dialogue between Jean-Pierre Changeux and Paul Ricœur (Changeux and Ricœur [17]).

[37]Wiener Kreis [19], Hahn, Neurath and Carnap [36], Soulez [111].

[38]See Joelle Proust's book on "questions of form" of logic and of analytical statements, from Kant to Carnap (Proust [103])

[39]Reichenbach [105,106].

[40]Carnap [10,11], Jeffrey [39].

[41]Bunge [8]. Alberto Cupani (Cupani [20]) rightly reminds us, however, that, for Bunge, the method is not a "recipe" that would be mechanically or automatically applied.

[42]Reichenbach [104,105,106,107], Popper [99,100].

[43]Popper [100], p. 154.

an impersonal universe of ideas is supposed to be that of pure rationality, cleared of affect as well as of chance, and even of materiality (these being sent back, respectively, to the second world and to the first one). This world of created forms, without the acts of creation, is akin to a reservation or a museum: a "museum of ideas", "virtual"—before the recent common use of this term—where one can draw prime matter from for formulating other ideas, is conceivable in fact only as organically related to the two other Popperian worlds. Karl Popper's third world seems to harbour a rather Platonian desire to purify the world of ideas from perishable elements such as the matter, flesh, affects, feelings and aims of which the individual subject is made, while also protecting a "logic of reconstruction" that is not very different from an algorithmic function.

This being said, we shall however recognize that an algorithm devoted to a logical application cannot be identified *per ipse* to a machine, because it does not necessarily involve the condition of reproducing the totality of the cognitive operations. Furthermore, one also should take into account a certain widening of the concept of machine, to include the possibility of new, *emergent* forms or properties concerning material systems as well as those from the space of ideas. But invoking a powerful algorithm or a machine, even with organic properties, as being able to reproduce or to describe the process of acquisition of a fundamentally new representation, gives rise to reserves, without any need to invoke some dualism, but on the contrary by holding an ontologically monist position: these reserves are similar to those which one can soundly oppose to reductionism and to a naturalistic conception of knowledge and of values. It is possible to emit them without for that denying any interest to the concept of epistemic operations.

It is possible to conceive an epistemic operation to be the source of an algorithm, such as the invariants considered above, or—another example—such as the Leibnitz differential calculus; and, once the algorithm has been invented, to permit to reconstitute or to reorganize a representation of it inside some chosen referential of meanings, with all the ascertained or predictable properties of such a representation. The algorithm could demonstrate its fruitfulness in the resolution of many problems, and it even could perhaps contribute to formulate new problems and to solve them. But can it be also conceived to generate, by itself, an essentially, a *qualitatively* new property or knowledge? This seems to be possible only if the algorithm contained in itself this break relative to its antecedents, that makes the new. But would we not, thereby, have already deserted the bounded framework inside which epistemic operations can be *formulated*?

So far as we know, the machine for producing *conceptual novelty* still remains to be invented. Does this entail that such a possibility is absolutely unthinkable for the future? The answer to this question, which artificial intelligence would like to be positive, depends on what can be said to be "qualitatively new": this characteristic certainly escapes the strictly *internal* content of a given knowledge, i.e., what in it can be formulated by reference to exclusively its *own* conceptual framework and inside the universe of meanings in which it is immersed. It is difficult to imagine the existence, or even only the possibility of a "machine for producing meaning", in the common acceptation of these terms, in the absence of a mind that would be at the origin of this meaning or that could "read" it.

Evaluating concepts and their possible character of novelty is of the order of signification, and as for now it is the human mind, tied with a brain inseparable from a body and from a practice of life, and setting aims for itself, by will or by desires, which imposes its meanings upon the machines. A "machine for produce sense" would have to possess the properties mentioned above, and certainly also others, including psychology and feelings: such a machine would astonishingly look like a man in society, whose generation by nature is the result of a very long—and maybe improbable—history the origin of which is lost in the night of times: a history of maturation, renewals, transmissions and exchanges, resulting from the diversification of the human phylum, biological, social and cultural, and from the accidents brought forth by chance. So the fundamental question is the following one: is it possible to conceive of an algorithm able to generate, for knowledge, meanings that would differ from all the already available ones and would appear to us to be legitimate, and even perhaps more certain or important? With questions of this type, we trigger, it seems, an unending chain of implications and an infinite multiplicity of open ways which a machine would have much difficulty to solve. While the human mind, issued from matter, does not calculate on all the possibilities as a machine does, it cuts across the available combinations and makes choices long before having exhausted them.[44] It simply posits the meaning which—according to its judgment that might be "subjective" in only a restricted sense—brings forth for it the sudden illumination of an intelligibility. This—be it Cartesian evidence, Spinozian knowledge of the third kind, illumination of the intuition in the sense of Poincaré, Einstein, and other contemporary thinkers—seems doomed to durably escape any reductionist representation or conception. This is so because a definite piece of intelligibility, in order to gain a ground, must call in other intelligibilities from an endless regressive chain. Like in Pascal's considerations on the situation of human intelligence in this world, that first lean on reasons which he believes to clearly understand by his experience but then, when questioned,

[44] As Poincaré noted when speaking of the "choice of the [significant] facts" (Poincaré [94]).

appear to draw down into a bottomless well.[45]

REFERENCES

1. A. Ashtekhar, "Non-perturbative quantum gravity: A status report," in M. Cerdonio and R. Rianci, eds., *General Relativity and Gravitation* (World Scientific, Singapore, 1989).
2. J. Audoze, P. Musset, and M. Paty, eds., *Les particules et l'univers* (Presses Universitaires de France, Paris, 1990).
3. F. Bailly, "About the emergence of invariants in physics: From 'substantial' conservation to formal invariance," in Mugur-Schächter and van der Merwe [56].
4. F. Bailly, "Remarks about the program for a formal epistemology," in Mugur-Schächter and van der Merwe [56].
5. B. Barnes and D. Bloor, 1982, "Relativism, rationalism and the sociology of knowledge," in Hollis, Luckes, 1982, pp. 21-47.
6. J. S. Bell, *Speakable and Unspeakable in Quantum Mechanics* (University Press, Cambridge, 1987).
7. L. M. Brown, A. Pais, and B. Pippard, eds., *Twentieth Century Physics, Vols. 1,2,3* (Philadephia Institute of Physics, New York, 1995).
8. M. Bunge, *Philosophy of Physics* (Dordrecht, Reidel, 1973).
9. M. Bunge, *Epistémologie* (Maloine, Paris, 1983).
10. R. Carnap, *Die Aufgabe der Wissenschaftslogik* (Vienna, 1934), translation of *Le problème de la logique de la science* (Hermann, Paris, 1935).
11. R. Carnap, *Logical Foundations of Probability* (Chicago University Press, Chicago, 1950).
12. R. Carnap, *Philosophical Foundations of Physics* (Basic Books, New York, 1966).
13. E. Cassirer, *Substanzbegriff und Funktionsbegriff* (Bruno Cassirer, Berlin, 1910); translation: *Substance and Function*, in Cassirer [14], pp. 1-346.
14. E. Cassirer, *Substance and Function and Einstein's Theory of Relativity*, English translation by W. C. Swabey and M. C. Swabey (Open Court, Chicago, 1923; Dover, New York, 1953).
15. E. Cassirer, *Philosophie der symbolischeen Formen*, 3 volumes (Berlin, 1923, 1925, 1929); transl. by O. Hansen-Love and J. Lacoste, *La philosophie des formes symboliques*, 3 volumes (Minuit, Paris, 1972).
16. J. P. Changeux, *L'homme neuronal* (Fayard, Paris, 1983.)

[45] Pascal [62].

17. J. P. Changeux and P. Ricœur, *La nature et la règle. Ce qui nous fait penser* (Odile Jacob, Paris, 1997).
18. R. Cohen, M. Horne, and J. Stachel, eds., *Experimental Metaphysics: Quantum Mechanical Studies for Abner Shimony, Vol. One* (Kluwer Academic, Dordrecht, 1997).
19. R. Cohen, M. Horne, and J. Stachel, eds., *Potentiality, Entanglement and Passion-at-a-Distance: Quantum Mechanical Studies for Abner Shimony, Vol. Two* (Kluwer Academic, Dordrecht, 1997).
20. A. Cupani, "Julgamento cientifico e racionalidade," in Dutra, ed., *Nos limites da Epistimolegia Analitica* (NEL, Florianopolis, 1999), pp. 139-159.
21. A. R. Damasio, *Descartes' Error. Emotion, Reason, and the Human Brain* (Grosset, 1994).
22. P. A. M. Dirac, *The Principles of Quantum Mechanics* (Clarendon, Oxford, 1930; 4th edn., 1958; revised 1981).
23. A. Einstein, "Spezielle Relativitätstheorie" (manuscript on the special theory of relativity, unpublished because of the war, prepared in 1912 for the *Handbuch der Radiologie*, E. Marx, ed.; copy at Einstein Archives), in Ref. [24], *Vol. 4*, pp. 3-108.
24. A. Einstein, *The Collected Papers of Albert Einstein, Vols. 1-8*, J. Stachel et al., eds. (University Press, Princeton, NJ, 1987-1998).
25. A. Einstein, H. A. Lorentz, H. Minkowski, and H. Weyl, *Das Relativitätsprinzip*, 4th edn., A. Sommerfeld, ed. (Teubner, Leipzig, 1922); translation by W. Perrett and G. B. Jeffery, *The Principle of Relativity* (Methuen, London, 1923; Dover, New York, 1952).
26. A. Einstein, B. Podolsky, and N. Rosen, "Can quantum-mechanical description of physical reality be considered complete?," *Phys. Rev.* **47**, 770-780 (1935).
27. B. d'Espagnat, *Le réel voilé. Analyse des concepts quantiques* (Fayard, Paris, 1994).
28. H. Focillon, *La Vie des Formes, suivi de Eloge de la main* (Presses Universitaires de France, Paris, 1981, 1990).
29. M. Foucault, *L'archéologie du savoir* (Gallimard, Paris, 1969).
30. R. Giere, *Explaining Science. A Cognitive Approach* (University Press, Chicago, 1988).
31. G. G. Granger, *Essai d'une philosophie du style* (Armand Colin, Paris, 1968; Odile Jacob, Paris, 1988).
32. G. G. Granger, *Pour la connaissance philosophique* (Odile Jacob, Paris, 1988).
33. G. G. Granger, *Formes, opérations, objets* (Vrin, Paris, 1994).
34. G. G. Granger, *Le probable, le possible et le virtuel* (Odile Jacob, Paris, 1995).

35. J. Hadamard, *An Essay on the Psychology of Invention in the Mathematical Field* (University Press, Princeton, 1945).
36. H. Hahn, O. Neurath, and R. Carnap, *Wissenschaftlische Weltauflassung der Wiener Kreis* (Vienna, 1929); reprinted in Neurath [1979], pp. 81-101; see [120].
37. W. R. Hamilton, "On a general method in dynamics," *Phil. Trans. Roy. Soc. (London)* 1934; reprinted in [38], Vol. 2.
38. W. R. Hamilton, *The Mathematical Papers of Sir William Rowland Hamilton*, 3 volumes (University Press, Cambridge, 1931, 1940, 1967).
39. R. C. Jeffrey, *Studies in Inductive Logic and Probabilities*, 2 volumes (Berkeley, 1980).
40. I. Kant, *Kritik der reinen Vernunft* (Hartknoch, Riga, 1781); 2nd modified edn., 1787; translation by Alexandre J. L. Delamarre and Franois Marty, "Critique de la raison pure," in *Emmanuel, Kant, Oeuvres philosophiques, Vol. 1* (Gallimard, Paris, 1980), pp. 705-1470.
41. J. Kouneiher, *Epistémologie et histoire récente de la gravitation quantique*, Doctoral Thesis in Epistemology and History of Sciences, Université Paris 7-Denis Diderot, 1998.
42. J. L. Lagrange, *Mécanique analytique* (Desaint, Paris, 1788; 2nd edn., 1811); reprinted in *Oeuvres complètes de Lagrange, Vols. 11 and 12*, published by J. A. Serret and G. Darboux (Gauthier-Villars, Paris, 1889).
43. I. Lakatos, *Philosophical Papers, Vol. 1: The Methodology of Scientific Research Programmes*, J. Worrall and G. Currie, eds. (University Press, Cambridge, 1978).
44. I. Lakatos, and A. Musgrave, eds., *Criticism and the Growth of Knowledge: Proceedings, International Colloquium in the Philosophy of Science, London*, 1965, Vol. 4 (University Press, Cambridge, 1970).
45. A. Lalande, *Vocabulaire technique et critique de la philosophie* (1926), 13th edn. (Presses Universitaires de France, Paris, 1980).
46. P. Langevin, "L'évolution de l'espace et du temps," *Scientia (Bologna)* **10**, 31-54 (1911) (conférence au Congrès de philosophie de Bologne, 1911); reprinted in Langevin [48], pp. 265-300.
47. P. Langevin, "Le temps, l'espace et la causalité dans la physique moderne," *Bulletin de la Société francaise de philosophie* **12**, 1-46 (1911); reprinted in Langevin [48], pp. 301-344.
48. P. Langevin, *La physique depuis vingt ans* (Doin, Paris, 1923).
49. C. Levi-Strauss, *Anthropologie structurale* (Plon, Paris, 1958).
50. H. Margenau, *Physics and Philosophy: Selected Essays* (Reidel, Dordrecht, 1978).
51. H. Minkowski, "Das Relativitätsprinzip," lecture in Göttingen, 5.11.1907; posthumous publication arranged by A. Sommerfeld, *Ann. Phys.* **47**, 927-938 (1915).

52. H. Minkowski, "Raum und Zeit," lecture at the 80th Meeting of German scientists and physicians, Cologne 21.9.1908; *Physikal. Z.* **10**, 104-111 (1909). "Space and Time," in Einstein et al. [25], pp. 73-91.
53. M. Mugur-Schächter, "From quantum mechanics to universal structures of conceptualization and feedback on quantum mechanics," *Found. Phys.* **23**, 37-122 (1993).
54. M. Mugur-Schächter, "Une méthode de conceptualisation relativisée: vers une épistémologie formelle apte à faire face aux complexités," *Revue Internationale de Systémique* **9** (2), 269-303 (1995).
55. M. Mugur-Schächter, "Les leçons de la mécanique quantique (vers une épistémologie formelle)," *Le Débat* **94**, 169-192 (1996).
56. M. Mugur-Schächter and A. van der Merwe, eds., this book (Kluwer Academic, Dordrecht, 2003).
57. J. von Neumann, *Mathematische Grundlagen der Quantenmechanik* (Springer, Berlin, 1932); transltion by Alexandre Proca, *Les fondements mathématiques de la mécanique quantique* (Librairie Alcan & Presses Universitaires de France, Paris, 1947).
58. E. Nœther, "Invarianten beliebiger Differentialausdrücke," *Nachr. Königl. Ges. Wiss. Göttingen, Math. Phys. Kl.*, 1918, pp. 235-257; in Noether [60], pp. 240-247.
59. E. Nœther, "Invariante variationprobleme," *ibid.*, pp. 235-257; reprinted in Noether [60], pp. 248-270; translation by M. A. Tavel, *Transport Theory and Statistical Physics* **1**, 183-207 (1971).
60. E. Nœther, *Collected Papers*, N. Jacobson, ed. (Springer, Berlin, 1983).
61. OULIPO (Ouvroir de Littérature potentielle), *Littérature potentielle: créations, re-créations, récréations* (Gallimard, Paris, 1977).
62. B. Pascal, "De l'esprit géométrique," in Pascal, *Oeuvres complètes*, Présentation et Notes de Louis Lafuma (Seuil, Paris, 1993), pp. 348-355.
63. M. Paty, "La non-séparabilité locale et l'objet de la théorie physique," *Fundamenta Scientiae* **7**, 47-87 (1986).
64. M. Paty, *La matière dérobée. L'appropriation critique de l'objet de la physique contemporaine* (Archives contemporaines, Paris, 1988).
65. M. Paty, "Sur la notion de complétude d'une théorie physique," in N. Fleury, Joffily, Sergio, J. A. Martins Simes, and A. Troper, eds., *Leite Lopes Festchrift: A Pioneer Physicist in the Third World;* dedicated to J. Leite Lopes on the occasion of his seventieth birthday (World Scientific, Singapore, 1988), pp. 143-164.
66. M. Paty, *L'analyse critique des sciences, ou le tétraèdre épistémologique (sciences, philosophie, épistémologie, histoire des sciences)* (L'Harmattan, Paris, 1990).

67. M. Paty, "L'endoréférence d'une science formalisée de la nature," in C. Dilworth, ed., *Intelligibility in Science* (Rodopi, Amsterdam, 1992), pp. 73-110.
68. M. Paty, *Einstein philosophe. La physique comme pratique philosophique* (Presses Universitaires de France, Paris, 1993).
69. M. Paty, "Mesure, expérience et objet théorique en physique," in J.-C. Beaune, ed., *La mesure, instruments et philosophies* (Champ Vallon, Seyssel, 1994), pp. 159-174.
70. M. Paty, "Le caractère historique de l'adéquation des mathématiques à la physique," in S. Garma, D. Flament, and V. Navarro, eds., *Contra los titanes de la rutina—Contre les titans de la routine* (encuentro en Madrid de investigadores hispano-franceses sobre la historia y la filosofia de la matematica Rencontre à Madrid de chercheurs hispano-franais sur l'histoire et la philosophie de la mathématique,18-22.11.1991) (Comunidad de Madrid/C.S.I.C., Madrid, 1994), pp. 401-428.
71. M. Paty, "Sur l'histoire du problème du temps: le temps physique et les phénomènes," in E. Klein and M. Spiro, eds., *Le temps et sa flèche* (Frontières, Gif-sur-Yvette, 1994; 2nd edn., 1995). (Collection Champs, Flammarion, Paris, 1996), pp. 21-58.
72. M. Paty, "The nature of Einstein's objections to the Copenhagen interpretation of quantum mechanics," *Found. Phys.* **25**, 183-204 (1995).
73. M. Paty, "Poincaré et le principe de relativité," in J.-L. Greffe, G. Heinzmann, and K. Lorenz, eds., *Henri Poincaré. Science et philosophie. Science and philosophy. Wissenscchaft und Philosophie. Congrès international, Nancy, France, 14-18 mai 1984* (Akademie-Verlag, Berlin/Albert Blanchard, Paris, 1996), pp. 101-143.
74. M. Paty, "Le style d'Einstein, la nature du travail scientifique et le problème de la découverte," *Revue philosophique de Louvain* **94**, 447-470 (1996).
75. M. Paty, "Predicate of existence and predictivity for a theoretical object in physics," in E. Agazzi, ed., *Realism and Quantum Physics* (Rodopi, Amsterdam, 1997), pp. 97-130.
76. M. Paty, "Philosophie et physique," in J.-F. Mattéi, ed., *Le Discours philosophique, volume 4 de l'Encyclopédie philosophique universelle* (Presses Universitaires de France, Paris, 1998), pp. 2104-2122.
77. M. Paty, "Universality of science: Historical validation of a philosophical idea," in S. I. Habib and D. Raina, eds., *Situating the History of Science: Dialogues with Joseph Needham* (Oxford University Press, New Delhi, 1998), Chap. 12.
78. M. Paty, "La place des principes dans la physique mathématique au sens de Poincaré," in J. Sebestik and A. Soulez, eds., "Actes du Col-

loque France-Autriche Paris, mai 1995, Interférences et transformations dans la philosophie francaise et autrichienne (Mach, Poincaré, Duhem, Boltzmann)," *Fundamenta philosophiae (Nancy)* **3** (2), 61-74 (1998-1999).

79. M. Paty, "La création scientifique selon Poincaré et Einstein," in M. Serfati, ed., *La recherche de la vérité* (Collection Ecriture des Mathématiques, Éditions ACL, Paris, 1999), pp. 241-280.
80. M. Paty, "Paul Langevin (1872-1946), la relativité et les quanta," *Bulletin de la Société Franaise de Physique* **119**, 15-20 (1999).
81. M. Paty, "Are quantum systems physical objects with physical properties?," *Eur. J. Phys.* **20**, 373-388 (1999).
82. M. Paty, "Cosmologie et matière quantiquea," in J. Seidengart and J.-J. Sczeciniarz, eds., *Cosmologie et philosophie. En hommage à Jacques Merleau-Ponty; Epistémologiques (Paris, Sao Paulo)* **1**, 219-249 (2000).
83. M. Paty, "Interprétations et significations en physique quantique," *Revue Internationale de philosophie* **54**, 199-242 (2000).
84. M. Paty, "Langevin, Poincaré et Einstein," in *Paul Langevin, son œuvre et sa pensée.* (Paris, São Paulo, forthcoming).
85. M. Paty, "The quantum and the classical domains as (provisional) parallel coexistents," in S. French, D. Krause, and F. Doria, eds. (festschrift in honour of Newton da Costa, on the occasion of his seventieth birthday), *Synthese* **125**, 179-200 (2000).
86. M. Paty, "The concept of quantum state: New views on old phenomena," in R. S. Cohen, and J. Renn, eds., *John Stachel Festschrift* (Boston Studies in the Philosophy and History of Science, Kluwer, Dordrecht, forthcoming).
87. M. Paty, "The idea of quantity at the origin of the legitimacy of mathematization in physics," in C. Gould, ed., *Conference on the Philosophy of Marx Wartofsky*, New School University, New York, March 6, 1999.
88. M. Paty, "L'analogie mathématique au sens de Poincaré et sa fonction en physique," in M.-J. Durand-Richard, ed., *Le statut de l'analogie dans la démarche scientifique* (Editions du CNRS, Paris, forthcoming).
89. M. Paty, *Einstein, les quanta et le réel. Critique et construction théorique*, in the press.
90. G. Perec, *La disparition* (Denoël, Paris, 1969; Gallimard, Paris, 1989).
91. H. Poincaré, "Sur les rapports de l'analyse pure et de la physique mathématique," *Acta Mathematica* **21**, 331-341 (1897); reprinted in Poincaré [98], pp. 17-30; "Les rapports de l'analyse et de la physique mathématique," *Revue générale des sciences pures et appliquées* **8**,

857-861 (1987); reprinted in Poincaré [93], Chap. 5 ("L'analyse et la physique"), 1970 edn., pp. 103-113.
92. H. Poincaré, "Sur la dynamique de l'électron" (received July 23, 1905), *Rendiconti del Circolo matematico di Palermo* **XXI** 129-176 (1906); also in Poincaré [97], Vol. 9, pp. 494-550.
93. H. Poincaré, *Science et méthode* (Flammarion, Paris, 1908, 1918).
94. H. Poincaré, "Le choix des faits," *The Monist*, 1909, 231-232; reprinted in Poincaré [93], Book 1, Chap. 1, 1918 edn., pp. 16-18.
95. H. Poincaré, "L'espace et le temps," *Scientia* **XXV**, 159-170 (1912); also in Poincaré [96], 1963 edn., pp. 97-109.
96. H. Poincaré, *Dernières pensées* (Flammarion, Paris, 1913, 1963).
97. H. Poincaré, *Oeuvres*, 11 volumes (Gauthier-Villars, Paris, 1950-1965).
98. H. Poincaré, *L'analyse et la recherche*, selection of texts with an introduction by G. Ramunni (Hermann, Paris, 1991).
99. K. Popper, *Logik der Forschung. Zur Erkenntnisstheorie der modernen Naturwissenschaft* (Springer, Wien, 1934); re-edited with additions, 1966; translation: *The Logic of Scientific Discovery* (Hutchinson, London, 1959, 1968).
100. K. Popper, *Objective Knowledge: An Evolutionary Approach* (Claredon, Oxford, 1972).
101. K. Popper, *Quantum Theory and the Schism in Physics, from the Postscript to the Logic of Scientific Discovery, Vol. 3* (Hutchinson, London, 1982; written in 1956).
102. K. Popper, *Realism and the Aim of Science, from the Postscript to the Logic of Scientific Discovery, Vol. 1*, W. W. Bartley III, ed. (Hutchinson, London, 1983; written in 1956).
103. J. Proust, *Questions de forme. Logique et proposition analytique de Kant à Carnap* (Fayard, Paris, 1986).
104. H. Reichenbach, *Experience and Prediction* (University Press, Chicago, 1938).
105. H. Reichenbach, *The Rise of Scientific Philosophy* (University of California Press, Berkeley, 1951; re-edited 1973).
106. H. Reichenbach, *Modern Philosophy of Science*, edited and translated by M. Reichenbach (Routledge & Kegan Paul, London, 1959).
107. H. Reichenbach, *Selected Writings*, 2 volumes, R. Cohen and M. Reichenbach, eds. (Reidel, Dordrecht, 1978).
108. D. Ruelle, *Chaotic Evolution and Strange Attractors* (University Press, Cambridge, 1989).
109. D. Ruelle, *Hasard et Chaos; Chance and Chaos* (Uniersity Press, Princeton, 1991).
110. A. Shimony, *Search for a Naturalistic World View*, 2 volumes (University Press, Cambridge, 1993).

111. A. Soulez, ed., *Manifeste du Cercle de Vienne et autres écrits: Carnap, Hahn, Neurath, Schlick, Waissmann, Wittgenstein* (Presses Universitaires de France, Paris, 1986).
112. W. Stegmüller, *Probleme und Resulate der Wissenschaftstheorie und Analytischen Philosophie. Band 2: Theorie und Erfahrung. Zweiter Halbband: Theoriestrukturen und Theoriendynamik* (Springer, Heidelberg, 1973).
113. W. Stegmüller, *The Structuralist View of Theories. A Possible Analogue of the Bourbaki Programme in Physical Science* (Springer, Heidelberg, 1979).
114. P. Veyne, *Comment on écrit l'histoire* (Seuil, Paris, 1971).
115. P. Veyne, "Foucault révolutionne l'histoire," in *Comment on écrit l'histoire et Foucault révolutionne l'histoire* (Seuil, Paris, 1978).
116. S. Walter, *Hermann Minkowski et la mathématisation de la théorie de la relativité, 1905-1915,* Doctoral Thesis in Epistemology and History of Sciences, Université Paris 7-Denis Diderot, 1996.
117. M. Wartofsky, *Conceptual Foundations of Scientific Thought: An Introduction to the Philosophy of Science* (Mac Millan, London, 1968).
118. H. Weyl, *Gruppentheorie und Quantenmechanik* (Hirzel, Leipzig, 1928, 2nd edn., 1931); translation by H. P. Robertson, *The Theory of Groups and Quantum Mechanics* (Methuen, London, 1931; Dover, New York, 1950).
119. J. A. Wheeler, and W. H. Zurek, eds., *Quantum Theory of Measurement* (University Press, Princeton, 1983).
120. Wiener Kreis, *Wissenschaftliche Weltaufflassung: der Wiener Kreis* (Wolf, Wien, 1929).
121. E. Zahar, *Einstein's Revolution. A Study in Heuristic* (Open Court, La Salle, IL, 1989).
122. E. Zahar, "Againts the (alleged) divorce between science and philosophy: Comments on M. Paty's "La matière dérobée," *Rivista di Storia della Scienza,* Ser. II, **2**, 191-208 (1994).

5

MATHEMATICAL PHYSICS AND FORMALIZED EPISTEMOLOGY: DEBATE WITH JEAN PETITOT*

Jean Petitot

Ecole Polytechnique
1 rue Descartes
75005 Paris, France

Interlocutors: Francis Bailly, Michel Bitbol, Mioara Mugur-Schächter, Vincent Schächter

This paper develops an updated transcendentalist perspective concerning the epistemological status of objectivity. The main point is that objectivity is neither an ontology nor a mere description of a phenomenal state of affairs. It is instead a principled (categorial) "legalization" and a mathematical reconstruction of the phenomena. As the very concept of a phenomenon is relational (relative to a receptive stance such as perception in classical mechanics or measure devices in quantum mechanics), the conditions for accessing them must be included in the very concept of objectivity. This paper emphasizes the transcendental content of symmetries in modern physical theories, from general relativity to gauge invariance. A great deal of the discussion focuses on epistemological key points in quantum mechanics.

Key words: physical objectivity, transcendental philosophy, mathematical models, symmetries, quantum mechanics.

Mugur-Schächter. Jean Petitot, in order to present us a first insight into your conception of "a formalized epistemology," you have accepted to summarize your views on mathematical physics.

Petitot. To begin with, I would like to stress that for me physics is a *factum rationis.* Mathematical physics is a scientific fact, and in spite of all its internal problems one has to accept that, in a sense that has to be precisely defined, it yields an objective knowledge concerning a very large set of phenomena. Physics works and requires therefore what I call a "plausible" operational epistemology, that is, an epistemology where physical effectivity

*See "Important Note" on p. xviii.

occupies the master-position. In this respect, I separate from most of my philosophers colleagues. I consider that philosophy does not have to rule the status of objectivity. So, if physics is taken as an object of philosophical investigations, it must occupy a leading position, and epistemology has to take it as a *datum.* From this factum rationis, and without any pre-judgment concerning the status of physical objectivity, it has to go on to problems of another kind, ontological, cognitive, etc.

I would like to stress also right away a second point, that is quite crucial: my theses are transcendentally oriented, in Kant's sense. This involves the key claim that physical objectivity is *not an ontology*: physics doesn't predicate on an "independent" reality but concerns exclusively phenomena. Now, phenomena are by definition *relational entities* that exist only with respect to a receptive device: no matter whether it is a measure instrument or a perceptive living organism, a receptive device must exist. This essential relativity of any phenomenon to a receptive instance has to be included in the principles of physics. In quantum mechanics (QM) for instance, a physical entity can be accessed only through convenient measuring apparatus. In classical mechanics the role of the instruments is hold by the perceptive system of the human observer (but this latter case is not so different from the former one, since the visual system, and especially the retina, is an extraordinary quantum instrument).

More generally, to uphold transcendental theses on physics amounts simply to require the conditions for accessing physical phenomena—space, time, instruments, etc.—to be included in the very concept of physical objectivity. I think that in the Bohr-Einstein controversy concerning this key point, Bohr was fully right.

So physics deals exclusively with phenomena but possesses also an undeniable objective content. But if physical objectivity is neither a subjective phenomenism nor a realist ontology, it must be something else.

The classical transcendental answer is that objectivity is a *"legalization" ruling phenomena.* This thesis persists from Kant on, up to the Vienna Circle. In fact, even if it was dogmatically anti-Kantian, the Vienna Circle was transcendentally oriented and developed what has been called a "grammatical" transcendentalism. Only the status of geometry in physics (the celebrated debate on "synthetic a priori" geometrical frames) launched a dramatic controversy against the Kantian tradition. But there was general consensus on the fact:

(i) That physics deals only with phenomena and must include the relational conditions for accessing them.
(ii) That objectivity stems from a legalization of phenomena and that a prescriptive and normative dimension is constitutive of objectivity.

Modern physicists have spontaneously rediscovered one of the main transcendental thesis, namely that there exists a specific legality of the phenomena as such. It is impossible to construct an objective explanation of physical phenomena by starting with an absolute ontology of an "an sich" independent reality. Objectivity can only be defined as an order of legality, and such a prescriptive definition distinguishes essentially objectivity from any ontology.

You see, objects of scientific knowledge are not given directly as such with their phenomenon. It is why they have to be constituted by their legalization, why, apart from its experimental descriptive dimension, any scientific knowledge presupposes in its very principle also a *prescriptive* (normative) dimension.

Therefore, besides the metaphysical "ontological" difference between phenomena and noumenal being (see Heidegger), there also exists the physical "objective" difference between phenomena and "object" of scientific knowledge. In contradistinction to phenomena, "objects" exist only if qualified according to norms, to eidetico-constitutive rules defining what Husserl called a "regional objective essence". The normative concept of an object is a condition of possibility of any scientific activity. It anticipates and predetermines what belongs *generally and typically* to the phenomena of the considered region.

The recurrent error of ontological realisms is to confuse the prescriptive dimension of objectivity with an underlying "an sich" ontological reality, which would at the same time exist in itself "behind" the phenomena, and be experimentally inaccessible, theoretically unknowable, and, in spite of all this, causally efficient.

If you agree with these preliminaries then the main question becomes: how can the legalization be performed?

In Kant's works concerning classical physics, the method consists essentially in interpreting the general "categories" of objectivity by starting from the "forms of givenness" of the phenomena, what he called the "pure forms" of phenomenal manifestation. But, since an interpretation of the categories of objectivity can be operational only if it is *mathematical*, these pure forms themselves must also be mathematized. So, in any objectivity conceived in the transcendental sense operates a *mathematical hermeneutic of the objective categories* – what Kant called a "schematization" and, moreover, a "construction" –, which eliminates the meaning of these categories (their metaphysical use): this is the mathematical aspect of the obligation to restrict the categories to empirical observables (what Kant called "transcendental deduction").[1] One is thus naturally led to a grammatical apriorism

[1]Here, the term "deduction" is not used in its logical sense but in its *juridical* sense of

according to which mathematics provide the pool of syntaxes for legalizing scientific objects. Moreover, those mathematics that do mathematize the forms of manifestation determine also the appropriate type of syntax. In short, transcendental philosophy is pre-eminently a philosophy of mathematical physics. As was emphasized by Jules Vuillemin, it can be defined as the philosophy taking into account the fact that the essence of objectivity is mathematical.

Of course, I go very much further than Kant concerning legalization. I hold that the ideal of mathematical science is, through the mathematization of the forms of phenomenal givenness and the correlative "schematization-construction" of the categories, to *reconstruct* the phenomena in *principled* mathematical terms, to reproduce, to simulate them as exactly as possible. This is conceivable precisely because phenomena are only phenomena. It would be absurd to claim to reconstruct an ontologically independent reality "in itself". But if what physics explains is strictly phenomenal, then it is fully legitimate to want to reconstruct what is given initially as a phenomenal datum, and to substitute for it by a mathematical constructed structure. And my thesis is that scientific explanation is precisely this. As Valéry said, we know only what we are able to produce. If, for instance, using Newton's law, I reconstruct in a mathematical principled way the trajectories of all the material bodies (from the falling bodies to the observed motion of the planets), ipso facto I understand them. But this objective understanding shares no ontological content. Its unique truth criterium is the accuracy of the simulation.

Bailly. Then what difference do you make between reproduction *via* Ptolemaic epicycles, and reproduction *via* the Kepler-Newton equations?

Petitot. Your question brings into play the core relation between algorithms of reconstruction and the categories and principles of objectivity. This relationship is the essence of the problem. Without adequate algorithms it is impossible to reconstruct the endless diversity of phenomena. And these algorithms, no matter of what could be their mathematical nature, must be precise and endowed with a strong internal generativity. But, on the other hand, in order to be explanatory, they must also be strongly connected to the objective categories. There is therefore a fundamental difference between algorithms reconstructing phenomena by *representing* them, and algorithms reconstructing phenomena by *deducing* them from law-like principles. The difference between the reconstruction of the motion of the planets by Ptolemaic epicycles and by Newton's equation is a very good example. In fact the epicycles algorithm is truly excellent. I have been told that spatial agencies make still use of it. Like Fourier series, it is highly

justification (of the restriction of the application of the objective categories to phenomena).

performant. But it is only a mere representation of trajectories, an approximation scheme. Its connection to the categories is missing and it lacks therefore any explanatory power. For each case it requires a more or less *ad hoc* analysis.

This is related to the problem of *algorithmic complexity*. The more an algorithm is generative and "simple", the more it is connected to basic categories and principles, the more it is a "good" efficient one. In this sense Newton's law represents a fantastic progress. When one follows up this transcendental line of thought it appears that epistemology of physics must essentially focus on the form of the main equations which connect deductively the categorial basis with the diversity of phenomena. If the equations of the physical theories are not taken into consideration directly and explicitly, then the whole transcendental conception breaks down and one must go back to a more classical conception of epistemology, such as it is nowadays generally practiced by most philosophers: one asks a lot of questions on ontology, experience, observation, induction, truth, judgement, etc., thus just applying general philosophy to physics, instead of working out a plausible epistemology specifically adapted to what physics really is. To place physics in a master position is to change our pre-conceived metaphysics in order to bring it to an inner agreement with physics, *not the converse.*

Mugur-Schächter. Do you consider that at this point the main contours of your view have been already expressed sufficiently for permitting a relevant first discussion?

Petitot. I would like to illustrate my views with some examples concerning both classical mechanics and quantum microphysics. But if you prefer we can already debate on some points.

Mugur-Schächter. All right. Then I would like to make a remark on "ontology". I thoroughly agree with your exclusion from physics of any attempt at the construction of an 'independent' ontology. This, in my view also, is simply a *self-contradicting expression* since 'ontology' means 'model of the things', any model involves qualifications, and 'independent' reality such-as-it-is-in-itself is *strictly* unqualifyable *by definition.*

However I would not exclude *any ontology.* I hold that each physical theory – besides the other sorts of only algorithmic mathematical models – should also involve a *'relative methodological ontology'* built in order to insure intelligibility (relative: including the ineluctable relativities of phenomena, to the receptive entities ; methodological: induced by the aim to achieve something; ontology: a model concerning also 'things', as opposed to chains of exclusively phenomena, or exclusively mathematical algorithms). Indeed, if such a relative methodological ontology is absent at the explicit level, as in nowadays quantum mechanics, the search for it never ceases before a satisfactory one is obtained.

A relative methodological ontology might come out to be very near to the connection required by Kant between the mathematized forms of 'donation' of the *phenomenal manifestations*, and the categories of objectivity. However, strictly speaking, it cannot be identified to this connection because modern physics contains exceptions with respect to the Kantian presupposition of a 'donation' of the phenomenal manifestations: in quantum mechanics, for instance, such a 'donation' does not exist, in general.

Now, relative methodological ontologies, in the specified sense, are not opposable to objectivity, quite the contrary. Often they are the *basic* place where invariances are first required.

Petitot. The term "ontology" has such a wide range of uses that one has to specify its meaning in each context. Personally, when I say "ontology" I have in mind not models of phenomena but a supposedly adequate conceptual description of an independent substantial reality. I think with Kant that "ontology" is a *scientifically ill-formed concept* (a category mistake). My rejection of ontology parallels that made by quantum physicists such Bernard d'Espagnat when they stress that quantum objectivity is only a "weak" objectivity, and cannot be, for principled reasons, a "strong" objectivity predicating about an independent reality in itself.

Now the question is: if we reject ontology are we necessarily committed to any sort of subjective idealism? If objectivity is divorced from substantial ontology and constitutively depends upon the forms of phenomenal givenness and accessibility, does this brings necessarily back to the subject and his cognitive processes? Not at all indeed. Let us take the case of QM. By construction, quantum objects can exist only in so far as they are measured and experimentally accessed. So the necessity of measuring apparatus must be included in the very concept of quantum object. However this does'nt entail that any specific theory of any specific sort of apparatus must be part of the axioms of QM. The relations to instruments making observable the phenomena is a sort of generalized principle of relativity. For instance, Heisenberg's principle is totally independent of any specific measuring device. It takes into account the unavoidable necessity of a "receptive" device, but at the same time it brackets all the peculiarities of such a device.

It is the same thing with classical mechanics. There, the restriction to observables means that mechanics can speak only of *space-time* trajectories, and not of any metaphysical "substance" that would exist outside space and time. One must not forget that at that time by ontology was meant for instance a Monadology in the sense of Leibniz. Philosophers thought that space and time were imaginary subjective entities existing only in our minds. Ontology was foreign to space and time, and so physics could'nt be "objective" in a strong sense. Kant's answer to this antinomy was to look at the way geometry operates in physics. Geometry concerns space as a *subjective* form. But at the same time there is absolutely no psychological element in it. Neurobiology, psychophysics, psychology *are not* part of the axioms of geometry. We can say that mathematical geometry is "ideal", completely *disembodied.* Space and time are *"pure"* forms. I say that they are universal *formats* for brain information processings. Once you have mathematized them, you can say that physical phenomena are spatio-temporal entities related with geometry. Then geometry *prescribes constraints* to objectivity, e.g., principles of relativity acting on positions, velocities, and motions.

To summarize, on the one hand, an objective theory has to incorporate the fact that phenomena presuppose a receptive instance, but on the other hand it must also bracket any particular theory of this receptive instance.

Mugur-Schächter. It is not canceled. It is maintained, but – exclusively – in its full generality with respect to the whole diversity of the receptive entities utilized by human investigators in their various investigating actions of the considered domain of phenomena. It is maintained in so far that it is 'objectively subjective', with respect to *human* investigators. This is a relative zero point, not an absolute one.

Petitot. Exactly, I agree.

Mugur-Schächter. So you don't cancel it.

Petitot. All right. It remains as a *trace.* For instance, general relativity reduces the "subject" to an "observer". But what is an "observer"? Just a mere reference frame.

Bailly. It is *inter-subjectivity.*

Petitot. It is a relativity group. "Subjectivity" has become a group of invariance. It is completely disembodied.

Mugur-Schächter. Not only one, indefinitely many. The 'observer' is the sum of all such known or potential human groups of relativity. (The 'conceptor' is still much more than this, but this is another topic). So the human 'subject' is not in the least canceled from physics. In particular he is not canceled from neither relativity nor quantum mechanics. He is *there*, and not as a 'trace', not as a sign of something that was there but ceased to be, he is there *actually*, as a quite positive, active, rich, dense, *specific* general extract. And it is noteworthy that the extract would be *different* in a science built by another sort of living beings. Another sort of being might experience other forms of phenomenal donation and also impose other categories of objectivity.

Petitot. All right, but I think that we must go still much further in this direction. I think that the role of relativity groups and symmetries raises a very deep philosophical question. As was emphasized by Daniel Bennequin, mathematical physics is "Galoisian" in the sense that what you can know is *negatively* determined by what you can't know. Symmetries express non-physical entities (an absolute origin of space or time, an absolute direction, etc.). They group such *indiscernable* entities in equivalence classes. But as far as they are the key of the mathematical reconstruction of the phenomena they are priors determining the positive contents of the theory. So the positive contents, what you can know, are deductive consequences of what you can't know. This is one of the most beautiful features of mathematical physics.

Bailly. What can be known does not belong to the same domain as that what cannot be known. What cannot be known bears on changes of the referentials and frameworks, while what can be known concerns what happens inside the frameworks.

Mugur-Schächter. What can be known concerning the studied system is modeled so as to stay consistent with the properties assigned by construction to the referential. But these last properties are defined so that the description of the studied system shall become as general and *synthetic* as possible. The referentials are very free conceptual constructs, not facts. The mathematical representations of directly perceivable phenomena, also, can be removed very far away from these perceivable phenomena. Thus there develops a to and fro process of representation, re-representation, etc., of the observable phenomena, in the course of which the referentials and the initial mathematical representation of the forms of phenomenal 'donation', are both radically changed, under constraints of global logico-mathematical coherence.

Petitot. Yes, but to grasp the point let us examine an example. In quantum field theory, Feynman's integrals provide an incredibly powerful algorithm. But what is their formal genesis? One starts from symmetries to construct (quite univoquely) a Lagrangian. According to the axioms of QM this Lagrangian leads to a path integral. Then using various algorithms (perturbative expansions, Wick's theorem, the stationary phase principle, the renormalisation group, etc.), one can construct models which can be confirmed up to many decimals! There, one really attend the mathematical genesis of physical contents from prior symmetries.

Mugur-Schächter. That is true. But the concept of a Lagrangian, and the Feynmann integral itself, contain already essences drawn from phenomena, *via* classical mechanics, elecromagnetism, quantum mechanics. How were all these prerequisites established? As far as I know there does not yet exist a worked out general method ruling the sort of operations that *lead* to the stage when it becomes possible to take maximal advantage of the symmetries which the nature of the involved referential and of the defined mathematical descriptors imposes upon a final mathematical representations of phenomena. To elaborate such a method could be a major aim for a formalized epistemology.

Petitot. I agree. But I think that the extraordinary efficiency of the constraints imposed upon physical phenomena by mathematical formalisms has not yet been given enough importance. Nor this idea that what can be known must be generated by what cannot be known, and is therefore negatively determined.

Mugur-Schächter. A rather Popperian sort of idea.

Schächter. The approach which you describe can be perceived *both* as negative or as positive: you start with some mathematical reconstruction of the forms of donation of the phenomenon, which is positive. Then you generalize maximally this first mathematical reconstruction by imposing all the symmetries required by the utilized referential, which amounts to introducing the 'negative' knowledge afterward, as a sort of correction or improvement. In this way, out of the initial 'positive' core of representation, one elaborates the final mathematical representation of the phenomena as the maximal class of equivalence constructible from the initial 'positive' representation, which again is 'positive'. This, moreover, is probably the order in which the representations do indeed evolve most frequently. A *posteriori* it is possible of course to reconstruct in the converse order also.

Petitot. Yes, but what I want to stress here is the very singular role of mathematics. The procedures of mathematical physics are radically different from those of common sense. Contrary to common wisdom, namely that there exists a continuity between common sense and mathematical physics, I think that the conceptual abstractions to which we submit the phenomena in current life are not at all of the same type as those from mathematical physics.

Bitbol. However a well known maxim asserts that *any* determination is "negative".

Mugur-Schächter. Expanding on the important remark of Francis Bailly that 'what can be known does not belong to the same domain as what cannot be know', and of Vincent Schächter's remark on 'negative' and 'positive' knowledge, I would like to submit a conjecture.

The formulations in terms of 'negative' and 'positive' knowledge have a paradoxical flavor, whereby they catch and mobilize the attention, which no doubt is useful. But I suspect that a thorough analysis might reveal at the bottom a matter of *purely logico-mathematical self-consistency* of the global representation.

Consider the simplest example, of position in space. One can *a priori* imagine this specification in only two sorts of contexts, either in connection with a space referential, or in connection with absolute space (in the Kantian sense of a form of pure intuition), *tertium non datur.* Now, in absolute space, the notion of 'absolute position' simply does not exist, it is not constructible, it contradicts the concept of absolute space because there – by definition – no origin is specified. Given that in the posited context an absolute position is a *logically impossible* concept, to say that in absolute space one cannot 'know' or 'distinguish' an 'absolute position' is somewhat misleading.

Whereas a relative mutual qualification of two positions is possible in absolute space, and even a quantitative relative qualification. Descartes introduced space-referentials only in order to organize *all* such relative quantitative characterizations, with respect to *one* common origin: a 'legalization' of reference!

Consider next the velocity of a particle. This mathematical concept is by construction independent of the spatial coordinates, in the utilized space referential, of the point where the velocity acts; it depends only on the numerical values of differences between spatial coordinates. If the initial referential is changed by translation, these differences do not change, though their notational expression does. So, if the velocity vector were to change, this would be *logically inconsistent* with the definition of the concept of velocity. Again, to indicate the identified constraint in terms of 'negative' and 'positive' knowledge, seems somewhat miseading: the conservation of velocity with respect to translation of the space-referential (hence, in Newtonian mechanics, also the conservation of momentum) is *just* a constraint of *inner logico-mathematical consistency* of the system of the concepts involved, namely space, space-referential, distance, velocity.

The general type of conceptual situation where considerations of symmetry are generative could reveal features that are much less immediately intelligible in terms of constraints of, exclusively, logico-mathematical self-consistency of the global mathematical representation. But in the trivial cases just examined, instead of saying that what can be known has been determined from what cannot be known, one can also only say that *a mathematical descriptive concept cannot, without contradiction, be conceived to change in consequence of changes of which, by construction, it is independent.* This is less mysterious, but also much less striking, of course.

Petitot. I don't understand how in absolute space there cannot be absolute positions. It is precisely when there is a relativity group acting transitively that all positions become equivalent. By definition, relativity negates absoluteness. In fact, the problem is much deeper than a mere question of logical coherence. What I have in mind is the determining role of Noether's theorem proving, in a variational context, the correlation between symmetries (negative geometrical limitations) and conserved quantities (positive physical contents). We will return to this point later.

Mugur-Schächter. Please let me go back to 'ontology', to try to come to a common conclusion. Is there some sort of ontology with which you would finally agree?

Petitot. Well, I think that we have to radically reject in science any ontology in the classical metaphysical sense. But there is another meaning of the word "ontology", philosophically tied with the concept of logical type, and refering simply to the general categories of object, whole, part, relation, set, function, event, process, etc., which we need to speak of something. Here "ontology" means in fact "categorial frame". It is the sense of the term, e.g., in Husserl's formal ontology, or in what is called in contemporary AI "ontological design".

Mugur-Schächter. We need? In what sense?

Petitot. In order to conceptualize the phenomena correctly.

Mugur-Schächter. Does 'correctly' mean so as to insure truth, or intelligibility, or both?

Petitot. Both.

Mugur-Schächter. In short, if instead of 'ontology' I say 'a correct model of intelligibility' (to distinguish it from the purely mathematical models'), I guess that we might finally agree. Only one more question in this context: by intelligibility, would you mean availability of any model, or only of a model that insures connection with the categories of objectivity in a simple 'explicative' way?

Petitot. As I explained before, for me scientific intelligibility is essentially linked with the possibility of an algorithmic simulation derived in a principled way, or, to make use of another terminology, of a *"computational synthesis" correlated to a categorial analysis.* Every time one succeeds to produce a very economic computational synthesis rooted in general and fundamental principles, then intelligibility is achieved.

Bailly. In usual computation one allows entirely *ad hoc* models, under strong technical restrictions, whereas, in the computation toward which you point, the condition of a relation to first principles, to *universal* categories, is very important. You do not explicate this in your assertions, which, in my opinion, weakens them.

Petitot. I strongly emphasized that the algorithms *should* be rooted into objective categories and principles.

Bailly. Together with 'simulation' you should then also require this explicitly and *from the start,* and declare the aim of intelligibility. But instead of pursuing the discussion now, I think that we should let you first finish your exposition and continue afterwards on a more complete ground.

Petitot. OK. Let's continue. I would like now to illustrate my general purpose with a couple of examples.

The problematic of the forms of phenomenal givenness and of their mathematization has been initiated by Kant in his *Transcendental Aesthetics.* The interpretation of the categories of objectivity starting from the "pure intuitions" is achieved at two distinct levels: first their "schematization", and then, at a much deeper level, their mathematical "construction", which radically transforms their traditional sense (therefrom the critical disjunction between scientific and common sense knowledge).

So, as we have already seen, in order to legalize phenomena, one must use paradoxical objectifying tools which take into account how phenomena hide their underlying being, and which, at the same time, bracket any "psychological" processing. They must break with both ontology and psychology.

Before considering the case of modern quantum microphysics, let me briefly indicate how this fundamental problem has been solved by Kant for classical mechanics in his *First Metaphysical Principles of the Science of Nature*, so deeply analyzed by J. Vuillemin in his *Kant's Physics and Metaphysics.*

In the *Phoronomics* (Kinematics) Kant studies the categories of *Quantity* and the associated transcendental principles called the "Axioms of intuition" that rule the function of extensive magnitudes. Two problems are treated there.

(a) The way in which *measure* becomes possible for the phenomenological forms of space and time. The introduction of coordinates allows an arithmetization of these continua, and then the introduction of a metrics allows to measure distances. Thereby space as a continuous medium for manifestation ("form" of intuition) becomes – in view of physics – a geometry ("formal" intuition) whose Euclidean structure is unseparable from mechanics (inertia principle and straight geodesics).

(b) The rectilinear uniform motions and Galilean relativity (that is the invariance group of the theory). This refers on the one hand to space-time symmetries: temporal translations, spatial translations and rotations, etc. (Kant was the first philosopher to strongly stress that spatial symmetries were constitutive of physical objectivity). On the other hand this refers to the properly kinematic Galilean transformations (rectilinear uniform motions). Wherefrom the constitutive role of the relativity principle. As J. Vuillemin notes,

"it is the relativity of motion which renders transcendentally

internal state cannot be identified with these dynamical descriptors.

Bailly. Certainly. But if you make use of a referential, you *can*: everything, then, is transferred in the referential relation.

Petitot. This is the relativity principle. As Fock emphasized, in this respect QM generalizes classical relativity.

Bailly. Yes. Concerning relativity I have a tendency to speak in terms of a sort of 'order', nearly an 'approximation process': Aristotelian theory was at the order zero, Galilean theory was at the first order, general relativity is at the third order, and quantum mechanics is a sum of all this up to infinity.

Mugur-Schächter. To go to the bottom of this line of thought, one should fully realize that no qualification *whatever* is 'predicable' in an *absolute* way. Any qualification presupposes some sort of point of view, so some sort of referential. A predicate, quite essentially, is a referential relation. No qualification can be conceived as 'a property *of* the qualified entity'. This expression is either a shorthand expression for 'a property of ... *with respect to* ... ', or, if not, it is a self-contradictory concept.

This is obscured by two facts. First, that very few classes of referentials have been reconstructed mathematically, like those from nowadays physics. (For instance, the referentials for the qualifications 'blue', 'big', 'intelligent', etc., are not even only formally constructed – in a 'legalized' way – so far they remained vague and implicit.) And, second, that sometimes the locution 'to *have something* ', where 'thing' stays for a substantive, is globally assimilated with a predicate, on the basis of linguistic characters (for instance, grammatically, 'position' is a substantive, so one tends to assimilate 'to have positions', with 'to have points', like absolute space, and to treat this like a predicate). A mixture between grammar and logic.

Petitot. Quite so. I agree. "To have a property" does not support conterfactuality and therefore there exists an irreducible conflict between physical objectivity and natural logic. Physics is not predicative in that sense.

Mugur-Schächter. Natural logic is not special in this respect. As far as I know, a logic cleaned of any trace of absolute predication does not yet exist? *A fortiori* there does not yet exist a *fully* relativized logic, taking into account – besides the relativity to the referential, involved in predication – also the unremovable relativity of any description (representation), to the process of generation of the described entity. This last sort of relativity is particularly obvious in quantum mechanics; but, more or less explicitly individualized with respect to predication, it is present in *any* process of representation. Nevertheless, concerning quantum mechanics as well as in general, it is still very weakly perceived. A certain type of analysis of quantum mechanics, if it is sufficiently deep, leads rather directively toward a full relativization of logic.

Bitbol. I would like to go over to another question. You say that objectivity in Kant's sense is a legalization *of phenomena.* I am not certain that this applies well to quantum mechanics. The laws of quantum mechanics, the equations of evolution, etc., apply to abstract objects, to state vectors in a Hilbert space, not to an entity immersed in physical spacetime where the phenomena are realized. Do you then think that this expresses an even more profound translation than that concerning ontology, to pass from classical physics to quantum mechanics?

Petitot. This, indeed, is a key point. In what I presented here, the term "phenomenon" meant "observable" and not merely "sensible phenomenon". I have already stressed that physics deals only with phenomena and that the axioms of a physical theory must include their relational essence. Now, how do the axioms of QM realize this condition? By substituting the concept of "observable" for that of sensible phenomenon. An observable is not a mere brute phenomenon, nor a phenomenon prepared in a definite way. It is already a pre-objectivized structure. Here appears a quite essential difference between classical and quantum mechanics. Classical mechanics is a theory of motion as pure exteriority. As we have seen, mechanical forces are "derivative" secondary forces produced by a noumenal "interiority" of matter on which classical physics can say nothing. If the observed phenomena are pure motions, then to convert them into observables means essentially to geometrize them by introducing spacetime, trajectories, and equations whose the latter are solutions. As we have seen, matter is reduced to mass and intensive magnitudes like velocities and accelerations. At each point of spacetime there are dynamical vectors, and what has to be reconstructed by the physical laws are the *fields* of such vectors. So the corresponding physical theories will necessarily be ruled by systems of differential equations.

From the transcendental point of view, QM represents an extraordinary comeback of "interiority". It introduces the presupposition that the interiority of matter can be described by internal quantum numbers. The Hilbert space of the states of a quantum system is in general not completely determined. Its relation to spacetime is not necessarily explicit. The only thing we know is that it must exist and that it is a good frame for mathematizing the results of measurements. But in fact the fundamental quantum structure is the algebra of observables. The Hilbert space can be retrieved from it. Moreover, the quantum internal degrees of freedom can depend on space-time and can be geometrized as in gauge theories by fibre bundles. Then the Hilbert space of quantum states becomes much more explicit geo-

rationality'. It could first concentrate upon the genesis of the agreement upon phenomena, which later can be reconsidered inside a further more general representation. Does there not exist something that is subjacent to the perfectly organized rationality which works in mathematics and in axiomatized science? Groping experimentation, comparisons....

Petitot. Certainly.

Mugur-Schächter. We should methodologize all this.

Bitbol. That would be the underground of the formal epistemology.

Mugur-Schächter. Its basis.

Petitot. To conclude, I would like to add a few remarks on the specific contents of the main categorial moments in modern physics and on how they could drive the research programm of a formal epistemology.

Kant studied the categorial moments of classical mechanics. We have to continue his transcendental research programm. We can first examine the categorial moments of the classical theory *of fields* (the theory of continuous media from the beginning of the last century up to now). We can then do the same for general relativity and QM. I think that I have already shown that it is possible.

Secondly, we can analyze the transcendental *trend* that drives the progressive *geometrization* of physical theories. In classical mechanics force is the physical expression of the categories of causality and reality. It is categorialy "dynamical" and not "mathematical". Forces are invariant with respect to the Galilean relativity group and are therefore endowed with a physical reality. In *general relativity*, the physical content assigned previously to forces, is transferred into the (pseudo-Riemannian, locally Lorentzian) metric of space-time. The symmetry group of the theory becomes incredibly larger (it is now the infinite dimensional Lie group of diffeomorphisms) and, as a consequence, all the forces become "inertial". This means that Mechanics is absorbed in a generalized Kinematics. In terms of the categorial structure, it is as if causality and reality were converted into pure "phoronomical" *a prioris*. This is possible because, conversely, the previously purely "mathematical" moment of the metric of space-time (Axioms of intuition) is converted into *a metric moment endowed with physical content.* The physical categories of causality and reality are lifted up into the geometrical ones, whereby correlatively geometry becomes laden with a physical content. By this sort of double movement *the structure of the categorial system is completely changed.* But of course it is still constitutive. In fact, as was deeply stressed by Ernst Cassirer, general relativity is one of the most achieved examples of transcendental physics. In a similar way, Noether's theorem, via the principle of least action, links the symmetries of a theory to its physical contents. This can be applied as soon as a variational formulation is available. This fact expresses the Galoisian quintessence of physics: what can be known and measured is determined from what cannot. I think that the philosophical deepness of this theorem, is far from having been fully recognized and understood.

Bailly. What is needed in Noether's theorem is just a metric, since a geodesic is researched. The approach is *equivalent* to that one which you mentioned before.

Petitot. Not really. You need only:

(i) the canonical symplectic structure of the cotangent bundle of the configuration space,

(ii) a symplectic group action on it, and

(iii) the invariance of the Hamiltonian.

And again there is a deep transformation of the categorial structure. The category of substance, already converted in conservation laws, becomes now a direct consequence of symmetries. Once again, the "mathematical" (geometrical) categories become physically laden, while at the same time the "dynamical" ones are geometrized.

Gauge theories deepen still this trend. Hermann Weyl, who was not only a giant of mathematics and physics (inventor of the concept of gauge symmetry) but also a specialist of transcendental philosophy and phenomenology, said explicitly that his aim when creating the gauge concept was to transform relativity principles into dynamical ones, that is to identify enlarged groups of symmetries with principles endowed with physical content.

This is a sort of "teleological" trend toward a *unification* of the categorial moments. The rigid categorial Kant's hierarchy is no longer operative, according to which one first specifies the geometrical framework, then prescribe *a prioris* upon what happens inside the framework, then builds laws. In modern physics the categorial moments are at the same time "mathematical" (geometrical) and "dynamical" (physical). This is the transcendental sense of geometrization. As a consequence, physical contents can be more and more reduced to enlarged symmetry groups and generalized relativity principles.

Mugur-Schächter. I suppose that in every *new* investigation one has to achieve first the previous phases. One cannot find directly the most economic algorithms. These can be identified only in the end. But they certainly deserve a thorough investigation. As *also* does the whole previous genesis. Without an explicit and methodologized knowledge of the genesis one cannot methodically achieve intelligibility, I think.

So, your program, Jean ...

Petitot. ... is to revisit the difficult problems of categorial structure and to work out a series of detailed case studies. To begin with "exercises of applied epistemology", then to elucidate the way in which, in each case, the categorial structure is mathematically schematized and constructed, and finally to show how the computational synthesis of phenomena can be algorithmically deduced from such constructions.

Bailly. Do you not think that the movement of geometrization that you so well described, might lead to confusion concerning *our* relation to the world, since the human need for explanation is tied to the 'natural' distinction between categories, such as Kant described them. Would you think that there emerges now a new sort of explanation, in terms of unification of the categories in structures of cognition, that will be able to reconstruct otherwise all that has been done before in philosophy in terms of separate categories?

Petitot. I don't know how to answer your question. My second domain of research deals with human cognition conceived of as a *natural* phenomenon. On the other hand, I have explained how and why my epistemology of physics is transcendental. Of course, the time must come, some day, for working out a relation between transcendental epistemology and natural cognition. But for the moment I make strong reserves concerning a naturalized epistemology of mathematical physics. I fear a vicious circle. Cognition is a natural neuro-biophysical phenomenon to which it is possible to apply sophisticated mathematical models. On the other hand, my epistemological views are strongly focused on the problem of mathematizing physical phenomena. If I tried to apply directly cognitive models to them, I would need, to be consistent, a cognitive theory of mathematics. But how could such a theory be obtained? Only by the help of mathematical models of the brain. This is a vicious circle.

Mugur-Schächter. I think that it is not possible to have in advance a general scheme for a not-yet-achieved development. One has just to begin, hoping that the development will progress along a spiral, and will also permit reflexive returns along the spiral, but without ever leading to circular *face à face* collisions on a plane level.

Bitbol. There always are blind spots in knowledge, which change with knowledge.

Petitot. Yes. This reserve is for me an only temporary one.

Mioara Mugur-Schächter. In any approach one has to start from posited concepts and assertions. At the basis of a representation of epistemology, these *a prioris* might always remain different from those needed for a representation of cognition regarded as a biophysical phenomenon. Between these two sorts of representations there probably acts like a sort of two-mirrors relation.

Petitot. Perhaps. This reminds me of a remark by Husserl: he said the categories of our living world and those of the physical objectivity *share the same names*, but are nevertheless completely different because the first ones are not concerned with any mathematical "substruction". This remark is very deep indeed.

Bitbol. I propose a metaphor: The mirrors are not exactly parallel.

Mugur-Schächter. Well, we all thank you, Jean, very much indeed.

Petitot. I thank you all. That was really a very kind and exciting exchange, and I want to thank particularly Mioara for having organized it.

REFERENCES

1. R. Abraham and J. Marsden, 1978, *Foundations of Mechanics* (Benjamin Cummings, New-York).
2. H. E. Allison, 1983, *Kant's Transcendental Idealism. An Interpretation and Defense* (Yale University Press, New-Haven).
3. V. Arnold, 1976, *Méthodes mathématiques de la mécanique classique* (Mir, Moscow).
4. D. Bailin and A. Love, 1994, *Supersymmetric Gauge Field Theory and String Theory* (Institute of Physics, London).
5. D. Bennequin, 1994. "Questions de physique galoisienne," in *Passion des Formes, à René Thom*, M. Porte, ed. (E.N.S. Editions, Fontenay-Saint Cloud), pp. 311-410.
6. M. Bitbol, 1996, *Mécanique quantique: Une introduction philosophique* (Paris, Flammarion).
7. N. Bohr, 1935, *Phys. Rev.* **48**, 696.
8. G. Brittan, 1978, *Kant's Theory of Science* (University Press, Princeton).
9. E. Cassirer, 1918, *Kants Leben und Lehre*; translation by J. Haden, 1981, *Kant's Life and Thought* (Yale University Press, New Haven).
10. G. Cohen-Tannoudji and M. Spiro, 1986, *La Matière-Espace-Temps* (Fayard, Paris).
11. H. Duncan, 1984, "Inertia, the communication of motion, and Kant's third law of mechanics," *Phil. Sci.* **51**, 93-119.
12. B. d'Espagnat, 1985, *Une incertaine réalité* (Gauthier-Villars, Paris).
13. B. d'Espagnat, 1994, *Le réel voilé* (Fayard, Paris).
14. H. J. Folse, 1978, "Kantian aspects of complementarity," *Kant-Studien* **69**, 58-66.
15. R. J. Gomez, 1986, "Beltrami's Kantian view of non-Euclidean geometry," *Kant-Studien* **77** (1), 102-107.

16. J. Honner, 1982, "The transcendental philosophy of Niels Bohr," *Stud. Hist. Phil. Sci.* **13**, 1, 1-29.
17. C. Itzykson and J. B. Zuber, 1985, *Quantum Field Theory* (McGraw-Hill, New York).
18. M. Jammer, 1974, *The Philosophy of Quantum Mechanics* (Wiley, New York).
19. M. Kaku, 1988, *Introduction to Superstrings* (Springer, New York).
20. I. Kant, 1781-1787, *Kritik der reinen Vernunft (Kants gesammelte Schriften, Band III,* Preussische Akademie der Wissenschaften (Georg Reimer, Berlin, 1911).
21. I. Kant, 1786, *Metaphysische Anfangsgründe der Naturwissenschaft (Kants gesammelte Schriften, Band IV,* Preussische Akademie der Wissenschaften (Georg Reimer, Berlin, 1911); translation by J. Gibelin, 1971, *Premiers Pincipes métaphysiques de la Science de la Nature* (Vrin, Paris).
22. I. Kant, 1796-1803, *Opus Postumum,* translation by F. Marty (Presses Universitaires de France, Paris, 1986).
23. Y. Manin, 1988, *Gauge Field Theory and Complex Geometry* (Springer, New York).
24. J. Marsden, 1974, *Applications of Global Analysis in Mathematical Physics* (Publish or Perish, Berkeley).
25. C. W. Misner, K. S. Thorne, and J. A. Wheeler, 1973, *Gravitation* (Freeman, San Francisco).
26. R. Omnès, 1994, *Philosophie de la science contemporaine* (Gallimard, Paris).
27. J. Petitot, 1991, *La Philosophie transcendantale et le problème de l'Objectivité* (Entretiens du Centre Sèvres), F. Marty, ed. (Osiris, Paris).
28. J. Petitot, 1992a, *Physique du Sens* (Editions du CNRS, Paris).
29. J. Petitot, 1992b, "Actuality of transcendental aesthetics for modern physics," 1830-1930, in *A Century of Geometry,* L. Boi, D. Flament, and J.-M. Salanskis, eds. (Springer, New-York).
30. J. Petitot, 1994b, "Esthétique transcendantale et physique mathématique," *Neukantianismus: Perspektiven und Probleme,* E. W. Orth and H. Holzhey, eds. (Königshausen & Neumann, Würzburg), pp. 187-213.
31. A. Philonenko, 1972, *L'oeuvre de Kant* (Vrin, Paris).
32. C. Quigg, 1983, *Gauge Theories of the Strong, Weak, and Electromagnetic Interactions* (Benjamin-Cummings, Reading, MA).
33. J. M. Souriau, 1975, *Géométrie symplectique et physique mathématique* (Collection Internationale du CNRS **237**, Paris).
34. B. C. Van Fraassen, 1991, *Quantum Mechanics: An Empirist View* (Clarendon, Oxford).

35. J. Vuillemin, 1955, *Physique et Métaphysique kantiennes* (Presses Universitaires de France, Paris).
36. A. Weinstein, 1977, *Lectures on Symplectic Manifolds* (Conference Series, American Mathematical Society **29**, Providence, RI).
37. C. F. von Weizaäcker, 1979, *Die Einheit der Natur* (Hauser, Munich).
38. H. Weyl, 1922, *Space-Time-Matter* (Dover, New-York).
39. J. E. Wiredu, 1970, "Kant's synthetic a priori in geometry and the rise of non-euclidean geometries," *Kant-Studien* **61** (1) 5-27.

6

ON THE POSSIBILITY OF A FORMALIZED EPISTEMOLOGY*

Robert Vallée

Université Paris-Nord
Paris, France

We present an attempt to a *mathematical epistemology* valid at the macroscopic level. Do similar perceptions correspond to similar objects? How can we recompose two perception processes? The use of mathematical *observation operators* provides an answer. Identical perceptions may correspond to different objects perceived, due to the lack of an inverse of an observation operator. Consequently, there is a process of *inverse transfer* of structures inherent to the observing system on the observed world. Moreover, multiplication of observation operators gives a representation of the composition of perception processes. So we have at hand an algebra opening the way to a mathematical and thus formal epistemology. Also, the intervention of decision, composed with perception, allows the introduction of considerations analogous to those valid in the purely epistemological case, giving rise to a formal *epistemo-praxiology*. The possibility of a *formal epistemology*, even completed by praxiology, at the macroscopic level provides arguments in favor of a general formal epistemology acceptable at all levels.

Key words: algebra, epistemo-praxiology, formal epistemology, inverse transfer, mathematical epistemology, observation operator.

On the paths of knowledge, for which there is no royal way, or maybe on the banks of the unknown, the human mind has met two temptations: direct, inexpressible apprehension and systematic, formal process. Between these opposite poles, these two attractors, of enlightenment and some "ars magna", mind is still hesitating. But if the first type of knowledge is strictly individual and subjective to an extreme, the second one, on which focuses here our interest, should be valid for all. Is the attractor corresponding to the second type so strange that it leads just to a mirage or, on the con-

*See "Important Note" on p. xviii.

trary, does it indicate a goal really accessible or which we can approach asymptotically?

As a preliminary, it may be interesting to consider the acquisition of knowledge, or much more simply the process of perception, at the macroscopic level. If this non-trivial process seems, at least partly, formalizable, we shall be encouraged to approach, in an analogous spirit, some wider aspects of epistemology. The angel of topology and even more the demon of algebra, in Bourbaki's words, seem able to participate in a formalization of macroscopic perception. Do similar perceptions correspond to similar objects? What can be said about the composition of two perceptive processes? This second question is perhaps the most interesting one and we are inclined to think that macroscopic perceptive processes may be formalized in algebraic terms. Examples are given by many physical or even biological devices such as signal transmission (frequency filtering, temporal limitation or sampling), and optical observation (spatial frequency filtering, framing). It is possible to elaborate an algebra of what we call *observation operators* [3,4].

This type of formalization clarifies many aspects of macroscopic perception, not only deformation but also non-inversibility. For some reason or another real perceptive processes have no inverse, it is not possible to reconstruct the observed object from the knowledge of its perceived image nor to separate two distinct objects having the same image. We have here a limitation of epistemic nature, expressed formally by the non-inversibility, in the mathematical sense, of the observation operator representing the process. This observation operator, in the most simple cases, has to do with the geometric concept of projection which reminds us of Plato's cave [7]. According to this well known allegory, the shadows, observed at the rear of the cave, are the only knowledge enchained men can acquire concerning statues moving between a bright fire and themselves. Not only do they obtain thus a distorted, impoverished image of the statues but furthermore they also attribute to the perceived shadows properties of the screen constituted by the rear part of the cave, for example its bi-dimensionality, its irregularities. This part of Plato's metaphor gives a poetical expression of the non-inversibility of perception and also of the subjective attribution, to the observed universe, of structures inherent to the perceiving device, whether purely physical or of biological nature. The use of non-inversible observation operators yields a mathematical formalization of this subjective *inverse transfer of structures* [6,7] which has to do with the tendency to project our inner world upon the external one, making it, even if abusively, more familiar, endowing it with an appearance of pre-established harmony [9].

The above considerations on perception plead in favour of the possibility of a general formal epistemology [1,2] or even of a *mathematical epistemology*, restricted to the macroscopic level [5].

In a less artificial way, it is possible to distinguish, for a physical or biological structure, between an act of observation and decisional process. In a formalized language, these steps can be represented by, respectively, an observation operator and a decision operator, both of a same mathematical nature. But, globally, we can define only one *pragmatic operator*, product (to the left) of the observation operator by the decision operator, giving access to an epistemology based upon decisions taken instead of perceptions felt. Here again a formalization seems at hand, opening the way toward a mathematical *epistemo-praxiology* [8,9].

So the cognitive aspect of epistemology seems at least partly formalizable in algebraic terms, at the macroscopic level. This is the encouragement we hoped to find. Is it possible to go further? A true formal epistemology should not confine itself to the macroscopic world; it must also include quantum aspects, thus encompassing both the macroscopic and the microscopic levels. Could a general method be formed, able to help us find very basic structures, more fundamental than those at hand. Let us try to proceed along this way imperfectly delineated. We have, for example, an algebra of macroscopic observation and also an algebra of microscopic observation. There are between these structures at least some analogies which give some hope to find a structure weaker than each of them, able to retain fundamental traits of both of them.

Other approaches seem possible. We can imagine an order, not necessarily total, concerning natural systems. This hierarchy could be generated by the process of observation itself, from the coarsest to the most delicate resolution level, from universe to particles. Epistemological analogies of structures could be seen, being obviously more and more difficult to cope with when approaching the microscopic level where they might become meaningless, unless some fundamental formal aspect, to be discovered, should still hold.

The construction of a formal epistemology, fundamental enough to be relevant at all levels of knowledge, from the macroscopic to the microscopic, is obviously a difficult task. The above considerations on a possible mathematical epistemology valid at the macroscopic level, the indications given about apparently plausible ways of research, are positive factors in favour of the feasibility of a formal epistemology.

REFERENCES

1. M. Mugur-Schächter, "From quantum mechanics to universal struc-

tures of conceptualization and feedback on quantum mechanics," *Found. Phys.* **23**, 37-122 (1993).
2. M. Mugur-Schächter, "Une méthode de conceptualisation relativisée: vers une épistémologie formelle apte à faire face aux complexités," *Revue Internationale de Systémique* **9**, 963-994 (1995).
3. R. Vallée, "Sur deux classes d'"opérateurs d'observation'," *C.R.A.S. (Paris)* **233**, 1350-1351 (1951).
4. R. Vallée, "A note on algebra and macroscopic observation," *Information and Control* **1**, 82-84 (1957).
5. R. Vallée, "Sur la formalisation mathématique en théorie de l'observation," *1ère partie* in *Actes du 7ème Congrès Internationale de Cybernétique* (Association Internationale de Cybernétique, Namur, 1974), pp. 225-232.
6. R. Vallée, "Observation, decision and structure transfers in systems theory," in *Progress in Cybernetics and Systems Research, Vol. 1* (Second European Meeting on Systems Research, Vienna, 1974), R. Trappl and F. Pichler, eds. (Hemisphere Publishing, Washington, 1975), pp. 15-20.
7. R. Vallée, "La caverne de Platon revisitée," in *Praxis et Cognition* (Colloque International Praxis et Cognition, Cerisy, 1988), E. Bernard-Weil and J. C. Tabary, eds. (L'Interdisciplinaire, Limonest, 1992).
8. R. Vallée, "Perception, decision and action," *Biological Systems* **2**, 43-53 (1994).
9. R. Vallée, *Cognition et Système, Essai d'Epistemo-Praxéologie* (L'Interdisciplinaire, Limonest, 1995).

Part Two

Constructive Contributions

7

QUANTUM MECHANICS *VERSUS* A METHOD OF RELATIVIZED CONCEPTUALIZATION

Mioara Mugur-Schächter

Centre pour la Synthèse d'une
Épistémologie Formalisée (CeSEF)
47 Bd Georges Seurat
92200 Neuilly-sur-Seine, France

A general representation of the processes of conceptualization, founded upon a descriptional mould drawn from fundamental quantum mechanics, is outlined. The approach is called the *method of relativized conceptualization*. This stresses that the representation is not researched as a "neutral statement of facts" but, from the start, as a *method* subject to definite descriptional aims, namely an *a priori* exclusion of the emergence of false problems or paradoxes as well as of any gliding into relativism. The method is characterized by an explicit and systematic relativization of each descriptional step to all the descriptional elements involved in this step, namely: the epistemic action by which the object-entity is generated, the object-entity itself, and the epistemic action by which the object-entity is qualified. Successive steps which complexify progressively a given initial relative description, form an unlimited chain of *cells* of conceptualization where the very first cell necessarily is *rooted in as yet strictly unconceptualized physical factuality*, while the subsequent cells consist of increasingly abstract descriptions that are connected hierarchically. The chains interact at nodes where they branch, thus generating an indefinitely evolving, complexifying web of relativized conceptualization, free of ambiguities, and where each element stays under control.

The method contains the posited assertion of a realism of which a definite sort of minimality *follows* then inside the method. This generates a clear distinction between illusory qualifications of "how-a-physical-entity-is-in-itself" and *models* of this physical entity. Thereby a worked-out connection with philosophical thinking is incorporated in the method. The method is shown to entail *a relativized genetic logic* and *a relativized genetic theory of probabilities*, more extended, respectively, than the classical logic and the classical theory of probabilities. Both are rooted in physical factuality whereby they merge in a unified representation of the logico-probabilistic conceptualization.

The relations between the general method of relativized conceptualization and the relativistic approaches in the sense of modern physics are specified. These last ones, in contradistinction to the method exposed in this work, are shown to concern *exclusively* the ways of constructing qualifiers of object-entities so as to insure intersubjective consensus among corresponding classes of observers, while the ways of generating the qualified object-entities and the consequences entailed by these ways, are not considered: Like in classical logic, as in all of classical thinking, the object-entities are simply presupposed to always pre-exist available.

Traditionally, the emergence and elaboration of knowledge has always been studied from a point of view founded on psychological and neurobiological data and in the spirit of a neutral account of the "natural" phenomena; the modern cognitivistic approaches continue this tradition. The approach exposed in this work is probably the first one in which a systematic representation of the processes of creation of knowledge is founded on *strategic* data drawn from physics and, correlatively, is constructed from the start as a *method* for the optimization of these processes *themselves*, accordingly to definite aims.

Key words: quantum mechanics, method, descriptional relativities, conceptualization, epistemology.

> "That a higher integration of science is needed is perhaps best demonstrated by the observation that the basic entities of the intuitionistic mathematics are the physical objects, that the basic concept in the epistemological structure of physics is the concept of observation, and that psychology is not yet ready for providing concepts and idealisations of such precision as are expected in mathematics or even physics. Thus this passing of responsibility from mathematics to physics, and hence to the science of cognition ends nowhere. This state of affairs should be remedied by a closer integration of the now separate disciplines."—
> *E. P. Wigner [1].*

Contents

1. INTRODUCTION

This work is submitted here as an illustration of how a formalized epistemology can be researched accordingly to the principles expressed at the beginning of this volume. Indeed, what I call *the method of relativized conceptualization* can already be regarded, I think, as a first but rather firm construct on the way toward a fully satisfactory formalized epistemology. Though initially induced by—specifically—the cognitive strategy brought forth by the analysis of fundamental quantum mechanics, this construct, by the universality of the epistemological essence drawn therefrom and by the way of elaborating it, possesses an unrestricted relevance concerning any process of conceptualization.

In a certain sense, the way in which the method of relativized conceptualization is offered here is highly artificial. This method developed in my mind very slowly, while periodically, year after year, in the course of my University lectures on elementary and advanced quantum mechanics, on probabilities and on information theory, I was once more scrutinizing the formalisms of these three theories. This recurrence, by a process of integration, produced the method of relativized conceptualization at the *same* time with what I now call *meta[quantum mechanics]*[1] and which—*a posteriori* - appears as a major illustration of the method, belonging organically to it. But meta[quantum mechanics] is far too technical to fit into this volume. So I chopped it off and healed the scars by a brief informal preliminary exposition of—strictly—only the essence of the considerations on quantum mechanics which triggered the method of relativized conceptualization. The result might appear somewhat strange due to restricted access to the structure of mathematical features which determined it from inside fundamental quantum mechanics and which, together with the emerging method itself, guided the modified reconstruction of quantum mechanics which I call meta quantum mechanics, which in its turn illuminates the method. But, on the other hand, any method, once constructed, should be able to convince by itself. So, insofar as a method, such as the one exposed here, fails to do that, it simply is devoid of a genuine inherent conceptual and operational value.

Inside the community of physicists, this work will appear as exterior to all the present-day main streams. Of course, there have been many famous physicists who have tried to understand how quantum mechanics works, what it really asserts, and what it leaves open. But, as far as I know at least, no physicist as yet has tried to work out explicitly, specifically, and systematically, the universal epistemological implications of the quantum mechanical formalism. The novelty of this aim imprints a peculiar character upon the approach. This is why physicists might feel disconnected while reading what follows. In order to nevertheless gain their attention and fix it upon the epistemological problems dealt with in this work, I take the liberty to claim once more that, inside meta[quantum mechanics], the results established in this work entail a clear optimization of the formalism of fundamental quantum mechanics—with respect to its own descriptional aims—and that they furthermore yield a thorough *intelligibility* of this formalism, which cannot but enhance the efficiency in dealing with the basic

[1] Partial indications on meta[quantum mechanics] can be found in various other works (Refs. 11 to 17). A complete final account is not yet available but I hope will be published soon.

problems of modern physics in general.

The philosophers, with respect to their own knowledge and criteria, will certainly find insufficiencies in this work. I apologize to them in advance: With the means available to me, I have tried to build a solid bridge between physicists and philosophers. Others might want to improve on it in various ways.

Of course, a formalized epistemology, in the full sense assigned to this term in the introduction to the present volume and in the contributions from the first part, should incorporate methodological procedures explicated also from other modern disciplines besides quantum mechanics, in particular from mathematics, informatics, biology, cognitive and neurological sciences, linguistics, and philosophy. Some steps in this direction can be found in other contributions to this volume (cf., in this volume, the contributions of Robert Vallée, Élie Bernard-Weil, Giuseppe Longo, Evelyne Andreewsky, and Vincent Schächter).

2. RETRO-PERSPECTIVE

Before entering upon the exposition of the method of relativized conceptualization, I shall briefly sketch out in what historical retro-perspective it fits in.

2.1. Objectivity and Descriptional Relativities

The concept of scientific objectivity is undergoing a revolution. The classical concept of objectivity was tied with the posit that science just *discovers* truths that are independent of any human aim-and-action, pre-existing "out there" such as they appear when discovered. But throughout the last century this view kept receding. It became increasingly clear that objectivity in the classical sense was an illusion; that scientific knowledge is *constructed* under certain constraints which characterize the epistemic situation and the epistemic aim of the acting observer-conceptor and imprint upon the result non-removable descriptional *relativities* to this situation and this aim. More or less implicitly, awareness of quite essentially involved [(epistemic situation)-(epistemic aim)] structures, developed steadily, perturbing the classical conception about objectivity while instating a new concept of objectivity in the sense of inter-subjective consensus.

So far however, only few have already gained an explicit and clear awareness of this evolution. Correlatively, on a metalevel, a fully organized and general view on the epistemic actions by which scientific inter-subjective consensuses are achieved is still lacking. What, exactly, in scientific consensus, insures subjection to *also* what is called reality and truth, thereby transcending mere conventionality and withstanding relativism? How, in what a sense and to what a degree, is reference insured? How, exactly, do the involved human aims and features come into play? What particular sorts of strategies are put to work in order to construct scientific inter-subjective consensuses? While such questions struggle for definite answers, the inertial forces that work inside language bring forth again and again the same old word—objectivity—to designate indistinctly either the emerging new concept, or the classical one. This favours the persistence of many circularities and confusions.

Let us now consider physics. The employed cognitive strategy *varies* radically as one shifts from fundamental quantum mechanics to the theory of relativity and to relativistic approaches in general.

Fundamental quantum mechanics incorporates—implicitly—a peculiar type of "basic" descriptional relativities which insert the very *first* stratum of conceptualization, deep into purely factual physical reality. The descriptional relativities of this basic type, when entirely explicated and then generalized, lead toward a recasting of epistemology. The main lines of this major consequence of the quantum mechanical strategy for constructing knowledge are captured in the method of relativized conceptualization. This method, while it strongly connects modern physics with philosophy, will be shown to entail also a non classical unification between logic, probabilities, and set-theory.

On the other hand, inside the theory of relativity and more generally inside the whole class of relativistic approaches, *another* sort of methods for constructing inter-subjective consensuses have been developed. These, much better recognized than those involved by fundamental quantum mechanics, are only very indirectly and loosely connected with physical factuality. They are quasi exclusively dominated by abstract constraints of a logico-mathematical nature imposed upon the representational features tied with "states of observation". The formal constructs entailed by this sort of constraints manifest a vertiginous growing of the degree of conceptual freedom displayed by modern physicists in the representations of physical reality. In these constructs one can again identify forms of the general tendency, in modern physics, to merge with epistemology and philosophy.

So in modern physics, objectivity, quite generally, means constructed inter-subjective consensus founded on descriptional relativizations that point toward an underlying trend of unification of physics, with epistemology and philosophy.

The method of relativized conceptualization, which is the core of this work, was crystallised out of this trend.

2.2. "Existents" or "Reality", and Objectivity

The existence, for each human being, of an inner psychical reality, probably has never been doubted by any normal person. Following Descartes, Berkeley, Kant, Husserl, the philosophers place it explicitly at the bottom of any knowledge. Physicists have never denied it. Nor did common sense. And nevertheless, paradoxically, for most people the quintessence of what is called reality, of what is hold to be "genuinely" existent, is *the exterior and physical* reality; even if this or that marginal individual happens to perceive the exterior physical reality as less certain than his own inner reality, or even—at the solipsistic limit—as wholly illusive.

This entangled hierarchy has multiple manifestations. For instance, it is striking that concepts, and more generally knowledge, languages, science, are seldom explicitly taken into account as constituents of reality, strictly speaking. It is true that Teilhard de Chardin did so (this is his major specificity); that Karl Popper [2] asserted "three worlds", the physical reality, the states of consciousness, *and* knowledge, arts, cultural facts; and that, no doubt, other important examples can be found. But, on the other hand, up to this day the debate on the existents (do the unicorns exist?) still continues among logicians [3]. Platonism has enemies as much as adepts, etc. And, more or less implicitly, a general tendency can be observed to *set aside* the word reality for designating exclusively what is posited to exist outside any psychism and moreover is physical. A larval form of this tendency is present in particular in the reductionist view according to which anything which at a first sight seems not to consist of exclusively physical entities, in fact is strictly deducible—without any loss—from the existence and laws of the physical reality *alone*. This view, favoured by a loose contact between philosophers and scientists, is still quite active in many eminent minds, notwithstanding that most philosophers perceived it as naïve and illusive already since Descartes, while since Kant they almost unanimously banished it explicitly and radically.

On the other hand, Einstein relativity and then—otherwise—quantum mechanics, induced a stream of change into the content assigned in physics to what is called truth and objectivity. The main contribution to this stream consists of deliberate *constructions* of symmetries concerning the processes of qualification of the considered object-entities, symmetries tied with groups of operations of transformation of the state of observation. But furthermore other modern developments of the "exact" thinking, logical, mathematical, informatical, also contribute to this stream, by direct elaboration of grammars (syntaxes) admitting of models (interpretations), by algorithms for reconstructing phenomena by simulation instead of representing them by assertions and proofs, etc. Now, all these new approaches are methods for constructing inter-subjective consensus concerning results of manners of conducting descriptional actions in order to reach a definite aim of knowledge. They all involve an explicit teleological dimension where factors of various natures—psychical or biological or physical, factual or abstract-conceptual—co-operate inside an organic whole. This amounts to an implicit deletion of the classical belief that consensus manifests a pre-existing objective truth which has to be just learned, apprehended.

This evolution induces the scientific thinkers into rediscovering by themselves certain basic features of Kant's constructivist view on objectivity [4,5]. This, among those who work in the foundations of science, generates an increased receptivity with respect to the philosophical thinking sedimented since millennia. While on the other hand the philosophers tend more and more to concentrate upon the methods and languages that emerge inside the sciences, trying to bring forth the new philosophical implications of these.

Globally, philosophy and the sciences are meeting in a process of re-elaboration of the concepts of reality and objectivity.

I shall now go to the bottom of this process, but specifically from the point of view of a physicist. I shall focus upon the content of the *very first* layer of the emergence of the inter-subjectively known, such as it can be characterized when the involved biological processes, though fully recognized to play a key role, are *not* themselves the object of investigation (as in the modern researches on cognition and consciousness [6,7,8]) but are regarded as only a datum to be explicitly taken into account.

2.3. The Polarity of Realism[2]

Kant stated explicitly that exclusively phenomenal appearances are known in a non-mediated way. The word phenomenon designates here a conscious event from an individual mind, already cast in the *a priori* forms of human intuition, time and space. This conscious event can be conceived by the man who experiences it as reflecting, or not, some object-entity; but in any case it somehow bears the mark of the acting human body-and-mind structure, in a non removable and inextricable way. This is the foundation of the well-known Kantian postulate of impossibility to know reality such-as-it-is-in-itself, i.e., independently of any structure interposed by the observer-conceptor.

It is curious to note that this famous Kantian impossibility concerns exclusively the reality that is exterior to the mind. Indeed, if one chooses to point *via* this same term, reality, toward any sort of existent, no matter whether assigned to the exterior universe or to some interior universe, this rather natural extension of language generates an exception to Kant's pos-

[2] This section has benefited from precious remarks made by Hervé Barreau.

tulate, a huge one. For on the one hand this extension of language entails that also a phenomenon from an individual mind is an element of reality. But on the other hand a phenomenon, by definition, *is* just that what *appears* to the mind where it emerges. So, for the sake of self-consistency, a phenomenon, as such, has to be posited to be known by the mind where it emerges *precisely* such-as-it-is-in-itself. To assert the contrary would simply be a logical contradiction in the construction of the whole consisting of [language and what it is posited to refer to]. Later the considered phenomenon might be perceived differently by the person who experienced it, or if it is communicated to another mind its description might there be variously interpreted, in psychoanalytical terms, or biological ones, etc. But in all such cases one is in fact speaking of another (meta)object-entity that is related with the initial phenomenon but is not identifiable with it. And this new (meta)object-entity, in its turn, again must be posited to be known by the mind where it emerges, such-as-it-is-in-itself, etc. (This same point has been made also by other authors, e.g., Goodman [9].) This characteristic of the inner phenomena, however, is not in the least a "problem". On the contrary it seems to be in deep harmony with the Cartesian cut.

Indeed, the fact that an entity from an inner individual universe has to be considered to be precisely such as it is perceived, can be considered to mark a polarity of reality with respect to knowledge, by which, while the exterior reality *never* can be known such-as-it-is-in-itself, any piece of interior reality—at the time when it emerges in this or that individual mind—can *only* be known by that mind such-as-it-is-in-itself, whereby its "truth" is beyond any doubt (or is a qualification devoid of pertinence, which amounts to the same thing), so it is endowed with the Cartesian sort of pre-eminence.

2.4. Knowledge and Communicability

But let us come back to the fact that a phenomenon, by definition, can only exist inside an individual mind. At the time when a given phenomenon emerges in an individual mind, it is known there without being also communicated. The subject can even know it without having ex-pressed it for himself: it can remain an unexpressed, a-symbolic individual psychical fact, chained to, and somehow melted to a certain degree in the interior universe where it happened. On the other hand, according to thinkers who know Kant's work deeply, in the Kantian view any scientific objectivity is constructed by a method of "legalization" of the primary phenomenal appearances. In this respect, Jean Petitot ([4]) writes:

> "*The object of experiment, of scientific knowledge, is not given in the donation of the phenomenon.* It emerges by objectual legalization of phenomena. So, apart from a descriptive dimension, any scientific

knowledge presupposes in its very principle also a prescriptive, a normative dimension, that is *constitutive of objectivity* .... In Kant's work—so concerning classical mechanics—*the method consists essentially in interpreting the categories of objectivity* [3] by starting from the instances of donation of the phenomena, that is, by starting from the forms of phenomenal manifestation. Since the interpretation of the categories of objectivity is operational only if it is *mathematical*, the forms of phenomenal manifestation themselves must be mathematized."

But such a legalization involves communicability. So, *how is the transposition of a phenomenon into communicable symbolizations to be set up?*

Here, at precisely this point, one is confronted with an obscure zone where is located—undefined—the structure of the very first stage of *inter*-subjective conceptualization, that on which the whole subsequent inter-subjective conceptualization is founded, so also objectivity in general and in particular scientific objectivity. Kant did not deal with this question.[4] And as far as I know, up to now the philosophical thinking did not yet concentrate constructive efforts upon this zone. But it produced already important "negative" developments. The whole question of *reference* on which Quine [10] and Putnam [11] for instance achieved so deep and compelling analyzes in order to establish the frontiers of the domain inside which language confines knowledge, takes its sources precisely in the above mentioned obscure zone.

Now, in so far that one agrees that any transposition of a phenomenon, in communicable terms, amounts to a description, the content of this obscure zone can be more narrowly pointed toward by the following formulation:

> *Nothing else but descriptions can be known in an inter-subjective way, neither exterior factual entities "themselves", nor non-described phenomena.*

This specification is far from being trivial: it focuses the attention upon the primary importance of the emergence of communicability. Communicability in general as a larger basis for the particular sort of communicability that is normed scientifically. By way of consequence it establishes

[3] The "dynamical" (physical) categories of substance, of causality and of interaction, the categories of quality and quantity, and the "modal" categories of possibility (potentiality, virtuality), of reality (actuality) and of necessity.

[4] As Hervé Barreau puts it, in Kant's view the phenomena seem to emerge *directly* Newtonian, already cast in scientific Euclidean spacetime. Any concern about geneses of the type of those examined later by Husserl, Bergson, Piaget, and so many others, is absent in the Kantian work.

the interest of defining a *canonical* structure for what is called a description, a normed form of the descriptions, a mould into which to pour in an agreed way any transposition of a phenomenal appearance, in communicable terms. It establishes the *inadequacy* of a notion of—directly—a "scientific legalization of phenomena" which omits, hides into the non-analyzed and non legalized, the more basic stage of accomplishment of descriptions. Indeed, only a conveniently structured general norm for accomplishing descriptions could act as a *universal* inter-subjective reference permitting to gauge against it any procedure for describing, the natural descriptional procedures, as well as, in particular, the various procedures for a "scientific" legalization of the descriptions of object-entities of any kind, so also of conscious phenomena. These procedures could then be all qualified, compared, understood, inside a common frame where a certain unity is set in advance beneath the specificities tied to this or that descriptional approach.

But how, according to which criteria, shall we identify the canonical form to be required for any description?

It is quite remarkable that the answer to a question of this nature and of such generality can be drawn from a physical theory. For it is quantum mechanics which shows the way, if the descriptional aim chosen in it and the strategy practised in order to reach this aim, are thoroughly explicated.

3. THE COGNITIVE SITUATION THAT LED TO THE QUANTUM-MECHANICAL DESCRIPTIONS

3.1. Historical Remarks

A cognitive situation like that one involved in the quantum mechanical formalism, *so extreme*, had never been dwelt with systematically before the construction of quantum mechanics. A cognitive attitude like that one induced by the mentioned cognitive situation, so radically creative, had never before been organized. But when a theory of "microstates" started being researched, the involved cognitive situation *acted*, without getting explicit for that. The various well-known contributions from Plank, Einstein, Bohr, de Broglie, Schrödinger, Heisenberg, Born, Pauli, von Neumann, Dirac, etc., led to a coherent whole because they all had to satisfy, more or less implicitly, the *same* strong and peculiar constraints, those imposed by the involved cognitive situation. (But, and it is curious to find this out, none among the so numerous and eminent contributors did fully grasp the new epistemological essence of the emerging construction). There has been no equivalent, for quantum mechanics, of a Newton, a Maxwell, a Carnot, a Boltzmann, or an Einstein.

The construction of the quantum mechanical formalism has been orchestrated by an impersonal, very peculiar *cognitive situation*.

This might explain why the formalism, notwithstanding its remarkable efficiency, is up to this very day thought to possess a cryptic character and to involve problems. These problems however, over and over again, are much more referred to the formalism itself than to the cognitive situation which commanded the form of the algorithms. While, as far as I know at least, the cognitive situation has never been explicitly and thoroughly reconsidered for itself. So, hidden beneath increasingly complex formal developments and surreptitious mutations of the theory as a whole, its seminal epistemological implications could remain for ever devoid of contour, their substance anonymously absorbed and assimilated in the process of the evolution of physics. This would be a big loss. Only what is named and described explicitly gets contour and can act sharply and deeply.

In what follows, I withstand this decay. In a very synthetic and simple way I shall outline the main specific epistemological features of the cognitive situation involved in the quantum mechanical formalism. Thereby, in fact, I achieve a first step in the direction of what I call meta[quantum mechanics]. Indeed, as already mentioned, this re-formalization of fundamental quantum mechanics, of which certain rather elaborate and much more technical elements (but never the whole so far) have been exposed in other works [12,13,14], is founded on—both—the basic considerations exposed in Sec. 3.2 *and* on the fully elaborated method of relativized conceptualization. Here, however, the aim is exclusively to bring into evidence the source of the method. The very simple exposition that follows (which summarizes two earlier non-specialized presentations ([15,16]) should suffice.

3.2. The Cognitive Situation Involved in Quantum Mechanics and the Strategy Induced by It

A description involves a definite object-entity (object-of-description) and qualifications of it. The basic object-entities of quantum mechanics are what is called states of microsystems (microstates).[5] These are *hypothetical* entities that no human being (in the present-day sense) will ever perceive. The

[5] The stable microsystems themselves (electrons, protons, neutrons, etc.) have first been studied in atomic and nuclear physics where they have been characterized by specific "particle"-constants (mass, charge, magnetic moment). Changes of stable microsystems (creation or annihilation) are studied in nuclear physics and in field-theory. States of stable microsystems—microstates—are specifically studied in fundamental quantum mechanics where they are characterized by probabilistic distributions of values of state-"observables" (for Dirac the word "sate" is short for "way of mouvement" of a dynamical system (microsystem)).

obtention for them of qualifications endowed with some sort of stability, raises difficult and deep questions. Nevertheless quantum mechanics exhibits a very performing description of microstates. This manifests a descriptional strategy that has succeeded to overcome the epistemological difficulties. We want to explicate this descriptional strategy.

Let us consider first the basic object-entities of the quantum mechanical descriptions, microstates. Since they cannot be perceived, such object-entities cannot be made available for study by just selecting them inside some ensemble of pre-existing entities. Nor can one study them by just examining observable marks spontaneously produced on macroscopic devices by admittedly pre-existing natural microstates: no criteria would then exist for deciding which mark is to be assigned to which microstate. The unique general solution, then, is to first accomplish a known and repeatable macroscopic operation posited to generate a given though unknown microstate, and to try afterward to somehow manage to "know" the generated microstate.

Consider the hypothetical microstate produced by a given operation of state-generation. The plan is to acquire concerning it informations cast in certain pre-established terms, involving what is called "position", or "momentum", or "energy", etc. The grids for the desired sorts of qualification are conceived beforehand, quite independently of the generated object-microstate, and with respect to *these* grids the object-microstate emerges in general still entirely unknown, still strictly non-qualified. This assertion is not in the least weakened by the fact that the presuppositions of the existence of microstates and of the emergence of a given sort of microstate when a given operation of state-generation is realized, insert already the generated microstate into a net of pre-conceptualization, so of a kind of pre-*posited* knowledge: the generated microstate emerges non-perceptible, so *a fortiori* still entirely non-singularized from the specific points of view expressed by the definitions of the grids of its desired further qualifications. But on the other hand it emerges also relative, in a non removable way, to the employed operation of state-generation, and this permits to *label* it: it is a result of *this*—known—macroscopic operation of state-generation. Let us immediately embody this possibility. Let us symbolize by G the considered operation of state-generation and by ms_G the corresponding generated microstate. Though in this incipient stage the symbols G and ms_G are devoid of any mathematical representation, their introduction is very important. Indeed it instates inside the realm of the communicable, the fact that the generated microstate, though unknown, is nevertheless captured, in the peculiar sense that one can now produce as many copies of it as necessary and subject each copy to some subsequent operation of examination, while communicating clearly what one does, by words and signs. This amounts to having achieved a sort of a-conceptual definition of an infinite set of replicas of the object-entity called a microstate generated by G and symbolized ms_G. A purely factual and nevertheless communicable definition. This is very remarkable because *it circumvents the lack, for defining ms_G, of any predicate: G is not a qualification of ms_G, it is the way of producing it.*

Thereby one of the extremities of the chain of information that was to be started, is now fixed.

Once the first stage, of production of a "given" object-entity, has thus been achieved, one can enter upon the second stage, of construction of a certain new knowledge concerning specifically the generated object-entity. Now, the object-entity denoted ms_G, such as it emerges from the operation G that generates it, in general does not reach the level of what is observable by man. So it has now to be *brought* to trigger on this level some observable manifestations. Furthermore these manifestations have to be endowed with significance, namely with precisely the researched kind of qualifying significance. In order to reach this new aim, *measurement interactions* $M(X)$ with macroscopic measurement devices are organized for measuring the *quantum mechanical dynamical quantities* X; X runs over the set of dynamical quantities—position, momentum, energy, etc.—that are mathematically defined inside quantum mechanics and $M(X)$ designates the process by which X is measured. The formal representations of the measurement-interactions $M(X)$ are mainly conceived in a peculiar sort of prolongation of the classical mechanics. Thereby, implicitly, history and *models* come in ([15], [16]). The practical realizations of the measurement interactions M(X) are planned such as to produce a perceptible set of marks $\{\mu_X\}$ upon a convenient X-registration-device of an apparatus $A(X)$ "good" for measuring X on microstates. What this means is quite non-trivial. In fact the processes $M(X)$ are produced by what is called the apparatus $A(X)$. Each set $\{\mu_X\}$ of observable marks, once realized, is interpreted, it is *coded* in terms of a value X_j of the quantum mechanical dynamical quantity X (X_j is called an *eigenvalue* of X); j is a discrete or continuous index. Which j corresponds to which sort of mark has to be specified so as to define a stable code-language. The coding-rules are determined by the formal quantum mechanical definition of X and by the specification of the interaction chosen as a measurement process $M(X)$.

> Codability in this sense—a rather complex operation—is a central condition for $M(X)$ to be acceptable as a "measurement" process of X, so for $A(X)$ to be acceptable as a "good" apparatus for measuring X.

In this manner—by a complex interplay of inherited pre-conceptualizations, of assumptions, implicit models, macroscopic operations theoretical representations, and of calculi and codings—are achieved the basic quantum mechanical qualifications of microstates.

Of microstates, indeed? Let us avoid inertial steps in the way of speaking, and check the pertinence of each verbal expression. For it seems clear that in general a measurement interaction must be imagined to *change* the microstate initially created by the employed operation of state-generation, possibly quite radically in certain cases; so the observable marks emerge indelibly *relative* to the employed measurement process. Which means that these marks characterize globally the measurement interaction, not separately the supposed object-microstate. One can however cling to the fact that the observable marks are relative to *also* the initially created microstate, while the type of change undergone by this microstate during a measurement interaction is ruled in an admittedly known way by what is called a measurement process $M(X)$. One has then to take furthermore into account that two distinct processes of change of the initially produced object-microstate, corresponding to two distinct measurement interactions $M(X)$ and $M(X')$ of two different quantum mechanical dynamical quantities X and $X' \neq X$, in general *cover two different spacetime domains.* When this happens, the measurement-processes $M(X)$ and $M(X')$ cannot be both simultaneously achieved starting from *one* single replica of a microstate ms_G: in this sense these two measurement interactions are mutually *incompatible.* So, if one wants to obtain observable qualifications involving the microstate ms_G, in terms of eigenvalues X_j and X'_k of *both* X and X', one has in general to generate *more* than only one replica of ms_G because one has to achieve two sorts of successions [(a given operation G of state generation),(a measurement process $M(X)$ on the supposed result ms_G of G)] (in short successions $[G.M(X)]$), namely $[G.M(X)]$) and $[G.M(X')]$), (the chronometer being re-set at the same initial time-value before the realization of each pair). Furthermore even the measurement on a microstate ms_G, of only *one* quantum mechanical dynamical quantity X, when repeated *via* the corresponding succession $[G.M(X)]$, in general does not yield systematically one *same* eigenvalue X_j; in general the results are distributed over a whole spectrum $\{Xj, j \in J]$ of possible eigenvalues of X (J: an index set, discrete or continuous). Moreover a given eigenvalue X_j can in general be obtained also with other microstates $ms_{G'} \neq ms_G$ corresponding to other operations of state-generation $G' \neq G$. In short, a stable information—if it can be obtained—*cannot concern isolately one individual microstate* ms_G. It necessarily concerns some pair $[G.M(X)]$ in which the measurement interaction M(X) is also involved. And, furthermore, in general a pair $[G.M(X)]$ has to be repeated in order for us to become able to assert a stable result. So a whole—big—set of replicas of the microstate generated by G is involved.

This means that the observational invariants that can be obtained by the help of pairs $[G.M(X)]$ can only consist of *probability laws* $\mathbf{p}(G, X)$ defined on the spectra $\{X_j\}$ of the quantum mechanical observables X. Now, nothing insures *a priori* the existence of such probability laws. This existence is not a logical necessity. And if no probability laws associated with the various pairs $[G.M(X)]$ were found, one would be obliged to finally give up the aim to construct some stable observable knowledge concerning microstates. But in fact it turns out that probability laws $\mathbf{p}(G, X)$ *do* arise, for each pair $[G.M(X)]$. So, by a very big number of repetitions of pairs $[G.M(X)]$ where X runs over the set of all the dynamical quantities defined inside quantum mechanics, classes $\{X_j, j \in J\}$ of eigenvalues are obtained, coding for sets of registered marks that are mutually incompatible in the sense specified above, and over these probability laws $\mathbf{p}(G, X)$ are found. These probability laws, like also the concerned observable events X_j and their individual probabilities $\mathbf{p}(G, X_j)$, are *relative* to both the involved operation G of state-generation and the involved dynamical quantity X[6].

But thereby knowledge of the studied object-entity itself, the hypothetical microstate labelled ms_G, remains non-extracted from inside the observable results of the pairs of operations $[G.M(X)]$. The descriptional strategy imposed by the cognitive situation leads to observable qualifications that can be posited to *involve* this object-entity, but cannot be assigned to *it* alone, separately from the macroscopic operations G and $M(X)$. This is a serious hindrance when one wants to think and speak about—specifically—"microstates". To overcome this handicap one can make use of a sort of an *ad hoc* conceptual construct. Instead of speaking of the probability $\mathbf{p}(G, X_j)$ of this or that observable event X_j tied with a pair $[G.M(X)]$, one can, equivalently, speak of the *potentiality* of the microstate ms_G itself to produce with probability $\mathbf{p}(G, X_j)$ the observable manifestation X_j *if* a measurement $M(X)$ is performed on this microstate ms_G. *Which centres the thought-and-locution upon the microstate ms_G itself.* In this way the concept of relative potentialities of observable manifestations permits to found upon the observable marks μ_X obtained by measurement interactions $M(X)$, a

[6] The fact that repetitions of pairs $[G.X]$ are necessarily involved in the construction of an observable knowledge concerning the hypothetical microstates, entails quite non-trivial conceptual questions. These, because no specific language for dealing with them conveniently has been constructed, have led to what is called the "problem of the completeness of quantum mechanics" [15, 16]. Here we slip over these questions because inside the method of relativized conceptualization we shall deal in detail with the sources of a generalalized equivalent of this problem. Let us only note that one probability law $\mathbf{p}(G, X_j)$ is not considered to be sufficient for an unambiguous characterization of the involved microsystem ms_G; to achieve such a characterization it is necessary to exhibit at least two such probability laws corresponding to two mutually incompatible observables X and X'.

standard way of speaking about the microstate ms_G itself, namely in terms of potential and relative hypothetical "properties" which are "possessed" by it alone, before the changes undergone during the measurement interactions that led to observable marks μ_X characterizing these interactions as a whole. But, mind that, what is achieved in this way is not more than just a model that should by no means be confused for an *impossible* specification of how-ms_G-really-is-in-itself. A very remote and poor, minimal sort of model, in fact, because of the non removable double relativization, to G *and* to $M(X)$, and of the only retroactive, hypothetical, potential and relative character of the assigned potential "properties". But nevertheless a model that introduces a standard way of speaking of the posited microstate itself. Which is a precious alleviation for thinking of it.

The spacetime incompatibilities between different measurement interactions $M(X)$ achieved on distinct replicas of the microstate ms_G generated by a given operation of state-generation G, entail, in terms of the minimal model specified above, that:

> The set of *all* the physical processes of actualization of the various relative potentialities of observable manifestations X_j assigned to a micro-state ms_G generated by a given operation of state-generation G, falls apart into a set of mutually incompatible classes of actualization. This brings forth a probabilistic whole of a new type, with a tree-like spacetime structure, and involving triadic chains with potential-actualization-actualized links.

I called this structure *the quantum mechanical probability tree of the operation of generation* G [12-14]. By systematic reference to the quantum mechanical probability trees, the quantum mechanical formalism can be understood clearly and in full detail. This sort of reference constitutes the key-procedure for the construction of what I call meta[quantum mechanics] (note 1).

The preceding account, brief and simple as it is, contains, I think, the whole essence of the quantum mechanical descriptional strategy and of the type of results brought forth by it. It shows clearly how epistemic aims, physical epistemic operations (required for acquiring the desired knowledge), hypotheses concerning the contents or the results of the epistemic operations, observable facts, codings of these in terms which assign meaning to them, and more or less explicit modelizations which come constantly in, are brought to form together coherent wholes called "states of microsystems".

What sort of objectivity do such descriptions insure? The knowledge constructed by the quantum mechanical descriptions is endowed with objectivity in the following sense. All the physicists who, working at different spacetime locations, are in states of observation constantly devoid of acceleration with respect to one another, obtain the *same* probabilistic distributions $\mathbf{p}(G, X)$ if they apply the quantum mechanical prescriptions for obtaining observable results concerning a given pair $[G.M(X)]$ as well as the galilean transformations of spacetime coordinates when a passage from an inertial referential to another one is involved. This means that the quantum mechanical probability distributions $\mathbf{p}(G, X)$ are *invariant* with respect to Newtonian changes of the spacetime coordinates, they are physical "Newtonian laws" associated with the considered pairs $[G.M(X)]$. That is, they are pieces of inter-subjective consensus involving physical operations and facts, insured inside a particular but "sufficiently" large class of different observer-conceptors.

3.3. Epistemological Universality

It appeared above that the quantum mechanical descriptions are the result of a deliberate construction of communicable knowledge, a construction founded on the systematic relativization to pairs of operations $[G.M(X)]$. In order to achieve a quantum mechanical description of a microstate it has been necessary:

(a) to achieve the epistemic action denoted G that introduces the object-entity, independently (in general) of any epistemic action by which this object-entity could be qualified;
(b) to achieve the epistemic actions that lead to qualifications of the object-entity;
(c) to realize both these distinct sorts of epistemic actions in a radically creative way, by *first* generating— physically, in spacetime—an object-entity that did not pre-exist, instead of just selecting it among already available physical objects, and by *then* generating, again physically, in spacetime, also observable manifestations of the previously generated object-entity, instead of just detecting pre-existing properties possessed by this entity;
(d) to realize a big number of replicas of the pair $[G.M(X)]$ for each quantum mechanical dynamical quantity X, in order to construct invariant probabilistic qualifications (because in general no individual invariants are found).

Now, this is a maximally displayed and creative way of achieving descriptions, where all the involved relativities are active. It is crucial to realize clearly that such a degree of display and creativity is absent in most of our current classical conceptualizations such as they are reflected by the natural languages as well as by logic, probabilities, physical theories, Einstein relativity included. In the classical conceptualizations it has always been possible to suppose more or less implicitly that the considered object-entities *pre-exist* to the descriptional process, that they are "defined" in advance by properties which they possess already actualized and independently of any act of examination. As long as the peculiar aim of describing states of microsytems had not yet been conceived, this supposition never led to noticed difficulties. Therefore, classically, a description is conceived to consist exclusively in the *detection* of one or more among the actual properties of the pre-existing object-entity. *The question of how the object-entity is introduced is entirely skipped.* As for the dynamical evolution that creates knowledge of a qualification, it is shrinked into one static act of mere detection. With respect to the quantum mechanical descriptional scheme, this last classical contraction is the source of the nowadays most explicitly known differences between quantum logic and probabilities, and classical logic and probabilities. While in fact the—*ignored*—consequences of the explicit consideration of the way in which the object-entity is generated, are still much deeper.

It is however noteworthy that, while in classical logic and classical probabilities—the two most fundamental classical syntactical structures—the quantum mechanical descriptional scheme is not apparent, this scheme nevertheless is explicitly involved in many classical and quite current epistemic situations and procedures. Indeed, once one has clearly perceived the peculiar and very difficult epistemic situation dealt with in quantum mechanics, as well as the descriptional strategy that permitted to dominate it, a very paradoxical inversion arises, by a sudden variation that reminds of those which make appear certain drawings of a cube as sometimes convex and sometimes concave. What first, in the quantum mechanical approach, had seemed to be fundamentally new and surprising, abruptly appears on the contrary as endowed with a certain sort of universality, so of normality. It leaps to one's mind that:

* Any explicit account of a process of description, in so far that it is self-contained, always includes a full specification of the action by which the object-entity is introduced, as well as a full specification of the action by which a qualification is obtained for this object-entity.
* Often these two actions are mutually independent.
* The introduction of the object-entity is sometimes achieved by *creation* of this entity, while the operation of qualification, if it is a physical process, *always*—in principle at least—changes the object-entity, and sometimes radically, in which cases the relativizing consequences of one or the other or both these epistemic actions, upon the development of the process of description, have to be explicitly taken into account and thoroughly analyzed.

For instance, think of a detective who is searching for material indications concerning a crime. What does he do? He usually focuses his attention on a convenient place from the physical reality, say the theatre of a crime, and there he first operates extraction of some samples (he cuts out fragments of cloth, he detaches a clot of coagulated blood, etc.); or he might even entirely create a test-situation involving the suspects, and insure registration by hidden apparatuses, of their behaviours. Only *afterward* does he examine the gathered samples or the behaviours registered during the test-situation. One can equally think of a biopsy for a medical diagnosis, or an extraction of samples of rock operated by a robot on the surface of another planet, and the *subsequent* examinations. In all these cases the observer-conceptor—more or less radically—*generates* an object-entity that did not pre-exist in the desired state or quantity, in order to qualify it later by operations that are quite independent of the operation which generated these entities. And in certain cases the operation of examination so radically changes the object-entity, that, if several different examinations of this object-entity are necessary, also several replicas of it must be produced. Furthermore, the obtained qualifications arise indelibly marked by a double relativity: relativity to the way of generating the object-entity (this way can simply exclude certain subsequent examinations), and also a relativity to the sort of examination that was achieved.

The preceding considerations prompt the following two correlated remarks:

In the first place, the nature and realm assigned by classical thinking, to the genesis of communicable knowledge, are misleading and shrinked. The whole zone where mind *actively constructs*, out of pure factuality, the very first forms of new communicable knowledge, is so deep-set that it remained hidden beneath the two basic building blocks of all the current occidental languages, namely subjects and predicates. These do both suggest available, pre-existing states of fact. Furthermore, the primordial creative zone of conceptualization remained cut off also from most classical *scientific* representations. Notwithstanding the well-known analises of Husserl, Poincaré, Einstein, Piaget, and many others, not only classical logic and probabilities, but also the set theory (hence most domains of modern mathematics), modern linguistic and semiotic, etc., take their start from a level organized *above* language, by use of—quasi-exclusively—language: physical operations are not considered. And factuality is widely supposed to spontaneously imprint—*via* language—upon passively receptive minds, information concerning already existing, actual properties of pre-existing objects. The active role is assigned quasi exclusively to the exterior factuality, not to the mind. This attitude, in fact, is stronger and more general concerning object-entities (typical grammatical subjects) than concerning qualifications (predications). Anyhow, globally, an attempt at an integrated and systematic representation of the emergence of individual object-entities and of qualifications of these, by deliberate epistemic actions, and the way in which these products get integrated into communicable concepts-and-language, is still lacking. From "the other part of the mirror" where the biological structure of the man's body is placed, the cognitive sciences are trying to initiate a representation of the sensorial bio-physiological processes involved in phenomenal appearances and in conceptualization, by including into the domain of investigation the inner volume delimited by a man's skin. But if this inner volume is excluded, then it is quantum mechanics which—for the first time—suggests the possibility of, and the method for a most deep-set attempt at a purely psycho-operational representation of the processes of conceptualization: an attempt founded on the very first interplay of what is called mind, with unknown factuality, and involving explicitly the descriptional aims, the physical operations and devices, and the evolving stratum of pre-existing conceptualization.

In the second place, the descriptional scheme explicated from the epistemic strategy involved in quantum mechanics, is paradigmatic. It has captured in it a certain sort of epistemic universality. Quantum mechanics involves a particular embodiment of an extreme epistemic situation, namely that which is realized when a communicable conceptualization is researched concerning non pre-existing physical entities of which—*a priori*—only the possibility is conceived, and which, if then effectively generated, emerge non-perceivable. In such extreme circumstances one has been compelled to a *radically* active, constructive attitude, associated with a maximal decomposition of the global process. All the stages of the desired description have had to be *built* out of pure physical factuality, independently of one another, each one in full depth and extension: the severity of the constraints revealed the most complete and explicit descriptional scheme where any other more particular description *must* find lodging. In this sense the quantum mechanical descriptional scheme possesses a universal epistemological value.

As soon as this universal value has been understood, one finds oneself in possession of a starting point for specifying a convenient canonical form of any description. Indeed such a canonical form must be precisely a complete abstract structure with a maximally carved out capacity. It must be a void form, a mould, able to offer an available, specific, and sufficiently large location, for *any* possible stage of *any* possible descriptional process. In this or that given description, one or more locations offered by this canonical form might remain partially or totally non utilized. But this, if it happens, will be known since the form will exhibit a labelled void of estimated ampleness. For instance, if I say "I consider what I see just in front of my eyes and this is a red surface", with reference to the maximally complete descriptional mould drawn from quantum mechanics it will appear that in this case the two canonically distinct descriptional actions, of generation of the object-entity, and of qualification of this entity, have coalesced in the unique act of "looking just in front of my eyes", which *both* delimits and qualifies the object-entity. So the location reserved for the stage of independent generation of an object-entity remains entirely void in this case. It will also be possible to estimate the magnitude of only partial voids and to draw consequences. For instance, imagine the assertion "I plucked this flower, I examined its morphology with a microscope, and the result is this". Comparison with the canonical mould brings forth that this amounts to a description where the object-entity—as such—is introduced by an only partially creative action—plucking a flower—while the act of examination might only very little change the object-entity initially introduced in this way. So in this case the two distinct locations reserved in the canonical mould in view of a possibly radical creativity in both the stage of production of an object-entity and in that of qualification of it, are both made use of, but each one to only a very reduced degree. It follows that a classical treatment (assuming the pre-existence of the object-entity as well as its invariance with respect to the process of qualification) can be posited to produce a very good approximation to the result that would be obtained by a complete canonical treatment.

4. NORMS FOR DESCRIBING: THE METHOD OF RELATIVIZED CONCEPTUALIZATION (MRC)

4.1. Preliminaries

Since 1982 I never ceased developing the method of relativized conceptualization ([14,17,18])—let us denote it as MRC—founded on the generalization of the descriptional scheme which I explicated from the quantum mechanical descriptions. This method can be regarded as an attempt at a certain "normation" of the processes of description of any sort, or in other terms, *a normation of the processes of communicable conceptualization.*

Because of the descriptional relativitisations that are explicitly built into it at each descriptional step, MRC withstands by construction the insertion of false absolutes, thus warding off false problems or paradoxes. And because it roots its constructions in physical factuality, at the lowest descriptional level that can be reached, MRC furthermore withstands any gliding into relativism:

MRC stands in polar opposition to what is called relativism.

It means confined, delimited, but strict precision of each descriptional step, associated with free though guided choices of the way of connecting the descriptional steps accordingly to the evolution of the descriptional aim. Which insures controlled rigor throughout a progressive construction of freely decided trajectories and nets of conceptualization, always indefinitely open.

The main difficulty has been to find a way of escaping the imprisonment inside the forms which current language, surreptitiously, imposes upon thought. In all the preceding publications concerning MRC, in order to achieve this liberation I made use from the start on of certain ideographic symbolizations, but I never tried to achieve a mathematical formalization. The ideographic symbolizations, however, have been felt by many to stay in the way of a natural and full access to meaning. Therefore in this work I adopt a different strategy. In a first stage I expose the nucleus of MRC in usual language, trying to get through the stubborn implicit forms of thought induced by the current usage of words, with the help of exclusively the resources of the associations of words themselves (and of abbreviating literal notations of words). In a second stage I give a summary of the ideographic symbolization utilized in all the previous expositions of MRC, because it permits a more suggestive and economic expression of certain basic concepts and assertions. Finally, in a third stage I sketch out a *mathematical* formalisation of the nucleus of MRC in terms of the theory of categories.[7]

This section is devoted exclusively to the nucleus of MRC. The way in which the nucleus works will be illustrated in the subsequent Sec. 5, by showing how it generates a deep and fully relativized unification between the logical conceptualization and the probabilistic one.

[7] The possibility of also another sort of mathematical formalization, more fit for calculations permitting numerical estimations—namely in terms of Hilbert-Dirac "individual" vectors (i.e., not belonging to a vector-space)—will be found in the exposition of meta-[quantum mechanics] (note 1). While in Sec. 5 it will become clear that the probably most natural vocation of MRC is to yield a non-mathematical formal system comparable to Russel and Whitehead's *Principia Matematica*, but concerning conceptualization in general instead of only logic.

4.2. The First Stage: a Presentation of MRC in Usual Language

In what follows I formulate definitions (D), a postulate (P), principles (P), conventions (C), and assertions which are called propositions (π) because they are justified by "natural deductions" (indicated by the word "proof" written between quotation marks in order to distinguish from deductions inside a formal system). Each step is labelled by the symbol of its nature—D, P, P, C, or π—followed by the ordinal of the step. There are 19 steps, namely 15 definitions, 1 postulate and 3 principles. When a step is splitted in sub-steps a sub-ordinal is added for each sub-step. A step is often followed by comments.

I proceed by enumeration of the steps and sub-steps. The sequence is interrupted by several intermediary titles which break the progression in small groups each one of which concentrates upon a given purpose.

4.2.1. Preparation of the concept of relative description

D1. ***Consciousness functioning.*** The activity of an observer-conceptor's mind—called here *consciousness functioning* and noted CF—is conceived to play a central generative role, acting on the exterior universe and on the interior universe where it belongs, and there, in particular, also on itself. This activity is regarded as the quintessence of the epistemic actor, irrepressibly anterior and exterior to any specified epistemic action. It is an (the?) invariant among all the epistemic actions the observer-conceptor is aware of, it is the tissue of his continuity, and each one of its products becomes exterior to it as soon as it has been produced. It marks a mobile, permanent and non removable cut—a ultimate cut—between itself and the rest.

Comment. The Cartesian cut between *res cogitans* and *res extensa* is second with respect to this mobile cut.

Throughout what follows CF is explicitly incorporated in the representation. Thereby, from the start on, this approach breaks openly and radically with the classical concept of objectivity. It introduces basically, in a declared and systematic way, the supplementary representational volume that is necessary for a non-amputated expression of the new concept of objectivity in the sense of inter-subjective consensus, such as this concept emerged from modern physics, from quantum mechanics and Einsteinian relativity. That is, inter-subjective consensus founded on systematically extracted fragments of pure factuality (quantum mechanics) and qualified by qualificators explicitly constructed in order to express definite classes of relative *obser-*

vational invariance (Einsteinian relativity). Indeed both these constraints, that are the core of modern physics, involve CF in a quite essential way.

D2. ***Reality.*** What is called reality is posited here to designate the *evolving* pool—always considered such as it is available at the considered time—out of which any given consciousness functioning either radically creates, or delimits, or only selects, object-entities of *any* kind whatever, physical or psychical or of a mixed kind. This pool will be indicated by the letter R.

Comment. This non-restricted definition of "reality" refuses the disputes on "existence" (do unicorns exist? does the number 3 exist? does a class exist? etc.). It will appear that inside the present approach the indistinctions entailed by this absence of restrictions generate no difficulties; that, on the contrary, they permit *a posteriori* clear definitions which so far could not be reached in the approaches in which more specification has been introduced at the start.

P3. ***The realist postulate.*** Throughout what follows is explicitly postulated the *existence*—independently of any mind and of any act of observation—of also a *physical* reality.

Comment. In the formulation of P3, as also in D1 and D2, the specific designatum of the expression "physical reality" (that implies that a sub-realm of what is called reality is considered), is assigned the status of a primary datum. This however is *only a starting point.* In what follows the general reflexive character of MRC will manifest itself, in particular, by the fact that, progressively, a more constructed distinction between "physical" reality and reality in general will constitute itself inside MRC.[8]

---

[8] This specification takes into account concurrent remarks by Jean-Louis Le Moigne, Michel Bitbol, Jean-Blaise Grize, and Gérard Cohen-Solal who—independently of one another—argued that the concept of "physical reality" seemed to them neither clear *nor necessary* in a context of the nature of MRC; that inside such a context this concept should *emerge.* Furthermore, on H. Barreau's opinion, speaking of "physical" reality might erroneously suggest some confusing necessary connection with Physics, which the word "empirical" would avoid. It will however appear that the crucial definition D14.3.1 of a basic transferred description, as well as the preparatory points 8 to 13, are endowed with significance *exclusively* with respect to what is usually called physical reality, while with respect to reality in the general sense of D2—which includes, for instance, *empirical* economic or cultural data, empirical aspects or components of what is called art, etc.—the formulations from the points 8 to 14 are *meaningless.* So I simply do not know how to avoid the assertion ab initio of P3 such as it is expressed above: such is the force of language. On the other hand, throughout the points 8 to 14 the concept of physical reality keeps acquiring constructed specificity. In this sense, a progressive specification of P3 *does* emerge from the evolving MRC-context, as desired by the above-mentioned colleagues, but it emerges on the basis, also, of P3 itself. So my final option is to conserve [D2+P3].

The posit P3 of existence of a physical reality might seem to be entailed by D2, so redundant, but in fact it is not. Indeed, though everybody agrees that what is called physical reality does contribute to the pool out of which the consciousness functioning extracts object-entities to be studied, nevertheless the various disputes concerning "existence" of this or that sort of object-entity (does Jupiter exist?) continue steadily. The association [D2+P3] is intended as *(a) a memento* of the fact stressed most by Descartes and recognized by the majority of the philosophers, that, in the order of the emergence of knowledge, the assertion of the existence of physical reality *cannot* be considered to be *primary* with respect to the assertion of the existence of subjective psychical universes (as classical physics might seem to suggest) : the word "also" in the formulation of P3 is intended to provocatively remind of this; *(b)* an explicit refusal of solipsism, on the other hand; *(c)* an inclusion in what is called reality, of the concepts and systems of concepts, of the behaviours, beliefs, social and economical facts, etc. (the third world of Popper).

D4. ***Generator of object-entity and object-entity.*** The epistemic operation by which a consciousness functioning introduces an object-entity will be regarded as an action upon R achieved by CF by the use of a *generator of object-entity* denoted G. The spot (or zone, or the sort of domain) from R where a given generator G acts upon R, is considered to be an essential element from the definition of that generator, and which has to be explicitly specified; it will be denoted R_G. The object-entity introduced by a given generator G will be denoted $œ_G$. For methodological reasons, a one-to-one relation is *posited* between a given definition of a generator G and the corresponding object-entity $œ_G$: *that* which emerges as the product of a given G-operation, *whatever it be*, is called "the object-entity produced by G and is labelled $œ_G$.

Comment. Any description involves an object-entity. Usually it is considered that it suffices to name or to label this object-entity thus just directing the attention upon it before it is more thoroughly examined. This "linguistic" attitude is restrictive since not any conceivable object-entity pre-exists available for examination. Therefore throughout what follows it is required that the primordial epistemic action accomplished upon R which brings into play the considered object-entity—as such—no matter whether this action is trivial or not, be *always* indicated explicitly and fully.

For the moment it is sufficient to understand the qualification "physical" as pointing toward anything involving an in principle definible amount of mass-energy. Then certain non-physical entities, like "art", etc., can involve physical *aspects*, while others, like the concept of the number 3, do not.

A generator G of object-entity can consist of any psycho-physical way of producing out of R an object for future examinations. Such a way involves systematically some psychical-conceptual component, but which can combine with concrete operations. A generator G can just *select* a pre-existing object or on the contrary it can radically create a new object. If I point my finger toward a stone I select a physical entity by a psycho-physical selective gesture that acts in a non creative way on a physical zone from R (here R_G is the volume where the stone is located). If I extract from a dictionary the definition of a chair I select by a non creative psycho-physical act, an abstract conceptual entity materialized by symbols in a physical zone from R consisting of the dictionary (so here $R_G \equiv$ dictionary). If I construct a program for a Turing machine in order to examine the sequences produced by this program, I bring into play a creative, instructional conceptual generator of object-entity that acts on a zone from R containing subjective and inter-subjective knowledge as well as material supports of these. If, in order to study a given state of an electron, I generate it by using some macroscopic device that acts on a place from the physical space of which I suppose that it contains what I call electrons, I delimit a physical object-entity, by a psycho-physical creative action. If now I apply the *same* operation upon a mathematical theory, or upon a place from the physical space where the vibrations of a symphony can be heard but the presence of electrons is improbable, according to the definition D4 I am making use of *another* generator, since it involves *another zone* R_G, and, in consequence of the one-one relation posited between G and $œ_G$, I delimit another object-entity (interesting, or not, probably not, in this case). When I define by words a new concept, as I am doing now, in order to later specify its behaviour, I produce a conceptual object-entity, by working, with the help of a psycho-conceptual-physical creative generator, upon the spot from R consisting of the reader's mind.

The inclusion, in the definition of G, of the "zone" R_G from R where G is supposed to act, requires two important specifications:

(a) R_G is *not* a qualification *of the produced object-entity* $œ_G$, obtained by examining *this* object-entity in order to learn about it. It is a condition imposed upon the operation of generation G in order to insure the location of *all* the products of G, inside a pre-decided conceptual volume indicated by some verbal label, "microstate", "chair", "program", etc. In the particular case of a selective generation like for instance pointing toward a stone, this pre-posited conceptual volume where G has to act, might degenerate inside our mind into an identification with the physical location of the object-entity $œ_G$, which has to be avoided). The methodological necessity of such a pre-decided conceptual location will be fully understood later, in

the comment of the definition D14.3.1.

(b) The "zone" R_G from R where G is supposed to act permits of uncontrollable fluctuations concerning what is labelled $œ_G$. The physical region from R where I act in order to generate a given microstate of an electron, can contain non perceptible and uncontrollably variable fields, etc.; the reader of these lines can happen to be a 16 years old boy, or a mature intellectual. These fluctuations entail an unavoidable non-predictability concerning the effect labelled $œ_G$ of an operation of generation of an object-entity. However one should clearly realize that it simply is *inconceivable* to "entirely" immobilize *a priori* the effect of G denoted $œ_G$: this would require to specify "completely" R_G. But such a requirement is both impossible (circular) and unnecessary. One simply cannot *start* a process of representation of the way in which descriptions, i.e., qualifications of any object-entities, emerge out of R, by specifying, so qualifying R itself everywhere and for any time, and also from any point of view. Such a circle cannot be realized. While the *a priori* non-determination concerning the effect of the individual operations of generation of an object-entity, is by no means an insuperable problem or a difficulty. It simply is an unavoidable constraint that MRC is obliged to recognize, include and control. The recognition of this constraint plays an essential and very original role in the dynamics of conceptualization from MRC. It brings into evidence one of the roots of human conceptualization and it comes out to be intimately tied with a reflexive character of MRC, of maximal *a priori* freedom, followed by *a posteriori* controls and restrictions. It opens up the way toward a constructive incorporation (*via* the sequence D14 of definitions of relative descriptions) of the fundamental fact called "non-determination of reference" established by the deep analyzes of Quine [10] and Putnam (cf. [11]), which marks the breaking line between factuality and mere language.

Consider now the one-one relation posited between a given definition of an operation G of object-entity generation and what is labelled $œ_G$. This relation is intimately tied with the above mentioned *a priori* non-determination involved by R_G, so also with the non-determination of reference. It is important to realize that *no other relation could be uphold ab initio*. Indeed in general the object-entity labelled $œ_G$ emerges still non qualified from the standpoint of the subsequently intended examinations, if not, in general its generation would be unnecessary for this aim. It can even emerge still entirely inaccessible to direct knowledge of any sort, if G is a radically creative and physical operation of generation (as in the case of the microstate generated by most quantum mechanical operations of state-generation). In these conditions what we called a one-one relation between a given definition of an operation G of object-entity generation, and the *mere label* $œ_G$, obviously *cannot* mean that the still unqualified replicas of $œ_G$ are all "identical" in some inconceivable absolute sense.

The one-one relation posited between G and $œ_G$ amounts to just a *methodological pre-organization of the language-and-concepts*, unavoidable in order to be able to form and express a beginning of the desired representation of a human conceptualization. Such a methodological pre- organization is, by its nature, a `formalizing` step, like an algebraic rule.

Indeed if from the start on we imagined that G might produce sometimes this and sometimes something else, how would we *speak* of what it produces, or think of it? We would have to re-label in only *one* way the product entailed by a given definition of G, whatever it be, and thus we would come back to precisely our initial choice of language and notation. On the other hand, if we asserted *a priori* a "real" one-one relation between G and what is labelled $œ_G$, we would thereby assert the sort of view that is sometimes called metaphysical realism (a God's Eye view, as Putnam puts it), which would directly contradict the very philosophical essence of the present approach. In the sequel, each time that some definite consequence of this *a priori* choice of language will appear, we shall deal with it for that definite case.

The explicitly methodological character of this constructive strategy adopted in the definition D4, is a quite crucial step. It saves premature, void, illusory questions and paradoxes that simply cannot be solved *a priori*. Instead, as it will appear, it brings forth *a posteriori* a clear, fully *relativized* operational concept of "identity" that emerges progressively in π12, π13 and D14.1 and then is specifically defined in π18.1; which suppresses inside MRC one of the most noxious false absolutes induced by current language. And the relativization of the qualification of identity permits then immediately to show by π18.2 and π18.3 that MRC, inside its *soma* progressively structured from the precedingly posited definitions, postulate and principles, eventually entails a well-defined sort of minimality of the realist postulate P3, initially posited without any further qualification. By this *minimality* the "metaphysical realism" will appear to by organically rejected by MRC.

D5. *Qualificators*

D5.1. ***Aspect-view.*** Consider a grid for examination which, *via* certain operations of examination performed on an object-entity $œ_G$, can be *a priori* imagined to produce qualifications of this entity. Such a grid will be called an *aspect-view* and will be denoted V_g. By definition V_g is structured as follows:

- The qualifications that can be generated by V_g are contained inside a semantic dimension called *aspect* and labelled globally by some index g (which can take on any graphic form: another letter, a group of letters, some other sign).

- The qualifications that can be generated by V_g are called *g-qualifications.* The set of all the possible g-qualifications is allowed to be arbitrarily rich but it is required to be finite, so discrete. Each g-qualification is called *a value k of the aspect g*, in short a gk-value, where gk—in one block—functions as only *one* index. The aspect g is conceived to *contain* the corresponding finite set of gk-values, not to identify with it.
- A gk value is permitted to be of either a physical or an abstract nature, but it is required to be directly perceptible by the involved observer-conceptor, *via* his biological senses and his mind.
- The aspect g is considered to be defined *if and only if* the specification of its values gk is associated with also the explicit specification of an effectively realizable modality—physical, or conceptual (in particular formal), or mixed—for:

  * Accomplishing the examinations—physical, or psychical or conceptual—from the semantical dimension consisting of the aspect g.
  * Expressing the results of these examinations in terms of "values gk of the aspect g", which amounts to the explicit specification of certain *coding-rules.*

Any object, device or algorithm involved by the modality required above, is to be included in the definition of the aspect g.

Comment. So, in contradistinction to the grammatical or logical predicates, an aspect-view V_g is endowed by definition with a *structure*, and with coding-rules which fix a finite "gk-language" consisting of operations, signs, names, referents, and the stipulation of the relations between these.

This structure exhibits explicitly all the restrictions to which is subjected an *effectively* realizable operation of qualification, that can be made use of without incurring ambiguities. If these restrictions are not all satisfied we simply are not in presence of an aspect-view in the sense of D.5.1.

Let us note that an order between the values gk of an aspect G is *not* required but is permitted.

The distinction between an aspect g and the set of all the gk values contained inside that aspect, takes into account the remarkable psychological fact that any set of gk-values, even only *one* such value, as soon as it is "conceptualized" (i.e., as soon as it ceases to be a mere "primeity" in the sense of Peirce), generates in the consciousness a whole semantic *dimension g* (a genus) that exceeds this set and constitutes a ground on which to place its abstract feet: every gk-value determines a location (a specific difference) on this semantic domain g that grows spontaneously beneath it (for instance, if gk labels the interior event toward which the word "red" points, this event, when conceptualized, generates the carrying semantic dimension toward which the word "colour" points). We are in presence of a fundamental law of human conceptualization that moulds logic, language, and even metaphysics (the concept of "substance" is the semantic ground on which are located the ways of existing of material systems, etc.). The adopted definition reflects this law, on which it tries to draw the attention of the cognitivistic approaches (what are the corresponding bio-functional substrata?).

Finally let us also note that, by definition, an aspect-view V_g acts like a qualifying *filter*: it cannot yield qualifications different from any corresponding gk-value.

D5.2. ***View.*** A grid for examination that consists of a finite but arbitrarily large set of aspect-views, is called a view and is denoted V.

Comment. The complexity and the degree of organization of a given view V are determined by the number of aspect-views V_g from V and by the structures of the various sets of gk-values introduced by the various involved aspect-views from V (number of gk-values, "position" (central, extreme) of each set of aspect-values on the corresponding semantic dimension g, existence or not of an order among the gk-values of a fixed aspect g, a reference-gk-value (a gk-zero), etc.). In particular a view can reduce to only one aspect-view or even, at the limit, to one aspect-view containing only one gk-value on its semantic dimension g. There is nothing absolute in the distinction between an aspect-view and a view: an aspect-view can be transformed in a view by analysis of its aspect in two or more sub-aspects, and *vice-versa* the set of distinct aspects from a view can be synthesized into a unique aspect. This stresses that a view, like also a generator of object-entity, is just a construct freely achieved by the acting consciousness-functioning CF, in order to attain a definite epistemic aim.

D5.3. ***Physical aspect-view and view.*** Consider an aspect-view V_g where the aspect g is physical and requires physical operations of examination of which the results consist of some observable physical effects. Such an aspect-view will be called a *physical aspect-view*. A view containing only physical aspect-views will be called a *physical view* (concerning this language, see note 8).

Comment. This definition can be best understood per *a contrario*.

A mathematical or a logical view is not a physical view, though the involved examinations do involve certain physical actions (writing, drawing, etc.), because what is called the results of the examinations (not their material expression) consists of concepts, not just of physical entities (marks on a measuring device, for instance). (And of course, a physical view does not in the least necessarily involve physics).

D5.4. ***Spacetime aspect-views.*** One can in particular form a *spacetime aspect-view* V_{ET}. Accordingly to Einsteinian relativity the double index ET can be considered as *one* aspect-index $g = ET$ where E reminds of the current Euclidian representations and T stands for time. However the partial aspect-indexes E and T can also be considered separately from one another, setting $g = E$ or $g = T$. The space-aspect E is associated with space-values or "positions" that can be denoted $E\vec{R}$ (setting a position vector $\vec{R}$ in the role of the index k introduced in D5.1) and the time-values can be denoted Tt (setting a time parameter t in the role of k). Indeed though in general the *numerical* estimations indicated by $\vec{R}$ and t are not mutually independent, nothing interdicts to symbolize separately the spatial position-value and the time-value.

Infinitely many spacetime views can be constructed (by varying, in the representations, the choice of the origins of space and time, of the units for measuring intervals, the form and direction of the involved reference-axes). Any spacetime aspect-view introduces an *ordered* grating of spacetime values. This is a specificity with highly important epistemic consequences ([17] and Sec. 5.2 of this chapter), because it endows the spacetime views with the power to strictly singularize an object-entity.

D6. ***Epistemic referential and observer-conceptor.*** A pairing (G, V) consisting of a generator G of object-entity and a view V, is called an *epistemic referential.*

A consciousness functioning CF that endows itself with a given epistemic referential is called an *observer-conceptor* and can be denoted $[CF, (G, V)]$.

Comment. A pairing (G, V) is permitted to be entirely arbitrary *a priori.* This is a methodological reaction to an unavoidable constraint: the capacity of a pairing (G, V) to generate meaning, can be examined only after having considered that pairing. This particular methodological reaction is a new manifestation of an already mentioned general reflexive strategy practised in MRC, of a tentative *a priori* approach that is entirely non restricted, but is systematically followed by *a posteriori* corrective restrictions.

An observer-conceptor $[CF, (G, V)]$ is the minimal epistemic *whole* able to achieve epistemic actions in the sense of MRC: by itself an epistemic referential (G,V) is not yet a closed concept, nor does it designate an active entity. This concept becomes closed and activated only when it is associated with the consciousness functioning CF that generated and adopted it.

D7. ***Relative existence and inexistence.*** Consider an *a priori* pairing (G, V_g). If an examination by the aspect-view V_g of the object entity $œ_G$ generated by G, never reveals to the involved observer-conceptor some value gk of the aspect G, we say that the *object-entity* $œ_G$ *does not exist (is not pertinent) with respect to the aspect-view* V_g (or equivalently, that V_g does not exist with respect to $œ_G$, or that $œ_G$ and V_g do not mutually exist)).[9]

Suppose now, on the contrary, an act of examination by the aspect-view V_g of the object entity $œ_G$ generated by G, that does reveal to the involved observer-conceptor one or more values gk. In this case we say that *the object-entity* $œ_G$ *exists with respect to the aspect-view* V_g (or that V_g exists with respect to $œ_G$, or that V_g and $œ_G$ do mutually exist).

Comment. The definitions of relative inexistence or existence can be transposed in an obvious way to one single value gk of an aspect g, or to a whole view V.

The concepts of mutual inexistence or existence concern, respectively, the general *impossibility* or *possibility* of the emergence of meaning, as well as the intimate connection between meaning and descriptional aims, which are induced by a tentative pairing (G, V_g) or (G, V). These concepts are essentially semantical. They express the general fact—*previous to any qualification*—that a given object-entity can be qualified only *via* the views to the genesis of which it can contribute by yielding matter for abstraction. Furthermore, the concepts of relative inexistence and existence permit to cancel *a posteriori*, among all the initially only tentative pairings (G, V_g) or (G, V) that an observer-conceptor has introduced, those which appear to be non-significant; while the other pairings can be kept and put to systematic descriptional work. The possibility of such a selection illustrates again the general reflexive strategy of MRC: maximal *a priori* freedom followed by *a posteriori* controls and restrictions.

The concepts of relative inexistence and existence have quite fundamental consequences, but with respect to which the classical conceptualizations are more or less blind. This generates various sorts of false problems and paradoxes.

---

[9] If one examined with the help of a voltmeter, a symphony by Beethoven, the operation might never produce an estimation of a difference of electrical potential (accidents being neglected). Of course during a more realistic sort of tentative research a mutual non-pertinence can be much less apparent *a priori* than in this caricatured example.

P8. ***The Frame-Principle.*** I posit the following principle, called *frame-principle* and denoted by FP.

Consider a *physical* object-entity $œ_G$ that can be (or is conceived to have been) generated by some definite physical generator of object-entity, G. The frame-principle FP asserts the following:

- This entity $œ_G$ does exist in the sense of D7 with respect to at least one physical aspect-view V_g (D5.3) (if not the assertion of a physical nature of $œ_G$ would be devoid of foundation (content)).
- If the physical object-entity $œ_G$ does exist in the sense of D7 with respect to the physical aspect-view V_g, then *ipso facto* $œ_G$ exists in the sense of D7 with respect to also at least one view V formed by associating V_g with a convenient *spacetime view* V_{ET} (it cannot exist with respect to *any* such association, if only because the values gk of a given aspect g can appear or disappear with respect to a given spacetime view when the spacetime units are changed). But the object-entity $œ_G$ is non-existent in the sense of D7 with respect to any spacetime view that acts *isolated* from any other physical aspect-view V_g where $g \neq ET$: *The spacetime views are frame-views which, alone, are blind, they cannot "see" anything.*
- According to what precedes what is called "physical spacetime" *cannot* be regarded as a physical object-entity $œ_G$. Indeed the assertion posited in the first part of this principle does *not* apply to what is called "physical spacetime": The designatum of this expression *itself*, considered strictly *alone*, is non-existent in the sense of D7 with respect to *any* physical aspect-view V_g where $g \neq ET$, and it is equally non-existent with respect to any association of such a physical aspect-view, with a spacetime aspect-view. In this sense: If spacetime were regarded as a physical object-entity we would need spacetime where to locate it and thus we would be drawn into indefinite regression.

What is called "physical spacetime" is—itself—only the locus of all the possible spacetime frame-views (referentials), the genus of these. It is the conceptual volume where physical entities, facts or aspects, can be assigned spacetime specifications which, if this is desired, can be numerically defined by the use of spacetime referentials.

Comment. The frame principle FP adopts, transposes in terms of MRC, and specifies, the Kantian conception according to which man is unable to conceive of physical entities outside physical spacetime, that he introduces as *a priori* "forms of the intuition inside which he casts all his representations of physical entities. FP isolates and stresses certain particular implications of this Kantian conception which so far seem to have remained insufficiently noticed by physicists. Namely that any mature and normal human being, by the nature of his consciousness functioning, as soon as he perceives or even only imagines a phenomenal appearance which he connects with what he conceives to be a physical entity $œ_G$, *ipso facto* introduces more or less explicitly:

(a) A spacetime frame-aspect-view V_{ET} (the observer-conceptor's body tends to yield—vaguely—the intuitive origin, the units, and—variable—directions of the axes, whereas in the technical or scientific approaches these are explicitly and freely specified in a precise and stable way, in mathematical, integral or differential terms).

(b) At least one aspect-view V_g, where g is a physical aspect *different* from V_{ET}, relative to which the considered physical entity $œ_G$ does exist in the sense of D7, and the values gk of which he combines with the value-indexes $E\vec{R}$ and Tt of the spacetime aspect-view V_{ET} (in mathematical terms, with the spacetime coordinates yielded by V_{ET}). J. Petitot ([4], p. 216) writes concerning Kant's conception of space and matter:

> "As quality (not as quantity any more), matter is *filling* of space. This filling is very different from a mere 'occupation' (anti-Cartesianism). It is a dynamical and energetical process characteristic of the substantial 'interiority' of matter."

In P8 the necessity of the presence of at least one physical aspect g different from the space or time aspects is a way of expressing *the presence of the matter that fills spacetime and of asserting that any phenomenal manifestation to human minds, if it does not stem from the inner universe, stems from this matter, not from spacetime itself.*

(c) With the help of a spacetime frame-view *alone*, in the strict absence of any other sort of physical aspect-view V_g (colour, texture, whatever), man is unable to perceive or even imagine a physical entity. He simply is unable to extract it from the background of exclusively spacetime frame-qualifications which, by themselves, act only as elements of a *grid of reference* inserted in an abstract, void *container* labelled by the words "physical spacetime". By themselves these elements of a grid of reference act exclusively as potential land-marks that can be "activated" only by the values of some other aspect $g \neq ET$.

The assertion that the designatum of the words "physical spacetime" cannot be treated itself as a physical (object-)*entity*—probably obvious for most physicists—is introduced here explicitly mainly in order to emphatically block certain very confusing ways of thinking induced in the minds of non-physicists by the verbal expressions by which the physicists use to accompany their relativistic formalizations: these verbal expressions suggest that what is currently called spacetime would itself *possess* this or that *metric*; while in fact any spacetime metric is just assigned by construction to this or that spacetime frame-aspect-view, on the integral level or *on the infinitesimal differential level*, on the basis of some definite (even if implicit) descriptional aim (this is discussed in the last chapter of this work).

The frame principle is endowed with a strong formalizing power which imprints its marks upon all the scientific representations of physical reality, inside physics as well as inside "abstract" mathematics.

C9. ***Conventions.*** In order to take explicitly into account the frame principle FP we introduce the following conventions:

- Any view V considered in order to examine a *physical* object-entity will contain a spacetime aspect view V_{ET} and one or more physical aspect-views V_g.
- The aspects denoted g are always *different* from the spacetime aspect ET.

P10. ***The principle of individual spacetime mutual exclusion.*** Consider a *physical* object-entity $œ_G$ corresponding to a physical generator G. Let V be a physical view with respect to which $œ_G$ does exist in the sense of D7, involving two distinct physical aspect-views V_{g1} and V_{g2} as well as a spacetime view V_{ET} (accordingly to C.9). The *principle of individual spacetime mutual exclusion* posits the following:

- Any physical examination involved by V quite systematically *changes* the state of the examined physical object-entity $œ_G$, even if only to a degree which in this or that context can be neglected: the state of a physical object-entity is not a stable datum with respect to an act of physical examination (in informatics one would say that it is a "consumable" datum).
- If, when performed separately on different replicas of $œ_G$, the examinations involved by V_{g1} and V_{g2} can be shown to cover different spacetime domains—the referential and the origins for spacetime qualifications being kept the same—which involves that they change differently the state of $œ_G$—then it is not possible to perform both these two sorts of examinations simultaneously upon a *unique* replica of $œ_G$ produced by only *one* realization of G (the word "individual" from the denom-

ination of P10 refers to this crucial unicity of the involved replica of $œ_G$).

If the type of impossibility specified above manifests itself, the two physical aspect-views V_{g1} and $V_{g2} \neq V_{g1}$ are said to be mutually *incompatible*. In the alternative case V_{g2} and V_{g1} are said to be mutually compatible.

Comment. It is probably possible to draw back P10 to other still more basic spacetime mutual exclusions (an attempt has been made in [17]B, p. 290). But here, for simplicity, we start from the formulation P10 because it is more immediately related with the consequences pointed out in the sequel.

The quantum mechanical principle of "complementarity" can be regarded as the realization of P10 for the particular category of physical object-entities consisting of states of microsystems. This brings into clear evidence the often only obscurely perceived fact that complementarity in the sense of quantum mechanics has an—exclusively—*individual* significance: indeed two mutually incompatible quantum mechanical measurements *can* be simultaneously realized on two distinct replicas of a given microstate (object-entity), and if this is done two distinct and useful pieces of information are obtained in a quite compatible way [13]. But this brings already up on a statistical level, and there what is called the mutual incompatibility of two physical aspect-views *is not manifest any more.* What *is* impossible indeed is only the simultaneous realization of two mutually incompatible quantum mechanical measurements upon *one* given replica of the considered microstate.

> *The concept of incompatibility of two physical aspect-views is defined only with respect to one individual replica of some given object-entity: it is not intrinsic to these physical aspect-views.*

This is of crucial importance from a logical point of view; cf. Sec. 5.1.2.

Π11.*Proposition.* Consider a physical object-entity $œ_G$ corresponding to a generator G and a physical view V with respect to which $œ_G$ does exist in the sense of D7. In general, in order to perform upon $œ_G$ *all* the operations of examination corresponding to all the different aspect-views V_g from V, it is necessary to realize a whole set of *successions* [(one operation of G-generation of $œ_G$), (one operation of V_g-examination of that replica of $œ_G$)] (in short $[G.V_G]$) containing (at least) one such pair for each physical aspect-view V_g from V.

"Proof". In order to achieve examinations of $œ_G$ *via* mutually incompatible physical aspect-views V_g from V, the operation G of generation of $œ_G$ has to be *repeated* (the time parameter being re-set to its initial value

t_0, as in sport-measurements, in the repetitions of chemical or physical experiments, etc.) and paired *successively* with these incompatible aspect-views.

Comment. This, though an obvious consequence of P10, is highly *non-trivial* by itself. It is important to know explicitly that the achievement of complex examinations of an object-entity involving "consumable" characters, entails in general the condition of reproducibility of all the involved pairs $[G.V_G]$ (either in succession or in simultaneity), thus involving a whole set of replicas of the involved sort of object-entity $œ_G$. (The proposition $\pi 11$ and its "proof" admit of generalization to also certain *conceptual* referentials (G, V)).

Π12. ***Proposition.*** Consider a physical object-entity $œ_G$ corresponding to a given generator G, and one given physical aspect-view V_g with respect to which $œ_G$ exists in the sense of D7. When a succession $[G.V_g]$ is repeated a big number N of times (the time parameter being re-set for each pair to its initial value $t_{œ}$) or when it is simultaneously realized on a big number of replicas of the object-entity $œ_G$, it is not impossible that the same observable gk-spacetime-values be found in each instance; in such a case one can say that an *individual qualificational N-stability* has been obtained. But in general this does not happen; in general the N obtained gk-spacetime values are *not* all identical, notwithstanding that in each realization of a pair $[G.V_G]$ the operations G and V_g obey strictly the same defining conditions.

"Proof". This follows *per a contrario*: to posit *a priori* that the results produced by repeated realizations of a given succession $[G.V_g]$ are all identical "because" in each pair both G and V_g obey the same specifications, neither follows with necessity from the previously introduced definitions and principles, nor could it be found *a posteriori* to be always *factually* true. To show this last point it is sufficient to produce a counter-example. Consider an object-entity generator G which acts by definition on a zone R_G from R consisting of a piece of land, and that delimits there the object-entity $œ_G$ consisting of a definite area of one square kilometer. Let V_g be an aspect-view (structured accordingly to D5.1 and C9) that permits to establish the aspect $G \simeq$[association of mean-colour-value-and-space-position over a surface (any one) of only one square *meter*]: inside the epistemic referential (G, V_g), two distinct realizations of the succession $[G.V_g]$ in general yield two different results, even though both G and V_g satisfy each time to the same operational commands.

Comment. Notice that if an individual qualificational N-stability *is* found for a given succession $[G.V_G]$, this by no means excludes the possibility that in another series of N' repetitions (with N' bigger than, or equal to, or smaller than N) no individual stability will be found any more.

Furthermore, and this is more important, if for a given object-entity $œ_G$ corresponding to a given generator G, an individual N-stability with respect to the examinations by a given aspect-view V_g *is* found, this does by no means involve that for the same object-entity $œ_G$ but another aspect-view $V_{g'}$ with $g' \neq g$ one will find again some individual stability for some big number.

> The individual stability of the qualifications of an object-entity $œ_G$ or the statistical character of these, are relative to the qualifying aspect-view V_g.

It is of the utmost importance to realize that—quite generally—a generator G of a physical object-entity being fixed by some operational definition of it, it would even be *inconceivable* that for *any* association of G with some aspect-view V_g, the results of repetitions of the corresponding sequence $[G.V_G]$ shall all be identical: that would be a *miracle* in so far that absolute identity—independent of the considered aspect-view V_g, i.e., for any tried aspect-view V_g—has never been observed concerning a *physical* object-entity which—factually—is always endowed with strict singularity (this probably holds even for a conceptual object-entity, like, say, the number 5, if its mental correspondent in a given mind is considered). As for "identity" in *absence* of *any* view—which, as many do in fact surreptitiously and vaguely imagine, would mean identity of $œ_G$ with *itself* from one realization of G to another one, not of the qualification of $œ_G$ *via* V_g when the succession $[G.V_g]$ is repeated—it is but an illusory concept tied with the quest for an impossible absolute objectivity of the thing-in-itself. (The psychological difficulty encountered for realizing this stems from the physical, "exterior" nature supposed for $œ_G$, which surreptitiously inclines to posit that—like $œ_G$ itself—the *qualifications* of $œ_G$ also exist independently of any observer-conceptor, as "intrinsic properties" of $œ_G$).

The above considerations bring back to the only methodological meaning which can be *a priori* assigned to the one-one relation posited between G and $œ_G$, and, correlatively, they bring back to also the roots of the non-determination of reference.

(Notice how all the preceding assertions acquired inside MRC a "natural" deductive character (i.e., outside any formal system) manifesting the formalizing essence of the features with which we progressively endow this approach. Which is a quite non-trivial feature of MRC).

Π1 **3.*Proposition.*** Given an epistemic referential (G, V_g) where both G and V_g involve *physical* operations, in general no stability at all is insured for the gk-spacetime values obtained by repeated or multiple realizations of the succession $[G.V_g]$, neither on the individual level of observation, nor on the statistical one.

"Proof". If only a maximal, an individual N-stability is considered, i.e., identity of all the N groups of observable gk-spacetime values corresponding to N realizations of a succession $[G.V_g]$, then $\pi 13$ becomes a mere repetition of $\pi 12$, hence the "proof" of $\pi 12$ still works. But suppose that no individual N-stability has been found, i.e., that a whole statistical distribution of dispersed triads of gk-spacetime-values has been found. Then it still remains *a priori* possible that a big number N' of repetitions of a series of a big number N of repetitions of the succession $[G.V_g](N' \neq N$ in general) shall bring forth, when N' is increased toward infinity, a convergence in the sense of the theorem of big numbers, of the relative frequencies of occurrence, in the mentioned statistical distribution, of the dispersed triads of gk-spacetime-values. In this case one can speak of a *probabilistic* (N, N')-*stability*. However, up to some given arbitrary pair (N, N') of big numbers, it might appear by experiment that in fact this second possibility does not realize either, even though G and V_g have been found to mutually exist in the sense of D7. Nothing excludes such a situation, neither some previous MRC-assumptions, nor the empirical experience. If this negative situation does realize indeed, then only two solutions are left: either one continues the search with pairs of increasingly bigger numbers N,N', or one stops at some given pair (N, N') and announces *a posteriori* that, even though G and V_g do mutually exist in the sense of D7, their pairing (G, V_g) has nevertheless to be (N, N')-cancelled from the subsequent conceptualization, because, while no individual N-stability has been observed, this pairing does not generate a probabilistic (N, N')-stability either; *tertium non datur* because apart from an individual or a probabilistic stability, no other sort of still weaker stability has been defined so far (in Sec. 5.2 this question is treated more thoroughly). In short, for any given pair of big numbers (N, N'), it is quite possible that no stability at all be found for the results of repeated successions $[G.V_G]$. Which establishes $\pi 13$.

Comment. The "proof" of $\pi 13$ does by no means exclude the possibility that, if the succession $[G.V_G]$ does produce a probabilistic (N, N')-stability, another succession $[G.V_g']$ with G the *same* but with $V_g' \neq V_g$, shall produce qualifications that are endowed with some *individual* N-stability, or with no stability at all, neither probabilistic nor individual:

> *The existence of a probabilistic stability of the qualifications of a given object-entity* $œ_G$ *is relative to the qualifying aspect-view* V_g *just like the existence of an individual stability. The nature—individual or probabilistic—of the stable qualifications of a given object-entity* $œ_G$, *is relative to the qualifying aspect-view* V_g *just like the existence of*

stable qualifications.

4.2.2. The normed concept of relative description

D14. ***Relative description***

D14.1. ***Relative description of a physical object-entity.*** Consider an epistemic referential (G, V) where: G is a *physical* generator that generates a corresponding *physical* object-entity œ$_G$; V is a *physical* view with m aspect-views V_g with respect to each one of which œ$_G$ does exist in the sense of D7; and, as required by P8 and C9, V contains also a spacetime view V_{ET} introducing an ordered spacetime grating (D5.4). Furthermore consider, for *each* V_g from V, a big number N of realizations of the corresponding sequence $[G.V_G]$, in simultaneity or in succession, the time parameter being re-set at the same initial value t_0 for each realization of a sequence $[G.V_g]$.

Suppose first that, when the succession $[G.V_g]$ is realized N times, for *each* aspect-view V_g from V, *identical* outcomes of the corresponding configuration of gk-spacetime-values are obtained, i.e., only *one* same "individual" result appears N times. We shall then say that an N*-individual* outcome has been obtained (the reference to this N is necessary because nothing excludes that for another sequence of successions $[G.V_g]$ some dispersion be found). The set of N-individual configurations of gk-$E\vec{R}$-Tt-values corresponding to *all* the m distinct aspect-views V_g from V, constitutes in the abstract representation space of V ordered by the spacetime grating introduced by V_{ET}, a definite "form" of gk-$E\vec{R}$-Tt- values. This "form" will be called an N-*individual relative description, with respect to V, of the physical object-entity* œ$_G$, (in short an individual relative description) and it will be indicated by the notation D_N/G,œ$_G$,$V/$ to be read "the description relative to the triad G, œ$_G$,V and to N" (in current usage the index N, supposed to be big, will be dropped). The individual relative description D/G,œ$_G$,$V/$ defined above can also be regarded as the set of all the individual one-aspect-relative-descriptions D/G,œ$_G$,$V_g/$ with $V_g \in V$.

Suppose now that, when the various successions $[G.V_g]$ with $V_g \in V$ are realized N times, *not* all the successions $[G.V_g]$ are found to reproduce identically one same configuration of gk-$E\vec{R}$-Tt-values; that at least for one $V_g \in V$ (not necessarily for all) the corresponding succession $[G.V_g]$ produces a whole set $S_{gi} = \{c_{gi}\}$ of mutually distinct, dispersed configurations cgi of gk-$E\vec{R}$-Tt-values, (with $i \in I$ and I a finite index-set, to preserve the finitistic character of this approach); but that, for *any* succession $[G.V_g]$ which produces dispersed results, when N is increased toward infinity, the relative frequency $n(c_{gi})/N$ of occurrence of each configuration $c_{gi} \in S_{gi}$

converges toward a corresponding probability p_{gi}. In these conditions each configuration $c_{gi} \in S_{gi}$ will be called an *elementary-event-description* corresponding to the succession $[G.V_g]$ with $V_g \in V$ and it will be denoted $D_{p(gi)}/G,œ_G,V_g/$. The epistemic referential (G, V) will be said to produce a *probabilistic relative description* of the physical object-entity $œ_G$ which will be denoted $D_p/G,œ_G,V/$.[10]

Comment. The definition D14.1 is the core of MRC. It finally assigns a *significance* to what has been called a physical object-entity $œ_G$. A significance which, though it is relative to a view V and in certain "basic" conditions that will be specified in D14.3.1 is far from being fully "satisfactory", nevertheless is now quite definite *and* endowed with communicability. Whereas G alone cannot systematically insure for "$œ_G$" a significance distinct from just the conventional label "effect of a realization of G", because the results of G might emerge still entirely non perceptible.

D14.1.1. ***Reference and relative meaning.*** In any case of qualificational stability, individual or probabilistic, we shall say that $œ_G$ *is the reference (or referent or designatum) of* $D/G,œ_G,V/$ *while* $D/G,œ_G,V/$ *is the meaning of* $œ_G$ *relatively to* V.

Comment. It thus appears that the initial methodological assertion of a one-one relation between a given definition of an operation G and its result labelled $œ_G$, does not hinder the subsequent construction of all the necessary specifications. On the contrary, it founds them.

The following is noteworthy:

The condition of existence of individual or probabilistic stability of the outcomes of the successions $[G.V_g]$, *with respect to repetitions of these, pre-supposes the possibility to achieve arbitrarily many successions* $[G.V_g]$, *for all the* $V_g \in V$.

This is a strong restriction. But when it is insured it extracts out of temporality the concept of "description" founded upon it and it puts it directly on highways of communicability where reference, meaning, and objectivity in the sense of intersubjective consensus, can most immediately be attained. Furthermore, it sets a standard with respect to which relaxing generalizations can be now defined.

[10] This definition of a probabilistic description is incomplete and simplifying. It will be thoroughly reconstructed and completed in Sec. 5.2. A more ancient but full treatment can be found in Ref. [18]. At this stage of the development of MRC we are obliged to introduce it in this unachieved form, as a provisional support for essential distinctions that cannot be postponed.

D14.2. *Two generalizations of D14.1*

D14.2.1. ***Relative description of a non-physical public object-entity.*** Let us suppress in the definition D14.1 the restriction to physical generators, while *excluding* generators that act on only one individual inner universe (there, in general at least, the sequences $[G.V_g]$ cannot be repeated (in succession or in simultaneity) and so the condition of stability of their results cannot be insured). Thus relaxed, the definition D14.1 enlarges to object-entities from the non physical but public, exterior reality (economical, social) for which the repeatability of sequences $[G.V_g]$ and the condition of stability of their results still may happen to make sense. The new sort of description obtained in this way will be called a *relative description of a non physical and public object-entity* and it will be indicated by the notation $(NPP).D/G$,œ$_G$,$V/$, in short $(NPP).D$.

Comment. The generalization D14.2.1 holds in particular concerning any already accomplished *description* in the sense of D14.1, selected itself as a new, always *conceptual* object-entity, to be examined in a subsequent description *via* some new view. Thereby:

> The definition D14.2.1 opens up specifically and explicitly the whole crucially important sub-realm of R consisting of constituents of a stabilized communicable *conceptual* reality where, in particular, "logic" is located.

In the case of non-physical object-entities that admit of a description in the sense of D14.2.1, any reference to the frame-aspect of ("physical") *space* can obviously be dropped, and so the obtained relative description amounts to a "form" of only *gk*-time values. If moreover it appears that the considered description can be regarded to be independent also of time values, (as for instance in the study of a fixed formal system), the reference to the frame-aspect of time can be equally dropped. (For instance, the dependence on time cannot be dropped for the relative description pointed toward by the verbal expression "this theory is true": The truth-value yielded by the examination of the object-entity consisting of a theory, *via* the aspect-view V_g where g = truth, *does* depend on the structure of knowledge (informations, understanding, modalities of verification, etc.) available to the acting observer-conceptor at the considered time; on the contrary, for the relative description indicated by the verbal expression "the sum of the angles of a Euclidean triangle is 180°", the time dependence can be dropped). Consider then a relative description where both the space qualifications and the time-qualifications can be dropped. If no one among the involved aspects g introduces by its own definition an order (cf. D5.1), this description consists of one or several *non-ordered* but *stable* configurations of gk-values. What does this mean? It means that the involved non-ordered configurations are characterized by some *correlations*, which are stable with respect to repetitions of the sequences $[G.V_g]$ permitted by the view V, i.e., a given gk-value is found to be associated with this or that other $g'k'$-value ($g' \neq g$ or $k' \neq k$ or both), always, or never (which is as strong a correlation as always), or with this or that probability.

D14.2.2. ***Relative testimony.*** Take again as a starting point the strong definition D14.1, and suppress now in it both the restriction to only a physical generator of object-entity and the condition of repeatability of the sequences $[G.V_g]$ for the V_g from V. What becomes of D14.1? It reduces to a mere set of "qualifications" generated by a definite epistemic referential. Indeed as soon as an epistemic referential (G, V) is given and the condition D7 of mutual existence is satisfied for the pair (G, V), qualifications *via* V can arise for the object-entity $œ_G$ produced by the generator G. From now on any structure of such qualifications will be called a *relative testimony* and will be denoted $\theta/G, œ_G, V/$, in short θ.

Comment. The generalization D14.2.2 of D14.1 gives a definite status inside the MRC-language to all the qualifications of *unique* object-entities of *any* nature. In the case of physical object-entities, uniqueness is often intimately connected with spacetime singularity, in particular with the principle P10 of individualizing spacetime mutual exclusion. This will come out to have a surprising importance in the identification of the characteristics of the deepest stratum of an MRC-logic (Sec. 5.1.2).

Furthermore, D14.2.2 introduces in the MRC-language all the qualifications of psychical events from the inner universe of a conceptor-observer. This is a huge inclusion that lays down a foundation for the future research of a clear connection *in MRC-terms*, between introspective reports and neurological facts. Which might lead to comparability of the MRC requirements on this sort of connection inside the framework of important new views on body *versus* mind, like those of Edelman [7], Changeux [6], Damasio [8], and more generally, in connection with the whole avalanche of results continually produced in the cognitive sciences. Thereby the problems of reference and truth that haunt this vast recent domain might find the conceptual framework for a guided approach.

Finally, the relative testimonies in the sense of D14.2.2 permit to take into consideration the historical descriptions, the poetical ones, etc. For these the fundamental concepts of reference and truth still remain wide open for discussion and for methodological organization.

D14.3. ***Basic transferred relative descriptions.*** In what follows we finally shall touch and transpose in quite explicit and generalized terms, the fundamental epistemological innovation specifically implied by quantum mechanics.

D14.3.1. ***Basic transferred relative descriptions of a physical object-entity.*** Consider a relative description *in the sense of D14.1 where:*

- The generator consists of a repeatable physical operation and it produces a physical object-entity that cannot be perceived directly by man. Such a generator will be called a *basic* generator and will be denoted $G^{(o)}$.
- The object-entity produced by a basic generator $G^{(o)}$ will be called a *basic object-entity* and will be denoted $œ^{(o)}$ (a simplified notation standing for $(œ_{G^{(o)}})^{(o)}$).
- The view able to draw phenomenal manifestations out of a basic object-entity is necessarily such that the phenomenal content of each gk-value of each involved aspect g consists of features of a material device for gk-registrations, biological or not, but which always is *different from the studied object-entity*, these features emerging as "marks" produced by the interactions between the registering-device and replicas of the considered basic object-entity. These marks acquire *significance* by their coding in terms of values gk of the aspects from the acting view. A view of the just specified kind will be called a *basic transfer-view* (in short a basic view) and will be denoted $V^{(o)}$. The aspect-views from $V^{(o)}$ will be called basic aspect-views and will be denoted by $V_g^{(o)}$.
- The epistemic referential $(G^{(o)}, V^{(o)})$ will be called a basic epistemic referential.
- A relative description in the sense of D14.1, individual or probabilistic, achieved with a basic generator and *one* basic transfer-aspect-view $V_g^{(o)}$, will be called a *basic transferred relative aspect-description* and it will be denoted $D^{(o)}/G^{(o)}, œ^{(o)}, V_g^{(o)}/$.
- A relative description in the sense of D14.1, individual or probabilistic, achieved with a basic generator $G^{(o)}$ and a basic transfer-view $V^{(o)}$ involving at least two mutually incompatible basic aspect-views $V_{g1}^{(o)}$ and $V_{g2}^{(o)}$, will be called a *basic transferred relative description* (also, in short, a basic description or a transferred description) and it will be denoted $\mathcal{D}^{(o)}/G^{(o)}, œ^{(o)}, V^{(o)}/$ (in short $D^{(o)}$).
- A basic transferred description $D^{(o)}/G^{(o)}, œ^{(o)}, V^{(o)}/$ is posited to *characterize* observationally the involved basic object-entity $œ^{(o)}$, which means by definition that it is posited that no other operation

of generation $(G^{(o)})' \neq G^{(o)}$ can be found which, associated with the same basic view $V^{(o)}$, produces the same basic transferred description.

Comment. It is difficult to fully grasp the meaning and the importance of the concept of basic transferred relative description. But it is crucial to grasp it fully. Indeed it is by *this* concept that MRC penetrates beneath natural language and the forms of thought involved by it, establishing a definite and rather complex relation between conceptualization and physical factuality. Therefore I shall comment on it in detail, even redundantly.

To begin with, let us stress that a basic physical object-entity produced by a basic physical operation $G^{(o)}$, *if* furthermore this sort of object-entity has never before been qualified *via* any transfer-view $V^{(o)}$ whatever, emerges still entirely unknown *in terms of the knowledge researched concerning it specifically*, notwithstanding that the operation of generation $G^{(o)}$ *does* singularize it out of the whole of physical reality. Indeed—factually – the result labelled $œ^{(o)}$ is entirely "specified" by $G^{(o)}$, it is "defined", since it is made available for any possible subsequent examination and, according to D14.3.1 and to the posited one-one relation between the operation $G^{(o)}$ and its result $œ^{(o)}$, it can be deliberately reproduced. More. Factually, each such result emerges from the operation $G^{(o)}$ that produced it, *fully* individualized, it lies on a level of zero-abstraction, still filled with its whole untouched concrete singularity. Which no language whatever could never do because we generalize as soon as we speak: full singularity is unspeakable. But—consequently in fact—this result produced by $G^{(o)}$ alone, not yet followed by an operation of examination, is individualized in *another* manner than that in which knowledge concerning it—specifically—is researched; namely in an only factual, physical sense, not an already conceptualized, qualifying sense. It is true that the specification of the generation operation $G^{(o)}$ involves necessarily some position of a pre-decided *conceptual* space of qualification (tied with the "zone" R_G from R where $G^{(o)}$ is supposed to act (D4 and comment on it). By its definition $G^{(o)}$ drops its products inside this pre-decided conceptual volume. That what is labelled $œ^{(o)}$ is pre-constrained to emerge inside this or that spacetime domain where $G^{(o)}$ acts, it is produced so as to correspond to some definite verbal designation ("a manifestation of stellar life", or "a state of a microsystem", etc.). In this sense $G^{(o)}$ and its result labelled $œ^{(o)}$ might be considered to never be "purely" factual. But:

> The preliminarily *posited* conceptual volume where the operation $G^{(o)}$ drops its products, cannot be equated to the new knowledge that is *researched* concerning these products. The elaboration of this new researched knowledge is the task left by construction for examinations

to be achieved *subsequently* upon the already produced $œ^{(o)}$, by this or that basic aspect-view $V_g^{(o)}$ that exists in the sense of D7 with respect to, non-specifically, anything lying inside the pre-decided conceptual volume where $G^{(o)}$ drops all its products.

It is important to realize that the specification of the operation $G^{(o)}$ of generation of an object-entity *must* contain a conceptual receptacle attached to the physical action involved by $G^{(o)}$; a conceptual receptacle to be lowered with this action into the depths of pure as yet non-conceptualized physical factuality, in order to receive inside it the results of the operation $G^{(o)}$ so as to be able to hoist them up into the stratum of the concepts-and-language. This is an unavoidable condition because *only* a receptacle made of concepts-and-language can hoist up into the thinkable and speakable a lump of pure factuality. A macroscopic operation $G^{(o)}$ can be itself shown, taught, repeated, and also said, pointed towards by words. But if nothing thinkable and speakable were posited concerning what $G^{(o)}$ produces, which by hypothesis is not perceivable, then this, the product, even if factually it has been produced, would simply stay out of conceptualization. While human mind, in order to be able to *think* about a non perceivable thing, needs, not only to have labelled it by a repeatable operation of generation and by a notation, but furthermore to have endowed it with some initializing conceptual status, with at least some approximate preliminary speakable location inside the unending and infinite-dimensional space of concepts.[11]

But of course a basic description $D^{(o)}$ does not *indefinitely* produce an object-entity $œ^{(o)}$ that is still unknown, specifically and precisely in the desired terms. Knowledge about $œ^{(o)}$ is a subjective and moving character. Think of a basic operation of generation $G^{(o)}$ that is repeated by the observer-conceptor X after it has founded for him the desired knowledge concerning $œ^{(o)}$, via some basic operations of examination $V_g^{(o)}$: Then even though $œ^{(o)}$ is generated by the same generator $G^{(o)}$ and emerges beneath the level of the directly observable by man, it is nevertheless already known to a certain extent by X (while for another observer-conceptor it can be strictly unknown, even if the knowledge acquired by X has been made socially available in public registration devices (apparatuses, catalogues, books, etc.). The only objective (inter-subjective) and perennial features of a "basic" description $D^{(o)}$ and of what is here called a "basic" object-entity $œ^{(o)}$ stem from the constant character of the involved referential, a "basic" referential $(G^{(o)},V^{(o)})$ where $G^{(o)}$ works on the physical factuality

[11] It was Evelyne Andreewsky who, by repeated questions and remarks, motivated me to specify how, exactly, the *pre-existing* conceptualization and the descriptional *aims* act upon the extraction of new knowledge out of as yet unconceptualized physical factuality.

and $V^{(o)}$ is a transfer-view as specified in the definition D14.3.1: it resides in the fact that what is called a basic description $D^{(o)}$ consists by definition of *exclusively* features imprinted upon registering devices that are all different from the studied object-entity œ$^{(o)}$.

Consider now the following question which is fundamental for the MRC treatment of reference: Does indeed the definition D14.3.1 of a basic description open up a way toward a communicable characterization of—specifically—the basic object-entity œ$^{(o)}$? The final posit from D14.3.1 concerns this question. Consider a basic *aspect*-description $D^{(o)}/G^{(o)}$,œ$^{(o)}$,$V_g^{(o)}/$ (the basic view consists of only one basic aspect $V_g^{(o)}$). In this case it seems clear that $D^{(o)}$ does *not* yield a characterization—individual or probabilistic, no matter, but specifically and isolately—of what is labelled œ$^{(o)}$, since it points toward observable manifestations brought forth by interactions between œ$^{(o)}$ and a material device for *gk*-registrations. Which changes what was labelled œ$^{(o)}$ (P10) and produces perceivable results that depend on the device for *gk*-registrations as much as of œ$^{(o)}$. But what about a "binocular" basic description $D^{(o)}$ where the basic view $V^{(o)}$ consists of two mutually incompatible basic aspect-views $V_{g1}^{(o)}$ and $V_{g2}^{(o)} \neq V_{g1}^{(o)}$? In quantum mechanics, for the particular case of a basic object-entity that is a state of a microsystem, it is (implicitly) admitted that, together, two quantum mechanical descriptions of a same microstate *via* two mutually incompatible quantum mechanical views, characterize that microstate. Which means only that no other operation $(G^{(o)})' \neq G^{(o)}$ of generation of a microstate can be assumed to yield both these same two quantum mechanical descriptions. The final posit from D14.3.1 generalizes inside MRC the above-mentioned quantum mechanical implication. It would be satisfactory of course to base this posit upon a general constructed argument (for instance *a reductio ad absurdum*). But so far I have failed to find one. So I introduce the condition as just a supplementary security for the solidity of MRC). This completes now on the *observational* level the methodological posit from D4 according to which a given operation of generation of an object-entity is assumed to always produce the same object-entity. The necessity of a complement of this type can be best understood *per a contrario*. In the absence of any phenomenal, specific, normed, communicable set of qualifications associated specifically with what has been labelled œ$^{(o)}$, one would have to regard "œ$^{(o)}$" as just a label that labels nothing distinct from this label itself. Then speaking and thinking of "what has been labelled œ$^{(o)}$" would be only a void sophistic trick, amounting to arbitrary implicit postulations.[12] We would be obliged to admit that pure factuality and human communicable knowledge stay for

[12] Putnam's thought experiments concerning the non-determination of reference [11] are very suggestive.

ever apart from one another. But this just does *not* happen. Quite on the contrary, our capacity to adapt to the environment and the technical powers that we are able to acquire manifest continually the astonishing, even miraculous agreement between human knowledge and factual being, attesting intimate couplings and transmissions which somehow manage to emerge between them.

The posit from D14.3.1 incorporates into the MRC-representation the assertion of a definite *way* in which a basic object-entity produced by a basic generator $G^{(o)}$, *can* be conceived to be captured inside pure physical factuality and then hoisted up into the conceptual net of inter-subjective knowledge: it is that what produces a pair of sets of mutually incompatible observable manifestations which - accordingly to the final posit from D14.3.1—cannot be obtained by the use of any other operation $G^{(o)\prime} \neq G^{(o)}$.

At a first sight the concept of a basic transferred description might seem very particular, and too radical. But in fact it *possesses absolute priority and non restricted generality inside the order of cognitive elaborations.* Quite universally, any object-entity corresponding to any generator, if it did reach the consciousness of an observer-conceptor, then it reached it *first* by some transferred descriptions. We remain unaware of this because usually the phenomenal appearance of the gk-values involved in these transferred descriptions stems from marks imprinted directly upon the biological domains of sensitivity of the observer's body which act at the same time as generators of object-entity and as views in the sense of MRC. So the involved epistemic referentials are of a nature which, with respect to the general MRC-descriptional mould, is particular and degenerate (cf. the global comments on D14 and the comments on D19.4, 5.1.1, and 5.1.2). This entails the following effects which occur all at the same time and beyond any control of logical consistency:

(a) It hides the transferred character of the marks.

(b) It inclines toward assigning systematically a passive role to the mind, in its interactions with physical factuality. The mind is supposed to just *receive* marks irrepressibly imprinted upon the sensitive apparatuses of the body by incessant streams from the physical factuality. (How far one is thus kept from realizing the possibility and the universal methodological value of two radically distinct epistemic stages which, in general, have to be both active during a deliberate achievement of "unnatural" transferred descriptions, like those on which quantum mechanics throws light!).

(c) It pushes surreptitiously toward ontological absolutizations. Indeed one encounters severe difficulties to realize that the (various) transferred descriptions of this chair, which my consciousness functioning achieved spontaneously by the help of my biological views (involving the eyes, the nervous system, the ears and fingers, etc.), cannot, without contradiction, be identified with "the-way-in-which-the-chair-in-itself-really-is"; that nothing, never, will be able to prove that this or that model of a chair "exists" independently of any perception, of any view. More, that such an instinctive hope contradicts both philosophy and logic, since in the absence of any view the very concept of description, and even that of merely an isolated qualification, simply vanishes (cf. D14.2.2, D19.1, D19.2). It is really hard to withstand the irrepressible trend toward identification of our spontaneous modelizations stemming from descriptions transferred on the human biological registering devices, with ontological credos that float on self-contradicting assemblages of words, alike to Magritt's tree that floats with its roots in the air. Kant, Poincaré, Einstein, Husserl, Quine, Putnam, have founded famous analyses on the explicit recognition of this fact.

But, and this is noteworthy, as soon as the transfer-view from a considered basic transferred description $D^{(o)}$ does not directly involve the biological human terminals—the nearest and which *in fine* cannot be eliminated—as soon as the transfer-view $V^{(o)}$ from $D^{(o)}$ involves marks registered on devices that are exterior to the observer's body (as it happens indeed for microstates), it suddenly becomes quite clear that a basic description $D^{(o)}$ itself constitutes a constructed intermediary object-entity which relays the access of the basic a-conceptual object-entity œ$^{(o)}$, to the observer-conceptor's consciousness-functioning; that phenomena are not always independent of aimed volition, that they are not always just psycho-physical facts which emerge spontaneously, but might have to be planned and produced by method. Then, like in quantum mechanics, the two distinct and mutually independent stages involved in a transferred description—the stage of generation of an object-entity œ$^{(o)}$, and the subsequent stage of creation of observable manifestations drawn from œ$^{(o)}$ by interaction with gk-registering devices—appear as obvious. Their active and deliberate character strikes the mind, and the invaluable normative value of the concept of basic transferred description can be fully understood.

The basic object-entity œ$^{(o)}$ from a transferred description $D^{(o)}$ roots this description directly into the physical factuality. Correlatively the transferred description $D^{(o)}$ achieves for the involved basic object-entity œ$^{(o)}$ a very first passage from pure physical factuality, into the domain of communicable knowledge. It yields for it a first communicable form, a first observable expression that points communicably toward the involved object-entity. So the basic transferred descriptions are the *local zero-points* of the chains of conceptualization, in the following sense. Each basic transferred description $D^{(o)}$ starts from a conceptual situation where, though some conceptual environment of the basic object-entity œ$^{(o)}$ (genus, etc.) always is more or less explicitly posited *a priori* (at least *via* the definition D4 of $G^{(o)}$), nevertheless nothing is known concerning œ$^{(o)}$ *specifically.*

The very first stratum of communicable knowledge available at any given time consists of the basic transferred descriptions achieved up to that time, not of just phenomenal appearances in the Kantian sense.

The transferred descriptions are the channels through which as yet non semantized but semantizable factual matter, is adduced into the domain of the inter-subjectively semantized. The "scientific legalization of phenomenal appearances" in Kant's sense (2.3) begins by the construction of transferred descriptions, of which $D^{(o)}$ yields a form that is normed. Which amounts to a formalization of the structure of the connections between knowledge and Being.

This is a quite fundamental contribution of MRC to epistemology. It separates the volume of the known in two essentially different strata. Indeed the whole rest of the available knowledge consists only of subsequent developments of this first—evolving—stratum of transferred descriptions which operate the very first connections between Being and knowledge. Namely developments consisting of spacetime modelizations which endow the basic transferred descriptions with the features required by the frame-postulate P8, thus insuring for them an "intelligibility" of which initially they are devoid; and then, a non limited succession of complexifications (or generalizations, etc.) of these spacetime modelizations (cf. D.19 and all the involved discussions).

I add a last remark concerning the concept of basic transferred description. From the viewpoint of MRC the quantum mechanical descriptions of microstates appear now as just particular instances of transferred descriptions of physical entities: the strategy of quantum mechanics, once identified explicitly, brings into evidence an example of the universal way in which the conceptualizations are rooted into pure physical factuality, and, for this example, it displays all the stages of the rooting. MRC recognizes the universality of this rooting and extends it to any sort of physical factuality, re-expressing it in general and *normalized* terms.

D14.3.2. ***Basic description of a psychical object-entity?*** Notwithstanding important difficulties (the non-repeatibility of the successions $[G^{(o)}.V^{(o)}]$, it might turn out to be possible to forge a useful concept of basic description of "psychical basic object-entities œ$^{(o)}$" (by some combination of testimonial descriptions θ in the sense of D14.2.2, with "biological basic transferred descriptions"). Thereby I mean a *conscious* but *not* yet conceptualized psychical object-entity, a *primeity* in the sense of Peirce that emerges in the acting observer-conceptor's *interior* universe, and, though perceived, is still entirely *un*known, non-qualified (A. Damasio ([8]) has elaborated a very subtle structure of concepts-and-facts concerning events of this sort). Think for instance of all the feelings of mere *existence* of an inner fact of which one becomes suddenly aware strictly without knowing as yet explicitly what and how they are, so *a fortiori* without understanding them; think of the genuine *research* conducted by Proust in order to identify the subjective *meaning* of such feelings; think also of the psychoanalytic methods which deal with features as if transferred upon behavioural "devices" (reactions, ways of acting, feelings) by interactions between a hypothetical entirely unknown inner configuration, and various accidental or systematically arising exterior circumstances; this hypothetical inner configuration is precisely what the therapies try to first somehow delimit "operationally" (by analyses of dreams, associations, etc.)—even if by creating it—and then to interpret, qualify, and control or suppress. The obtained description is then in a certain sense precisely what seems to deserve being called a basic relative description of a basic psychic object-entity.

It is however clear that for the moment these are just conjectures. The central concept of basic transferred description has an indisputable pertinence only with respect to physical object-entities.

Global comment on the definitions D14. Finally, let us now consider globally the whole set of definitions D14 and make some comments on the general concept of relative description.

The general notation $D/G,œ_G,V/$ stresses that any description that is normed in the sense of MRC brings into play a triad $G,œ_G,V$ to which it is essentially relative: this is the general descriptional mould induced from quantum mechanics and required now for any description, whether it is transferred or not. The first location from this triad is the place reserved for an epistemic action, the generation of an object-entity, which up to now has quasi systematically been ignored, because the canonical basic transferred descriptions where the generation of an object-entity plays a separate and active key role, were ignored. Indeed for a description that is not transferred, the operation of generation of the desired object-entity is often accomplished without any difficulty, in a natural or even implicit way (think of descriptions of conceptual entities, for instance, the definition of the concept of "table" that pre-exists implicitly inside my mind). While when the transfers occur in—directly—the biological sensorial apparatuses (views, in the sense of MRC), the involved view V *acts also like a generator* G which just selects out of R an object-entity, namely the field of perceptibility of V, and—simultaneously—also qualifies this object-entity: we can symbolize by $G(V)$ such a generator of a view and by $(G(V), V))$ the corresponding epistemic referential. In this case the existence of a generator of object-entity is still deeper hidden than in the preceding case. This highly degenerate and so wide-spread natural situation contributed strongly to the lasting occultation of the fundamental role of principle of the operations of object-entity generation. Quantum mechanics, for the first time and only implicitly, made a separate use of the operations of generation of object-entity, which permitted to become aware of their general and fundamental epistemological importance.

> *The generator of object-entity remained the big omission of the grammars, the logic, and of all the approaches that involve the processes of conceptualization.*

This is why the question of reference has raised insuperable problems: the basic object-entities are only *surreptitiously* drawn into the natural basic descriptions (the *degenerate* ones being produced in a reflex way *via* the biological sensorial apparatuses), with the status of a present but non specified reference. The problem of identifying *a posteriori* of what this reference consists, starting from the already achieved description, has stubbornly resisted solution.

But accordingly to MRC, an operation of generation of object-entity is *always* involved, even if in a non separated and implicit or reflex way.

By construction, any relative description $D/G,œ_G,V/$ is, itself, distinct from the generator, the object-entity and the view involved by it, to all of which it is conceptually posterior; it qualifies only the object-entity which it concerns, not also the generator and the view of which it makes use, nor itself, globally. As for the generator and the view, these are by definition distinct from one another, often by their content, but in *any* case by the *role* held during the process of description.

> *In the definition of a relative description the three notations $G,œ_G,V$ designate three descriptional roles, three descriptional functions,* ***not*** *the nature of the entities to which these roles are assigned in the case of this or that particular relative description.*

And all these three roles are systematically played in any relative description, even if an actor cumulates distinct roles, or plays a role superficially, or both. For instance, if I say ""red" is a too poor expression, better say "colour of blood"", the first proposition expresses verbally a relative description $D/G,œ_G,V/$ where "red", though grammatically it is an attribute, holds the role of the object-entity $œ_G$ (generated by use of a generator G which is a selector acting upon the spot R_G from R indicated by the word "colour"), while "poor" is placed in the role of the view V. But if I say "my cheeks are red", "red" plays the role of the view. So the structure required by the definition D5.1 of an aspect-view, is only a necessary condition for acting as a *view*, but this condition does not hinder a view in the sense of D5.1 to act also in the role of an object-entity (like in the above first example) or in the role of a generator $G(V)$ of object-entity that generates its field of perceptibility by interaction with R.

> *According to MRC no operation or concept possesses intrinsically a fixed descriptional role.*

In each descriptional act, the descriptional roles are assigned by the acting consciousness functioning, and in general these roles change from one description to another one. When a natural description is examined in order to compare it to the MRC norms, the first step is to examine what plays the role of object-entity, what the role of generator, and what that of view. A description $D/G,œ_G,V/$ is a piece of constructed normed meaning which, essentially and explicitly, is relative to the epistemic actions that achieved the semantization asserted by it. *Any* asserted meaning bears inside it the genetic structure designated by the sign $D/G,œ_G,V/$, but it can include this structure in a more or less implicit, truncated, malformed way. Whereas in the normed form $D/G,œ_G,V/$ all the three involved roles $G,œ_G,V$ are explicitly indicated, each one at its own location and following the genetic order of the corresponding epistemic actions. They are to be treated as void, available, labeled rooms that have to be filled up in a check questionnaire to which any achieved or envisaged description must be subjected.

The distinction, inside a relative description $D/G,œ_G,V/$, between the relativity to the operation G of object-entity generation of which the role is to produce an object-entity, and the relativity to this object-entity $œ_G$ itself of which the role is to bear subsequent qualifying examinations, is one of the most subtle and important features of MRC. In particular it preserves from the very strong inertial tendency induced by classical thinking, to forget that as soon as an entity is regarded as playing in a description the role of object-entity, *ipso facto* a corresponding epistemic action of generation of object-entity has produced it as such, implicitly or explicitly, even if this entity somehow pre-existed and so has only had to be selected as object-entity, not to be radically created as such. The importance of a normed *memento* of this fact will fully appear in Secs. 5.1 and 5.2.

The association, in any relative description $D/G,œ_G,V/$, between a one-one relation G-$œ_G$ and the requirement for D of, indifferently, either a strong individual stability or an only probabilistic one, is intimately related with the impossibility, for mere language as well as for mere notations, to grasp and capture the factual individualities, neither in an absolute sense *nor* in only a relativized sense (cf. $\pi 12$, its "proof" and the comments). Umberto Eco remarks: "The tragedy comes from this that man speaks always in a general manner about things which always are singular. Language names, thus covering the non transcendable evidence of individual existence" [19]. Indeed each predicate (view) is general, and no conjunction of a finite number of predicates can ever exhaust the open infinity of the possible qualifications of a physical object-entity.

The concept of relative description is selective. It does not admit inside the class delimited by it, illusory descriptions where one of the three roles G, $œ_G$, V is not played at all. Consider for instance the famous illusory description "this is a lie" (or "I am a lie")" where the word "this" (or "I") masks the absence of specification of the operation G of generation of object-entity, so also *the absence of specification of the object-entity* $œ_G$ *itself.* In fact, there simply is no object-entity at all. This blocks any further conceptual development. Indeed, previously to any research of a truth-qualification of the description, one finds oneself in a situation of impossibility to decide concerning the mutual existence in the sense of D7 between the involved object-entity $œ_G$—nonspecified—and the involved view V. If this primary non-decidability concerning the *a priori* possibility of meaning were somehow (unimaginably) overcome, it would manifest itself later in the form, also, of a paralysis of any attempt at a metaqualification of the *relative proposition* founded on this illusory description *via* the values gk =true or gk =false of a meta-aspect-view g = empirical truth (cf. DL.2 and DL.3 in Sec. 5.1.2).

When descriptions that violate the MRC norms, are reconstructed in a normalized way, the paradoxes stemming from them disappear. There is no need for this to introduce levelled languages of logical types, the illness is cured *locally* by the normed reconstruction of only the considered description.

But nothing hinders to generate (select) as an object-entity any natural description excluded by MRC, and to characterize its incapacities or specificities by reference to the MRC-norms. In this sense the methodological selectivity of the concept $D/G,œ_G,V/$ by no means constitutes an *a priori* pauperisation of the ensemble of descriptions that can be studied inside MRC.

Finally, the general concept of relative description, by its various realizations, permits to discern definite categories inside the realm of the problem of reference and of meaning, and a *dégradé* of proposed solutions: the definitions D14.1, D14.2.1, and D14.3.1 introduce, for the corresponding circumstances, what might stand as a solution or be completed to become one; the definition D14.2.2 suggests a possible approach concerning some of the circumstances to which it applies, while others are isolated as the most problematic; finally, the non achieved definition D14.3.2 concentrates in it definite questions and suggestions.

Like the one-one relation between a given generator of object-entity and the corresponding object-entity, like the definition of relative existence and then the frame principle P8, the concept of relative description with the three roles involved by it, is an act of (qualitative) formalization involving a methodological essence.

4.2.3. Cells of relative description. Chains of descriptional cells. Non-reducible complexification of the conceptualization

P15. ***The Principle of separation.*** Since any one relative description $D/G,œ_G,V/$, whatever its complexity, involves by construction one generator of object-entity, one object-entity, and one view, all well defined, as soon as some change is introduced in the actor designated for holding one of the roles from the triad G,$œ_G$,V, another description is considered.

By a methodological principle called the *principle of separation* and denoted PS, this other description must be treated `separately`.

Comment. Any human observer-conceptor, in presence of reality, is condemned to parcelling examinations. The successivity inherent in human mind, the spatial confinements imposed by the bodily senses—whatever prolongation is adjusted to them—and the absence of limitation of what is called reality, compose together a configuration which imposes the fragmentation of the epistemic quest. MRC reflects this situation in the relativity of any one description, to one triad $G,œ_G,V$. Indeed the relativity to one triad $G,œ_G,V$ specifies, but also limits the capacity of one given relative description to generate further information.

Relativization, limitation, and *precision*, are tied to one another in an unseparable way. *They constitute together an indivisible whole that withstands relativism.*

On the other hand, any fragment generated out of reality in order to play the role of an object-entity, admits of an infinity of kinds of examinations. Moreover any examination achieved on this object-entity, raises the question of the appearance of its result *via* this or that view with respect to which this result exists in the sense of D7, or the question of the relations of this result to descriptions of other object-entities, etc., thus multiplying the conceivable subsequent object-entities and examinations. These confinements and these endless and changing vistas call forth haste and panics of the mind which entangle in knots of "paradoxes" and block the understanding. So they also block the further development of the started conceptualization. The limitations imposed by each specified description are flooded by the implicit fluxes of the rush toward more conceptualization. Without being aware of this, mind yields to whirls of implicit interrogations which generate a subliminal tendency to fluctuate between different operations of generation of an object-entity and different views; a tendency to work out simultaneously several different descriptions. But as soon as the elaboration of several different relative descriptions is simultaneously tried, the various involved generators of object-entity, object-entities and views, are offered a ground for oscillation. And then the oscillations actually happen, because it is very difficult to perceive them, so *a fortiori* to hinder them. So the different descriptions that are simultaneously entered upon, get mixed, and in general none of them can be achieved. Their interaction coagulates nonsense that stops the conceptualization.

The principle of separation hinders such coagulations. It requires the conceptualization, by method, to be achieved by explicit separation in mutually distinct, successive, closed, cellular descriptional steps.

In particular the principle of separation PS surveys the saturation of a description. It rings the bell as soon as the descriptional capacities of a started description must be considered to have been exhausted, because all the qualifications *via* the view chosen for acting in that description, of the object-entity corresponding to the generator chosen for acting in that description, have been already realized by performing a big number of repetitions of all the successions $[G.V_g]$available in that description. PS announces that once this has been done, the descriptional cell potentially delimited by the chosen epistemic referential (G, V) has been saturated with actualized qualifications; that from now on any attempt at obtaining new information inside this same epistemic referential, either is useless or it manifests the surreptitious intrusion of another generator of object-entity, or of another view, or both; that—to avoid stagnation, paradoxes or infinite regressions—one has to stop this intrusion or mixture, by identifying the new epistemic referential that weighs with subliminal pressure upon the consciousness functioning, and by putting it explicitly to work in its own turn, separately.

The systematic application of the principle of separation plays, in the development required by MRC for a process of conceptualization, a role similar to that conveyed by the sign "." or the word "stop" in the transmission or writing down of a message; or else, a role similar to that played in algebra by the closure of a previously opened parenthesis. It is a formalizing requirement of the nature of a rule of calculus.

Any process of conceptualization that is normed accordingly to MRC, is clearly divided in a sequence of localized descriptional cells, and thus it develops by systematically renewed local frameworks, under systematically renewed *local* control.

While the tests of mutual existence (D7) detect the *a priori* impossibilities to construct meaning, the principle of separation permits to avoid any stagnation—illusory paradoxes, infinite regressions—throughout the processes of development of meaning. The concepts of mutual inexistence and the principle of separation cooperate to prevent sources of unintelligibility, and also to detect and suppress them.

The principle of separation possesses a remarkable capacity of organization of the conceptualization. This assertion will find many illustrations in the sequel of this work.

D16. ***Relative metadescription.*** The principle of separation requires descriptional closures and new starts. These entail the necessity of an explicitly and fully relativized concept of metadescription prescribing how to transcend "legally" an already saturated description.

Consider a precedingly achieved relative description to which the *order* 1 is assigned conventionally: $D^{(1)}/G^{(1)}$,$œ^{(1)}$, $V^{(1)}/$ (in short $D^{(1)}$; and instead of $œ_G$ we write œ, to simplify the graphism). Consider a generator that selects $D^{(1)}$ as a new object-entity $œ^{(2)}$, denote it $G^{(2)}$ and call it *a metagenerator (or a generator of order 2 relative to* $D^{(1)}$. So we have $œ^{(2)} \equiv D^{(1)}$. Consider also a view involving *aspects of order 2* with respect to which $D^{(1)}$ does exist in the sense of D7 (for instance the aspect of empirical truth of $D^{(1)}$, or else some aspect of relation inside $D^{(1)}/G^{(1)},œ^{(1)},V^{(1)}/$, between the various *gk*-spacetime qualifications produced by the examinations of $œ^{(1)}$ by the initial view $V^{(1)}$, etc.; call it a *metaview (or a view of second order) relative to* $D^{(1)}$ *and denote it* $V^{(2)}$. The description which is relative to the triad $G^{(2)},œ^{(2)},V^{(2)}$ will be called a *metadescription (or a description of order 2) relatively to* $D^{(1)}$ and it will be denoted $D^{(2)}/G^{(2)}, œ^{(2)}, V^{(2)}/$ (in short $D^{(2)}$).

The same denomination and notation are conserved if ***(a)*** $G^{(2)}$ selects as a new object-entity $œ^{(2)}$ not only $D^{(1)}$ considered globally, but furthermore it includes in $œ^{(2)}$ also separate elements from $D^{(1)}/G^{(1)}, œ^{(1)}, V^{(1)}/$ specified explicitly ($G^{(1)}$, or $œ^{(1)}$, or $V^{(1)}$, or two or all three of them) which permits then to introduce in $V^{(2)}$ aspects of relation between such an element, and the global result $D^{(1)}$ to which it has contributed. Or if ***(b)*** $G^{(2)}$ selects a whole set $\{D_1^{(1)}, D_2^{(1)}, ...D_m^{(1)}\}$ of previously achieved relative descriptions (with an explicit reconsideration, or not, of elements from these descriptions), in which case $D^{(2)}$ is relative to all these descriptions. In this way a very free and rich concept of normed relative metadescription is introduced.[13]

Comment. The definition D.16 can also be applied to $D^{(2)}$ thus leading to a metadescription $D^{(3)}$ of order 3 relatively to $D^{(1)}$ and of order 2 relatively to $D^{(2)}$, etc. In this way it is possible for any consciousness-functioning CF to develop unlimited descriptional chains $D^{(1)}, D^{(2)}, ...D^{(j)}...D^{(n-1)}, D^{(n)}$ of hierarchically connected relative descriptions of successive orders $j = 1, 2, ...n$—with an arbitrary origin denoted $D^{(1)}$—in each one of which the involved metaview can contain all the desired pertinent new meta-aspects of order n.

So *in general the order of a description is not an absolute*, it labels the place where this description emerges inside the considered chain of conceptualization, while a chain can be started *conventionally* by these or those previously achieved descriptions to which the order 1 is assigned.

> *But a basic transferred description can only have the minimal conceivable order, no matter in which chain it is involved.* Therefore this non-conventional minimal order will be denoted by 0, to distinguish it from any conventional initial order 1.

And any chain, if it has first been conventionally started with already previously achieved descriptions to which the order 1 has been assigned, can always be later completed downward until a basic transferred description

[13] Here we can go back to the important distinction from the note 5 between "objectual" qualifications—call them "objectities"—and "state"-qualifications. The objectities are (relatively) stable qualifications that apply in an invariant way to a whole class of evolving states, thereby definig the "object", in the current sense, that assumes this or that state. So according to this language *the term object-**entity** labels only **a descriptional role** in the sense of the general comment of D14, while "object" in the current sense means "endowed with some objectities"*: inside MRC these two words should not be confounded. For instance, the state-qualifications called position, momentum, energy, etc., can vary or evolve from one state to another one, thereby introducing an infinite class of states of a definite sort of "object" labelled, say, "electron", which, inside a convenient metadescription (with respect to this whole class of states), is characterized by the *meta*qualifications consisting of the numerical values obtained (with some given system of unities) for objectities like mass, charge, spin, that are the same inside the whole class of what is called "states of electrons". These objectities however can themselves change by creation or annihilation of the corresponding object, and when the conditions for such changes are realized they can be regarded as states of some more general object (at the limit, of what is called field or energetic substance). In this way the language introduced here can organize conveniently various hierarchies of degrees of abstraction.

is identified which roots the chain into pure factuality. Thereby the chain hits an absolute end (or equivalently, it finds an absolute beginning), which entails a corresponding re-notation upward, of all the successive orders of the involved descriptional cells. But a given relative description can belong to different chains that meet in it (it can be a node of the web of chains of conceptualization); so, regarded as a cell from distinct chains, a same description can have different orders: though all the basic descriptions are absolute zeros of descriptional chains, a given non-basic description from a chain possesses only relative orders with respect to its various zero-points.

> Nevertheless, since the zero order of a transferred description is an absolute, the feature of being a metadescription, or not, is an absolute *if* transferred descriptions constitute the origin used as reference.

This amounts to the remark (rather obvious *a posteriori*) that:

> The (open) set of all the possible relativized descriptions falls apart in just *two* (evolving) layers: ***(a)*** the layer of transferred descriptions of physical basic object-entities which, by definition, are not themselves previously achieved descriptions, and ***(b)*** the layer of metadescriptions in the absolute sense, i.e., of descriptions of object-entities consisting of previously achieved descriptions.[14] Both layers have an evolving content.

Through the first layer, the prime matter for the elaboration of meaning is drawn into conceptualization, and inside the second layer the basic meaning produced in the first layer undergoes abstract transformations which progressively elaborate indefinitely complexified meanings.

It is essential to note that in any chain, for each passage from a descriptional level n to the following level $n+1$, the new epistemic referential to be used $(G^{(n+1)}, V^{(n+1)})$ is freely decided by the acting

[14] However it is curious to note that there are various sorts of rooting of a basic object-entity, into pure factuality: the *objectual* manifestations of a basic object-entity, in the sense of the note 13 can be *conceived* (not known, just imagined) to be tied with pre-existing "own" features of this basic object-entity (cf. D19) which, though unknown, are always the *same.* In this sense, a basic object-entity which is a priori researched as located inside the genus labelled micro-*object* (i.e., is researched exclusively *via* objectual manifestations) is thereby *a priori* endowed with a rooting into pure factuality which is less hidden than that of a basic object-entity researched *a priori* as located inside the genus labelled micro*state*, because it is posited to reach the level of observability by just a time-invariant coding transposition, not by the coding of the effects of a (measurement) evolution produced by the processes of examination. These remarks amount to the assertion of various possible deliberately chosen *depths* of the rooting of a transferred description, into physical factuality.

consciousness-functioning CF, as an expression of his own (evolving) descriptional curiosities-and-aims, such as these emerge at any given time from his own biological, temperamental, and social-cultural background: it is the consciousness-functioning CF who, step by step, chooses the "direction" of the descriptional trajectory drawn by the succession of the cellular but connected descriptional closures $D^{(n-1)}, D^{(n)}, D^{(n+1)}, \ldots$. which, according to [P15+D16], produce the indefinite progression of a hierarchical chain started by conventionally initial conceptual descriptions $D^{(1)}$ or by absolutely initial basic descriptions $D^{(o)}$.

So—as long as no method or algorithm is found for determining automatically, as a function of some definite parameters, a new epistemic referential, when a passage from a description to a metadescription (with respect to it) has to take place—a descriptional chain remains a concept that cannot be absorbed in the concept of computation. And even if such an algorithm were specified, furthermore also the determination of the parameters on which the new referential depends should emerge automatically: accordingly to what criteria? Etc. The subjective successive descriptional aims play a decisive role in the representation of the processes of conceptualization offered by MRC. But on the other hand, the structure assigned by MRC to the conceptualized, namely the structure of a web of chains of increasingly complex relative descriptions, is a (qualitatively) formalized structure, involving definite methodological rules and conventions.

> This brings clearly into evidence that this formalized epistemological method is quite fundamentally distinct from a computational reduction.

Furthermore, MRC excludes also the conceptual reductions:

Π17. ***Anti-reductionist proposition.*** Inside MRC the "reduction" of a metadescription of order n (D.16) to the descriptions and elements of descriptions of order $n-k, k=1,2,\ldots,n-1$ involved in it, is in general impossible.

"Proof". Consider the metaobject-entity $œ^{(n)}$) from a metadescription which, inside the considered chain, is of order n, $D^{(n)}/G^{(n)}, œ^{(n)}, V^{(n)}/$. An isolated element from $œ^{(n)}$ (a description $D_j^{(n-1)}$ of order $n-1$, or some other descriptional element of order $n-1$ from such a description (generator, object-entity, view)) in general simply *does not exist in the sense of D7 with respect to the new meta-aspects of order n from* $V^{(n)}$. For instance, a metaview $V^{(2)}$ of order 2 from the metadescription $D^{(2)}/G^{(2)}, œ^{(2)}, V^{(2)}/$ relatively to $D^{(1)}/G^{(1)}, œ^{(1)}, V^{(1)}/$, can contain the aspect of *distance* between two space-gk-qualifications of order 1 involved by $D^{(1)}/G^{(1)}, œ^{(1)}, V^{(1)}/$, with respect to which these qualifications *themselves* do not exist in the sense of D7. Or else, $œ^{(2)}$ can contain two previously achieved descriptions of physical object-entities, $D_A^{(1)}$ and $D_B^{(1)}$ involving both a same view $V^{(1)}$ (so qualifications of a same nature) while $V^{(2)}$ contains a meta-aspect of order 2 of *comparison* of these qualifications, whereas neither $D_A^{(1)}$ alone nor $D_A^{(1)}$ alone, nor descriptional elements from these, do exist in the sense of D7 with respect to this meta-aspect of comparison. In general terms now, the new qualifications of order n that can be involved in a metadescription $D^{(n)}$ while they cannot be involved in the descriptions of order $n-1$ contained in $D^{(n)}$, consist of *global* or *connective* metaqualifications of order n concerning two or more descriptional entities of order $n-1$ from the object-entity $œ^{(n)}$ from $D^{(n)}$ (consisting of whole descriptions of order $n-1$, or generators of object-entities, or object entities or views, of order $n-1$). These, when considered separately inside the descriptions of order $n-1$, *do not exist in the sense of D7 with respect to any of such new metaqualifications of order n involved by $D^{(n)}$.*

So in general $D^{(n)}$ is not reducible to the descriptions or descriptional elements of orders n-k from the same chain.

The biologists should pay particular attention to this circumstance:

Comment. On each descriptional level of a given order n from a descriptional chain (D.16), the descriptional cell $D^{(n)}$ placed on this level introduces, *via* the condition of relative existence D7, the possibility of new qualifications, of which the very definibility and meaningness are conditioned by the previous achievement of the descriptions from all the previous levels $n-1, n-2, \ldots, n-n$:

> Throughout the development of a process of conceptualization normed accordingly to MRC one can literally watch the creative complexifying work of cognitive time: *One can literally see what "emergence" means.*

It is remarkable that inside MRC this conclusion follows from the system of basic definitions, postulate and principles, in a way that permits a clear perception of the nature of each contribution to the conclusion. One can distinguish between contributions of a factual nature as for instance those brought in by a basic description $D^{(o)}$, and on the other hand contributions of psychological nature like the choices of epistemic referentials for the successive descriptional cells, or of methodological nature like the condition D7 of mutual existence and the principle of separation P15:

> There is no need any more for pleading, arguments, etc. in order to draw attention upon the specific character, the mechanisms and the

features of what is labelled by the words "complexity", "complexification", "emergence".

So, by normed complexification, the transferred descriptions that start from the inside of pure factuality and by which phenomena acquire a first communicable form, are then developed in unlimited chains of hierarchically connected metadescriptions of increasing order. These chains can meet and interact variously at various levels and thus they weave indefinitely compexifying and non predictable forms of communicable significance.

The consequences of the association between the principle of separation and the concept of relativized metadescription, are innumerable and always important. But in the absence of a normed descriptional structure to which any description be referable, they cannot be systematically identified and controlled.

4.2.4. Reference and minimality of the MRC-realism

At this stage of the elaboration of MRC it is already possible to entirely elucidate *a posteriori* the *a priori* somewhat obscure features introduced by the definition D4 of a generator of object-entity (the posited one-one relation G-$œ_G$) and by the realist postulate P3 (cf. note 8). We shall now achieve this by a succession of three propositions. Thereby also the reflexive character of MRC will gain new illustrations, while the formalized character of MRC will become still clearer.

Π18. *Propositions on reference and minimal realism*

Π18.1. ***(On comparability, identity, and the relation* G-œ$_G$)**. A basic object-entity is inexistent in the sense of D7 with respect to any "comparison-view": such a view is a metaview with respect to which only descriptions exist in the sense of D7, never basic object-entities.

"Proof". What is not already pre-qualified cannot be compared. Only two (or more) previously achieved *descriptions* D_1 and D_2 can be compared, and only concerning some definite aspect-view or view with respect to which these descriptions do both exist in the sense of D7. One can for instance ask: are D_1 and D_2 identical or different with respect to this or that gk-value of the aspect-view V_g? If V_g is absent in one or in both considered descriptions, the question is *meaningless* because D_1 and D_2 constitute together a meta-object-entity $(D_1, D_2)^{(2)}$ that does not exist in sense of D7 relatively to a metaview of g-comparison, say $V_{gc}^{(2)}$, so *a fortiori* a gk-identity can be neither established nor refuted. If on the contrary both D_1 and D_2 do make use of V_g, then $(D_1, D_2)^{(2)}$ and $V_{gc}^{(2)}$ do satisfy D7 and so one can research whether, yes or no, D_1 and D_2 do possess some gk-identities. In this example, I have brought into play a most simple metaview of comparison with respect to only one aspect g. Nevertheless this view is already, quite essentially, a *metaview*. One can form much richer metaviews of comparison. But all are metaviews relative to definite views with respect to which only previously achieved descriptions can exist in the sense of D7.

A basic object-entity—a bulk of pure a-conceptual factuality—is not a previously achieved description. Therefore it cannot be compared, neither to "itself" nor to something else.

Comment. So the whole stratum constituted by the very first products of the epistemic actions—the stratum of basic object-entities introduced by basic generators—is not reachable by the concept of comparison and by the qualifications derived from it, identity, difference, degree of similitude. For basic object-entities these qualifications cannot be established by investigation, *they can only by posited by method* (like in the definition D4 of a generator of object-entity). When a given *basic* operation $G^{(o)}$ of generation of object-entity is repeated, it simply is meaningless to ask whether yes or not the object-entities $œ^{(o)}$ produced by this operation are all identical: this finally founds *"deductively" inside MRC the impossibility to assign a* **general** *meaning to the question whether yes or not the repetition of a given operation G of generation of an object-entity* $œ_G$*, produces identical results* $œ_G$. So the posit of a one-one relation G-$œ_G$ appears *a posteriori* to be necessary indeed in order to be always able to speak and think fluently concerning the products of G; while the significance of this posit, already specified to a certain degree in the comment on $\pi 12$, becomes now fully clear.

The one-one relation G-$œ_G$ founds a methodological strategy according to which the referent $œ_G$—*independently defined from the start*, namely by the operation G, and posited to be unique—associates coherently with, both, the *a priori* condition of possibility in the sense of D7 of an as yet non-defined meaning of $œ_G$ with respect to a given view V, and with a subsequently constructed specified meaning of $œ_G$ with respect to V (while for another view $V' \neq V$, the relative existence D7, or a meaning of $œ_G$, or both, might fail to exist).

Thus the question of reference obtains a self-consistent and effective solution.

Π18.2. *"Local" proposition on the realist postulate.* Consider a *physical* object-entity $œ_G$. This is a fragment of physical reality generated by a given physical operation of generation G. The fact that any communicable knowledge is description, and the relativity of any basic description to a basic view, entail that the sequence of words "knowledge of how $œ_G$ is in itself" is void of significance.

"Proof". Consider a physical object-entity $œ_G$. Any communicable knowledge concerning $œ_G$ amounts to some relative description $D/G,œ_G,V/$. Any relative description $D/G,œ_G,V/$ belongs to some net of descriptional chains that is rooted in pure factuality *via* a (finite) number of *basic* transferred descriptions $D^{(o)}/G^{(o)}, œ^{(o)}, V^{(o)}/$ the basic object-entities $œ_G^{(o)}$ from which somehow contributed to $œ_G$, have hereditarily transmitted into $œ_G$ some of their own semantic substance. Now, in each one of these basic transferred descriptions, the transfer-view $V^{(o)}$ acting there yields for the involved basic object-entity $œ_G^{(o)}$ a very first access to observability. But the principle P10 of individual mutual exclusion, the propositions $\pi 11$, $\pi 12$, $\pi 13$, and the definition D14.3.1 of a basic description, show that, and how, the basic transfer-view $V^{(o)}$, while it yields this first access, also *inserts a non removable opaque screen between the acting consciousness-functioning CF and "$œ_G^{(o)}$-in-itself"*, it bars the way of human knowledge toward "$œ_G^{(o)}$-in-itself". So the unavoidable and non removable descriptional relativities explicated inside MRC, and the fact that any communicable knowledge is description, entail inside MRC that [knowledge-of-the-physical-reality-as-it-is-in-itself] is nothing more than a meaningless combination of words, devoid of any designatum.

Comment. Since Kant the impossibility to know how a physical entity "is-in-itself", is accepted as an obvious postulate inside philosophy. But many physicists still are reluctant to fully realize this definitive limit of human rational knowledge. So it seems worth mentioning explicitly that inside MRC this limit follows from the posited assumptions without being one of these. So that there is no need to assert it as a logically independent assumption. Then those who contest this limit should specify which posited assumptions they contest.

Π18.3. ***"Global" proposition on the realist postulate: minimality.*** Inside MRC the realist postulate P3 can only be given a *minimal* significance: it can only be understood to assert *exclusively* the credo of the *existence*, apart from the interior reality from my own mind, of also a physical reality independent of any act of observation; but an existence which is strictly *non-qualifiable* "in-itself", beyond the mere trivial and non-informative, idempotent assertion of its relativized qualifi*ability*, if acts of observation of it do take place in the conditions D4-D7 (in the absence of which P3 would be aimless).

***"Proof"*.** According to the definition D2 "the physical reality", globally considered, is just a *posited* substratum wherefrom all the *basic* object-entities $œ_G^{(o)}$ considered in π18.1 and in the proof of π18.2, are conceived to be extracted. Only this and nothing more. It would then be an arbitrary conceptual discontinuity, a leap, a kind of spontaneous generation, of *Deus ex Machina*, and even an inner inconsistency, to assign to this substratum posited by *us*, properties that transcend the very descriptional essence of *all* the fragments $œ_G^{(o)}$ that we extract from it, namely the impossibility shown by [π18.1+π18.2], to know any qualification whatever concerning a basic object-entity $œ_G^{(o)}$ in-itself.

Comment. It is quite non-trivial that inside MRC this minimality of the realist postulate P3 is a feature that emerges as a consequence—in the weak sense that marks all the "proofs"—of the non removable descriptional relativities. So much more so that the forces which withstand the distinction between mere existence of something, and *knowledge* of *how* this something is, are huge.

Final global comment on the realist postulate (cf. note 8). By now, I think, the specificity of the concept of "physical reality" with respect to the general concept of reality introduced by D2, has come out with satisfactory definiteness, mainly *via* the frame principle P8, the principle P10 of individual mutual exclusion, the propositions π11, π12, π13, the concept D14.3.1 of basic transferred description, and the propositions from this point 18. Thereby, retroactively, the necessity of the postulate P3 as well as its significance should have become clear. This necessity lies in the fact that the *formulations mentioned above would not have been possible without P3.* As for the significance of P3 inside MRC, it can be best grasped *per a contrario*: it is that which inside MRC makes no sense, or no clear sense, when one considers elements of reality consisting of *concepts*, social facts, etc.

As for the minimality of the realism asserted here, I suppose that notwithstanding the proposition π18.3 many will tend to continue to nurture in their minds a non-minimal realism. But reconsider in full light the quasi irrepressible hope that, in spite of all, some model or "only some invariants", might some day transpierce the obstacle generated by the descriptional relativities and inform us definitively, even if only in a coded way, on *how* the physical reality is-in-itself, independently of any perception. And on the other hand, consider the necessarily fragmenting character of the knowledge that human mind can construct, the indefinite and evolving multiplicity of the possible basic object-entities $œ^{(o)}$ as well as of the basic transfer-views $V^{(o)}$ which—now or in the future—could be found to exist in the sense of D7 with respect to a given basic object-entity $œ^{(o)}$: these stress even more, if still possible, the illusory character of such a hope for non-minimality. Indeed, given the non removable dependence of thought on perception, given the non removable dependence of perception on fragmenting descriptional relativities, given the unpredictable and incessant complexifications brought forth by the so various, and unbounded, hierarchical chains of metadescriptions that are growing everywhere, given the unpredictable changes of "viewpoint" (of epistemic referential) which these complexifications might bring forth—certainly radical from time to time—on *what* a rational basis could one uphold the postulation of some convergence toward a definite, definitive, terminal, absolute descriptional structure (supposing that this succession of words were endowed with some meaning)? What a sort of invariants, magically stabilized against all the changes brought forth by the growth of thought, and magically freed of any descriptional relativity, could, thus stripped, nevertheless carry knowledge of the way of being of physical reality in-itself, beyond the posit of its mere existence? When knowledge is nothing else than qualifications *via* some view, of a somehow delimited object-entity, so qualifications relative to some view and some generator of object-entity? Obviously one ends here up in a whirl of circularity.

4.2.5. Relative models versus minimal realism

But if any knowledge-of-how-physical-reality-is-in-itself, is indeed an illusory self-contradicting concept, *why do our minds so stubbornly keep to this concept?* This is a question which deserves being examined.

So I close now this exposition of the nucleus of MRC as follows. First I shall show why the illusory belief in the possibility to reach knowledge of how physical object-entities are in-themselves, is quasi irrepressibly generated by human mind, in consequence of the frame-proncipле P8. And then I shall show how, once identified, the fallacy vanishes and leaves place to dimensions of conceptual liberty.

I proceed by defining a last group of four concepts which specify entirely the philosophical status of the minimal realism asserted here.

On the insufficiency of the basic transferred descriptions. Consider first an *individual* transferred description $D^{(o)}/G^{(o)}, œ^{(o)}, V^{(o)}/$ of a physical basic object-entity $œ^{(o)}$ (i.e., for any aspect-view $V_g^{(o)} \in V^{(o)}$, when the succession $[G^{(o)}.V_g^{(o)}\ ]$ is repeated, always the *same* value gk is obtained). In this case, by hypothesis, the epistemic referential $(G^{(o)}, V^{(o)})$ insures for the transferred results the *strongest* possible sort of qualificational stability (π12, π13, D14.1). While furthermore, according to D14.3.1 the basic transferred description $D^{(o)}$ *characterizes* observationally the involved basic object-entity œ$^{(o)}$. So one finds oneself already in possession of an observational invariant that associates a quite definite *meaning* to what has been labelled *a priori* "œ$^{(o)}$" (cf. the comments on the final posit from D14.1.3). It might then be argued that this "suffices", that in such conditions there is no reason for researching further specifications concerning what has been labelled œ$^{(o)}$. But the fact is that in general such a "sufficiency" simply is not experienced by the observer-conceptors: in presence of even an *individual* transferred description $D^{(o)}$ that produces a most immediately manifest observational stability, many thinkers (if not most)—quite modern thinkers, and even physicists—experience an irrepressible tendency toward a subsequent epistemic elaboration that shall produce a better, a clearer meaning assignable to what has been labelled œ$^{(o)}$. But a "better, clearer meaning of *oe*", in what a sense, exactly?

When one tries to answer this question it appears that what is researched is a representation of œ$^{(o)}$ that shall endow it with an *own* form of spacetime-*gk-values*, *separated* from any process of observation and any registering device; and moreover a form of spacetime-*gk*-values possessing "unity", i.e., covering a connected space-domain obeying some definite dynamical law.

Furthermore a global and explicit spacetime representation is (vaguely) desired for also the processes that have led from the basic object-entity œ$^{(o)}$ with its own spacetime location, to its basic transferred description. The frame-principle P8 is here at work.

The requirements of the frame-principle cannot be violated definitively. One can at most postpone dealing explicitly with them. The frame-principle expresses a psychical fact which is as irrepressible as the physical fact that masses are tied with gravitation. If a basic transferred description of a basic object-entity is asserted, then one should be able to imagine some possible own form of spacetime *gk*-values of this object-entity, as well as some possible own structure of spacetime *gk*-values of the process that has generated the description. If not, the frame principle will keep active and upset us.

A basic transferred description $D^{(o)}$, though, yields no hint for satisfying these requirements. It is expressed *exclusively* in terms of observable features of registering devices which are all *distinct* from what is labelled œ$^{(o)}$. It yields no representation whatever concerning the spacetime location of the basic object-entity œ$^{(o)}$ itself. Inside a basic description $D^{(o)}$ the involved basic object-entity œ$^{(o)}$ is not represented as an autonomous individuality endowed with an own form, it still floats behind as a mere labelled nebula suggested by the words basic object-entity and their notation $œ^{(o)}$. And even if, for a moment, we suspend any question concerning specifically $œ^{(o)}$ and we consider $D^{(o)}$ as a whole, again we find ourselves in presence of an absence of spacetime intelligibility. Indeed, given that each registered mark gk involved by $D^{(o)}$ is found on a g-apparatus and that the transfer-view $V^{(o)}$ must involve at least two different g-apparatuses for measuring two mutually incompatible basic aspect-views, the "form" of spacetime gk-values involved by the basic transferred description $D^{(o)}$ *itself* is found to cover a scattered domain of space, tied with different registering devices that can lie arbitrarily far from one another. And given that the time-origin t_o has to be re-established after each realization of a succession $[G^{(o)}.V_g^{(o)}]$, it is not even clear whether it is possible to somehow associate this form with some continuous evolution (or persistence) ordered by a unique increasing time-parameter.

In short, by $D^{(o)}$ alone one cannot "understand" intuitively, neither how the basic object-entity can be conceived to "be", nor in what a sense, exactly, $D^{(o)}$ is a "description" of this basic object-entity. This situation is tiring for the mind. Therefore an individual basic transferred description $D^{(o)}$ is not perceived as an *achieved* descriptional action. It is not felt to have reached a conceptual stage of epistemological equilibrium. It is obscurely felt as if loosely fixed on a steep conceptual slope where a conceptual force draws it toward a separated representation of $œ^{(o)}$ in terms of own *gk-spacetime* aspect-values. This sort of need might be regarded as a methodological instinct tied with the frame-principle, induced by the adaptive biological evolution of our minds.

All the preceding remarks hold also concerning a probabilistic transferred description. The now seventy years old debate on the interpretation of quantum mechanics proves this enough.

So one is led to consider the following question: is it possible to elaborate, out of a previously achieved basic transferred description $D^{(o)}$, a separated description of the basic object-entity $œ^{(o)}$ involved in $D^{(o)}$? Not a description of "how $œ^{(o)}$ really is"—by now such naïve epistemic quests can be supposed to have been entirely transcended inside MRC—but a specification of just a possible modus of *thinking* of $œ^{(o)}$ in a self-consistent, transparent, intellectually operational way that be naturally insertable into the current language-and-conceptualization. The answer to this question is positive and it is brought forth by the following three new definitions.

D19. ***Intrinsic metaconceptualization. Intrinsic model***

D19.1. ***Intrinsic metaconceptualization of a basic transferred de-***

scription. Consider a basic transferred description $D^{(o)}$ of a physical object-entity œ$^{(o)}$, individual or probabilistic.

- Let $G^{(1)}$ be a metagenerator of object-entity consisting of a conceptual selector (D4) that selects for examination the meta-object-entity consisting of œ$^{(1)} \equiv [D^{(o)} +$ œ$^{(o)}]$.
- Let $V_I^{(1)}$ indicate an *intrinsizing metaview* (I: *intrinsizing*) which, starting from the initial, purely observational, transferred description $D^{(o)}$, works out *intrinsic* qualifications of the basic object-entity œ$^{(o)}$ involved in $D^{(o)}$ (intrinsic: word used in order to distinguish from the philosophical term "in itself"). This, inside the new epistemic referential $(G^{(1)}, V_I^{(1)})$, is achieved as follows:

  * Let $V_{Ig}^{(1)}$ (I fixed, $g = 1, 2, \ldots, m$, Ig functioning as *one* compact index) be a set of m intrinsizing meta-*aspect*-views which, together, constitute the intrinsizing metaview $V_I^{(1)}$.
  * Each intrinsizing meta-aspect-view $V_{Ig}^{(1)}$ involves an *abstract*, conceptual $V_{Ig}^{(1)}$-operation of examination of the *metaobject-entity* $[D^{(o)} +$ œ$^{(o)}]$, namely an examination constructed in a way such that its possible results – necessarily values $(Ig)k$ of $V_{Ig}^{(1)}$, accordingly to the definition D.5.1—are all *conceivable* as separate intrinsic qualifications $(Ig)k$ of the basic object-entity œ$^{(o)}$, which are compatible with $D^{(o)}$.
  * The values $(Ig)k$ of the intrinsizing metaview $V_{Ig}^{(1)}$ are furthermore constructed as: ***(a)*** intrinsic qualifications of œ$^{(o)}$ *at the time* t_o which is the time-origin re-established at the beginning of each succession $[G.V_g]$ having contributed to the elaboration of $D^{(o)}$; ***(b)*** qualifications located inside a connected space-volume $\partial_{\mathbf{r}}$ ($\mathbf{r}$: space-position vector inside the spacetime referential involved by $D^{(o)}$ according to P8 and C9) which œ$^{(o)}$ is posited to occupy at the time t_o.

The relative metadescription $D^{(1)}/G^{(1)},$ œ$^{(1)}, V_I^{(1)}/$ constructed as specified above will be called *an intrinsic metaconceptualization of the basic (individual or probabilistic) transferred description* $D^{(o)}/G^{(o)},$ œ$^{(o)}, V^{(o)}/$ and it will be also assigned the alternative more specific symbol $D_I^{(1)}/[D^{(o)}, V_I^{(1)}]$.

Comment. We speak of "an" (not "the") intrinsic metaconceptualization of $D^{(o)}$, because in general many different intrinsizing metaviews can be constructed, and each one of these yields a corresponding and possibly specific intrinsic metaconceptualization.

An intrinsic metaconceptualization of a basic transferred description $D^{(o)}$ realizes a *retro-active* localizing projection of the scattered form of $D^{(o)}$, onto a connected and instantaneous spacetime domain $[\partial_{\mathrm{r}} t_o]$. The uniqueness of the temporal qualification t_o, even though it is retro-active, suffices now for permitting to posit, starting from it, an intrinsic time-order that is hidden to observation. This permits now to assign a law of intrinsic evolution to what has been labelled œ$^{(o)}$, underlying any evolution of the observable transferred description $D^{(o)}$. As for the transferred description $D^{(o)}$, it can now finally be *explained*. The basic object-entity œ$^{(o)}$ can now be conceived to have "possessed" at the time t_o—on the connected spatial domain ∂_{r}—the features assigned to it by the intrinsic metaconceptualization $D_I^{(1)}/[D^{(o)}, V_I^{(1)}]$. These, one can now think, were *own* features of œ$^{(o)}$, separated from those of any measurement device, independent of them, but features which $D^{(o)}$ has been able to transpose into observable manifestations, only by disorganising the form of intrinsic gk-spacetime aspect-values constituted by them. The scattered form of spacetime-gk-values involved by $D^{(o)}$ can now be thought of as the result of a bursting and change of the initially integrated intrinsic features of œ$^{(o)}$ itself. A bursting produced by the mutual incompatibility of certain aspect-views $V_g^{(o)}$ from the transfer-view $V^{(o)}$ which has obliged us to perform a set of *different* successions $[G^{(o)}.V_g^{(o)}]$, $V_g^{(o)} \in V^{(o)}$ in order to obtain the global transferred description $D^{(o)}$ (according to D14.1 at least two such incompatible aspect-views $V_g^{(o)}$ are necessary in order to characterize œ$^{(o)}$).

In short, by the assumptions from D.19.1 the basic object-entity œ$^{(o)}$ has acquired the specification of an *own* form of gk-spacetime aspect-values, and the process of emergence of the basic, transferred description $D^{(o)}$ has been *causalized*: the categories of space, time and form have been restored for œ$^{(o)}$, so $D^{(o)}$ has now become *intelligible*. But thereby the frontier between the two strata of our conceptualizations—the primoridal stratum of basic transferred descriptions $D^{(o)}$ which draws in fragments of pure factuality, and the second stratum consisting of a web of indefinite chains of *meta*descriptions $D^{(n)}, n = 1, 2, \ldots$ (see the comment on D16)—is now crossed: from now on we find ourselves inside the unlimited depth of verbal-symbolic developments of the basic transferred descriptions.

D19.2. ***Intrinsic model of a physical basic object-entity.*** So the intrinsic metaconceptualization $D_I^{(1)}/[D^{(o)}, V_I(1)]$ constructs "explanatory" *relations* between its global meta-object-entity œ$^{(1)} \equiv [D^{(o)}+$œ$^{(o)}]$ and the basic object-entity œ$^{(o)}$ involved by $D^{(o)}$, as well as an own spacetime representation of this basic object-entity $œ^{(o)}$. Once this construction has been achieved it is possible to extract from it exclusively the representation of the basic object-entity $œ^{(o)}$, in the following way.

The set of intrinsic qualifications of the basic object-entity $œ^{(o)}$ produced by the intrinsic metaconceptualization $D_I^{(1)}/[D^{(o)}, V_I(1)]$, when considered *alone*, severed from all the other elements with which it is tied inside the intrinsizing metadescription $[D_I^{(1)}/D^{(o)}, V_I^{(1)}]$, will be called an (intrinsic) *model of* $œ^{(o)}$ and will be symbolized by $M(œ^{(o)})/[V^{(o)}, V_I^{(1)}]$ in order to remind explicitly of the non-removable relativity of this model to the pair of views $[V^{(o)}, V_I^{(1)}]$ which determined its genesis and its characters.

Comment. It is important to realize clearly that an intrinsic model $M(œ^{(o)})/[V^{(o)}, V_I^{(1)}]$ is *not* a relative description of $œ^{(o)}$ in the sense of the definitions D14.

> The intrinsizing meta-aspect-views from $V_I^{(1)}$ that produced the qualifications assigned to $œ^{(o)}$ by the intrinsic model $M(œ^{(o)})/[V^{(o)}, V_I^{(1)}]$, have examined the *meta-object-entity* $œ^{(1)} \equiv [D^{(o)} + œ^{(o)}]$, `not` *the basic object-entity* $œ^{(o)}$.

The model $M(œ^{(o)})/[V^{(o)}, V_I^{(1)}]$ occupies finally a position of full saturation and equilibrium of the meaning assigned to what had been initially labelled $œ^{(o)}$. Its genetic compatibility with the transferred description $D^{(o)}$, as represented by the intrinsizing metaconceptualization $[D_I^{(1)}/D^{(o)}, V_I^{(1)}]$, detached it from $D^{(o)}$ like a mature fruit that has been plucked from its tree. The model $M(œ^{(o)})/[V^{(o)}, V_I^{(1)}]$ superposes now to the initial purely observational basic description $D^{(o)}$, a pragmatic, economic and stable conceptual *closure*. Namely a closure consisting of *an invariant with respect to the group of transformations from one succession* $[G^{(o)}.V_g^{(o)}]$, $V_g^{(o)} \in V^{(o)}$, *which contributed to the elaboration of* $D^{(o)}$, *to any other such succession with a different aspect-view in it,* $G^{(o)}$ *being* `fixed` : the observable effects of all these different successions $[G^{(o)}.V_g^{(o)}]$, $V_g^{(o)} \in V^{(o)}$, are now all assigned one common and definite "causal" ancestor $M(œ^{(o)})/[V^{(o)}, V_I^{(1)}]$ which produces various perceptible manifestations, in a "normal" way, which means in a way that is understandable accordingly to the frame-principle P8.

When the basic transferred description $D^{(o)}$ on which the model $M(œ^{(o)})/[V^{(o)}, V_I^{(1)}]$ is founded involves exclusively the human biological sensorial apparatuses, this sort of closure emerges in an unconscious, non-mediated, genetically wired way: It *is* precisely what we believe to perceive, and *this* we automatically assign to, exclusively, the involved object-entity....*in-itself* . The stage of a transferred description $D^{(o)}$ remains unknown and unsuspected. And even when fabricated apparatuses are connected to the biological ones, if the whole apparatus thus obtained still offers a directly intelligible form of spacetime *gk*-values, this form, again, is irrepressibly felt to reveal how the perceived object-entity is *in-itself* (think of perceptions *via* a microscope or a telescope). Moreover, when, as in quantum mechanics, the observable basic transferred data do not themselves offer a directly intelligible form of spacetime *gk*-values, so if an intrinsic model $M(\text{œ}^{(o)})/[V^{(o)}, V_I^{(1)}]$ has to be explicitly constructed from these data treated as mere coding signs, still, once a model has been constructed, it usually is felt to be satisfactory and *necessary* to such a degree that its only *hypothetical, retro-active, and relative* character tends to be skipped. Implicitly and fallaciously the intrinsic models $M(\text{œ}^{(o)})/[V^{(o)}, V_I^{(1)}]$ conquer inside our minds a primary and absolute status.

> This is the fallacy that instates the irrepressible belief that physical object-entities can be *known* "such as they are in themselves".

The unavoidable dependence of any intrinsic model of $\text{œ}^{(o)}$, on *both* an initial transferred description $D^{(o)}$ that has had to be achieved first and has involved some particular transfer-view $V^{(o)}$, and a subsequent process of intrinsic metaconceptualization $D_I^{(1)}$ involving a particular intrinsizing metaview $V_I^{(1)}$, tends to be overlooked. In particular, it tends to remain unnoticed that `another` *pair* $(V^{(o)}, V_I^{(1)})$ *would have led to a* `different` *model of* $\text{œ}^{(o)}$.

> These occultations mark all the classical descriptions, in physics, in mathematics, etc., as well as in the current thinking expressed by the current language: they are the opaque fictitious platform that floats above the physical factuality and on which is erected the classical concept of objectivity. The roots which insert the conceptualizations into physical factuality, with the relativities involved by them, are hidden beneath this fictitious platform.

Starting from the transferred data that are available for it and on which it takes support without trying to express them, human mind always rushes as rapidly and as directly as it can toward a representation of the involved object-entity by an intrinsic spacetime-*gk*-values model. As soon as such a representation has been attained, it is spontaneously felt to be "true" in a primary, certain and absolute way, without reference to the initial transferred data on which it is founded and forgetting that it is just an economic, hypothetical, retro-actively imagined *construct.* While the initial transferred data, even though they are the sole certainties, because of their dispersed unintelligible phenomenal appearance, are implicitly and irrepressibly perceived as nothing more than "subjective" tools for finding access to the "objective truth": *a fallacious, illusive inversion.* We systematically commit what Firth [20] called "the fallacy of conceptual retrojection". Simplicity, invariance, and what we tend to call "truth" and "objectivity", have coalesced in a knot imprinted upon our minds by ancestral processes which, by implicit pragmatic causalisations, optimizes the efficiency of our behaviour, but blocks and botches the reflexive knowledge of our fundamental epistemological functioning. The interpretation as ontological assignments, of the results of our instinctive human adaptive constructs involving the frame-principle, is one of the worst and most stubborn pathologies of thought.

But in quantum mechanics this process has hit an obstacle. Up to this very day a type of intrinsic model $M(œ^{(o)})/[V^{(o)}, V_I^{(1)}]$ fitting satisfactorily the quantum mechanical transferred descriptions of what is called a microstate, has not yet been found. So it has been necessary to stop the attention upon these transferred descriptions themselves such as they have emerged, and to embody these transferred descriptions in mathematical expressions able to yield, if not understanding, at least numerical predictions. And then, like a tireless insect when its instinctive constructive actions are hindered, human mind came back again and again upon these quantum mechanical transferred descriptions that resist modelization. And so it has become possible to discern more and more explicitly their specificity, which inside MRC has been redefined in quite general terms and has been called a "basic transferred" character. In this way we finally become aware of the unavoidable necessity of a quite universal first phase of conceptualization in terms of basic transferred descriptions.

> Inside MRC the distinction between illusory ontological assertions concerning an absolute way in which $œ^{(o)}$ "really-is-in-itself", and relative methodological intrinsic models of $œ^{(o)}$, is quite radical, elaborate and clear cut. And *the genetic order of the descriptional steps is re-constructed correctly and is fully displayed.*

Under these conditions the irreplaceable pragmatic and heuristic power of intrinsic models can be put to work without triggering any more insoluble philosophical pseudo-problems. Correlatively, the vain and exhausting battle between positivists and defenders of modelization, evaporates. The transferred descriptions are the unavoidable first stage of our processes of conceptualization, while the intrinsic metaconceptualizations of the initial transferred descriptions and the relative models extracted from these are a stabilising subsequent stage which, if realized, brings us down onto a (local, psychological and provisional) minimum of our potential of conceptualization.

There is no choice to be made. There is just an unavoidable *order of elaboration* to be observed, in a normed way, or to be recognized when it occurs implicitly.

D19.3. ***Minimal intrinsic metaconceptualization. Minimal intrinsic model.*** Consider a basic transferred description $D^{(o)}$ of a physical basic object-entity. The effect labelled $œ^{(o)}$ of the basic operation $G^{(o)}$ of generation of an object-entity can *always* be trivially metaconstructed accordingly to D19.1 so as to be conceivable as:

> A bulk of potentialities of future observable manifestations, determined by $G^{(o)}$ on a finite space-domain ∂^3_{r}, at the time t_o when $G^{(o)}$ comes to an end, each one of these potentialities being relative to an aspect-view $V_g^{(o)}$ from the basic view $V^{(o)}$ operating in $D^{(o)}$.

For this it suffices to posit in D19.1 the *minimal intrinsizing view corresponding to* $V^{(o)}$—let us denote it $[min.V_I^{(1)}/V^{(o)}]$—defined as follows. For each basic aspect-view $V_g^{(o)}$ from the basic view $V^{(o)}$, $[min.V_I^{(1)}]$ contains a corresponding intrinsizing minimal meta-aspect-view $[min.V_{Ig}^{(1)}]$ possessing a unique *minimal meta-aspect-value* denoted Ig_{min} that consists of the intrinsic *potentiality*, assigned to what has been labelled $œ^{(o)}$, to produce at a time $t_g > t_o$, any one among the transferred observable aspect-values gk of the basic aspect-view $V_g^{(o)}$, iff $œ^{(o)}$ is subjected at t_o to an $V_g^{(o)}$-examination ($t_g - t_o$: the duration of a $V_g^{(o)}$-examination, characteristic of the considered aspect g) (I recall that "intrinsic" means here assigned to $œ^{(o)}$ itself as an own feature, the word having been chosen in order to distinguish from the meaning of the philosophical term "in itself").

The trivial realization of the definition D19.1 specified above will be called the *minimal intrinsic metaconceptualization of the basic transferred description* $D^{(o)}/G^{(o)},œ^{(o)},V^{(o)}/$ and it will be denoted by $[min.D_I^{(1)}/D^{(o)}]$ (the relativity to the acting intrinsizing view $V_I^{(1)}$ is now included in the definition of the minimal intrinsizing view $[min.V_I(^1)/V^{(o)}]$, so it is absorbed in the proper "min."). The intrinsic model of $œ^{(o)}$ extracted from $[min.D_I^{(1)}/D^{(o)}]$ will be called the *minimal intrinsic model of* $œ^{(o)}$ and will be denoted by $[min.M(œ^{(o)}/V^{(o)}]$.

Comment. The following consequence of the final posit from D14.3.1 is quite worth being noticed. Any basic view $V^{(o)}$ that involves two mutually incompatible basic aspect-views $V_{g1}^{(o)}$ and $V_{g2}^{(o)} \neq V_{g1}^{(o)}$ entails a minimal intrinsic model $[min.M(œ^{(o)}/V^{(o)}]$ which now characterizes $œ^{(o)}$ *conceptually* (by predication). It yields a conceptual definition of $œ^{(o)}$ that can now be added to the purely factual definition of $œ^{(o)}$ insured initially by the operation $G^{(o)}$ alone (whereby $œ^{(o)}$ still remained outside knowledge) and to the subsequent purely observational description of $œ^{(o)}$ offered by the basic description $D^{(o)}$ (whereby $œ^{(o)}$, though characterized observationally, nevertheless was still devoid of an own conceptual representation). MRC brings forth *degrees* of characterization of a basic object-entity $œ^{(o)}$, which compose the complexifying sequence [purely factual→ purely observational→ conceptual]. From that stage on, chains of *non* minimal intrinsic metaconceptualizations can indefinitely increase the degree of conceptual characterization of $œ^{(o)}$. This illustrates the reflexivity of the method and its unlimited character.

As any intrinsic metaconceptualization and any intrinsic model, the trivial minimal models also may be perceived as "opportunistic" constructs where what is actually observed is posited to stem from an *a posteriori* imagined *ad hoc* explanatory potentiality. This however does not in the least diminish the pragmatic importance of the fact that a minimal model of what is labelled $œ^{(o)}$ is a representation that permits a most natural, easy insertion of $œ^{(o)}$ into the conceptualization. Moreover it is always and automatically realizable. It is however useful to remember again and again that inside MRC this sort of representation is accepted as just an unavoidable strategic step that must be carefully distinguished from an ontological credo: nothing whatever is naively asserted concerning the impossible question of how the basic object-entity $œ^{(o)}$ "really-is-in-itself". It is only stated how this object-entity can be most simply conceived in order for us to become able to speak and think of it in structured, consistent, fluent terms.

4.2.6. Final comment on the realism involved in MRC

> " ... Thus the aim of the book is to draw a limit to thought – not to thought, but to the expression of thoughts: for in order to be able to draw a limit to thought, we should have to find both sides of the limit thinkable (i.e., we should have to be able to think what cannot be thought)."
> —Ludwig Wittgenstein, in the Preface to his *Tractatus*.

The concept of *minimal* realism possesses, I think, an essential philosophical importance. Imagine an abstract surface on which are displayed all the grammatically correct structures of words that human mind can compose about the physical reality. On this surface, the concept of minimal realism is delimited by a boundary which coincides strictly with the boundary that separates the domain of communicable *knowledges*, from the domain inside which can be found only expressions that are grammatically correct, so communicable, but *devoid of reference*: This boundary defines the extreme limit that expressions of communicable knowledge can reach. The communicable knowledges cannot transcend this frontier. They can just advance toward it and eventually hit it by this or that basic transferred description which acts like a small squad carrying a local net of pre-conceptualization inside which it captures a small load of as yet unknown physical factuality which it hoists up on the very first level of speakable, communicable knowledge. But thereby the progression of the squad from inside the zone of knowledge, toward the physical reality, is stopped. The squad is reflected *back* like an elastic ball toward the inside of the realm of relative *descriptions*, where it delivers its load which, from that moment on, can indefinitely be elaborated along innumerable branches of complexification by intrinsic metaconceptualizations and/or by extraction of corresonding intrinsic models. But each one of these complexifying elaborations introduces new descriptional relativities which thicken the screen between physical reality in-itself, and our mind's representations of it, they thicken this screen so as to *improve intelligibility* and thereby the capacity to think and to act. Such is the paradoxical relation between physical reality and mind.

It is crucial to become aware, intensely, of the surreptitious advent of this *inversion* in our direction of conceptualization, of these unavoidable rebounds in the opposite direction each time that the extreme frontier of the domain of communicable knowledge is hit by a basic description. If not, we remain imprisoned in the inertial illusion that by modelizing more and more we *approach* more and more the knowledge of how the physical reality "is-in-itself". The grammatically correct associations of words which express this illusion are founded upon a self-contradicting concept of reality-in-itself, namely the concept of a *qualifiable* reality-in-itself. *Whereas reality-in-itself—by definition— is precisely what cannot be qualified more than by its mere quali-fiability.* By these words, "in-itself", what is pointed toward deliberately is *nothing* more than a *posited* existence, *posited* also to be qualifi-*able* but to be devoid of any other more specifying qualifi-*cation.*

> *This is not a matter of* fact; *it is a matter of organization of language-and-concepts.*

The words "description" and "physical reality in itself" have to be somehow endowed with a definition (even if only implicitly). And when this is done in a coherent way what is called description is opposed by construction to what is called "physical reality in itself".

One might perhaps believe, for instance, that it is possible to gain one more inch by specifying that the reality-in-itself is "such" that the qualifications which it admits from our part are precisely those which are elaborated by our senses and our investigations. But when we focus attention upon this supposedly supplementary specification, trying to capture an element of positive novelty added by it to the minimal realist postulate, we find only nothingness. We find ourselves placed on exactly the same content of information as before. Any qualification added to that of mere qualifiable but unqualified existence, even the most feeble one, the most vague, is either idempotent, or generates contradiction. Any attempt to superpose some nuance expressible in terms of approximations or of asymptotic apprehension of *how* the physical reality is in itself, would only manifest a misunderstanding of the nature of what is here involved, namely an optimized organization of concepts-and-words. One can reasonably try to fight against a physical circumstance, even if it is a "physical law", trying to master it so as to realize some technical aim. But trying to fight against the limitations entailed by a conceptual-linguistic organization, manifests a confusion concerning essence: what meaning would that have, for instance, to fight against the limitations imposed by a previously constructed formal system, say arithmetic, which one does not criticize and inside which one has placed oneself? "The-way-of-existing-of-reality-in-itself" is a self-contradicting notion stemming from a confusion between empirical circumstances and conceptual organizations of which on the other hand one makes use.

In his Conference on Ethics, Wittgenstein said (concerning the more or less similar confusion between value and truth): "It is perfectly, absolutely hopeless to thus bump our forehead against the walls of our cage" (my own retro-translation in English, from French translation). One can apply the same assertion to the confusion between an impossible ontological quest, and an organization of language-and-concepts constructed by man. This confusion entails chimerical aims and fictitious problems, like in the quantum mechanical orthodoxy the arbitrary positivistic interdiction of intrinsic metaconceptualizations and intrinsic models *because these are confounded with impossible qualifications of reality-in-itself*. This mythic fauna that spouts from the bursting of an inertially oriented impetus to understand more, against the barrier placed by thought itself between communicable knowledge, and a just posited and denominated rest, must be exorcised.

So the minimal realism involved by MRC has a composite logical status. While the feature of minimality follows "deductively" inside the method (π18), the main term, realism itself, is just a posit, the postulate P3. It is a declaration of metaphysical belief, wholly subjective. Any question of truth or objectivity is meaningless concerning it. But this metaphysical belief plays a fundamental role for MRC: It seats the method on a unifying ground. It asserts that beneath the endless proliferation of branching relativities which mark the contents of descriptions, there exists a substratum of non referred absolute, wherefrom the relativities emerge together with the conceptualizations. I say "beneath" in order to stress that the thesis of realism draws *out* of the domain of language and descriptions. By the mysterious powers of self-transcendence of language, this thesis acts like a verbal directional indicator, pointed from inside the volume of the expressed, but which points toward an existence from outside this volume. It grasps the attention, displaces it, and installs it at the very core of the non expressible. There, inside this background of unverbalisable which it succeeds to designate, the realist thesis fixes the ends of the threads with the help of which the basic transferred descriptions web to one another the two regions that stretch out on the two sides of the ghostly but insuperable wall between what is by construction devoid of legal communicable expression, and the legally formulated and communicable. In spite of the fact that we cannot "find both sides of the limit thinkable". This is the fundamental, the huge epistemological innovation hidden in the quantum mechanical formalism, which inside MRC is explicated, generalized, and organized in detail by radical and systematic relativizations. Whereby *all* the false absolutes are suppressed, not only those which vitiate esthetics, ethics and metaphysics; for *everywhere* thought is invested by hosts of false absolutes that generate pathological tissues of illusory problems and paradoxes that blur out the sound limit between the thinkable and mere nonsense.

It might seem that this background of non referred, because it is absolute, is incompatible with the method of relativized conceptualization. But, and it is important to stress this, MRC by no means banishes *any* absolute. It banishes exclusively the false absolutes, those which hide descriptional relativities of which the presence can be identified, and which, if ignored, can generate illusory problems. But it is clear that when one constructs, it is unavoidable to posit certain absolutes. All the definitions from MRC, principles, etc., as *such*, have nothing relative about them. They are absolutes of the method, by the help of which the descriptional relativities are defined. And the existence of a physical reality posited in P3 is also an absolute of the method. This concept is introduced as just the absolute reference without which thought would get lost in an unexplained profusion of diversity; an absolute reference which unifies in one closed coherent whole all the indefinitely evolving descriptional relativities defined by the method.

I confess that the beauty which, to my eyes, emanates from this unification, appears to me irrepressibly as a sign of pertinence. Man and "reality" form a whole, and the feeling of beauty that can emerge in a human mind, intimately tied with coherence, has for me the significance of an announcement that certain slopes of the real have been embodied without having been violated. Whatever the unimaginable designatum of the succession of words which I just aligned, I want to align them, for we must somehow speak in order to communicate—paradoxically and in spite of all—concerning the unspeakable.

4.2.7. Global remarks on the nucleus of MRC

MRC is:

* ***Explicitly founded upon the functioning of human mind, with its cognitive aims.*** The choices of the epistemic referentials that generate the relativized descriptions, stem from the consciousness functioning of the acting observer-conceptor. Each such choice expresses a curiosity, a descriptional aim of this consciousness functioning. The descriptional aims expressed by the successive choices of an epistemic referential, inside a chain of conceptualization, mirror the evolution of the descriptional aims of the acting consciousness functioning, and thereby they determine the "direction of conceptualization", step by step. Inside MRC, in its present stage at least, the descriptional aims do not follow from methodological prescriptions. This means the following:

  No AI-machine could, by applying MRC, work like a human being, without being directed by a human being. But an AI-machine endowed with an "MRC-program" (if this were possible) and drawn by a man, *would work exactly like* that *man.*

  This specifies the difference between AI and MRC as well as the particularity of an "MRC-program".

* ***Explicitly rooted in pure factuality***, which entails the possibility of a systematic and constructed distinction between potentiality of an infinity of processes of actualization of relative observable manifestations, and this or that actualized observable manifestation (cf. the concept of "genset" in Sec. 5.2.2). *Thereby it brings in the modal dimension potential-actualization-actualized.*

* ***Radically relativizing.*** The whole approach bears the seal of the relative mutual existence of object-entities and views (or, equivalently, of generators of object-entity, and views) and of the relativities of descriptions to the triads G,$œ_G$,V.

* ***Methodological, normative, legalizing.*** MRC is not an attempt at describing the natural processes of conceptualization. Though data (introspective, linguistic, etc.) concerning these natural processes are strongly taken into account, nevertheless MRC recognizes the *impossibility* of a "purely" descriptive account on the processes of description. So, deliberately, it takes distance with respect to such an aim, by constructing definitions and principles conceived in order to optimize

the processes of conceptualization in compatibility with definite goals, namely the *a priori* elimination of any false absolutization, reflexivity, construction of a conceptual structure with respect to which it be possible to "localize" any other descriptional structure, natural or not, etc. *Thereby MRC is formalized.* Not yet mathematically and quantitatively formalized, like a physical theory, but already formalized, qualitatively formalized.

* ***Finitistic, cellular, local.*** The fact that the construction of knowledge requires parcellings, steps, is taken into account quite fundamentally throughout MRC, *via* the principle of separation P15 and the concept D16 of relative metadescription.
* ***Globally unlimited.*** Though everywhere there are strict *local* delimitations of the descriptional quest, which withstand any gliding into relativism, globally nowhere a boundary is pre-imposed: the finalized finitism of MRC generates infinities.
* ***Hierarchical.*** MRC generates *hierarchical* ***trajectories*** of conceptualization, in contradistinction to the theory of logical types, or that of levels of language, which introduce extended hierarchical *strata.*
* ***Directional and reflexive***, endowed with a capacity for an *a priori-a posteriori* double way progression. Before starting a given descriptional cell, a free choice of the direction of conceptualization desired by the observer-conceptor is expressed in a corresponding choice of an epistemic referential. Later the results of this choice can be rejected or kept and developed, on the basis of explicitly defined criteria.

The various features enumerated above are not exhaustive. Nor, by no means, are they mutually independent. Quite on the contrary, they all stem from one core-structure that induces an innumerable host of connections between these features. This core-structure is dominated by the systematically recurrent role of the consciousness-functioning which introduces the epistemic referentials. And along the whole hierarchy of distinct descriptional cells of increasing order from each chain of conceptualization from the web of such chains, the same fundamental MRC-requirements for a relativized normed conceptualization manifest themselves with a sort of fractality: Each time that an epistemic referential has been chosen—no matter on which level of conceptualization—the generator of object-entity, the object-entity and the view from it entail non removable descriptional relativities to them.

4.2.8. On the conceptual status of MRC.

To what class of conceptual beings does MRC belong? *Any* representation of "natural facts" is more or less normative, never purely descriptional as the classical myth of objectivity involves.

> In the case of MRC the explicitly and resolutely *methodological* character is a major feature of the approach. Any confusion between ontological assertions or implications, and methodological constructs, is most carefully avoided.

Nevertheless MRC can also be regarded as:

> An attempt at a finitistic representation of the natural processes of generation of meaning where both relativism and false absolutizations are excluded *ab initio* by explicit rooting into pure factuality and by deliberate systematic relativizations.

The fact that throughout the process of constructing MRC one acts "logically", is neither a circularity, nor does it involve that MRC is reducible to a logic. It only illustrates the general reflexive, (*a priori*)-(*a posteriori*) character of any approach and in particular of this one: *a priori* the logical criteria are supposed to be fulfilled and they are utilized implicitly [21], but later, at a convenient level of development of the approach, the logical criteria—as it will be shown in 5.1— become *a posteriori* explicitly expressible in MRC-terms. (This sort of inner evolution partakes of the general reflexive character of MRC that has permitted to admit *a priori* the possibility of any pairing (G, V) and to introduce only *a posteriori* criteria concerning the relevance of a given pairing (G, V): first became expressible the criterion of mutual existence D7, and then the subsequent criterion of stability involved in the definition D14.1).

So probably the best characterization is as follows.

> MRC is a strongly normative representation of the processes of conceptualization, of which the major specificities are: the place explicitly reserved to the consciousness functioning; the radical descriptional relativizations; and the fact that it explicates the structure of the very first step in the construction of objectivity, in the course of which intakes of a-[conceptual-linguistic] fragments of pure factuality adduce into language and thought the hard core of scientific objectivity.

4.3. The Second Stage: an Ideographical Symbolization of MRC

In all the expositions of MRC that preceded the present one, I included in a presentation made in usual language, an ideographic symbolization which—without being neither a formalization *stricto sensu* nor a mathematical representation—permits certain suggestive and economic expressions. In this work I present it simplified and separately. In this way the symbolizations are made available while the drawbacks as well as the advantages appear clearly:

- *A consciousness functioning* CF is represented by the sign $\square\!\!\rightarrow$ suggesting the whirling place from D1 that acts on both the Exterior Universe and the Interior Universe where it belongs, and in particular also on itself.
- *Reality* is again symbolized by the letter R.
- *A generator G of object-entity* will be represented by the sign Δ and will be re-named a delimitator of object-entity, in order to stress that, whatever the nature of G, the final result is a delimitation, out of R, of a corresponding object-entity. Thereby however one *looses* the accent placed by the term "generator" upon a (possibly) *radically creative* character of an operation of object-entity generation. Then:
- The "place" from R where Δ works will be denoted R_Δ.
- The object-entity produced by Δ will be denoted by $œ_\Delta$.
- *The process of delimitation* by Δ, of an object-entity $œ_\Delta$, will be represented indifferently by

$$\Delta R_\Delta \Rightarrow œ_\Delta, \qquad \text{or} \qquad œ_\Delta \Leftarrow \Delta R_\Delta,$$

where the arrows do *not* have a logical meaning and cannot be considered separately, they are cemented into the global symbolizations which read respectively: "the delimitator Δ, acting on R at the place R_Δ, produces the object-entity $œ_\Delta$", and "the object-entity $œ_\Delta$ produced by the delimitator Δ that acts on R at the place R_Δ". Notice that the introduction of these symbolization permits us to *distinguish* between:

* Δ: an epistemic operator (in the sense of usual language, *not of mathematics*);
* $\Delta R_\Delta \Rightarrow œ_\Delta$: a *process*, that mentions its beginning and its result;
* $œ_\Delta \Leftarrow \Delta R_\Delta$: an explicit specification of an object-entity *via the process that produced it*, which permits to specify an unobservable object-entity, by the way of producing it.

Thereby the expressivity concerning this zone from MRC is considerably increased.

- An *aspect-view* will be symbolized by the same sign V_g as before;

- The *operation of examination of* $œ_\Delta$ *by* V_g will be represented by $V_g œ_\Delta$.

Notice that the introduction of these symbols permits us to *distinguish* between:

* the epistemic operat*or* V_g (in the sense of usual language, not of mathematics) and
* the operat*ion* of examination $V_g œ_\Delta$,

which again is an increase of expressivity.

- A *view* will be symbolized as before by V.
- The global *operation of examination of* $œ_\Delta$, *by* V (achieved accordingly to $\pi 11$), will be represented by $V œ_\Delta$.

The remarks concerning V_g hold also concerning V.

- An *epistemic* referential is represented by (Δ, V).
- The representation of an observer-conceptor $[CF, (G, V)]$ becomes $[\sqsupset, (\Delta, V)]$.
- *The mutual inexistence* between an object-entity $œ_\Delta$ and a view V will be symbolized by

$$\nexists œ_\Delta / V \quad \text{or} \quad \nexists V / œ_\Delta,$$

  which reads, respectively, "the object-entity $œ_\Delta$ does not exist with respect to the view V", "the view V does not exist with respect to the object-entity $œ_\Delta$".
- The *mutual existence* between an object-entity $œ_\Delta$ and a view V will be represented by

$$\exists œ_\Delta / V \quad \text{or} \quad \exists V / œ_\Delta,$$

  which reads "the object-entity $œ_\Delta$ does exist with respect to the view V", "the view V does exist with respect to the object-entity $œ_\Delta$". (All these symbolizations can also be used, in particular, with the symbol of an aspect-view V_g instead of V, which changes the meaning correspondingly).
- A *spacetime view* is represented as before by V_{ET}.
- The *frame-principle* can be symbolized in the following way:

$$[\exists œ_\Delta / V_g] \rightarrow [\exists V_{ET} : \exists œ_\Delta / (V_{ET} \cup V_g)]$$

$$[\nexists œ_\Delta / V_{ET}], \quad \forall V_{ET}, \forall œ_\Delta.$$

  Here the arrow, quite independently of any connotation suggesting *formal* logic, reads "entails that" (in the sense of natural logic); $\exists$ and

$\nexists$ —outside any *formal* system, just in the sense of usual language or of "natural logic"—read, respectively, "there exists" and "there does not exist"; V_g denotes a *physical* aspect-view; $V_{ET} \cup V_g$ considered as a one-block symbol, reads "the view formed with a spacetime view V_{ET} and another *physical* aspect-view V_g". The global reading of this symbolic picture is the verbal formulation of P8.

- The symbol of a *relative description* $D/G,œ_G,V/$ becomes $D/\Delta,œ_\Delta,V/$, and the symbol for a *basic relative description* $D^{(o)}/G^{(o)},œ_G^{(o)},V^{(o)}/$ becomes $D^{(o)}/\Delta^{(o)},œ_\Delta^{(o)},V^{(o)}/$; a relative metadescription of order n, $D^{(n)}/G^{(n)},œ_G^{(n)},V^{(n)}/$, $n = 0, 1, 2, \ldots$, is symbolized by $D^{(n)}/\Delta^{(n)}, œ_\Delta^{(n)}, V^{(n)}/$.

Together, these symbolizations constitute *the ideographic representation* $[\text{⊏}^{\to}_{\downarrow}, \Delta, œ_\Delta, V, (D^{(n)}, n = 0, 1, ...)]$ *of MRC.*

4.4. The Third Stage: a Scheme of a Mathematical Representation of MRC in Terms of the Theory of Categories

The verbal formulation of MRC conveys a methodology by which the activity of constructing knowledge, though exposed with the help of words, nevertheless is extracted from *mere* language by a peculiar sort of qualitative and non-mathematical formalization. The above ideographic symbolization increases the degree of this extraction. But in order to increase this liberation still more, it seems important to achieve now a mathematization. Indeed current language inextricably incorporates hosts of surreptitious false absolutizations, of insidious obscurities, a pullulation of sonorities and implications that arouse unpredictable philosophical suspicions, refusals, passions. Furthermore, it is devoid of a clearly defined calculus. Neither a verbal extraction *via* a system of definitions, postulates and axioms, nor an ideographic symbolization associated with such an extraction, cannot sufficiently remedy to these lacunae. A transposition of the definitions and principles which form the nucleus of MRC, in *mathematical* terms, would re-produce the essence of MRC in a still more unambiguously defined *form*, a more synthetic form, more purified of uncontrolled philosophical harmonics. It would also open up the possibility of calculatory treatments.

On the other hand, the full content conveyed by the verbal presentation should be kept in mind: It certainly points best toward the whole wealth of the singular conceptual being symbolized "MRC" which, like any singular designatum, escapes any sort of language, but, if touched and grasped by the mind in prolongation of a "direction" of thought well materialized by associations of words from current language, acts as a guide and a fertiliser of the process of understanding.

4.4.1. Preliminary summary

The first target of a mathematical expression is a synthetic re-expression of the skeleton of the nucleus of MRC. So we begin by extracting this skeleton.

Imagine a consciousness functioning CF in interaction with the reality R.

- This interaction induces inside CF epistemic *aims* that generate there the conception of corresponding epistemic referentials, i.e., *a priori* non restricted pairings (G, V_g) or (G, V).
- The epistemic aim synthesized by (G, V_g) (or (G, V)) leads to a first epistemic *action*, namely of G upon the corresponding "spot" R_G from R, that generates out of R the object-entity $œ_G$.
- Consider now the definition D7 of mutual existence of G and V_g (or V). If G and V_g (or V) do not mutually exist in the sense of D7, then the *a priori* pairing (G, V_g) or (G, V) must be *a posteriori* dismissed; but if G and V_g (or V) *do* not mutually exist in the sense of D7, then the action of V_g (or V), upon $œ_G$—to be accomplished accordingly to the principles P8 and P10 and to the proposition $\pi 11$ when spacetime is involved—produces observable results.
- Concerning these results consider now the condition of *stability* from D.14.1 (cf. also $\pi 12$ and $\pi 13$). If this condition does not obtain, neither on the individual level of description nor on the probabilistic one, then the *a priori* pairing (G, V_g) or (G, V) must be *a posteriori* dismissed, even though it has resisted the first test D7 of mutual existence. But if the condition of stability does obtain either on the individual level of description or on the probabilistic one, then hierarchical chains of relative descriptions $D^{(n)}/G^{(n)}, œ_G^{(n)}, V^{(n)}/$, $n = 0, 1, 2, \ldots$ involving (G, V_g) or (G, V) can be constructed accordingly to the principle of separation P15, the concept D16 of metadescription, and to the concepts D19 of intrinsic metaconceptualization; these enrich the content of an *evolving* net of chains of relative descriptions.

This is the essence of the skeleton of MRC.

4.4.2. Mathematical framework in terms of the theory of categories

We seek now a mathematical representation of the skeleton of MRC. It is crucial to begin by making use of the weakest possible mathematical structure, i.e., which introduces a minimum of formal restrictions not stemming from MRC itself. Only in this way can it be hoped to avoid a too amputating

stract concept of a category, the semantics associated with the involved objects and arrows will be given as much importance as the syntactical constraints imposed by the theory of categories.

$Ob(C_{\mathbf{MRC}})$

The objects from the class $Ob(C_{MRC})$ are called *epistemic sites* (in short, sites) and are denoted S. A site is posited to designate a definite sort of conceptual ground—just a semantic receptacle similar to an axis in a graphic representation, or, more generally, to a multidimensional representation space—available for lodging inside it an evolving and unlimited content to which no general structure is pre-imposed (for the representation of particular MRC-problems one can pre-impose a particularly adequate structure, for instance an order). This content, however, is required to have a nature consistent with the general definition of the considered semantic receptacle (to "fit" into it, as, for instance, the red of this flower or the dark of this cat do fit into the semantic dimension labelled by the word "colour", but not into that labelled by the word "form"). The most important feature of the content of a site is that it is not required as given from the start on (though it is permitted such): in general it is conceived of as being created progressively and *indefinitely* in the same way that the "population" of a point from an axis can indefinitely be increased by adding entities located at that point.

The distinction itself between a stable pre-existing conceptual receptacle (a genus, an axis, a multidimensional conceptual space), and a corresponding sort of content of which any constituent or part can always be lodged inside this receptacle, indefinitely, at this or that definite "location" (specific difference, point), is by no means new. Quite on the contrary, more or less explicitly it underlies the whole classical organization of thought (linguistic, logical, mathematical; it was already quite explicit for Aristotle), and it includes also the basic notion of a referential. But neither classical logic nor nowadays mathematics do represent in general and explicit terms the most complete possible process of *generation* of the content of a pre-posited conceptual receptacle, as it is specified inside MRC by the generator $G^{(o)}$ or $G^{(n)}$ from, respectively, the concepts of a basic transferred description and of subsequent intrinsic metaconceptualizations and modelizations. And very often this content is tacitly supposed to somehow be entirely "given" from the start on, to somehow pre-exist all done, "out there", in a Platonic manner. Only if *ab initio* this hypostatic view is systematically replaced by a genetic one, will it be possible to mimic in the terms of the theory of categories, the fundamental MRC-concepts of basic transferred description and of intrinsic metaconceptualization. This is why here a specific definition of the concept of "site" is needed.

The sites from $Ob(C_{MRC})$ are:

- S_R, which represents formally the location of the evolving content of the reality R, as defined in D2;
- S_{CF}, which represents formally the *location* of the evolving content of the consciousness-functioning CF, as defined in D1;
- $S_{œ}$, where have to be located all the *formal representations* of the object-entities $œ_G$ defined in D4, as these emerge;
- S_D, where have to be located all the *formal representations* of the relative descriptions $D/G,œ_G,V/$ (Def. D14.1) or metadescriptions $D^{(n)}/G^{(n)}, œ^{(n)}, V^{(n)}/$, $n = 0, 1, 2, ...$ (Def. D16), as these emerge.

As already stressed, the explicit distinction between a permanent site determined by a static definition, and the (in general) evolving content located on this site, is quite essential for $Ob(C_{MRC})$. Furthermore, according to MRC it is necessary to posit explicitly that $S_R \supset [(Ob(C_{MRC})]$ (see Def. D2), which will induce *reflexive* features into the formalization [22].

In a future elaboration of particular MRC-problems, $S_{œ}$ and S_D will have to be assigned structures. $S_{œ}$ will have to become a mathematical space lodging in it an evolving content of some sort of specified mathematical beings (real or complex functions, kets, sequences of signs, etc.) generated one by one and in general independently of one another and offering a convenient representation of the considered sort of object-entities (for instance, in the particular case of the Hilbert-Dirac formulation of quantum mechanics $S_{œ}$ becomes the Hilbert space of non-referred state vectors $|\psi\rangle$). S_D will have to become another kind of mathematical space, lodging in it an evolving content of some other sort of mathematical beings, again generated one by one and in general independently of one another and representing conveniently the considered type of achieved descriptions (in the case of quantum mechanics S_D consists of the space of column-matrixes that represent any state vector in some given basis). These spaces will have to be endowed with general structures *such* that the formal behaviour of the elements from the space, if tied with *physical* object-entities $œ_G$, when combined with the other elements of the mathematization, shall permit to reflect conveniently the spacetime restrictions imposed by the principles P8 and P10, as well as the propositions $\pi 11$,$\pi 12$, $\pi 13$. Moreover the two structures posited on $S_{œ}$ and S_D will have to be connected with one another consistently from both a mathematical and a semantic point of view. In order to reflect formally this or that particular class of object-entities and/or of descriptions, further more specific structural restrictions can be added.

Ar($C_{\mathbf{MRC}}$)

Consider now the class of arrows, $Ar(C_{MRC})$. The arrows from this class will be called *epistemic arrows.*

Inside the theory of categories, given some category C, an arrow from $Ar(C)$ is currently conceived to represent an already constituted morphism that pre-exists in a Platonian manner. This sort of semantics, however, is not coherent with our previous definition of $Ob(C_{MRC})$ as containing sites with evolving content. For consistency with the definitions from MRC and with our previous definition of $Ob(C_{MRC})$, any arrow from $Ar(C_{MRC})$ will be posited to represent a *process* of which the action is unlatched inside the source-site, at a definite "content-point" which in certain cases is *itself* created by that process, as its source-point; then the process develops in time (and sometimes in spacetime) always ending by the creation at its head of a local contribution to the evolving content of the target-site. In this sense an CMRC-arrow is posited as a *local genetic arrow.*

The epistemic arrows from $Ar(C_{MRC})$ themselves are generated inside the consciousness functioning CF *by its* `free choices`, in consequence of its interactions with the contents of S_R and with itself. So:

> Though it does not belong to $Ob(C_{MRC})$, the generic concept $Ar(C_{MRC})$ can be best described by making use again of the concept of site, namely as a site bearing an evolving content of arrows.

The set of arrows $Ar(C_{MRC})$ can be split in two sub-classes of epistemic arrows, a sub-class of *primitive epistemic arrows* $PAr(C_{MRC})$, and a sub-class of *composed epistemic arrows* $CAr(C_{MRC})$.

$PAr(C_{MRC})$. The primitive epistemic arrows from $Ar(C_{MRC})$ are:

- *Data-arrows* $d\rightarrow$ denoted d, with $s(d)=S_R$ and $t(d) = S_{CF}$ (so belonging to $Hom(S_R, S_{CF})$), that represent the generation of data inside CF, by the influxes from the reality R.
- Endomorphic *aim-arrows*, of two kinds:
    * *(Object-entity-generation-aim)-arrows* $GA\rightarrow$ (in short GA) with $s(GA)=S_{CF}$ and $t(G_A)=S_{CF}$ (so belonging to $Hom(S_{CF}, S_{CF})$, that represent the process of constitution inside CF of the aim to know specifically about a somehow pre-figured sort of object-entity $œ_G$.
    * *(Qualification-aim)-arrows* or, in short,*view-aim-arrows*, of two kinds, $V_gA \rightarrow$ or $VA \rightarrow$, indistinctly short-noted VA, with $s(VA) = S_{CF}$ and $t(VA)=S_{CF}$ (so again belonging to $Hom(S_{CF}, S_{CF})$, that represent the process of constitution inside CF of the aim to qualify (some object-entity) *via* an aspect-view V_g or, respectively, a view V.

- *Operational-arrows* of two kinds:
  * *(Object-generation)-operational-arrows* or, in short, *generation-arrows* $G \rightarrow$ (in short G) that represent the epistemic operations of *effective* generation of an object-entity. By definition $s(G) = S_R$ and $t(G) = S_{œ}$, so $G \rightarrow$ belongs to $Hom(S_R, S_{œ})$.
  * *Qualification-operational-arrows* of two kinds, *aspect-view arrows* $V_g \rightarrow$ or *view-arrows* $V \rightarrow$, indistinctly called *view-arrows* (in short V), with $s(V) = S_{œ}$ and $t(V) = S_D$ (so belonging to $Hom(S_{œ}, S_D)$). The view-arrows represent the elaboration of relative descriptions by operations of qualification of an object-entity *via*, respectively, examination by an aspect-view or a view. Mind that a view-arrow $V\rightarrow$ represents *globally* all the processes of examination that establish the *one* corresponding relative description, so it has to be *constructed* from aspect-view-arrows $V_g \rightarrow$ by taking into account the proposition $\pi 11$.

[-] *Aim*-`activating`-*arrows* $Aa \rightarrow$ (in short Aa) of three kinds, that represent the passage (decided by the working consciousness functioning) from a given epistemic *aim*, to the corresponding effective epistemic *operation* :

  * *(Generation-aim)*-`activating`-*arrows* $GAa \rightarrow$ (in short GAa) with $s(GAa) = S_{CF}$ and $t(GAa) = S_R$, so $GAa \rightarrow$ belongs to $Hom(S_{CF}, S_R)$;
  * *(View-aim)*-`activating`-*arrows* $VAa \rightarrow$ (in short VAa) with $s(VAa) = S_{CF}$ and $t(VAa) = S_{œ}$, so $VAa \rightarrow$ belongs to $Hom(S_{CF}, S_{œ})$);
  * *(Descriptional-aim)*-`activating`-*arrows* $DAa \rightarrow$ (in short DAa), that just initiate *globally* the whole descriptional program involved in the choice of an epistemic referential. (An arrow $DA \rightarrow$ itself, a descriptional-aim-arrow, is a *composed* arrow and as such it will be defined below. Nevertheless the corresponding aim-`activating`-arrow $DAa \rightarrow$ is a monolithic primitive arrow with $s(DAa) = S_{CF}$ and $t(DAa) = S_{R \cap D}$, so $DAa \rightarrow$ belongs to $Hom(S_{CF}, S_{R \cap D})$ (we have $S_R \supset S_D$, so $t(DAa)$, being in S_D, is also in S_R).
- *The unit-arrows* required by the theory of categories for each site from C_{MRC} could be introduced as purely formal arrows. However it is obvious that a fully satisfactory representation of MRC inside of the theory of categories should endow each unit-arrow, with an adequate semantics. This might be possible but it might involve quite non trivial epistemological considerations. It might even lead to certain deep and rigorous explicitations concerning the reflexive features to be assigned to the sites from C_{MRC}. (For S_{CF} the role of unit-arrow could be assigned to each one of the already defined endomorphic aim-arrows, which involves a problem of choice). So, for the moment, we leave open the question of a meanigful definition of the unit arrows.

Before continuing with the sub-class of composed epistemic arrows, let us note the following. An epistemic referential (G, V) as defined in D6 can be now represented formally by the corresponding pair of operational arrows $(G \rightarrow, V \rightarrow)$. In order to represent formally the *a priori* possibility of any MRC-pairing (G, V), inside CMRC any pairing $(G \rightarrow, V \rightarrow)$ will be permitted *a priori*. An observer-conceptor as defined in D6 can then be represented inside CMRC by the association $[CF, (G \rightarrow, V \rightarrow)]$ between the evolving content CF of the site S_{CF} and the representation of an epistemic referential.

$CAr(C_{MRC})$. The composed epistemic arrows from $Ar(C_{MRC})$ are:

- Given two aim-arrows $GA \rightarrow$ and $VA \rightarrow$, whatever they be, they are always composable in any order, since $s(GA \rightarrow) = t(QA \rightarrow) = s(GA \rightarrow) = t(VA \rightarrow) = S_{CF}$. However the MRC-semantics requires to take into consideration only the order $GA \rightarrow \circ VA \rightarrow$. So, denoting the result by $DA \rightarrow$ (in short DA), we have DA with $s(DA) = t(DA) = S_{CF}$. We call it a *descriptional-aim-arrow* and we write

$$DA = DA \rightarrow \; = GA \rightarrow \circ VA \rightarrow$$

This descriptional-aim-arrow $DA \rightarrow = GA \rightarrow \circ VA \rightarrow$, like a fragment of DNA, holds in it, still non-realized so still a-temporal, the whole descriptional program corresponding to the pairing $(GA \rightarrow, VA \rightarrow)$, whether realizable or not.[15]

Given a pair of arrows $d \rightarrow, DA \rightarrow$, the composition, in this order, is always possible formally. But it is MRC-significant iff $DA \rightarrow$ corresponds to the content of data supposed to be carried by $d \rightarrow$ (this, being a fundamentally semantic matter, cannot be established formally). The composition will be taken into account only when it is meaningful. We then call it an *induction arrow*, we denote it $ind.DA \rightarrow$ (in short

[15] The *selection*—among all the syntactical possibilities offered by a formalism—of exclusively those that translate the *semantic* features to be represented, is unavoidable when an interpretation of a formal system is built. In particular the procedure is quite current throughout mathematical physics. (For instance, in a quantum mechanical problem of square potentials, the general solution of the differential equation of the problem offers exhaustively all the possible formal terms; among these, those which have no physical correspondent in the data of the problem are dismissed, while the conserved expressions are specified as required by these data (limiting or initial conditions, etc.), *which cannot follow syntactically*. Another example can be found in Schrödinger's solution of the problem of a one dimensional harmonic oscillator where subtle and very constructed physical arguments are introduced in order to identify restrictions that are *not* imposed mathematically; etc.).

ind.DA), and we write

$$ind.DA\!\rightarrow\, = d\!\rightarrow\! \circ DA\!\rightarrow$$

$s(ind.DA) = S_R$ and $t(ind.DA) = S_{CF}$, which represents formally an induction of a descriptional aim from R into CF.

- Consider the representation $(G\rightarrow, V\rightarrow)$ of an epistemic referential. Formally the two operational arrows are always composable in this order. MRC also requires, for methodological reasons, to systematically admit the composability *a priori*, but to exclude it *a posteriori* if the condition D7 of mutual existence or the condition of individual or probabilistic stability involved by D14, appears not to obtain. So inside C_{MRC} we proceed as follows. First, systematically and tentatively, we do form the composition between $G\rightarrow$ and $V\rightarrow$, in this order, naming it a descriptional arrow $D\rightarrow$ (in short, D). Thus we write

$$D\!\rightarrow\, = G\!\rightarrow\! \circ V\!\rightarrow,$$

with $s(D) = S_R$ and $t(D) = S_D$(so belonging to $Hom(S_R, S_D)$. But if later it is found that no description arises because D7 or the condition of stability from D14 fails (which, being fundamentally a matter of semantics, cannot follow syntactically), then we cancel *a posteriori* the previously formed arrow $G\rightarrow \circ V\rightarrow$ and the corresponding epistemic referential $(G\rightarrow, V\rightarrow)$. Any epistemic referential considered in what follows is supposed to have been found to satisfy both D7 and D14. The composed arrow $D\!\rightarrow\, = [G\!\rightarrow\! \circ V\!\rightarrow]$ formed with such a "good" epistemic referential is the operational nucleus of C_{MRC}. It has to be constructed so as to yield a satisfactory formal expression of all the conditions relevant to the considered description, as required by D14 (so P10 and P11) as well as by (according to the case) P15, D16, D19:

> In consequence of P10 and π11, $D\rightarrow$ involves an (in general) non-commuting algebraic structure imposed upon the set of arrows $V\rightarrow$.

- Given an epistemic referential $(G\rightarrow, V\rightarrow)$, the following corresponding composition, called a *complete-description-arrow* (in short, CD) is always possible and significant:

$$\begin{aligned} CD\!\rightarrow\, &= CD = d\!\rightarrow\! \circ DA\!\rightarrow\! \circ DAa\!\rightarrow\! \circ G\!\rightarrow\! \circ V\!\rightarrow \\ &= indDA\!\rightarrow\! \circ DAa\!\rightarrow\! \circ G\!\rightarrow\! \circ V\!\rightarrow, \end{aligned}$$

with $s(CD) = S_R$ and $t(CD) = S_D$ (thus belonging to $Hom(S_R, S_D)$, which reads: Data from the reality R induce a descriptional aim into the consciousness functioning, this is activated, and so first an object-entity is generated out of R (which brings on the site of object-entities) and then this object-entity is qualified, whereby a description is obtained (which brings on the site of descriptions). The explicit "sites-trajectory" of a complete description CD is

$$S_R - S_{CF} - S_{CF} - S_{CF} - S_R - S_o - S_D.$$

The triplet $S_{CF} - S_{CF} - S_{CF}$ expresses satisfactorily the dominant role of the consciousness functioning in a descriptional process.

- Other compositions also are permitted by the introduced definitions (for instance $GAa \rightarrow oG \rightarrow$, $VAa \rightarrow oV \rightarrow$, etc.). But it seems not necessary to examine them exhaustively. Notice that the MRC-definition D2 of reality requires to extend now the previous assumption $S_R \supset [(Ob(C_{MRC})]$ by positing explicitly $S_R \supset [(Ob(C_{MRC}) + Ar(C_{MRC})]$.

***The axioms* $\mathbf{C}_1$ *and* $\mathbf{C}_2$.** They seem to raise no problems.

Representation of the evolving contents of the* $\mathbf{C}_{MRC}$*-sites. The theory of categories does not specify a general modality for expressing individualizations *inside* an object from $Ob(C)$, as being the source or the target of an arrow tied with that object. While MRC involves such individualizations quite essentially. So we construct the necessary individualizations as follows.

We consider only the operational arrows $G \rightarrow$ and $V_g \rightarrow$ that form the hard core of C_{MRC}. This will suffice.

Each arrow $G \rightarrow$ can be labelled by a pair of indexes $(R_G, œ_G)$ defining respectively its local start inside S_R (by the "spot" R_G where G has to be applied (D4)) and the element $œ_G$ from the evolving set $\{œ_G\}$ that constitutes the new contribution to the content of $S_{œ}$ by the creation of which the considered $G \rightarrow$ arrow ends. So for each definite arrow $G \rightarrow$ we shall write $(R_G, œ_G) \rightarrow$, which distinguishes it from any other arrow $G \rightarrow$. Thereby the set $\{(RG, œ_G) \rightarrow\}$ associated to the generation arrows $G \rightarrow$, itself also an evolving set, is now *connected with the evolving inner contents of the two sites* S_R *and* $S_{œ}$ represented, respectively, by the evolving sets $\{R_G\}$ and $\{œ_G\}$. This connection can be then organized more by putting mutually compatible structures on the sets $\{R_G\}$, $\{œ_G\}$ and $\{(R_G, œ_G) \rightarrow\}$ (physical operations of object-entity generation are subject to the frame-principle P8, which requires a convenient extension of the principle P10 of mutual exclusion, to operations of object-entity generation also).

Mutuatis mutandis, one can connect in a similar way each definite processual arrow $V_g \rightarrow$, with a "pair" of indexes $(œ_G, \{gk\})$, $k = 1, 2, ...$, by re-writing $(œ_G, \{gk\}) \rightarrow$, $k = 1, 2, \ldots$,where k takes on a unique value if the attempted descriptional process reveals an individual stability, or a whole set of different values if it reveals a probabilistic stability ((D5.1),

$\pi 12$, $\pi 13$, D14). In $(œ_G, \{gk\})$ the index $œ_G$ defines the element from the discrete evolving content of the source-site $S_{œ}$ where $(œ_G, \{gk\}) \rightarrow$ begins, and $\{gk\}$, $k = 1, 2, ...$ defines the element from the discrete evolving content of S_D by the creation of which $(œ_G, \{gk\}) \rightarrow$ ends. So the (evolving) set $\{(œ_G, \{gk\}) \rightarrow\}$ of aspect-view arrows is connected with the evolving content of the sites $S_{œ}$ and S_D, expressed respectively by the sets $\{œ_G\}$ and $\{gk\}$ (where $\{gk\}$, $k = 1, 2, ..., g$ fixed, amounts to the description of $œ_G$ *via* V_g, which is an element from $\{D\}$). The connection between the evolving sets $\{œ_G\}$, $\{(œ_G, \{gk\}) \rightarrow\}$ and $\{D\}$ can be then organized more, by putting on these sets mutually compatible structures obeying all the MRC-requirements and furthermore conveniently reflecting the particular considered class of descriptional processes (the nature presupposed for the object-entities and the aspect-view-examinations).

The procedure can be extended to the class of arrows $V \rightarrow$: in consequence of D5.2 each *definite* $V \rightarrow$ arrow is a *set* of arrows $\{(œ_G, \{gk\}) \rightarrow$, $k = 1, 2, \ldots, \}$, $g = 1, 2, \ldots m$, m finite.

Then a relative description $D/G,œ_G,V/$ from MRC becomes in C_{MRC} a complete-description-arrow

$$[CD \rightarrow = CD = d \rightarrow \circ DA \rightarrow \circ DAa \rightarrow \circ G \rightarrow \circ V \rightarrow],$$

where $G \rightarrow \circ V \rightarrow$ is indexed:

$$(RG, œ_G) \rightarrow \circ (œ_G, \{gk\}) \rightarrow, \; k = 1, 2, ...\}, \quad g = 1, 2, \ldots, m, \quad m \text{ finite}.$$

C_{MRC} versus quantum mechanics.

We consider the Hilbert-Dirac formalism of quantum mechanics. The Hilbert-space $\mathcal{H}$ of the state-ket-vectors $|\psi>$ of the studied microsystem corresponds to the C_{MRC}-site $S_{œ}$ where are lodged mathematical representations of the considered class of object-entities. The set $\{|\psi>\}$ of state-ket-vectors $|\psi>$ from $\mathcal{H}$ corresponds to the evolving set $\{œ_G\}$ from $S_{œ}$. *The vector-space structure assigned in quantum mechanics to* $\{|\psi>\}$ *is a particular feature entailed by the principle of superposition posited for quantum states, a principle justified by the* **wavelike** *features manifested by what is called quantum states.* So *in general* such a structure has no semantical counterpart, so it will have to be dropped. Furthermore

> *The* C_{MRC} *generation arrows* $(R_G, œ_G) \rightarrow$ *have no* **general** *correspondent in the quantum mechanical formalism: they are represented only in the particular case of microstate-generation by a measurement process.*

This is a striking lacuna (which is suppressed in meta[quantum mechanics]) (see note 1).

The quantum mechanical (in general) non-commuting linear differential "dynamical" operators defined on $\mathcal{H}$ correspond to the C_{MRC}-aspect-view arrows $(œ_G, gk) \rightarrow, k = 1, 2, \ldots$.

The quantum mechanical representation of a state-ket $|\psi>$ with respect to the basis of eigenvectors introduced by a given quantum mechanical operator A, namely as a column-matrix of which the elements are calculated with the help of $|\psi>$ and the considered eigenvectors, corresponds to a basic transferred description $D^{(o)}/G^{(o)}$,$œ^{(o)}$,$V_g^{(o)}$/ from S_D created for a basic object-entity $œ^{(o)}$ by a basic aspect-view-arrow $(œ_G, \{gk\}) \rightarrow, k = 1, 2, \ldots$ (which can be re-written $(œ^{(o)}, gk^{(o)}) \rightarrow, k = 1, 2, \ldots.$).

The set of all the column-matrix representations of a given state-ket $|\psi>$ with respect to all the bases of eigenvectors introduced by all the quantum mechanical dynamical operators, corresponds in C_{MRC} to a complete-description-arrow

$$CD\rightarrow = CD = d\rightarrow \circ DA \rightarrow \circ DAa \rightarrow \circ G \rightarrow \circ V \rightarrow$$

(with $G \rightarrow \circ V \rightarrow$ indexed: $(R_G, œ_G) \rightarrow \circ(œ_G, \{gk\}) \rightarrow, k = 1, 2, \ldots\}$, $g = 1, 2, \ldots m$, m finite).

> So it will be possible to attempt a systematic transposition of the Hilbert-Dirac formulation of quantum mechanics, in terms of the theory of categories, *via* MRC with its central concept of basic transferred description.

It is of course obvious from the start that the explicit C_{MRC}-representations of reality and of the consciousness-functionings have no correspondent in quantum mechanics where not even the actions of object-entity generation are represented mathematically, nor are they at least conceptually and verbally clearly distinguished from the *qualifying* actions *via* measurements. By comparison with C_{MRC} quantum mechanics appears as flawed by very flattening lacunae.

Nevertheless, once the main relations C_{MRC}-(quantum mechanics) have been established, the quantum mechanical formalism becomes a precious guide for a subsequent development of C_{MRC} (any non-necessary restriction suggested by the—particular—case of quantum mechanics having to be carefully avoided). One first important step in the mentioned direction will be the identification of the individualized *MRC-meaning* of Dirac's dual space of linear functionals defined on the Hilbert space of state-ket-vectors, and of the various sorts of scalar products from the Hilbert-Dirac formulation of quantum mechanics (see [13]). Then the C_{MRC}-transposition of these will have to be conveniently achieved.

4.4.4. Concluding comment on C_{MRC}

The outline indicated above needs development. For instance, the condition $S_R \supset [(Ob(C_{MRC})+Ar(C_{MRC})]$ imposed by MRC entails reflexive characters that might raise difficult syntactical problems connected with the definition of the categorial concept of a sub-object. The postulate, the principles and the propositions from MRC must systematically acquire inside C_{MRC} mathematical expressions, and the MRC-propositions should furthermore acquire mathematical proofs. Etc.

5. ILLUSTRATIONS OF THE FUNCTIONING OF MRC

In this section we illustrate by examples the functioning of MRC, thereby also developing the method. We shall first consider logic and then probabilities.

5.1. Classical Logic [23] *versus* the MRC-logic of Relative Classes of Cognitive Actions

Because logic is so particularly important when a method of conceptualization is proposed, we shall, by a brief sequence of remarks, try to convey a notion concerning the relations, and the gap, between MRC and classical logic. We shall then very briefly indicate along what lines an MRC-logic can be constructed and what novelties it introduces. It will appear that the MRC-logic achieves an explicit connection between physical factuality and formal structure, and that it disconnects the question of the consistency of a formal system, from the question of decidability (completeness) considered in Gödel's basic theorem, on which it yields a different perspective.

5.1.1. Critical remarks on Frege's basic definitions

Insufficiencies of the concept of Frege-class of a predicate

The logic of classes and predicates has first been developed by Frege. The starting remark is that a predicate "determines" a class of objects, namely those that partake of the meaning (sense, comprehension) of the considered predicate and hence constitute its extension. In order to identify these objects, first ***(a)*** it is remarked that a predicate, by itself, is neither true nor false, but that ***(b)*** its assertion concerning a given object-entity *can* be true or false *if* the predicate is "pertinent" concerning this object-entity. Then

(c) for each predicate P a propositional function $f_P(x)$ is introduced where f_P represents the predicate and x is an *object-variable*:

> "The expressions which.....include letters 'x','y','z', and are such that they become true or false propositions as soon as the objects designated by these letters are specified, are called propositional functions (J-B. Grize [23], p. 150)."

And ***(d)*** it is posited by definition that **any** *value of the object-variable* x for which $f_P(x)$ is `true`, belongs to the class determined by P. In short:

> The class of P is the set of values of the object-variable x for which $f_P(x)$ is true.

From the standpoint of MRC these very first steps call forth already the following remarks:

* In the first place, we are in presence of a qualification[16] of an object-entity—"the class of P"—of which the generator G is of a particular and so a restrictive type, namely a "generator $G(V)$ of a view V" (cf. final general comments on D14 and 5.1.2): V is supposed to act first in the role of a generator $G(V)$ that selects as object-entity the whole field of perceptibility of V ("any value of an object-variable..."), and then it furthermore acts in the role of a view, by qualifying isolately the "values of the object-variable" from this field of perceptibility, but qualifying them from inside the metapredicate "P is true" (cf. the sequel). This very particular sort of generator of object-entity, $G(V)$, produces either conceptual object-entities—i.e., already previously achieved descriptions—or basic object-entities that transfer directly on the sensitive biological apparatuses of the human beings, marks called "impressions". It has been already remarked that this last sort of cognitive situation produces basic transferred descriptions that are spontaneously and implicitly metaconceptualized during the very first period of a man's life, and are reduced to intrinsic models (D19.3) that seem to pre-exist independently of observation, "out there", available for examinations, in particular examinations of [truths of P's].

> Both sorts of descriptions mentioned above, perpetuate a full ignorance of the rooting of conceptualization, in physical factuality.

[16] I say "qualification", not "description", because no condition of stability of the qualificational result with respect to repetitions of the process of qualification, is required here, as it is in the set of all the definitions D14 (with the unique exception of relative testimonies D14.2.2) (cf. the comments on the generalization of D14.1).

* In the second place, the qualification "x is P" and the metaqualification of empirical truth of this first qualification, are combined in a sort of coalescence where fundamental MRC-conditions get lost. Indeed from the point of view of MRC the qualification "x is P" is just a piece of meaning, no matter whether true or false, and possibly not even that, if *a posteriori* it appears that no view of empirical truth can be constructed which exist in the sense of D7 with respect to the assertion "x is P". Whereas in Frege's approach such reservations are totally absent. Moreover the qualification "x is P" is first introduced in a quasi subliminal way, and then it comes into stable being together with, and indistinctly from the metaqualification "it is true (or false) that (x is P)". This conveys the illusory assumption that a truth-qualification is always possible for any qualification "x is P", whatever its semantic content. Which of course is not the case, as Tarski claimed much later ("the snow is white" is true iff the snow is white).

* In the third place, the involved predicate P, considered separately, is neither endowed with some structure, nor is it subjected to any conditions of effectivity of the examination which P is supposed to perform on x: a sort of ghost-predicate (compare with an aspect-view D5.1 or a view D5.2). Furthermore, as just mentioned, the so feebly formed significance of what is called a predicate P is immediately dissolved in the metapredicate of [truth of P]. While for the metapredicate of [truth of P], again, no structure whatever is specified, nor some condition of relative existence and of effectivity.

In sum, on the one hand, a predicate P and its truth qualifications are assigned the fundamental logical role of, together, producing always, automatically, a proposition, i.e., the tentative assertion of a description, that can then be found, *via* some definite procedure, to be true or false. But on the other hand:

> The classical predicates "P" are reduced to no more than shadows of *undefined* intensive extracts from factuality, just verbal labels which, while they are hypostatized, are also smuggled away by an immediate translation in terms of a purely extensive domain of correlates "x" inside the realm of object-entities on which they act, this correlation being subjected to another *undefined* meta-intension called truth. A vague but dense knot.

* Consider now the "values of x" in general—not only those selected in "classes of P"—and notice that these are the equivalents of MRC-object-entities $œ_G$. Now, no genesis whatever is specified for the "values of x". They

are simply posited to always be "out there", passively waiting to fall inside the field of perceptibility of the predicates P.

> *Classical logic implies in its foundations a hypothesis of universal actuality. The Boolean algebra of classes and predicates is constructed for the already actualized.*

(This, by isomorphy, holds also concerning the nowadays set-calculus on which classical mathematics are founded). Fundamentally, the modal dimension of existence (not to be identified with the "logical" modalities of necessity or possibility) along which potential existence is transformed into actualities by processes of actualization, remains exterior to the classical calculus of classes and predicates. When needed, this dimension has to be superposed by *a posteriori* manipulations. This is not disturbing in the usual language where everything is plethoric, contextual, minimally structured, which for the specific aims of current language is optimal. But in a fundamental formalized representation of thought operations, like logic, the absence of the modal dimension of existence is an imprisoning poverty comparable to what the absence of techniques for the representation of perspective must have been in painting. Only addiction to the traditional methods can hinder to perceive to what a degree such a lacuna is amputating, and that, in particular, it is an obstacle in the way of a basic and explicit connection of logic, to conceptual geneses, to aims (finality), to praxis.

* The fact that no genesis is specified for the "values of x" (the MRC-object-entities $œ_G$) has also another consequence, a radical one:

> The generators G of the object-entities $œ_G$ themselves (not the generators P of the [classes of some P]) are simply not considered.

This absence of an explicitly defined object-entity generator G, so, *a fortiori*, the absence of a generator G required to be in general independent of any qualification and permitted to be physical-operational, restricts *a priori* and arbitrarily the domain of object-entities to which the classical logic can be applied:

> In classical logic all the basic physical object-entities that have to be first radically generated by deliberate physical operations of object-entity-generation, independently of any subsequent qualification, and then might have to be transformed in order to draw from them observable manifestations, are simply eliminated *a priori* from consideration. Indeed "predicates", i.e., linguistic-conceptual qualificators, cannot "determine classes" among *basic* object-entities in the sense of

MRC. They cannot act upon such only factually singularized object-entities, *because they are not homogeneous in nature with these.*

Since the cognitive situation tied with basic physical object-entities, however, is endowed with a certain universality of principle (3.3), we are in presence here of a huge arbitrary amputation. Namely the massive amputation of the whole stratum of conceptualization where the structure of its rooting in physical factuality is specified. In such conditions one can, in particular, well understand why, for classical rationality, quantum mechanics seems unintelligible. Indeed one of the fundamental features of quantum mechanics is precisely the liberation (in general) of the operation of generation of object-entity from any view. And it is by this liberation that MRC transpierces the armoured platform of language and succeeds to build a representation of the processes of conceptualization that is rooted in physical factuality. But, and this comes as a surprise, not *exclusively* basic physical object-entities are eliminated because the object-entity generators are not explicitly considered. All the *conceptual* entities which are first constructed independently of any qualification and are only afterwards qualified—like many mathematical systems and formal systems of logic itself !—are equally eliminated from clear consideration *as* created, constructed object-entities. This leads to *false* problems, and to enormous unnecessary efforts to solve them (5.1.2). An amputation of such an extent, and which concerns logic itself, is not acceptable in a fundamental discipline like logic.

Let us now take a second step. By definition:

"Two classes of object-entities α and β are equal iff all the elements of α are elements of β and vice versa (iff α and β hold the same elements).

Two propositional functions f_α et f_β that determine two classes α and β are *equivalent* if the classes α and β are equal (cf. *op. cit.*, in continuation)."[17]

This prompts a new critical remark:

* How can one know, for instance, whether yes or not for any value of an x for which it is true that it is red, it is equally true that it is spherical? It is implicitly supposed that the answer to such questions can always be given. But this supposition is founded upon the same restrictive hypothesis identified above that any value of any object-variable x (an object-entity

[17] Hervé Barreau remarked that precisely these definitions have already been the object of basic criticisms opposed to Frege's logic. This might somehow be related to the remarks that follow in the main text. However here Frege's approach is examined exclusively by confrontation with MRC, and on a level of principle where technical features do not appear.

$œ_G$ in MRC terms) pre-exists out there, already accomplished, ready to be pointed toward with one's finger, certainly available for examination *via* the metapredicate [truth of P's], equally always available. But this time it is furthermore implied that a P-examination of a value of an x *never* changes the considered value of that x: if it did, *this* value of the considered x, after having been examined by $P \equiv \alpha$, would in general cease to stay available for an examination also by $P \equiv \beta$. Thereby, again but otherwise, are eliminated *a priori* all the basic transferred descriptions that are so deeply rooted in physical factuality that they have to be radically changed in order to draw from them observable manifestations. Now, in the constructive outline from Sec. 5.1.2 it will be shown that the basic descriptions, precisely because in general they have to allow for changes of the involved basic object-entity during its processes of qualification, entail certain consequences on logical *form*, consequences involving *strict physical singularity*. But when the rooting of logic in the as yet unknown physical factuality is obtruded, these consequences remain hidden with it, which restricts *a priori* our perception of logical form, to exclusively its plural, statistical aspects.

* This radical occultation, in classical logic, of the features tied with strict physical individuality, is what permitted to claim that logic is just formal structure; more, to *require* logic, for the sake of "maximal generality", to be a "pure" syntax, freed of any intension, cut from any semantic matter.

But in fact this severance is illusory. It has been possible to imagine it to be realizable precisely because the way in which unspeakable factuality loaded with semantic potentialities is drawn into descriptions at each local relative zero-point of a descriptional chain, remained so completely ignored. As soon as one becomes aware that any local zero-point contributing to the foundation of descriptional chains, consists of a (more or less canonical) *transferred* description, the illusion of the possibility of a complete elimination from a syntax, of any semantic content, is dissolved. It becomes clear that any syntax stems from numerous bulks of physical factuality, which is the prime matter for phenomenal appearances. It is out of these bulks that are drawn the observable manifestations of which the phenomenal appearances consist, while the whole conceptualization is founded on phenomenal appearances. Through these phenomenal appearances, semantic matter goes over into language-and-conceptualization, by primary codings, and then it irrepressibly diffuses up into all the levels of abstraction and complexification. Language is a circulatory-system *for* factual, semantizable prime matter. It emerged and got form in *order* to carry from mind to mind information about factuality, about semantics. If this were not so the societies of men would not have lasted. They would not even have started being. Our minds work with intensions. These, adduced inside language by the phenomenal effects of the interactions between pure factuality and mind, have then osmotically impregnated with semantic contents all the levels of abstraction. So, unavoidably, they have infused into logic also, where they generate its natural forms,[18] those which, more or less implicitly, command in real circumstances our choices, methods, and actions. If on the other hand in the theoretical formalized logic any connection between syntactical form and semantic content is first refused, this instils there by reaction lacunae and awkward features as well as difficult fictitious problems, like for instance those of the *a posteriori* connectivity of modern formal logic with modern semantics. (Try to design *in abstracto* the human circulatory system, strictly without using as a guide the condition that blood has to circulate in it in such a way as to nourish every tiny volume of living tissue: what is the chance to end up with the natural scheme?).

Criticism of the classical logical void ∅: the semantic relativities of ∅

A trivial but striking example of the way in which ignorance of semantic aspects induces syntactic insufficiencies, is that of equality of all the void classes. In consequence of the extensive (set-theoretic) definition of the class determined by a predicate, all the void classes are equal because they all contain "the same element", namely the null element ∅. So, if it is true that no immortal man exists, and it is equally true that no symphony lasting less than one minute does exist, then the class of immortal men and that of symphonies shorter than one minute, are equal. This argument induces a feeling of artifice, of twisting of what one would be prepared to accept as "meaningful". One feels a gliding. The trajectory of this gliding can be retraced:

When one wants to determine quantity, extension, number of elements, starting from only the quality—the predicate—that qualifies, a ground for ambiguity is surreptitiously inserted. So long as a class in the sense of Frege is not void, the quality specific of this class—the one expressed by the predicate P that determines the class—is present, it is held by each element of the class. But at the limit where the class becomes void, the specific quality P that characterizes the class is discontinuously transmuted into pure *qualitatively indistinct* quantity, into a purely numerical zero. This transmutation has been instilled as follows. The mathematicians, when they defined the number zero, in fact have extrapolated into nothingness a certain quality, namely the degree of "numericity" N of any (finite) number

[18] Cf. [21].

so the predicate $P' \equiv N$ of which mathematics studies the manifestations *via* the object-entities called "numbers", of which a quite general property is to be able to "measure", to quantify. It is the prolongation into nothingness, *of numericity*, of this particular predicate $P' \equiv N$, which, by definition, has been called "the number zero". Whereas the logicians, while they make use of numbers in order to measure by the help of $P' \equiv N$ the quantity of supports of a given quality $P \neq P'$—this time *any* quality P whatever, any predicate – did *not* take care to prolong into nothingness *also* this quality P, in order to dispose of a veil of quality P, specifically, to be co-extended, together with the mathematician's zero-of-numericity, over the void encountered at the limit where the *quantity of carriers* of this quality P comes to an end. So at that limit they are left with only a zero-of-numericity, uncovered, stripped of quality P. While the other numbers of carriers, 5, 100, etc., were *all* tied with also the quality P characteristic of the considered class: at this limiting point, the conservation of the way of representing a class breaks down, a *solution of continuity* inside the way of representing a class has been surreptitiously introduced. This is a heavy methodological error, comparable, for instance, to a dimensional inhomogeneity inside an equation. The non homogeneity of conceptual treatment inside a closed conceptual system is always the source of very slippery problems. Any two void classes are considered to be "equal" on the basis of a purely extensive estimation of the null content of a concept that has been first characterized in an exclusively intensive way, even if this characterization possessed also an extensive counterpart: a predicate P is *only* quality, and, by definition, it is P *alone* that determines the corresponding class f_P, not also the quantity of carriers of the quality P. It is then inconsistent, if one distinguishes clearly between quality and [quantity of supports of this quality] (in MRC terms between views V and object-entities œ$_G$ that exist in the sense of D7 with respect to this view), to permit the *defining* quality to disappear "because" all its *supports* disappeared, while *the class itself, defined by the quality, is still maintained.* The predicate P that defines the class f_P should subsist with the class, in spite of the vanishing-support-of-quality-P, i.e., when the set of numbers that label the supports reduces to the number 0. It is inconstant to end up in such a materialist idolatrous manner when one has begun by adoring an abstract God. One should act like the mathematicians, or like Lewis Carrol who leaves us with smile-of-cat-without-cat when the smiling- cat vanishes completely.

The logical void $\emptyset$, which is an element of the "purely" syntactical system called the classical logic of classes and predicates—is subjected to *semantic relativities* that require a *specific syntactical expression*: the asserted possibility of a radical separation between syntaxis and semantics is obviously contradicted in the case of the logical void.

Ferdinand Gonseth said that "logic is the physics of any object." But any given object has some semantic content, and the types of semantic content have to be mutually distinguished in a thoroughly worked out formal representation of safe derivational vehiculations of our knowledge concerning empirical truth-valuations involving "any" object.

Global critical conclusion

The classical logic of classes and predicates, which founds the whole modern classical logic, floats above language, inside the stratum of the already pre-verbalized-conceptualized. The rooting of the processes of conceptualization, in physical factuality, the creative cognitive actions which produce object-entities and qualificators of these, the modal dimension of existence where potentiality, actualization and actuality are located, remain hidden to it. By occultation of the genetic stages from the processes of conceptualization and by substitution to these of false hypostatizing absolutizations, it introduces arbitrarily restricted conceptual platforms that cannot withstand artificial and inadequate formal representations.

Only when all the involved descriptional geneses, with the descriptional relativities entailed by them, are explicitly taken into account, is it possible to dominate from a formal point of view any descriptional situation, whatever its complexity. This can be better understood *per a contrario* and on examples.

For instance, inside MRC where any descriptional relativity is taken into account explicitly as soon as it comes into play, the treatment of the logical void is preorganized in consequence of the way in which the very first levels of general conceptualization are structured. As soon as one considers an (independently defined) object-entity $œ_G$ and a view V (D4 and D5), the test of their mutual existence in the sense of D7 is methodologically required, before trying to perform the corresponding relative description. If this test is negative one finds oneself precisely in the case that can be designated as "the void class determined by V inside the set of object-entities $œ_G$", which means "absence of object-entities $œ_G$ admitting of the qualification V", i.e., absence of the possibility of a meaning generated by the pair (G, V). So a conceptual void, doubly relativized to the semantic features involved by the considered pair (G, V), comes into being *ab initio*. Later, once the possibility of meaning has been insured by mutual existence in the sense of D7 and then a first descriptional level has been insured by the existence of some stability of the qualifications in the sense of D14.1, comes furthermore into consideration, in its turn, the question of empirical truth:

given an already achieved description in the sense of one of the definitions D14, is this description a proposition, i.e., does *it* exist in the sense of D7 with respect to some view of empirical truth that can be effectively exhibited? The still higher and more particular level of "logical" characterizations concerns object-entities consisting of systems of propositions. A proposition from a formal system of propositions S, can be described by the logical views of provability inside S and of decidability inside S, while the system itself considered as a whole can be examined by the logical views of completeness and of formal consistency. All the mentioned sorts of logical description are related with the previously developed relativized conceptual-semantic voids (mutual exclusions, absence of descriptional stability). Indeed these *entail* the definibility of syntactical, calculational relativized voids (see Sec. 5.1.2) and thus they go over into the form of the logical descriptions. So in this specific case it is clear that, and how, inside MRC the semantic contents determine progressively aspects of logical form. And these, the calculational relativized logical voids, preserve from a whole category of false problems. Indeed the absolutization of the logical void is one of the most prolific sources of illusory problems. (Even in modern quantum logic there subsists much confusion concerning complementations tied to the logical void [24]; cf. also [13] as well as 5.1.2).

When instead of a system of propositions, a formal system in the most abstract sense is considered, either any connection between semantics and syntax has deliberately been suppressed by the process of conceptualization (which is difficult) and in this case one obtains just a Wittgensteinian "game" that resists any non distorting and useful interpretation in terms of some domain of natural facts, or some connections between semantics and the constructed formal system have been deliberately preserved, and then precisely these insure possibilities of useful interpretations of this system.

The corpus of relativizations required by MRC does not only insure a controlled penetration of semantics into the logical descriptions, it also exerts another crucial sort of control which classical logic cannot exert systematically because of the artificial separation between semantics and syntax. Namely, it insures automatically all the types of descriptional "homogeneity" amounting to the conservation of the method of representation inside a closed descriptional universe, i.e., throughout the work accomplished with a given epistemic referential. While, on the other hand, the principle of separation P15 regulates the passages from one set of homogeneous descriptional contents, to another one whatever their type.

This is important. Indeed the creation of sense, in all its stages, is ruled by the implicit imposition of methodological principles of homogeneity: physical operations can directly change only physical entities, concepts can directly change only concepts and can be localized only inside nets of concepts; in an equation the semantic dimensions from the first member must be the same as the semantic dimensions from the second member; statistical-probabilistic qualifications do not exist in the sense of D7 with respect to individual events, nor with respect to only statistical distributions of events, they exist only with respect to statistical-probabilistic distributions of events; and *vice versa*, individual qualifications do not exist in the sense of D7 with respect to statistical-probabilistic distributions, they are blind with respect to these; etc. When no matter which one among these various sorts of implicit principles of homogeneity is violated, paradoxes or false problems emerge. Inside MRC this is always expressed as a consequence of a violation of the principle of separation P15, i.e., of a non-explicit modification of the epistemic referential which is made use of.

The false absolutizations that flaw the classical logic of classes and predicates have prolongations in many domains of modern science, in particular in the theory of sets. Indeed the elements of a set are always supposed to somehow pre-exist already realized, and this, just like in the definition of the equality of two classes and of the equivalence of two propositional functions, entails arbitrary *a priori* restrictions. But the most noteworthy consequence might consist of the fact that *classical logic, because of its lack of explicit connection with strictly singular physical factuality, remains unaware of the spacetime specificities of the descriptions of physical object-entities.* This has favoured a surreptitious gliding

conceptualization→natural logic→ formal logic→calculus→computation,

whereby often, in computational simulations of physical processes, understanding disappears entirely into mere doing.

Globally, the apparently so clean-cut and rigorous classical logic, when scrutinized, reveals non-intelligibility tied with superficialities and arbitrary posits.

5.1.2. Outline of an MRC relativized genetic logic

The preceding critical considerations entail by contrast a constructive approach of which what follows conveys only an extremely synthetic notion. The aim is, inside MRC, to explicate the consequences upon logical descriptions, of the relativization to the cognitive actions from which these logical descriptions stem, so also to the semantic contents introduced by these cognitive actions. The main step is the introduction of the concept of genetic class.

Double extremity genetic classes

Let us recall that inside MRC what is called "object-entity" is just a descriptional role (see final comment on the general concept D14 of a relative description). No entity never pre-exists **as** *an object-entity.* It always has to be introduced in the role of object-entity by the explicit action of a definite operation G which either radically creates—physically or conceptually – an actor for this role, or only recruits some pre-existing entity for acting in this role. This, in general, is done independently of any pre-established qualification of, specifically, the object-entity introduced as such by the chosen operation G, so also independently of any "predicate P" (cf. comment on D14.3.1). Only after having been thus put, *via* G, in the role of object-entity, becomes the involved entity available for the action on it of a view (D5.1, D5.2)—any one—which, in its own turn, is chosen for acting in the role of a view. The necessity of an apparently so redundant and intricate way of saying can be best understood when the chosen epistemic referential has the particular degenerate form $(G(V), V)$ where $G(V)$ is the "generator of the view V" (cf. the general final comment on D14; V denotes here indistinctly an aspect-view or a view while if specifically an aspect-view is meant, we write V_g) which is precisely the form presupposed implicitly by the whole classical logic. Indeed in this case the view symbolized by V, though from the beginning on it is structured accordingly to the definitions D5.1, D5.2 of qualificators, nevertheless acts first in the role denoted $G(V)$ of generator of object-entity. Namely it acts first either by selecting as object-entity its own field of perceptibility, or by radically creating this field, like for instance in the case of the generation of a microstate by a given quantum mechanical measurement process. And afterward, on the product of this first action accomplished by itself but in the role $G(V)$ of generator of object-entity, V can furthermore act also in the role of a view or an aspect-view, for which its initial definition has been specifically intended. (Let us also recall that what is structured as an aspect-view or a view in the sense of D5.1 or D5.2, respectively, can be selected for the role of object-entity, by a convenient generator (here a conceptual selector) (cf. the final general comment on D14).) The existence of situations like those mentioned above requires indeed ways of speaking that distinguish clearly between the general descriptional roles, and the specific actors to which the roles are assigned. This distinction is quite essential because according to MRC, in order to describe, both the role G and the role V have always to be acted, even if in a reduced or a degenerate way and which reflects also the characteristics of the particular actor put to hold the role. So inside MRC it would be neither necessary nor sufficient to consider, as it might seem natural at a first sight, that the equivalent of a "predicate P" is just an aspect-view V_g. In order to achieve qualifications, MRC requires to make systematically use, instead of just a "predicate P", of some definite *succession* $[G.V_g]$ or $[G.V]$ of an actor put in the role G followed by an actor put in the role V.

So far the relative description D/G,œ$_G$,$V/$ produced by an epistemic referential (G, V), once obtained accordingly to one of the definitions D14, has been considered separately from its genesis. By the following definition DL.1 (L: logical) we shall now introduce a synthetic concept that takes systematically into explicit account, together with a given description, also the whole genesis involved by it.

DL.1. ***Double-extremity genetic classes.*** Consider an epistemic referential (G, V) where V is a view containing in general several aspect-views and which exists in the sense of D7 with respect to the generator G of object-entity.

DL.1.1. ***Double-extremity genetic class involving a physical object-entity.*** Suppose that (G, V) introduces a physical object-entity and that it does produce a relative description D/G,œ$_G$,$V/$ of it in the sense of the definition D14.1, individual or probabilistic. Then the repetitions of the succession $[G.V]$ of pairs of cognitive actions, constitute [the class of all the operational processes of gk-valuations involved by D/G,œ$_G$,$V/$] (in this context the term "operational" is intended to stress that no model whatever is asserted). The class specified above will be called a *double extremity genetic class involving a physical object-entity*, in short a physical genetic class, and will be labelled $C_{ph}[G.V]$.

If V consists of only one aspect-view V_g an aspect-description D/G,œ$_G$,$V_g/$ is obtained, and the succession $[G.V_g]$ produces [the class of all the operational processes of gk-valuations involved by D/G,œ$_G$,$V_g/$]. We name this a one aspect *double-extremity genetic class involving a physical object-entity*, in short a one aspect physical genetic class, and we label it $C_{ph}[G.V_g]$.

When a basic referential $(G^{(o)}, V^{(o)})$ is considered, a basic transferred description $D^{(o)}/G^{(o)}$,œ$^{(o)}$, $G, V^{(o)}/$ is obtained, and the corresponding genetic class will be called a basic genetic class; such a class will be denoted $C[G^{(o)}.V^{(o)}]$. This is [the class of all the operational basic processes of gk-valuations involved by $D^{(o)}/G^{(o)}$,œ$_G^{(o)}$,$V^{(o)}/$].

DL.1.2. ***Genetic class involving a non physical public object-entity.*** Suppose that (G, V) introduces a non physical public object-entity and that it does produce a relative description NPP.D/G,œ$_G$,$V/$ of it in the sense of D14.2.1. Then the repetitions of the succession $[G.V]$ constitute [the class of all the processes of gk-valuations involved by NPP.D/G,œ$_G$,$V/$]. The class specified above will be called *a genetic class involving a non physical public object-entity*, in short a non physical public class, and will be labelled CNPP[$G.V$]. In particular V can consist of only one aspect-view V_g and then we have a class CNPP[$G.V_g$].

DL.1.3. ***Testimonial double-extremity genetic class.*** Suppose that (G, V) does not insure the possibility to realize arbitrarily many repetitions of the successions $[G.V_g]$, for all the aspect-views V_g from V. So it produces an only testimonial description θ/G,œ$_G$,$V/$ in the sense of D14.2.2. Nevertheless according to MRC in this case also a certain set of known or unknown implicit qualifying processes $[G.V_g]$ have necessarily been involved: if not, there would be no qualification at all (cf. general comments on the definitions D14). These will be said to constitute the *double-extremity genetic class of the testimony* θ. Such a class will be indicated by the notation $C_\theta(G, V)$ where only the involved epistemic referential is specified.

Comment. The general concept of a genetic class is posited here as the MRC-equivalent of the Frege-class of a predicate P.

The distinction between a relative description D/G,œ$_G$,$V/$ (or a relative testimony)—i.e., exclusively the final global result of the processes of qualification produced with the considered epistemic referential (G, V)—and the corresponding genetic class, draws attention upon the absence, in the classical logic of classes and predicates, of any reference to the epistemic actions involved by an "object-variable x" or a "predicate P". Thereby is brought into full light the contrast between the active conception on knowledge involved by MRC, and the passive, hypostatizing and absolutizing implications of classical logic.

In a genetic class the undefined, hypostatic shadow-predicates P are replaced by the views V founded upon aspect-views V_g obeying the definition D5.1, which consist of effective operations and tests incorporated to a definite conceptual-operational structure. While the hypostatic object-variable x is replaced by a definite operation of generation G associated with the object-entity œ$_G$, an operational-conceptual pair (G,œ$_G$) that opens up the possibility to grasp and to draw up into conceptualization bulks of physical factuality of which the semantic matter nourishes with meaning all the levels and sorts of description, the logical one included. Any unnecessary absolutization is suppressed by the fact that a view is explicitly allowed to act in the role of generator of object-entity (labelled $G(V)$) while both views and generators are allowed to play in *other* descriptions the role œ$_G$ of object-entity. The specificity of the concept of generator introduced by the definition D4 is not in the least diminished by this absence of strict solidarity with the descriptional role G, nor is the specificity of the concept of a view V as defined by D5.1 and D5.2, diminished by the absence of a strict solidarity with the role of view. So inside a double-extremity genetic class, the classical object-variables and predicates—abstract, vague, hypostatic, absolute, as if out of reach of human action—transmute into a quite definite and complex operational-conceptual whole of relativized and constructive epistemic actions.

Outlook on a calculus with double-extremity genetic classes

In what follows we consider exclusively genetic classes involving stable relative descriptions in the sense of the definitions D14.1, D14.2.1, or D14.3.1 (the genetic classes of testimonial descriptions are too vaguely defined to be included in a calculus). Furthermore we drop the lower indices as well as the prefixes and write uniformly $C[G.V]$.

The logical operations, sum, intersection, complementation, must all be redefined step by step for the case of genetic classes, in a way fully relativized to the involved generator G and to the whole content of the acting view V. The reconstruction requires the definition of *laws of composition* of object-entity generators G, of object-entities $œ_G$, of views V, and of descriptions D (accordingly to P8 and P10), and it has to be carried out for all the possible *sorts* of compositions of genetic classes $C[G.V]$ (two classes with both G and V different, or with the same G and different V, or with different G's and the same V, or a basic class $C[G^{(o)}.V^{(o)}]$ and a non-basic one, or two basic classes, or two non-basic classes (of same order or of different orders), or an aspect-class $C[G.V_g]$and a general one $C[G.V]$, etc.). For instance:

Consider the two genetic classes generated by the successions $[G_1.V_1]$ and $[G_2.V_2]$, both not basic. Then the involved object-entities $œ_{G_1}$ and $œ_{G_2}$ are conceptual (previously achieved descriptions, or intrinsic metaconceptualizations, or intrinsic models) while the final global results are two descriptions D_1 and D_2. Suppose now $(G_1 \equiv G_2) = G$ and $V_1 \neq V_2$. Then only one object-entity $œ_G$ is involved and the intersection $[C[G.V_1] \cap C[G.V_2]]$ leads to an (absolutely) void result if V_1 and V_2 involve no common aspects; while if V_1 and V_2 do involve common aspects this intersection yields a description containing the qualifications present in both D_1 and D_2, so one can pertinently say that the resulting description is the intersection (or product) $D_1 \cap D_2$ of D_1 and D_2 which can be denoted $D_{\cap 12}$. In the same conditions the union $[C[G.V_1] \cup C[G.V_2]]$ produces a final description that can also be pertinently called the sum of D_1 and D_2 and can be denoted $D_{\cup 12}$. Suppose now on the contrary $G_1 \neq G_2$ and $(V_1 \equiv V_2) = V$. Then according to the nucleus of MRC the view V yields a (meta)description of the metaobject-entity

$œ_{G_1} \cup œ_{G_2}$ where all the qualifications from D_1 and all those from D_2 are contained, so one could speak, for instance, of the description of an *object-sum* and introduce the notation $D_{\cup œ12}$ with $D_{\cup œ12} \equiv D^{(2)}/G^{(2)},œ_G^{(2)},V^{(2)}/$ where $G^{(2)}$ selects the sum-object-entity $œ_G^{(2)} \equiv (œ_{G1} \cup œ_{G2})$ and $V^{(2)} \equiv V$. Etc.

> The last example entails that the classical definition of the class of a predicate P can be progressively approached inside the MRC-logic by composing additively an increasing number of genetic classes with distinct generators G of object-entity and identical views V.

In *any* case the global result of a permitted composition of genetic classes $C[G.V]$ is just a relative description. Furthermore, let us consider the logical voids.

> *Each class $C[G.V]$ introduces various semantically relativized voids tied with corresponding semantically relativized complements:*

Given a qualification (gk) from $C[G.V]$, it introduces its own relative void—let us label it $[\emptyset/(gk)]$—that sends to the corresponding relative complement consisting of the set $\{(gk)'\}$, $(gk)' \neq (gk)$, of all the other qualifications *from* $C[G.V]$; analogously an aspect-view $V_g \in V$ introduces the relative void $[\emptyset/V_g]$ that sends to the complement consisting of the set $\{V_{g'}]$, $g' \neq g$, of all the other aspect-views *from* V, so also to the set of all the qualifications $g'k$ from $C[G.V]$ produced by the aspect-views from V that are different from V_g. These were examples of relative voids internal to the genetic class $C/[G.V]$, i.e., which send to complements contained in $C[G.V]$. If now G and V are regarded as wholes, the genetic class $C[G.V]$ introduces three relativized metavoids $[\emptyset/G]$, $[\emptyset/V]$ and $[\emptyset/G,V]$ which send to complements from the outside of $C[G.V]$, namely to the three exterior metacomplements with respect to, respectively, G, or V as a whole, or the referential (G,V) as a whole (there is no difficulty to characterize these metacomplements by words).

So a genetic class $C[G.V]$ introduces a hierarchical organization of relative voids and of corresponding relative complements sending into definite domains of observation or epistemic action. We are already far from the connection between a hypostatized "object-variable x" and a hypostatized predicate P—always just a conceptual selector—*associated with only one absolute void.* Now, the qualifications gk, the aspect-views V_g, and the generators G, are all semantic descriptional elements which determine semantic relative voids and the corresponding semantic complements; but, *via* the symbols that represent them, these semantic relative voids and complements go into the calculus with genetic classes where they become "logical" voids and complements that imprint their mark upon a syntax.

We are in presence here of an example in which one can see how semantic features gain access toward a specific syntactical expression. What appears on the horizon is a *syntax of the extraction and elaboration of semantic matter*, a syntax of conceptualization where the artificial and illusory frontier between semantics and syntax is transcended.

The calculus with genetic classes it not yet elaborated, but nothing hinders to elaborate it. It will have to be worked out in compatibility with the whole content of the nucleus of MRC. In particular all the restrictions or methodological rules involved by the frame principle P8, the principle P10 of individualizing mutual exclusion, and the principle of separation P15 with the concept of relative metadescription D16 entailed by it, will have to be taken into account systematically. Inside the enlarged framework created by this calculus, the concept of proof will require reconstruction.

These brief indications suffice for conveying a first notion concerning the content and the degree of novelty and complexity of the calculus with genetic classes.

Views of empirical truth. Relative proposition

Consider a genetic class $C[G.V]$. It involves as its final global result a corresponding relative description $D/G,œ_G,V/$, i.e., some *specified* structure of spacetime-gk-values (where one or both frame-aspects of space and time can be absent), the aspect-index g running over the aspect-views $V_g \in V$ that are different from the frame-aspect V_E and V_T. Now, following Tarski in this respect, we note that the mere assertion of the description $D/G,œ_G,V/$ is *not* itself what is called a proposition. It generates a proposition if and only if $D/G, œ_G,V/$ *can* be asserted to be empirically true (this is the MRC way of saying like Tarski that ["the snow is white" is true iff the snow is white]). Indeed only a *previously* constituted *description* can be empirically true or false. For instance, a basic object-entity cannot *exist* in the sense of D7 with respect to an aspect of empirical truth, because it does not exist in the sense of D7 with respect to *any* view of comparison (π18.1) while an aspect-view of empirical truth is an aspect-view of comparison. Indeed it must somehow compare the mere assertion of the considered description, with some perceptions of empirical facts to which this assertion refers; it must somehow be a view of "verification" able to establish identities or non-identities concerning, on the one hand the *assertion*, for a definite object-entity, of definite aspect-values gk of definite aspects g, and on the other hand the *effective emergence* for that object entity, of precisely those asserted gk-values, when it is examined *via* that aspect g. Even not any description can be empirically true or false (think of the description of a minotaur).

So, quite essentially, each aspect of empirical truth is a meta-aspect which is *relative* to an aspect g involved in the description that has to be "verified". Like in any identity-valuation, the (two) meta-(aspect)-values of a meta-aspect of empirical truth, namely "true" or "not true" (false), are *inconceivable* in an absolute sense, they can be imagined only relatively to some definite gk-value of a definite aspect g. If D/G,œ$_G$,$V/$ is an individual description, then one can desire to establish for each gk-value asserted by D/G,œ$_G$,$V/$ whether it is true or false; and if D/G,œ$_G$,$V/$ is a probabilistic description one can desire to establish whether the probabilistic distribution asserted by it for the values gk of each aspect g is true or false; so one can also ask: "Is D/G,œ$_G$,$V/$ true with respect to all the gk-values or all the distributions of gk-values asserted by it?". But to research a valuation of empirical truth of D/G,œ$_G$,$V/$ concerning no specified gk-value or distribution of gk-values, would obviously be meaningless. So we introduce the following definition:

DL.2. ***Meta[aspect-view] or view of empirical truth.*** Consider a meta[aspect-view] consisting of one meta-aspect (τ/g) which is relative to an aspect g in the sense of D5.1. Let us designate by $[V_\tau^{(2)}/g]$ a corresponding meta(aspect-view). The meta-aspect (τ/g) from $[V_\tau^{(2)}/g]$ is posited to contain only two aspect-values, namely $(\tau/g)1 \equiv$ "true with respect to g" and $(\tau/g)2 \equiv$"false with respect to g". Accordingly to the general definition D5.1 of an aspect-view—which concerns any aspect of any order – each meta-aspect $(\tau)/g$ must introduce a *definite* and *effective* corresponding operation of $(\tau)/g$-examination, as well as an explicit *coding rule* for deciding which results of the $(\tau)/g$-examination are to be coded "true with respect to g" and which ones are to be coded "false with respect to g". A meta[aspect-view] of the specified sort will be called a *meta[aspect-view] of relative empirical truth* (τ: empirical truth). A view containing two or more meta[aspect-views] of empirical truth relative to two or more distinct aspects g will be called a *metaview of empirical relative truth* and will be symbolized by $V_\tau^{(2)}$.

Comment. Consider a previously achieved description D/G,œ$_G$,$V/$ and a metaview of relative empirical truth $V_\tau^{(2)}$. If $V_g \in V$ and $V_\tau^{(2)}$ contains a meta-aspect-view $[V_\tau^{(2)}/g]$ of empirical truth relative to g *that is effective with respect to* œ$_G$, then D/G,œ$_G$,$V/$ and $V_\tau^{(2)}$ do mutually exist in the sense of D7, with respect to g. In this case $V_\tau^{(2)}$ is able to qualify the empirical truth of D/G,œ$_G$,$V/$ with respect to that aspect g. If this is not the case, then with respect to that g the description D/G,œ$_G$,$V/$ and the metaview $V_\tau^{(2)}$ do not mutually exist in the sense of D7 and $V_\tau^{(2)}$ is not able to qualify the empirical truth of D/G,œ$_G$,$V/$ with respect to g. If D/G,œ$_G$,$V/$ and $V_\tau^{(2)}$ do mutually exist in the sense of D7 with respect to all the aspects g involved by V, then $V_\tau^{(2)}$ can yield for D/G,œ$_G$,$V/$ a *complete* valuation of empirical truth.

It follows that according to MRC the concept of empirical truth possesses some *meaning* relatively to D/G,œ$_G$,$V/$ only if it is possible to construct at least one metaview $V_\tau^{(2)}$ of empirical truth which exists in the sense of D7 with respect to D/G,œ$_G$,$V/$. But this condition is far from being always insured (as it often seems to be implied). It is a rather restrictive condition, because of the requirements of definiteness, effectivity and codability entailed by the general definition D5.1 for any (τ/g)-examination. Indeed, of what can a (τ/g)-examination consist? One possibility is that it consists of a mere repetition of the V_g-examination itself which inside the genetic class $C[G.V]$ leads to this or that aspect-value gk, or this or that probabilistic distribution of gk-values asserted by D/G,œ$_G$,V, followed by a comparison between the result obtained in the re-production and the result asserted by D/G,œ$_G$,$V/$ (the aim of the condition of re-producibility currently imposed in the "exact" experimental sciences like experimental physics, chemistry, molecular biology, is precisely to insure possibility of (τ/g)-examinations of the type specified above). But re-producibility is relatively *rare*, even for descriptions of physical facts,[19] and even for descriptions of physical facts that belong to what is called an exact natural science. In history, palaeontology, human biology, police researches, current life, etc., one is in presence of just testimonial qualifications in the sense of D14.2.2 with respect to which other sorts of definite, effective and codable (τ/g)-examinations must be invented, and in many cases this simply is not possible. As for religious, metaphysical, mythical, poetical testimonial qualifications, the meaninglessness of any relative metaview of empirical truth is entailed by the very content of the testimonies.

Consider now a description D/G,œ$_G$,$V/$ for which a complete

[19] For instance, if the considered assertion is "yesterday at 14h35, a grain of dust carrying on it a germ X has left my pillow," it seems highly improbable to be able to construct for it some meta[aspect-view] of empirical truth founded on reproducibility. So the *testimonial* descriptions are eliminated (which is why I did not call them "descriptions"). In experimental physics, in chemistry, biology, etc., the specification of metaviews of empirical truth founded upon reproducibility, that be acceptable from all the points of view, constitutes a basic part of the research.

metaview $V_\tau^{(2)}$ of empirical truth has been constructed. Then the valuations of empirical truth of D/G,œ$_G$,$V/$ achieved *via* the (τ/g)-examinations involved by $V_\tau^{(2)}$ are in a non removable way relative to these particular examinations. In general $V_\tau^{(2)}$ is not unique, and with another metaview $V_\tau^{(2)}$ involving other (τ/g)-examinations one obtains in general other truth-valuations.

In consequence of the relativizations specified above, the questions of empirical truth become precise and they admit of definite but only relative solutions.

This stands in polar opposition to relativism.

The conception on empirical truth exposed above can rather obviously shown to be in essential agreement with K. Popper's concept of "relativity of truth to theory", as well as with H. Putnam's views. While Quine, Kuhn, and many other important thinkers, put less or no accent on the definiteness and effectivity required for a (τ/g)-examination, so in their writings the question of empirical truth, like that of reference, seems to involve a general and irrepressible doom to relativism.

We can now define a relativized concept of proposition:

DL.3. *Relative proposition.* Consider a description D/G,œ$_G$,$V/$ for which it has been possible to construct a complete metaview $V_\tau^{(2)}$ of relative empirical truth. Consider the metadescription $D^{(2)}/G^{(2)}$,œ$^{(2)}$,$V_\tau^{(2)}/$ where: the metaobject-entity is œ$^{(2)} \equiv D/G$,œ$_G$,$V/$ (introduced by a corresponding meta-generator $G^{(2)}$, namely a conceptual selector); $V_\tau^{(2)}$ is the metaview of empirical truth that exists in the sense of D7 with respect to D/G,œ$_G$,$V/$, the results of all the involved (τ/g)-examinations being *a priori* asserted—*tentatively*—to consist only of the relative truth-values "*true* with respect to the aspect g" for all the $V_g \in V$, which remains to be validated or invalidated *a posteriori* by the effective realization of all the (τ/g)-examinations involved by $D^{(2)}/G^{2)}$,œ$^{(2)}$,$V_\tau^{(2)}/$. Because the specified tentative assertion is a "proposition" in the etymological sense, $D^{(2)}/G^{2)}$,œ$^{(2)}$,$V_\tau^{(2)}/$ will be called an *atomic proposition relative to* D/G,œ$_G$,V *and to* $V_\tau^{(2)}$and will be labelled $p(D, V_\tau^{(2)})$. It can consist either of the global, integrated formulation "D/G,œ$_G$,$V/$ is true (or false) with respect to $V_\tau^{(2)}$" or of the analyzed set of all the formulations "D/G,œ$_G$,$V/$ is true (or false) relatively to $V_\tau^{(2)}$ in its assertion concerning *that* value gk of *that* aspect g (or relatively to its assertion of the distribution of gk-values of g").

Comment. *Via* the MRC-concepts of metaview of empirical truth and of relative proposition, the calculus of genetic classes $C[G.V]$ leads to a corresponding relativized calculus of propositions, where the truth-value of the final description produced by a composition of genetic classes $C[G.V]$ has to be established as a function of: the nature of the composition; the involved metaviews of empirical truth; the values of empirical truth assigned *via* these to the descriptions produced by the classes involved in the considered composition of classes. So, while in the classical approach a truth-valuation is from the beginning on involved in the definition of the class of a predicate P, in the MRC genetic logic the genetic classes are clearly separated from the corresponding propositions, of which the truth-valuations require different, explicit, analyzed, non-trivial relative specifications. This, at first sight, might seem to be a huge complication, to be avoided at any price. But in fact it is a complexification of the treatment that can determine with any desired precision *the configuration of the channels along which semantic matter is adduced into logical syntax.* Too often, for the sake of simplicity, false absolutes are introduced, which block the growth of thought.

A non-classical logical stratum concerning strictly singular physical factuality

Consider a basic genetic class $C[G^{(o)}.V^{(o)}]$. Even if the basic description $D^{(o)}/G^{(o)}$,œ$^{(o)}$,$V^{(o)}/$ involved by this class is called "individual", in consequence of the condition of stability from the general definition D14.1 and its particularization D14.3.1, it involves nevertheless a big number of repetitions of the realization of each succession $[G^{(o)}.V^{(o)}]$.

> *The epistemic* `action` *leading to* $D^{(o)}/G^{(o)}$,œ$^{(o)}$,$V^{(o)}/$ *as a whole, no matter whether* $D^{(o)}$ *is individual or probabilistic, is always directly placed on the level of* `statistics`.

However, by penetrating *inside* $D^{(o)}$ and taking into account only two distinct successions $[G^{(o)}.V^{(o)}_{g1}]$ and $[G^{(o)}.V^{(o)}_{g2}]$, it is possible, by use of the concept D14.2.2 of testimonial description, to dig down to the level of the strictly individual qualifications, and to define for these a *semantical* character which determines a certain corresponding *logical form.* This is an innovation with respect to classical logic. We proceed as follows.

Consider two distinct successions $[G^{(o)}.V^{(o)}_g]$ and $[G^{(o)} \cdot V^{(o)}_{g'}]'$ that have been realized either with $g' \equiv g$ or with $g' \neq g$, no matter, but have led to two different aspect-values $gk \neq gk'$ or $gk \neq g'k'$, respectively. These two successions—with their outcomes included—do not insure a test of descriptional stability as required by D14.1 or D14.3.1. So they are not descriptions in the sense of the mentioned definitions, they are just two testimonies in the sense of D14.2.2, say θ_1 and θ_2. Now, *because* θ_1 *and* θ_2 *involve by hypothesis two* **distinct** *registered aspect-values, each one of these testimonial descriptions required its own realization of a replica of the object-entity* $œ^{(o)}$ *produced by* $G^{(o)}$. So, if we label $œ_j^{(o)}$ one given realization of a replica of the basic object-entity $œ^{(o)}$, the two testimonies $\theta_1(œ_j^{(o)})$ and $\theta_2(œ_j^{(o)})$ are mutually incompatible because they cannot both realize for that one given replica $œ_j^{(0)}$.

> As soon as a restriction to only *one* definite **replica** $œ_j^{(o)}$ of a basic object-entity $œ^{(o)}$ is posited—not only restriction to no matter how many replicas of one sort of basic object-entity $œ^{(o)}$ as defined by a given operation $G^{(o)}$, but furthermore restriction also to only one replica of that sort of basic object-entity—there arises a mutual incompatibility between the factual realizability of $\theta_1(œ_j^{(o)})$ and that of $\theta_2(œ_j^{(o)})$. *This holds even if the qualifications involved by* θ_1 *and* θ_2 *concern both one same basic aspect* g.

This is a mutual exclusion of a semantical nature. But *via* the concept of empirical truth it entails a formal logical consequence. To show this we proceed as follows. To begin with, we define:

DL.4. ***Basic relative atomic testimonial proposition.*** We call *basic relative testimonial proposition* and we label $p(\theta^{(o)})$ the tentative assertion of the empirical truth of a relative basic testimony $\theta/G^{(o)},œ^{(o)}, V_g^{(o)}/$ (with respect to some definite view of empirical truth $V_\tau^{(2)}/g^{(o)}$ supposed to have been constructed); which tentative assertion remains to be validated or invalidated *via* the $(\tau/g^{(o)})$-examinations involved by $[V_\tau^{(2)}/g^{(o)}]$.

Comment. A basic aspect of empirical truth concerning a basic testimony $\theta/G^{(o)},œ^{(o)},V_g^{(o)}/$ can consist, for instance, of the consensus concerning the genesis and the outcome of the testimony θ, among an arbitrarily big number of observers that have watched and witnessed together these non-repeatable phenomena.

So to the two testimonies $(\theta_1^{(o)}) = \theta_1(œ_j^{(o)})$ and $\theta_2^{(o)} = \theta_2(œ_j^{(o)})$ there correspond two testimonial propositions $p_1[\theta_1(œ_j^{(o)})]$ and $p_2[\theta_2(œ_j^{(o)})]$. Now since $\theta_1(œ_j^{(o)})$ and $\theta_2(œ_j^{(o)})$ cannot be both realized because they involve by hypothesis two different outcomes for one same replica $œ_j^{(o)}$ of the involved basic object-entity $œ^{(o)}$, *a fortiori* $p_1[\theta_1(œ_j^{(o)})]$ and $p_2[\theta_2(œ_j^{(o)})]$ cannot be both true. So:

A logical conjunction of $p[\theta_1(œ_j^{(o)}0]$ and $p[\theta_2(œ_j^{(o)})]$ is *devoid of factual counterpart.* It cannot be defined, which is a case different from that in which it can be defined but comes out to be false.

This can be better understood by the help of truth-tables: Given two propositions p and q, their logical product $p \wedge q$ is defined by:

p	q	$p \wedge q$
T	T	T
T	F	F
F	T	F
F	F	F

What happens if $p \equiv p_1[\theta_1(œ_j^{(o)})]$ and $q \equiv p_2[\theta_1(œ_j^{(o)})]$? In this case the top line "TTT" represents a combination which, factually, is *systematically impossible.* The factually possible cases are *only*

p	q	$p \wedge q$
T	F	F
F	T	F
F	F	F

But what this last set of possibilities claims, is that the logical product $p \wedge q$ simply does not "exist", factually, since it **never** *is factually true.*[20]

[20] Wittgenstein [25] made an analogous analysis related to another sort of factual mutual spacetime exclusion: "I have said elsewhere that a proposition 'reaches up to reality,' and by this I meant that the forms of the entities are contained in the form of the proposition which is about these entities .... For the sentence, together with the mode of projection which projects reality into the sentence, determines the logical form of the entities .... For if the proposition contains the form of an entity which it is about, then it is possible that two propositions should collide in this very form. The propositions "Brown now sits in this chair" and "Jones now sits in this chair" each, in a sense, try to set their subject term on the chair. But the logical product of these propositions will put them both there at once, and this leads to a collision, a mutual exclusion of these terms .... It is, of course, a deficiency of our notation that it does not prevent the formation of such nonsensical constructions, and a perfect notation will have to exclude such structures by definite rules of syntax. These will have to tell us that in the case of certain kinds of atomic propositions described in terms of definite symbolic features certain combinations of the T's and F's must be left out (T: true; F: false). Such rules, however, cannot be laid down until we have actually reached the ultimate analysis of the phenomena in question. This, as we all know, has not yet been achieved". Wittgenstein's propositions "Brown now sits in this chair" and "Jones now sits in this chair" are related with a *dual* spacetime mutual exclusion (two distinct sorts of object-entities are involved, not only one) and furthermore a spacetime mutual exclusion

It claims this in the amputating "purely syntactical" language of classical logic. But what is thus claimed is not a purely syntactical matter, it is a matter of syntax which directly expresses a matter of *fact*. If $p \equiv p_1[\theta_1(\text{œ}_j^{(o)})]$ and $q \equiv p_2[\theta_2(\text{œ}_j^{(o)})]$, the logical product $p \wedge q$ considered above is meaningless with respect to the value "true" of *any* aspect g of any constructible view of empirical truth with respect to which both p or/and q do exist in the sense of D7. This is so in consequence, not of the falsity of either p or q considered separately, but in consequence of the fact `prior` to such a falsity, that the realizability of the testimony $\theta_1(\text{œ}_j^{(o)})$ is incompatible with that of the testimony $\theta_2(\text{œ}_j^{(o)})$, so that p and q *cannot coexist*. To represent this new sort of situation by still saying in an inertial and non specific way that $p \wedge q$ is "false"—exactly as we say in the cases when p and q *can* coexist but one of them is false—amounts to a too loose formalization-and-language which by construction is unable to express the specificities of a whole definite category of cases. Obviously the aim of maximal formal "generality" cannot justify such a categorial non-specificity. In a well-adjusted logical formalization the situation from the last table requires an own syntactical sign that shall prevent void writings of logical products $p \wedge q$ that are *a priori* impossible factually.

This is the usually so fuzzily understood core of what is called "quantum logic", reflected there in such a truncated and distorting fashion.[21]

But as soon as two or more replicas of a given object-entity are allowed (so *a fortiori* if also two or more sorts of object-entities are allowed) the mutual exclusions founded on the unicity of the involved replica of object-entity vanish, and a factual counterpart can be defined for the logical con-

---

that can happen or *not* (if in the second proposition, instead of Jones, we set "Brown's bacterian flora" there is no exclusion any more). Therefore this kind of dual spacetime mutual exclusion cannot be expressed by a *principle* like P10. But it is very striking indeed that—without benefiting of guidance by quantum mechanics, which in the present work led toward "the ultimate analysis of the phenomena in question"—Wittgenstein as early as 1929 identified the decisive *individualizing* role played by spacetime in the factual mutual exclusions of two propositions (he labels propositions like those considered above by the group of letters PT where P means place and T means time: our condition of *one* same replica of the object-entity $\text{œ}^{(o)}$ corresponding to $G^{(o)}$ amounts to a labeling by the same values PT of space-and-time!). It is also striking that, notwithstanding Wittgenstein's work quoted above, the illusory belief of independence of syntax, on semantics, still is so strong up to this very day.

[21] In quantum mechanics the distinction between the individual level of description and the statistical one is not sufficiently clear, so the ways of speaking often seem to involve that qualifications by two mutually incompatible observables are *always* mutually exclusive, while qualifications by two compatible observables are never mutually exclusive: the decisive role of—`exclusively`—the restriction, or not, to only *one* `replica` of the involved object-entity $\text{œ}_j^{(o)}$, is not recognized.

junction of any two successions $[G^{(o)}.V^{(o)}]$, *even if they correspond to mutually* `incompatible` *basic views.* Then, however, one finds oneself already in the realm of statistics, and there, *grosso modo*, the "Boolean" logic, so the algebras from the classical probability spaces, do operate (cf. note 21).

> The classical assumption of a non-restricted possibility of logical conjunction *presupposes* statisticity. The classical Boolean logic is quasi systematically statistical. It overlooks the specificities of strict individuality.

By its "universals" (at least) classical logic usually begins above the level of strict individuality and then keeps floating over it, loose and dead, cut away from its unknown roots implanted in strict factual individuality. While `only` *a level of logical conceptualization where strict individuality is explicitly characterized can contain a* `common foundation` *for classical logic and classical probabilities* (cf. 5.2).

For the particular case of quantum mechanics [13], I have already introduced a logical conjunction restricted by a syntactical sign of factual mutual exclusion between two propositions reflecting the unicity of the involved replica of object-entity. This permits to deal with the question of quantum logic in a much deeper way than the usual one. Now, the mentioned approach can be generalized to any two testimonial propositions $\theta_1(œ_j^{(o)})$ and $\theta_2(œ_j^{(o)})$. When this is done it becomes possible to effectively construct an MRC-calculus with testimonial propositions which connects the level of strict factual individuality, with the statistical level of logic, *via* a very first stratum of logical form where the conjunction is not universally permitted.

> Like the relativization to semantic features of the syntactical logical void, the dependence of the domain of pertinence of the logical conjunction, on semantical features (the mutual incompatibility of two testimonial propositions $\theta_1(œ_j^{(o)})$ and $\theta_2(œ_j^{(o)})$) and so the mutual factual exclusion of the corresponding propositions $p_1[\theta_1(œ_j^{(o)})$ and $p_2[\theta_2(œ_j^{(o)})$, illustrates again how factuality, semantics, can determine logical form.

The MRC-status of the "objects" of the classical logic of classes and predicates

Inside the general category of genetic classes, the classical concept of class is re-obtained in only the following two cases:

(a) A basic genetic class of the type $C[G^{(o)}(V^{(o)}).V^{(o)}]$ is involved, where $V^{(o)}$ denotes a human biological sensorial view. In this case the generator $G^{(o)}(V^{(o)})$—i.e., $V^{(o)}$ *itself* but in the *role* of generator of object-entity—even though it is basic, is not explicitly perceived to create out of the physical reality the corresponding object-entity, namely the field of sensitivity of $V^{(o)}$; while the basic view $V^{(o)}$, again the view involved in the description but which now also plays the *role* of a view, can be assumed without inner contradiction to qualify the created object-entity *without* changing it. This particular sort of basic referential produces a very simplified version of basic description $D^{(o)}$ that can be, and indeed is, spontaneously metaconceptualized intrinsically, by an implicit process; and then it furthermore is immediately reduced implicitly to the corresponding intrinsic model $M(\text{œ}^{(o)})/[V^{(o)}, V_I^{(1)}]$, where the relativities to the basic view $V^{(o)}$ and to the intrisizing view $V_I^{(1)}$ remain hidden, only the model $M(\text{œ}^{(o)})$ itself is perceived, and so it is taken to be absolute. This model is what is illusively felt to somehow exist eternally and immutably, independently of any observer, in an abstract Platonian space where it stays available for passive perceptions of [truth's of P's] (cf. D19.1, and D19.2 with their comments, and Sec. 5.1.1). This—the models $M(\text{œ}^{(o)})$ cut from their relativizing ties with the basic transferred descriptions wherefrom they stem—is the basis of the Platonian realism (in the scholastic sense), which down to the present day grasps the minds with irresistible force. The logicians and mathematicians are particularly exposed to this force because they have found methods to distil consistent systems of very abstract models $M(\text{œ}^{(o)})$ which are so perfect that *a posteriori* they seem to be endowed with divine pre-existence and supreme intelligibility (think of Pean's arithmetic).

(b) A basic genetic class of the type $C[G^{(o)}(V^{(o)}).V^{(o)}]$ is involved where $V^{(o)}$ denotes *extensions* by apparatuses of the domain of human biological sensorial aspect-views. All the preceding remarks are valid for this case also. The intrinsic models elaborated in this somewhat enlarged framework belong to the realm of exact classical sciences (think of what is called atomic spectra, and the corresponding intrinsic models of atoms) to most of which the classical logic still applies.

In both cases mentioned above the *content* of the epistemic operator playing the role G, identifies with the content of the view V which plays the role of a view, and furthermore this view V is reduced to an undefined and structureless abstract "predicate P". So $G \equiv V \equiv P$, all the involved descriptional actors being identified to P. This point-like degeneration is what entails the loss of awareness of the ineluctable action, in any description and so in any proposition, of also a generator G of object-entity. Correlatively the "direction of conceptualization" defined by a double-extremity genetic class $C[G.V]$, gets lost also. The classical definition of a class determined by (the truth of) exclusively a predicate P is just tangential to the superficial level of the already verbalized-conceptualized intrinsic models represented by "object-variables x", a definition which is loose like the needle of a compass on the surface of the earth.

So we have recovered here in analyzed terms a conclusion already asserted in the preliminary critical comments from Sec. 5.1.1:

> Inside MRC, the domain of "objects" directly considered in the classical logic is found to consist of exclusively intrinsic models $M(œ^{(o)})/[V^{(o)}, V_I^{(1)}]$, always conceptual constructs extracted from spontaneously achieved implicit intrinsic metaconceptualizations of degenerate transferred descriptions $D^{(o)}$ produced by successions of the particular type $[G^{(o)}(V^{(o)}).V^{(o)}]$ where the human biological apparatuses cumulate the role of generator of object-entity and the role of view.

These "objects" never disclose the bulk of a as yet non-conceptualized physical factuality from their cores, wherefrom any conceptualization stems *via* basic transferred descriptions. The connection between an intrinsic model $M(œ^{(o)})/[V^{(o)}, V_I^{(1)}]$ and the corresponding basic description $D^{(o)}/G^{(o)},œ_G^{(o)},V^{(o)}/$ remains ignored because both the basic view $V^{(o)}$ and the intrinsizing metaview $V_I^{(1)}$ are wired into the morphology and the reflex functioning of our bodies, so the relativities to this pair of views $(V^{(o)},V_I^{(1)})$ remain hidden to the immediate natural perception of the human mind. The neurobiologists and the cognitivists are now studying them intensively from a psycho-biological standpoint. But among the sciences of non-biological domains of the physical reality, only quantum mechanics has succeeded to get down to these cores of a-conceptual physical factuality hidden inside the classical models, and it has *represented* their extraction as well as their very first transposition in communicable terms, by basic transferred descriptions. It has represented all this indeed, but *only cryptically*, mathematically from the start on, and without being able to formulate their descriptional status, nor to accomplish also the subsequent descriptional phase of intrinsic metaconceptualization. The integral conceptual trajectory that leads from the basic transferred descriptions to classical models $M(œ^{(o)})$, remained hidden to quantum mechanics also, and in consequence of this the universal significance of quantum mechanics itself remained hidden. So the possibility—always—of at least a *minimal* intrinsic model in the sense of D19.3, has not been pointed out, and the *universal* rooting of any model, in physical factuality, remained non perceived. And now, when eventually all this becomes apparent and so a general law of growth of the processes of conceptualization is brought forth, it will be tried for some time, no doubt, to ignore or even to deny it, because a positivistic philosophy has had time to constitute and to consolidate itself, and so now it opposes its own inertial resistance.

In sum, in classical logic we circulate swiftly on an aerial net of smooth highways for deduction, erected out of models drawn from a thick stratum of unsuspected hidden conceptualization that keeps us far from the nourishing background of an as yet non-conceptualized physical reality. The MRC genetic logical approach explicates the presence of this stratum and its whole morpho-functional structure into which the classical models $M(\text{œ}^{(o)})$ are fixed by innumerable genetic threads. This offers now this stratum, as well as the models, to control and deliberate use and also to confrontation with the neural biological processes.

Formal systems versus genetic classes

It might now seem that the integral domain of the MRC-logic of double-extremity genetic classes, can be obtained by simply adding to the sub-domain corresponding to classical logic as specified above, the domain of basic double-extremity genetic classes $C[G^{(o)}.V^{(o)}]$ with $G^{(o)} \neq V^{(o)}$ and where in general $V^{(o)}$ changes to a significant degree the object-entity created by $G^{(o)}$. But in fact such a juxtaposition would not exhaust the domain of the genetic classes $C[G.V]$. Indeed it would leave out all the double-extremity genetic classes produced by a *conceptual* epistemic referential (degenerate or not) that is creative and yields stable relative descriptions of type D14.2.1. While the natural representations, and even the scientific ones, *quite currently* do involve double-extremity creative conceptual genetic classes, notwithstanding that classical logic does not define them.

This is a paradoxical situation of which a massive illustration can be found in mathematics *as well as in the modern formal logic itself*! The central concept in these disciplines is that of a *formal system* S. A finite formal system consists of a finite list of *primitive symbols*, a finite list of terms formed with primitive symbols, a finite list of *well-formed expressions*, a sub-set of well-formed expressions called axioms, and a finite list of *rules of transformation* of a given well-formed expression, in another one. In a non finite formal system the list of primitive symbols can be indefinitely enlarged (as in Peano's arithmetic). The well-known concept of formal system needs no further specification in order to be reconsidered inside MRC, so we do not introduce a specifically MRC-definition. Let us simply note that a formal system is generated by the conceptor's mind *via* a generator of object-entity, say G_S, that consists of an epistemic action upon the zone of "reality" (in the sense of D2) consisting of the "conceptual reality" from the conceptor's mind, his knowledge included, say R_C (cf. D4). The process of generation of this object-entity is quite essentially creative.

We now try to specify what a *formal description* is accordingly to MRC.

Once constructed, a formal system S can be regarded as the abstract zone or domain from "reality" in the sense of D4 where all the formal descriptions permitted by S are carried out. It is a sort of conceptual platform, smooth, stable and solid, conceived in order to permit us to achieve on it particularly precise descriptional trajectories. So S itself has to be constructed in the first place; afterward one can elaborate also descriptions "in" S. This preliminary condition for the achievement of a formal description will have to be somehow explicitly expressed in the specification of the notations that characterize a description in S.

In all the mathematical or logical treatments it is assumed more or less implicitly that as soon as the formal system S is given, *ipso facto* one knows how to work with it because the rules are incorporated. But inside MRC one is obliged by method to always specify explicitly the epistemic referential (G, V) inside which a (relative) description $D/G,œ_G,V/$ is attempted, as well as the involved object-entity $œ_G$. So we ask: of what does a formal description consist, what plays in it the role of generator of object-entity, what plays the role of object-entity, and what plays the role of view?

A formal description from a given formal system S is a finite proof carried out inside S. Let us call it here a *proof-description* and label it D_j^S where j is an index that distinguishes between the various proof-descriptions from S. By classical definition, a proof-description D_j^S consists of a finite sequence of n well-formed expressions that are all permitted in S in consequence of the fact that the sequence always starts with an axiom or a well-formed expression known to follow from the axioms from S (theorem) which then, in a sequence of n descriptional steps D_{jk}^S, $k = 1, 2, \dots, n$, is progressively transformed by the combined use of rules of transformation from S and of "lateral" introductions of other axioms from S or of already proven theorems from S, the end of the sequence being reached when an "interesting" well-formed expression emerges which previously was not known to follow from the axioms of S and which now is listed as a new theorem in S.[22]

Let us denote by $œ_j$ this final well-formed expression from S: it can be regarded—at least *a posteriori*—as the object-entity of the considered

[22] Since a given theorem is the result of a definite proof, it might seem inconvenient to mix the definitions of distinct proofs by making use in the definition of a proof A, of theorems established in other proofs. But the use in A of a theorem established in another proof B is just a short-hand for the—equivalent—introduction of that whole other proof B. So the definitions of the various proofs are separable.

proof-description D_j^S.

So the involved generator of object-entity is by definition that which generates œ$_j$. For this the generator must dispose of S. Therefore it is pertinent to posit for the notation of the generator of œ$_j$ the form of a product of two successive operations of generation, say $G_j^S = G_j G_S$ where: G_S acts first on the zone R_C from reality consisting of the conceptor's mind, thereby producing the zone R_S of "reality" consisting of the formal system S; G_j acts subsequently, on $R_S \equiv S$, thereby producing the well-formed expression œ$_j$ to be proven. (Of course this analysis is only notational. Once S has been created by the epistemic action labelled G_S it remains indefinitely available, and there is no need to effectively re-produce its generation for each object-entity œ$_j$: only G_j has to be chosen and acted with in each case.)

The aim of D_j^S is to establish whether yes or not œ$_j$ is provable inside S. So œ$_j$ has to be examined by a formal view of provability inside S. (Retroactively it is always possible to represent the proof-process in this way, though in fact most often œ$_j$ emerges constructively *together* with its proof). We introduce now an explicit MRC-definition of the view that acts in a proof-description:

DL.5. ***View of demonstrability in S.***[23] Consider a formal system S. From the classical definitions of S and of the proof-descriptions D_j^S from S it follows that the view which acts in D_j^S must be able to qualify the object-entity œ$_j$ in terms of the aspect-values of two aspects g_1 and g_2, namely: ***(a)*** an aspect of *form* inside S, $g_1 \equiv \Phi$, endowed with two aspect-values, say, respectively, Φ-yes (well-formed inside S),Φ-no (not well-formed inside S); and ***(b)*** an aspect of *transformation* inside S, say $g_2 \equiv \Theta$, equally endowed with two aspect-values, say Θ-yes (correctly transformed inside S), Θ-no (not correctly transformed inside S). The view consisting of these two aspects is a formal view relative to S that will be called a *view of demonstrability* in S. It will be labelled V_d^S.

Comment. The upper index S stresses that the formal view V_d^S is extracted from S (remember that according to MRC this dependence between V_d^S and S, so also between V_d^S and G_S (which is involved as a "factor" in the global generator $G_j^S = G_j G_S$) is a *restriction* with respect to the most general situation of mutual in-dependence between the generation operators and the acting view). The aspect Φ from V_d^S qualifies accordingly to the list of well-formed expressions posited in S, and its aspect Θ qualifies

[23] We choose the word demonstrability only in order to index by d: the word provability would require the index p that might lead to confusion with indexes concerning the concepts of proposition or of probability.

accordingly to the transformation-rules posited in S. So in fact what V_d^S is able to ascertain for any expression from a proof-chain, is just that it is formally consistent with the requirements of well-formedness and of ways of transformation from S.

> V_d^S is—exclusively—a yes-no filter concerning well-formedness in S and transformation in S. Nothing else.

The fundamental but often obscure problems concerning the relations between demonstrability in S and "truth", will be discussed in the next paragraph. For the moment, in what follows immediately we speak *only* of demonstrability.

So the epistemic referential corresponding to D_j^S is $(G_j^S, V_d^S) \equiv (G_j G_S, V_d^S)$.

Now, how can we represent the *emergence* of a proof-description $D_j^S/G_j^S, œ_j, V_d^S/$? Can it be conceived as the result *directly* produced by a corresponding genetic class $C[G_j^S.V_d^S]$, i.e., as the result of re-productions of a set of successions $[G_j^S.V_d^S]$ defined *from the start on*? The structure posited for a proof-description D_j^S shows immediately that the answer is *negative.* Indeed in the course of the elaboration of D_j^S the view V_d^S does not work constantly on one same object-entity, namely the object-entity $œ_j$ generated by G_j^S which has to be proven in S. V_d^S works on *other* intermediary well-formed expressions produced by other generators that become possible progressively while the proof-description D_j^S is developing. So a more analyzed answer is needed here. It can be established as follows.

We have noted before that the integral description D_j^S emerges by n successive mutually different descriptional steps. Let us denote by D_{jk}^S, $k = 1, 2, \ldots, n$ the k-th step. This is a one-step "elementary" proof-description $D_{jk}^S/G_{jk}^S, œ_{jk}, V_d^S/$. It involves a generator of object-entity $G_{jk}^S = G_{jk} G_S$ which is *different* from the generator $G_j^S = G_j G_S$ from the epistemic referential (G_j^S, V_d^S) corresponding to the integral proof D_j^S and produces a "local" object-entity $œ_{jk}$ that is *different* from the object-entity $œ_j$ from $D_j^S/G_j^S, œ_j, V_d^S/$ (the index j which has been conserved in the notations D_{jk} and $œ_{jk}$ reminds that the designata of these notations are both referred to the object-entity j). In the first descriptional step $D_{j1}^S/G_{j1}^S, œ_{j1}, V_d^S/$ the object-entity j1 with which D_{j1}^S *ends* (which had to be ascertained) is produced by a generator G_{j1}^S that still acted on the zone from the conceptual reality consisting of S *itself*, like in the case of the global generator G_j^S from the integral proof-description D_j^S, but nevertheless this is already *another* generator because it produces the object-entity $œ_{j1} \neq œ_j$. And for $k > 1$ the corresponding generator G_{jk}^S does not even work on S any more. It works on $[S \cup \{œ_{jm}\}]$, $m = 1, 2, \ldots, k-1$ where $\{œ_{jm}\}$ is the set of well-formed expressions of which the demonstrability in S has been established by the sequence $D^S_{j1} D^S_{j2}, \ldots, D^S_{jk-1}$ of the previously accomplished elementary descriptional steps. (The set $\{œ_{jm}\}$, $m = 1, 2, \ldots, k-1$ has to be added to S because now it is explicitly available in the conceptor's mind and the k-th choice of an object-entity $œ_{jk}$ takes support on this set also, not only on S any more). So the integral proof-description D^S_j emerges by an "additive composition in succession" of the n elementary step-descriptions D^S_{jk}, $k = 1, 2, \ldots, n$ where the generator of object-entity and the corresponding object-entity *change*—in a way that is not prescribed by S – while the view remains the same. In these conditions we can show that:

ΠL6. *Proposition.* *A proof-description D^S_j can be considered to be produced by a non-degenerate double-extremity and* creative *genetic class.*

"Proof". A one-step description D^S_{jk}, for any k between 1 and n, can be non trivially regarded as the result of the corresponding double-extremity genetic class $C[G^S_{jk}.V^S_d]$. Indeed the succession of epistemic operations $[G^S_{jk}.V^S_d]$ is indefinitely repeatable and its result stays unchanged, namely it is the k-th final well-formed expression $œ_{jk}$ that has been shown to be provable. So we are strictly in agreement with the concept of an individual conceptual description produced by a double-extremity genetic class, as formed by the definitions D14.1, D14.2.1, and DL1.2. Furthermore, the *ordered* juxtaposition in succession of the elementary proof-descriptions D^S_{jk} brought forth by the elementary genetic classes $C[G^S_{jk}.V^S_d]$ with $k = 1, 2, \ldots, n$, yields a definite new description in the sense of D14.2.1, namely precisely the integral proof-description D^S_j as defined from the start on. So we can write

$$D^S_j / G^S_j, œ_j, V^S_d / \equiv \sum_k D^S_{jk} / G^S_{jk}, œ_{jk}, V^S_d /,$$

$k = 1, 2, \ldots, n$ where in the last descriptional step $D^S_{jn}/G^S_{jn}, œ_{jn}, V^S_d/$ the object-entity $œ_{jn}$ has the same content as $œ_j$ but is generated by the generator G^S_{jn}—heuristically different from G^S_j—which acts on $[S \cup \{œ_{jm}\}]$, $m = 1, 2, \ldots, n-1$, not exclusively on S like the (in general fictitious) generator G^S_j. So $D^S_j/G^S_j, œ_j, V^S_d/$ is the global result of the "sum in succession" $\sum_k D^S_{jk}|G^S_{jk}, œ_{jk}, V^S_d|$ (see "Outlook on a calculus with genetic classes"). Which means that we can write $\sum_k [C^S_{jk} \cdot V^S_d] \cong [G^S_j \cdot V^S_d]$. In *this* sense D^S_j can indeed be considered to be produced by the genetic class $[G^S_j \cdot V^S_d]$ (which, globally, can be repeated an arbitrary number of times, once it has been obtained from the sum $\sum_k [G^S_{jk} \cdot V^S_d]$, $k = 1, 2, \ldots, n$. This establishes

πL6.[24]

Comment. In the first place, the fact that $œ_{jn}$ has the same content as $œ_j$ while on the other hand $G^S_{jn} \neq G^S_j$, might seem to contradict the one-one relation $G - œ_G$ posited in D4. But in fact G^S_{jn} working on $[S \cup \{œ_{jm}\}]$, $m = 1, 2, \ldots, n-1$ amounts to an effective and explicit representation of precisely the global generator G^S_j, *iff* the proof of $œ_j$ succeeds (if not, the very concept of what is denoted G^S_j is discarded). So the effective expression of G^S_j is given by the definition

$$D^S_j/G^S_j, œ_j, V^S_d/ \equiv \sum_k D^S_{jk}/G^S_{jk}, œ_{jk}, V^S_d/, \quad k = 1, 2.....n.$$

What appears here is that, as already remarked, the *a priori* operation of generation G^S_j of the well-formed expression $œ_j$ to be proven is in general just an *a posteriori* fiction, that in fact $œ_j$ is obtained progressively, constructively, by trial and error, while $D^S_j/G^S_j, œ_j, V^S_d/ \equiv \sum_k D^S_{jk}/G^S_{jk}, œ_{jk}, V^S_d$, $k = 1, 2, \ldots, n$, is being sedimented. And once $œ_j$ and G^S_j have been settled—together—the one-one relation between them *is* insured: I postulate that two different proofs *never* have identical results, they can imply the same result, but each one also has specific entailments.

In the second place, the definition

$$D^S_j/G^S_j, œ_j, V^S_d/ \equiv \sum_k D^S_{jk}/G^S_{jk}, œ_{jk}, V^S_d/, \quad k = 1, 2, \ldots, n,$$

obtained above suggests that the to-be-established calculus with genetic classes will include a *general* definition of "additive composition in succession" of certain types of genetic classes. (Such a definition can appear to be important in an attempt at a mathematical formalization of MRC).

In the third place:

[24] It is noteworthy that in a certain sense the structure found for the process of emergence of a proof-description presents certain similitudes with the way in which the basic transferred quantum mechanical description of a microstate is brought forth. Indeed the quantum mechanical measurement-evolutions draw into the realm of the observable and communicable, aspects of the studied microstate that can be conceived *a posteriori* as relative potentialities possessed *ab initio* by the studied microstate which have been actualized by the measurement evolutions. While the provability in S of the studied well-formed expression $œ_j$ can also be conceived *a posteriori* as a potentiality of S to yield $œ_j$ that has been actualized by the proof-description $D^S_j/G^S_j, œ_j, V^S_d/$. The visibility of all the intermediary steps D^S_{jk}—devoid of equivalent in the quantum mechanical case—stems from the fact that here the cognitive situation is different, the object-entity as well as the whole descriptional process being conceptual, which permits a uniform perceptibility that cannot be realized for a physical microprocess.

While a formal system S itself is an object-entity generated by a *creative* abstract generator, i.e., it does not pre-exist like a "value" of a classical "object-variable x", furthermore the concept of formal proof inside S, in its turn, appears to have the nature of a relative description produced by *genetic* classes, not by pre-existing shadow-"predicates P": the concepts that are the very core of the modern classical logic stem from epistemic actions that are *not defined* inside modern classical logic.

This paradoxical situation illustrates strikingly how we currently act inside conceptual volumes that are not included in our explicit representations.

We sum up. The whole set of the researched MRC-terms concerning a proof-description is this. The object-entity generator $G_j^S = G_j G_S$ is a fundamentally creative conceptual generator consisting of, first the construction of the stable formal "ground" consisting of S itself, and then, out of S, of the choice or the construction of the object-entity $œ_j$ to be proven in S. The view is the formal view V_d^S of demonstrability in S, extracted from S, so a view that depends on S (or, equivalently, on G^S, so on G_j^S). So the epistemic referential where any proof-description D_j^S is achieved is (G_j^S, V_d^S). The explicit structure of a proof-description in S is

$$D_j^S / G_j^S, œ_j, V_d^S / \equiv \sum_k D_{jk}^S / G_{jk}^S, œ_{jk}, V_d^S /,$$

$$k = 1, 2, \ldots, n, \quad œ_{jn} \equiv œ_j, \quad G_{jn}^S \neq G_j^S.$$

The MRC-relations between empirical truth and demonstrability

We have shown that a view of demonstrability V_d^S in the sense of DL.5 has nothing to do with empirical truth as ascertained by a metaview DL.2. However it is quite currently said of an expression which has been proven in S *via* V_d^S that it has been shown to be "true". Those who want to be more specific make sometimes use of the expression "formally true" in S. Furthermore a well-formed expression that has been proven in S, is often referred to as a "proposition" which, "because" it has been proven in S, necessarily is also "true", not in S this time, just true in the sense of "empirical mathematical truth". Whereas inside the MRC-logic a (relative) proposition in the sense of DL.3 is a concept quite different from a well-formed expression from a formal system (which is consistent with the axioms from S *via* the transformation rules from S), it concerns empirical truth, not consistency. In fact *all the formulations of the sort mentioned above, where the word true is made use of, are related with the supposition that the axioms from S are empirically true.* If this is not made clear it might entail much confusion. So below I shall now explicate the MRC-relations between propositions tied with empirical truth, axioms, and demonstrability inside a formal system. For the logicians and mathematicians such specification are certainly trivial: I apologize for this.

According to MRC, a view $V_\tau^{(2)}$ of empirical truth (DL.2) is a metaview which can exist in the sense of D7 only with respect to a previously achieved relative description $D/G,œ_G,V/$. Only a piece of meaning that has been elaborated into a relative description previously, independently of any question of truth, can afterward be found to be empirically true or false; and this, if it can happen at all, can happen only with respect to a *specified* metaview $V_\tau^{(2)}$ of empirical truth. An absolute assertion of empirical truth is rejected inside MRC as devoid of significance. This is why the MRC-concept of proposition $p(D, V_\tau^{(2)})$ defined in DL.3 is a metaconcept, and is doubly relative. It must involve an independently constructed description and a metaview of empirical truth constructed for definite aspects.

A view V_d^S of demonstrability in a formal system S (DL.5) can exist in the sense of D7 only with respect to well-formed expressions from S. Though inside another formal system which is a metasystem with respect to S it might also be possible to construct a metaview of demonstrability in S, V_d^S is not *quintessentially* a metaview.

Now, in general a relative description $D/G,œ_G,V/$ is not a well-formed expression from a formal system S, so in general it does not exist in the sense of D7 with respect to a view of demonstrability V_d^S; in general well-formedness and correction of transformation inside some formal system have no relevance with respect to a relative description. And *vice versa*, in general a well-formed expression from S is not a relative description, it is just a sequence of signs permitted inside S, expressly posited to have been purified of any semantics, of any meaning; it is by construction "invisible" to the views of empirical truth which consist of procedures for testing assertions of values of empirically perceivable aspects g. So the relative descriptions, the metaviews of empirical truth and the propositions, form a group of essentially semantical concepts which simply have nothing to do with the well-formed expressions and the view of demonstrability from a "purely" formal system.

In these conditions, what is the reason why provability and truth are so readily coalesced with one another?

The main reason is the current assertion that the axioms from a formal system S are posited to be true. But in fact no formal system at all is—*stricto sensu*—concerned by this way of speaking. The axioms are posited to be true only in the *interpretations* of a formal system S, if these exist, or in the deliberate formalizations of this or that theory of a domain of empirical facts (physical or abstract) that has first been constructed quite independently of any formal system and afterward has been axiomatized, and then formalized. In both these cases the "axioms" are well-formed expressions from the formal system obtained in this way, which are explicitly constructed so as to translate relative propositions $p(D, V_T^{(2)})$ posited to obtain the empirical-truth-value "true" when the description D involved by $p(D, V_T^{(2)})$ is examined *via* the metaview $V_T^{(2)}$ of empirical truth relatively to which $p(D, V_T^{(2)})$ is defined (the form required by DL.3 can always be achieved). So *in these cases the axioms are double-faced.* On one hand they are just asemantical well-formed expressions from the considered formalized system, on the other hand they are meaningful propositions concerning empirical facts and posited to be true.

Now, one of the theories of a domain of empirical (conceptual) facts, namely deductions, is logic. Logic establishes *logical laws*, "tautological propositions" that are always true exclusively in virtue of their mere form: a composed proposition where the atomic propositions are laws of physics, can have a form *such* that exclusively the truth value 1 ("true") is assgined to it by its truth-table, which means that the composed proposition can be true even if some of the physical laws asserted by the atomic propositions, or *all* these laws, are false. In this sense the tautological logical axioms are *closed* with respect to non-logical domains of facts; they are isolated from the truth-qualifications of the atomic propositions which concern factual domains different from the logical one; they are endowed with an immutable truth-value "true" which concerns exclusively *logical* empirical truth, "logical form", being devoid of reference to any view of empirical truth different from the view of "logical empirical truth" (if such an expression is permitted). B Russell [26] wrote:

> "All the propositions that are demonstrable in any admissible logical system must share with the premises the property of being true in virtue of their logical form; and all propositions that are true in virtue of their logical form ought to be included in any adequate logic."

(Here "premises" stands for "axioms".) But a formalization of logic can introduce also axioms that are *not* tautologies (the axiom of infinity, the axiom of choice), whereby empirical truth of non-logical essence can be also injected into a formalization of logic: this, in Russell's view, is *a problem.* Anyhow the essential point in this context is that even the logical laws which by their tautological form express *logical* empirical truth, have been constructed such by man, with the deliberate aim to codify in a performing and method-offering way the domain of facts consisting of human deductive reasoning. *Logical systems are not purely formal systems, they are formalizations of a theory which legalizes, normalizes a domain of conceptual facts.* They build methods for the conservation and vehiculation of the empirical truth captured in the axioms. *In any acceptable formalization of logic this basic aim entails intimate relations between logical empirical truth and demonstrability* (still a rather unexplored domain).[25]

---

[25]**Note added in proof.** Now, what—exactly—is a "logical truth" (a tautology, a logical axiom) according to MRC?

Consider an atomic relative proposition in the sense of DL.3, $p(D, V_\tau^{(2)})$ where D stands for a relative description $D/G, œ_G, V/$ and $V_\tau^{(2)}$ represents a metaview of empirical truth supposed to exist with respect to D in the sense of D7. Furthermore, consider a metaproposition composed from atomic relative propositions with the help of the usual logical operators $(\vee, \wedge, \neg, \equiv)$. Call it a relativized composed proposition. The "logical form" of a composed proposition is determined exclusively by the way in which the propositions from it are connected *via* logical operators (and parentheses). We assert the following (non-numbered) proposition:

ΠL. According to MRC a "logical truth"—a tautology—is a metadescription of a set of two previously specified relative propositions (atomic or composed), with respect to a *metaview of comparison* **of the locations inside the global qualification space introduced by the views** V **from the involved descriptions**, this metadescription yielding uniformly the result "identical location" in consequence of—**exclusively**—the logical forms of the two compared propositions (so independently of the object-entities, views and metaviews of empirical truth involved by these). *Mutatis mutandis*, a similar conclusion holds concerning logical contradictions.

"Proof". Let us begin by an example. Consider the tautology $p \vee q \equiv [(p \wedge q) \vee (p \wedge \neg 1) \vee (\neg p \wedge q)])$. Inside MRC-logic p and q are two relative atomic propositions $p(D, V_\tau^{(2)})$ and $q(D', (V_\tau^{(2)})')$ where D and D' stand for two relative descriptions $D/G, œ_G, V/$ and $D'/G', œ'_G, V'/$, while $V_\tau^{(2)}$ and $(V_\tau^{(2)})'$ stand for two different metaviews of empirical truth in the sense of DL.2, supposed to exist in the sense of D7 relatively to, respectively, D and D'. Let us denote by *cp*1 and *cp*2 the two relativized composed propositions of which consist, respectively, the first and second member of the tautology. Both members involve the same qualification space, say S_Q, namely that one determined by all the aspect views involved by V or by V' or by both (each one counted one time). *No matter* what sort of object-entity $œ_G$ is considered in D, according to the definitions D14 of a relative description D finally sends into a certain subspace or "atomic location" $L(D)$ from S_Q, while D', for *any* object-entity $(œ_G)'$, sends into a certain atomic location $L(D')$ from S_Q.

In the first member *cp*1 of the tautology these two atomic locations are then composed accordingly to the logical form expressed there by the help of the operator $\vee$ of logical sum, thus yielding a global "composed location" $L1(D, D')$ (which, according to the well-known rules of the classical logic of propositions and its relation with the logic of classes determined by predicates, consists in this particular case of the set theoretic sum $L(D) \cup L(D')$).

In the second member *cp*2, D and D' from the arguments of, respectively, $p(D, V_\tau^{(2)})$ and $q(D', (V_\tau^{(2)})')$, act like in the first member *cp*1, while in the arguments of $\neg p(D, V_\tau^{(2)})$ and $\neg q(D', (V_\tau^{(2)})')$, D and D' send, respectively, into the complement of $L(D)$ with respect to S_Q, say $K[L(D)/S_Q$, and the complement of $L(D')$ with respect to S_Q, say $K[L(D')/S_Q$.

The preceding remarks hold also for the logico-mathematical systems. These are constructed as formalizations of this or that domain of spontaneously formed intrinsic mathematical models in the sense of D19.2 (the integer numbers, the geometrical objects, etc.). By insertion into such a formal-

---

These 4 atomic locations $L(D), L(D'), K[L(D)/S_Q, K[L(D')/S_Q$, are then composed accordingly to the logical form of *cp*2, as it is determined, accordingly to the rules from the classical logic, by the operations indicated by the symbols involved in *cp*2 and their succession. This, again, sends finally into some global composed location $L2(D, D')$ from S_Q.

Consider now the identity sign "≡". Since, once performed, *cp*1 and *cp*2 *reduce* to, respectively, $L1(D, D')$ and $L2(D, D')$, this identity sign can only assert that these two locations $L1(D, D')$ and $L2(D, D')$ from inside S_Q, coincide. So, in MRC-terms, *the identity sign "≡" from the considered tautology acts like a value "identical location inside S_Q" of a metaview $V_L^{(2)}$ of comparison of cp1 and cp2 with respect to an aspect of* location, say, L, inside the global qualification space S_Q determined by the views V and V' involved in the descriptions D and D' from the atomic relative propositions $p(D, V_T^{(2)})$ and $q(D', (V_T^{(2)})')$ on which the tautology is built. Furthermore, since the calculations that lead to $L1(D, D')$ and $L2(D, D')$ do *not* depend on the truth valuations of p and q that can be obtained by the use of the metaviews $V_T^{(2)}$ and $(V_T^{(2)})'$ from the atomic propositions $p(D, V_T^{(2)})$ and $q(D', (V_T^{(2)})')$, nor on the object-entities $œ_G$ and $(œ_G)'$ or on the views V and v' from these, the considered tautalogy asserts the identity between the locations $L1(D, D')$ and $L2(D, D')$ *independently* of—all—$œ_G$, $(œ_G)', V$ and $V', V_T^{(2)}$ and $(V_T^{(2)})'$. So this identity between the two locations $L1(D, D')$ and $L2(D, D')$ inside the global qualification space S_Q introduced by $p(D, V_T^{(2)})$ and $q(D', (V_T^{(2)})')$, emerges as a consequence of *exclusively* the logical forms of *cp*1 and *cp*2.

The preceding reasoning holds in its essence for *any* tautology.

A similar reasoning can be constructed concerning logical contradictions (logical impossibilities), and it leads to the following conclusion. The two members *cp*1 and *cp*2 of any logical contradiction reduce to two global locations $L1(D, D')$ and $L2(D, D')$ which, with respect to the involved global qualification space S_Q, cannot *both* realize concerning the *unique* set of atomic relative propositions from *cp*1 and *cp*2, because, in consequence of *exclusively* the logical forms of *cp*1 and *cp*2, one of these locations emerges as interior to S_Q while the other one emerges as exterior to S_Q, *independently* of the contents of the elements $œ_G, V, V_T^{(2)}$ introduced by each atomic relative proposition from the contradiction.

Which establishes the asserted proposition.

Comment. So a tautology or a contradiction filters out two different "formal calculational dynamics of occupation of a location" (accordingly to the rules of the logical calculus) relative to the one same given subspace from the global qualification space S_Q introduced by the views (in the sense of Secs. 5.1 and 5.2) from the relative descriptions from the relative atomic propositions from the tautology or contradiction, this dynamics being independent of the object-entities, views and metaviews of empirical truth introduced by these atomic relative propositions. In a tautology these different calculational dynamics end up with the same subspace of S_Q, while in a contradiction they end up with two mutually exclusive locations. This seems more precise a characterization than just asserting like Wittgenstein that tautologies and contradictions are not propositions, but "limiting cases—indeed the disintegration – of the combination of signs" (*Tractatus*, 4.06, 4.462, 4.466, ...). Instead of cryptic formulations, MRC brings forth explicit and crystal clear ones.

ization these models are much purified, accordingly to various requirements, and are organized in structures endowed with a strict formal coherence. The result is endowed with a power of rigorous deductive re-expression of the essential features of the initial spontaneous models wherefrom it stems, which often is so remarkable that it is perceived as if miraculous.

But in a genuinely "pure", non-interpreted formal system, the axioms are *not* also relative propositions, they are exclusively well-formed expressions from S selected as those by which a proof-description is permitted to *start*: this is the specificity of an axiom from a strictly formal system, *not* truth (think of formal games or of certain calculi). The axioms from a non interpreted formal system are simply not connected with the concept of empirical truth. This, however, is forgotten in the current ways of speaking, just because formal systems which are neither interpreted, nor interpretable, nor obtained from a theory of a domain of empirical facts by axiomatization and formalization, are devoid of interest. So the double-faced [axioms-propositions] are present in the mind as soon as one thinks of an interesting case, and therefore it is continued to think and speak in terms of truth of the axioms. Then, given that formal proofs start with axioms, furthermore the intermediary well-formed expressions are often called propositions, and the theorems, having been proven, are *ipso facto* considered to be also empirically true. Which amounts to a surreptitious fading away of the case of exclusively formal characters, and a fallacious substitution to these, of *semantic*-deductive characters.

Inside MRC this sort of gliding is refused by method. We are in possession of an explicit definition of each one of the involved concepts: *a priori* possibility of relative meaning in the sense of D7, piece of elaborated relative meaning in the sense of one or the other of the definitions D14 of a relative description, relative view of empirical truth in the sense of DL.2, relative proposition in the sense of DL.3, formal system S, view of demonstrability inside S in the sense of DL.5, proof-description inside S. We shall never say that the axioms from a purely formal system S are posited to be true; nor shall we say that a theorem from S is a well-formed expression that has been proven to be true in S, we shall only say that it has been proven to follow from the axioms in the way required in S. And we shall distinguish sharply between a formal system and the formalization, logical or logico-mathematical, of a theory of a domain of physical or conceptual facts, logic itself included.

These distinctions do by no means exclude the pertinence of the concept of empirical truth concerning the work with "formal facts". The metaviews of empirical truth are here recognized to have the major role in the intuitive pre-construction of formalizations of a theory of a domain of facts—logic included—as well as in the intuitive pre-construction of complex proof-descriptions. For instance, one can want to prove inside arithmetic that given any prime number there always exists a bigger one. The above expression of this assertion in terms of usual language can be without difficulty put in the canonical form of a relative proposition in the sense of DL.3, defined relatively to a specified metaview of "mathematical empirical truth" that introduces a case by case examination of truth-value consisting in each case of the exhibition of an example. With respect to this metaview of mathematical empirical truth one might then find that the assertion has never been found false in any of the examined cases. This sort of empirical (conceptual) research develops in the conceptor's mind the preliminary intuitions necessary for becoming able to attempt a proof-description inside, say, Peano's formalized arithmetic (how to start, what deductive trajectories to imagine tentatively, etc.). But this preliminary empirical work is not the researched formal proof itself, and this proof, if it can be achieved, *cannot* make an explicit, declared use of the metaviews of mathematical empirical truth that generated the intuitive knowledge of the conceptual situation: there is no *place*, in a formalized mathematical proof, for metaviews of empirical mathematical truth.

Gödel's proofs versus MRC

Let us now consider the properties of completeness or decidability of a *formalized* system S (not a purely formal one), and of consistency of this system[26]. This leads to Gödel's famous proofs [27]. These proofs establish that ***(a)*** if Peano's first order formalization of arithmetic, A_P say, is posited to be representable inside the formalization of logic achieved by Russell and Whitehead [28] in *Principia Matematica* (PM), then A_P is found not to be complete; and ***(b)*** from this first conclusion of non-decidability of A_P it follows (with a slight generalization of PM) that the consistency of A_P cannot be proven inside A_p either. These results hold for a large class of other formalized systems and other formalizations of logical thought.

We shall now show that MRC throws a new light on the questions of consistency and completeness. In the first place, it entails—quite *independently* of the question of completeness—that in general the consistency of a formal system cannot be formally examined inside this system. The same impossibility holds concerning the completeness of the system, this

[26] *Completeness* (or decidability) of S: the (presumed) property of S according to which any expression that one can exhibit, which is well-formed according to S, is decidable *in* S, i.e., either this expression or its negation can be proven in S. *Consistency* of S: the (required) property of S according to which, for any well-formed expression from S that can be exhibited, it is not possible to prove in S both this expression and its negation.

time quasi without reservations. So—according to MRC—examination of both consistency and completeness, but independently of one another, require the specification of a metasystem and are then explicitly relative to the utilized metasystem, not just properties of the studied formal system itself. In the second place, MRC suggests to *require by method* that a "good" metasystem, offering an optimized formalization of the logico-mathematical thinking, *shall not permit inside itself undecidable expressions that can be treated like propositions* (which *PM* does permit). Thereby MRC displaces the accent from a deductive point of view centered upon the studied formal system, to a constructive methodological point of view concerning the acceptable metasystems.

We begin by reproducing Sec. 1 from Gödel's work.[27]

ON FORMALLY UNDECIDABLE PROPOSITIONS OF *PRINCIPIA MATEMATICA* AND RELATED SYSTEMS[1]

by Kurt Gödel, Vienna

1

The development of mathematics in the direction of greater exactness has—as is well-known—led to large tracts of it being formalized, so that proofs can be carried out according to a few mathematical rules. The most comprehensive formal systems yet set up are, on the one hand, the system of *Principia Matematica (PM)*[2] and, on the other, the axiom system for set theory of Zermello-Fraenkel (later extended by J. v. Neumann[3]. These two systems are so extensive that all methods of proof used in mathematics today have been formalized in them, i.e., reduced to a few axioms and rules of inference. It may therefore be surmised that these axioms and rules of inference are also sufficient to decide all mathematical questions which can in any way at all be expressed formally in the systems concerned. It is shown below that this is not the case, and that in both the systems mentioned**[28]** there

[27] Lacking the German original, the English translation, found on the web, has been verified with the French one as published in E. Nagel, J. R. Newman, K. Gödel, and J.-Y Girard, *Le théorème de Gödel (Seuil, 1989).* Taking into account both the significance of the word and its French translation, we have substituted the word "true" to the word "correct", which in the English translation available to us introduced confusion. Gödel's notes are all reproduced—with their own numbering—after the quotation from his main text, in order to avoid confusion with our own notes. Those among Gödel's notes that are irrelevant here (bibliography) are not reproduced, only their existence is indicated, followed by dots. *Our* notes concerning the quotation from Gödel's text are inserted in the general series of our notes, but their numbers are written with Arial Black characters.

[28] These are (essentially) formalizations of logic, so involving meaning and empirical truth.

are in fact relatively simple problems in the theory of ordinary whole numbers[4] which cannot be decided from the axioms. This situation is not due in some way to the special nature of the systems set up, but holds for a very extensive class of formal systems, including, in particular, all those arising from the addition of a finite number of axioms to the two systems mentioned,[5] provided that thereby no false propositions[29] of the type described in footnote 3 become provable.

Before going into details, we shall first indicate the main lines of the proof, naturally without laying claim to exactness. The formulae of a formal system—we restrict ourselves here to the system *PM*—are, looked at from outside, finite series of basic signs (variables, logical constants and brackets or separation points), and it is easy to state precisely just which series of basic signs are meaningful formulae and which are not[6,30] Proofs, from the formal standpoint, are likewise nothing but finite series of formulae (with certain specifiable characteristics). For metamathematical purposes it is of course immaterial what objects are taken as basic signs, and we propose to use natural numbers[7] for them. Accordingly then, a formula is a finite series of natural numbers[8], and a particular proof-schema is a finite series of finite series of natural numbers. Metamathematical concepts and propositions thereby become concepts and propositions concerning natural numbers, or series of them[9],[31] and therefore at least partially expressible in the symbols of the system *PM* itself. In particular it can be shown that the concepts "formula", proof-schema", "provable formula" are definable in the system *PM*,[32] i.e., one can give[10] a formula $F(v)$ of *PM*—for example with one free variable v (of the type of a series of numbers), such that $F(v)$—

---

[29] Although he employs the word "false", Gödel means here *apparent* propositions, not untrue ones, as his note 4 shows: He explicitly says there that the "false" (apparent) propositions from the metasystem are *undecidable*. But he continues to make use of the word "proposition" in order to point toward these only apparent propositions. Though in the explicit conclusion of his proof as presented in the above-quoted section 1, Gödel did not assign a role to this fact, let us note that it was present to his mind. It will appear below that this fact is the crucial feature for understanding the MRC-significance of Gödel's work.

[30] Later in his proof Gödel re-expresses the meaningful sequences of signs which he wants to make use of, in terms of *defined* notations that point briefly toward the logical meaning of the considered sequence of signs (variable, proof- sequence, provable, etc.).

[31] "Isomorphic" in Gödel's note 5 means that the logico-mathematical meanings and the truth valuations are preserved.

[32] For instance, according to the PM definition no. 20, "x is an elementary formula" is the meaning of the writing $Elf(x)?(\exists y,z,n)[y,z,n \leq x \; \& \; Typ_{n+1}(z) \; \& \; x = z = Typ_n(y)$, and "$x$ is a provable formula" is the meaning of the writing (in the German original) $Bew(x) = (Ey)yBx$ from the definition no. 46.

interpreted as to its content—states: v is a provable formula. We now obtain an undecidable proposition of the system PM, i.e., a proposition A, for which neither A nor not-A are provable, in the following manner.

A formula of PM with just one free variable, and that of the type of the natural numbers (class of classes), we shall designate a *class sign*. We think of the class signs as being somehow arranged in a series[11], and denote the n-th one by $R(n)$; and we note that the concept "class-sign" as well as the ordering relation R are definable in the system PM. Let α be any class-sign; by $[\alpha; n]$ we designate that formula which is derived on replacing the free variable in the class-sign α by the sign for the natural number n. The three-term relation $x = [y; x]$ also proves to be definable in PM. We now define a class K of natural numbers, as follows:

$$n \in K = \sim (Bew[R(n); n])^{11a} \tag{1}$$

(where $Bew\ x$ means: x is a provable formula). Since the concepts which appear in the definiens are all definable in PM, so too is the concept K which is constituted from them, i.e., there is a class-sign S[12] such that the formula $[S; n]$—interpreted as to its content—states that the number n belongs to K. S, being a class-sign, is identical with some determinate $R(q)$, i.e.,

$$S = R(q)$$

holds for some determinate natural number q. We now show that the proposition $[R(q); q]$[13] is undecidable in PM. For supposing the proposition $[R(q); q]$ were provable, it would also be true;[33] but that, on the basis of what precedes, means that q would belong to K, i.e., according to (1), $\sim (Bew[R(q); q])$ would hold good, in contradiction of our initial assumption. If, on the contrary, the negation of $[R(q); q]$ were provable, then $\equiv (n \in K)$ would hold good. $[R(q); q]$ would thus be provable at the same time as its negation, which again is impossible.[34]

The analogy between this result and Richard's antinomy leaps to the eye[35]; there also is a close relationship with the "liar"

[33] This distinction is essential. It takes support on the logical theorem according to which a provable universal proposition is true (Gödel's proposition $[R(q); q]$ is a universal).

[34] PM is *supposed* here to be consistent.

[35] According to PM, Richard's antinomy is vitiated by the confusion between distinct logical types in the sense of Russell's theory of logical types, while Gödel's proof respects the Russellian stratification of distinct types.

III.7. ***Proposition on consistency.*** According to *MRC* the question of the consistency of a formal system *S* cannot, in general, be settled inside *S*, for reasons that are *independent* of any assumption concerning the completeness of *S*. In general this question can be settled only by formal examination inside a conveniently constructed metasystem *MS*. Then the solution established inside *MS* is relative to *MS*.

"Proof". The consistency of *S* is by definition the (required) property of *S* according to which, for any well-formed expression from *S* that can be exhibited, it is not possible to prove inside *S* both this well-formed expression and its negation (note 26).

Now, the *whole* qualificational power defined inside *S*, is concentrated in the view of demonstrability V_d^S from the proof-descriptions $D_j^S/G_j^S, œ_j, V_d^S/$. In each one of these the object-entity $œ_j$ consists by definition of just *one* well-formed expression from *S*: V_d^S does not exist in the sense of D7 with respect to "*any* well-formed expression from *S* that can be exhibited"—a potential meta-entity with respect to those, $œ_j$, from the achieved proof-descriptions $D_j^S/G_j^S, œ_j, V_d^S/$— so V_d^S cannot qualify this meta-object-entity as a *whole*. However, though *S* says nothing concerning the way in which one may "exhibit" well-formed expressions different from those enumerated *ab initio* in the definition of *S*, otherwise than by achieving proof-descriptions D_j^S, it might happen that somehow—with the help of projections from some metasystem *MS*, or by empirical research—a well-formed expression from *S* be first found without any proof D_j^S, that can then be proven inside *S*, *via* the view of demonstrability V_d^S, as well as its negation. Thereby the in-consistency of *S* would be proven inside *S*, by construction, and the question would be closed. But this is a particular circumstance which may stay indefinitely non realized; and as long as a proof of inconsistency by construction has not been produced, the question of the consistency of *S* stays open. Or otherwise, in the case of certain trivial finite systems *S*, it can be possible to produce one by one all the well-formed expressions permitted by *S*, and to study them by corresponding proof-descriptions $D_j^S/G_j^S, œ_j, V_d^S/$, thus concluding inside *S* concerning the consistency of *S*. But in general an assertion of consistency of *S* cannot be founded on a sequential production of well-formed expressions from *S*. In general such a process is not efficient because there is no way to ascertain that the production is finished, nor that, while it continues, inconsistency will never be found, nor that, if for a given well-formed expression $œ_j$ no proof of inconsistency is *found*, none is *possible*. So according to *MRC* the question of the consistency of *S* cannot—in general—be settled inside *S*. It follows that only a formal examination *of S as a whole*, achieved from the outside of S, could settle this question.

But this, according to MRC, requires *another* sort of description than the proof-descriptions D_j^S/G_j^S,œ$_j$,V_d^S/ *from* S, where *not* [S-as-a-whole] is the object-entity. Indeed the principle of separation P15 asserts that "Since any one relative description D/G,œ$_G$,V/, whatever its complexity, involves by construction one generator of object-entity, one object-entity, and one view, all well defined, as soon as some change is introduced in the content or the role designated by a term from the triad G,œ$_G$,V, *another* description is considered". And, by method, P15 posits that "this other description must be treated *separately*".

Now, since a formal proof is researched, it must be achieved inside some formal system, namely some convenient metasystem MS inside which S be somehow embeddable.

Suppose then that such a metasystem has been found and that inside it a proof of the consistency or the inconsistency of S has been achieved. Then nothing excludes that with another metasystem $(MS)' \neq MS$ the conclusion of this proof be contradicted: Though inconsistency *can* in principle happen to be provable inside S by an example—i.e., in an absolute way—in general a proof concerning the consistency of S is relative to some metasystem MS.

So πL.7 is entirely established.

Comment. The fact that in general a proof of consistency of a formal system, requires a metasystem, is well known. The new element here is only that (and how) this follows inside MRC, and quite independently of considerations concerning the completeness of the studied system.

We consider now Gödel's proof of undecidability.

ΠL.8. *Proposition about the expression* [R(q); q]. According to MRC the well-formed expression $[R(q); q]$, by construction, is not a proposition, so it cannot be true or false. So Gödel's reductio becomes impossible and aimless.

"Proof". According to MRC, a relative description D/G ,œ$_G$,V/is a piece of elaborated meaning where the three roles G, œ$_G$, and V have all to be defined, and played accordingly to their definitions, by *definite* epistemic actors.[40] Furthermore a relative proposition $p(D, V_\tau^{(2)})$(DL.3) involves a def-

[40] Even if in a degenerate way (i.e., two roles are held by one same actor) and/or without radicality (the generator G does not radically create the object- entity œ$_G$, the view V does not radically change this object-entity while qualifying it) (cf. the final general comment

inite relative description $D/G,œ_G,V/$ that has been previously established independently, and afterward is subjected to valuation by the truth-values of some definite metaview of empirical truth $V_T^{(2)}$ that exists with respect to $D/G,œ_G,V/$, in the sense of D7 (DL.3).

The metasystem PM dwells with well-formed expressions that can be "interpreted as to their contents" (meanings) and with respect to these can be *a priori* awaited to be found to be empirically true or false. So, though implicitly, descriptions and propositions *are* involved in PM. Then, according to MRC, what is the descriptional status of the formula $[R(q);q]$ from PM?

$[R(q);q]$ is not a relative description $D/G,œ_G,V/$. Indeed $[R(q);q]$ is first constructed by a succession of syntactical steps. Once obtained in this way, it is "interpreted as to its content" (cf. Gödel's text) and found to assert its own unprovability. But $[R(q);q]$ *consists* `exclusively` *of this self-qualifying assertion.* The roles of generator of object-entity and of object-entity are not defined, so they are not played. In "I am not provable" of what does "I" consist? $[R(q);q]$ asserts the unprovability *of nothing definite.* The fact that $[R(q);q]$ has been constructed in full agreement with all the syntactical requirements from PM (the stratification of distinct types included), does not change this situation.

Now, inside MRC the definition D14 of a relative description (cf. the final global comment) banishes explicitly self-referential constructs, on semantic-methodological grounds, whatever their grammatical or logico-mathematical correctness. So according to MRC there is no description corresponding to $[R(q);q]$. *A fortiori* $[R(q);q]$ is not a relative proposition in the sense of DL.3 either.[41] It is not even a proposition in the loose sense of the classical logic, (i.e., [(an assertion) (that can be true or false)]), since the first element, an assertion, *in its own right*, is lacking), its place is held by a sequence of signs which, though it is correct from a syntactical viewpoint, is devoid of any possible meaning by *its very inner structure.* As Gödel himself says, it is a false (apparent) proposition (cf. the end of the first paragraph from our quotation of Gödel's text, and our note 37 on Gödel's note 15). In these conditions $[R(q);q]$ is doomed not to be provable in PM: *there is no way to prove the truth of something that is not a proposition.*

---

of the definitions D14).

[41] In order for $R(q);q]$ to be a proposition in the sense of DL.3 it would have been necessary to *first* specify inside PM a well-formed expression, say X_{PM}, which, interpreted as to its content, be a definite description $D/G,œ_G,V/$; and *then*, in order to "propose" tentatively that the meaning (the content) carried by X_{PM} is true, to construct $[R(q);q]$ as a genuine (universal) proposition, i.e., such that, considered itself now as to its content, it be found to be a metadescription $D^{(2)}/G^{2)},œ^{(2)},V_T^{(2)}/$ with $œ^{(2)} \equiv D/G,œ_G,V/$ and $V_T^{(2)}$ a metaview of empirical (logico-mathematical) truth.

So Gödel's *reductio* becomes impossible (and aimless). The hypothesis "supposing the proposition $[R(q);q]$ were provable, it would also be true" is known *a priori* to be impossible, by construction, which dissolves the *reductio*.

Comment. In so far that one is *aware*, as it does happen inside MRC, that a linguistic construct like $[R(q);q]$ does not exist in the sense of D7 with respect to any view $V_\tau^{(2)}$ of empirical truth (be it logical or mathematical empirical truth), Gödel's *reductio* is settled in advance and acquires the character of a game of play pretend.[42]

ΠL9. ***On the MRC-significance of Gödel's proof.*** According to MRC, the main previously unknown result established by Gödel's proof is that the metasystem PM permits inside it well-formed expressions that are not decidable, so are not propositions, and that these can be injected into Peano's arithmetic AP by isomorphic projection. There, *via* examinations monitored by PM, they reproduce their non decidability.

"Proof". Obvious from Gödel's proof and the preceding *"Proof"*.

Comment. Strictly expressed, according to MRC Gödel's proof establishes a conclusion about the metasystem $PM + A_P$, not about Peano's arithmetic A_P considered independently of PM. The undecidability proven by Gödel is relative to the Russell-Whitehead metasystem PM. So the propositions πL.8 and ΠL.9 displace the accent from the studied formal system, on the metasystem which is made use of for the study. But thereby one is led to a further quite general question, analogous to the question concerning consistency examined in πL.7: is it conceivable to study the completeness of a formal system S from inside S?

ΠL.10. ***Proposition on completeness.*** According to MRC, in general the question of the completeness of a formal system S cannot be settled inside the system. It requires the use of a convenient metasystem MS. This entails that the result is relative to MS.

"Proof". Completeness of a formal system S is the (presumed) property of S according to which any expression that one can exhibit, which is well-formed according to S, is decidable *in* S, i.e., either this expression

---

[42] J-Y. Girard, in *Le champ du signe ou la faillite du réductionnisme*, in *Le théorme de Gödel*, by E. Nagel, R. Newman, K. Gödel, and J-Y. Girard (Seuil, 1989), writes: "Si l'on dégage les idées profondément novatrices—essentiellement la distinction vrai/démontrable—autour desquelles se charpente le théorme, la démonstration résulte d'une suite impitoyable de truismes—ou de 'provismes.'"

or its negation can be proven in S.

The argument from the "proof" of πL.7 can be transposed in an obvious way.

Comment. In these conditions, speaking of "the" completeness of S as if it were an absolute property of S, is in general misleading. In general the property of completeness of a formal system is radically dependent on the metasystem that is made use of for establishing its existence.

This leads us to ask whether the features of a metasystem MS, which permit to induce in a studied formal system S, *undecidable* well-formed expressions, are indeed *unavoidable* features. The statement πL.8 seems to indicate a negative answer. Indeed, since MRC—a nonformalized method—*does* avoid the emergence of "undecidable false propositions", *a fortiori* it should be possible to build also a formalization of logic which avoids such *emergences* (as well as any other features that can generate undecidability).

The pertinent question, in this respect, seems to consist of the specification of methodological rules for constructing "good" metasystems.

The above (very rapid and quasi informal) examination of the questions of consistency and completeness of a formal system illustrates well the fundamental difference between classical logic and MRC. In classical logic all the creative epistemological features are occulted by storage in the absolutized and hypostatized concepts of "values of an object-variable x" and of a shadow-predicate "P". The involved descriptional relativities are not apparent, hence their consequences also remain hidden, so they cannot avoid false problems, nor show the way toward the natural solution when problems do creep in. Whereas inside MRC the double-extremity genetic classes expose explicitly all the involved cognitive actions, so the relativizing consequences of these upon the produced qualifications are obvious. Furthermore, the limitations entailed by the descriptional relativities are explicitly tied to a methodological obligation P15 to interrupt the current descriptional process and to take a new start on a metalevel, which organizes in cells the conceptual progression and keeps it under control.

5.1.3. Conclusion on the MRC-logic

It is remarkable that MRC, such as it has been constructed by taking initial support exclusively on quantum mechanics, leads to the outline of a logical approach that is relevant not only for the basic, the physical creative genetic classes $C[G^{(o)}.V^{(o)}]$ of the type of those involved in quantum mechanics, but also for *conceptual* creative genetic classes $C[G.V]$ found to be involved in formal systems.

The quantum mechanical cognitive strategy, generalized inside MRC, has opened up a way of conceptualization that is not mute with respect to the most fundamental questions of nowadays abstract mathematical and logical thinking.

This is so because the canonical descriptional mould $(G, œ_G, V)$ drawn from quantum mechanics has been constructed at the lowest level of conceptualization which human mind has been able to reach, possibly the final one. There the most severe conditions that can be encountered in a process of conceptualization, are all active. So a basic structure of labelled receptacles for conceptualization which is constructed to fit these conditions, is sufficiently comprehensive for harbouring any descriptional possibility that might occur. Inside this structure, semantics and cognitive actions—which always involve *aims* —combine with the syntactic features, and this induces both intelligibility and control.

5.2. MRC *versus* Probabilities

One of the major successes of MRC is the representation of a deeper general concept of probability, which contains and explains the so cryptic quantum mechanical probabilities [14,16,17A,18]. Indeed, when Kolmogorov's classical concept of a probability space is examined inside MRC, the limitations and the absolutizations which flaw this concept come into striking evidence. By suppressing them, the concept of probability expands to the limits of its whole natural volume which rests on the most basic level of transferred conceptualization and extends up to very high descriptional levels.

Throughout the process of construction of the MRC-concept of probability, the methodological principle of separation P15 plays a key role. Therefore this process can also be regarded as a succession of illustrations of the very peculiar way in which the principle of separation works.

5.2.1. Komogorov's classical definition of a probability space

The fundamental concept of the modern theory of probabilities—in Kolmogorov's formulation [29]—is a probability space $[U, \tau, p(\tau)]$ where: $U = \{e_i\}$ (with $i \in I$ and I an index set) is a *universe of elementary events* e_i (a set) generated by the repetition of an "identically" *reproducible procedure* $\mathcal{P}$(called also an experiment) which, notwithstanding the posited identity between all its realizations, nevertheless brings forth elementary events e_i that *vary* in general from one realization of $\mathcal{P}$ to another one; τ is an *algebra of events* built on U,[43] an event, let us denote it e, being a subset of U

[43] An algebra built on a set S is a set of subsets of S—S itself and $\emptyset$ being always included—which is such that if it contains the subsets A and B, then it also contains

and being posited to have occurred each time that any elementary event e_i from e has occurred; $p(\tau)$ is a *probability measure* defined on the algebra of events τ.[44] A pair $[\mathcal{P}, U]$ containing an identically reproducible procedure $\mathcal{P}$ and the corresponding universe of elementary events U is called a *random phenomenon*.

On a given universe U, one can define various algebras τ of events. So it is possible to form different associations [[random phenomenon],[a corresponding probability space]], all stemming from the same pair $[\mathcal{P}, U]$.

With respect to the previous representations (Bernoulli, von Mises, etc.)—where only a concept of "probability law" (or "probability measure") was defined mathematically—Kolmogorov's concept of a probability space $[U, \tau, p(\tau)]$ has marked a huge complexifying progress.

5.2.2. Critical remarks

In Kolmogorov's classical theory of probabilities, the procedure $\mathcal{P}$ is neither formally defined, nor symbolized or otherwise represented. This theory contains no symbolic location reserved for the procedure $\mathcal{P}$, so *a fortiori* the random phenomenon $[\mathcal{P}, U]$ as a whole is not represented. The consequence is that the structure of the connection between the considered probability space $[U, \tau, p(\tau)]$, with the substratum wherefrom it is generated, is very rarely explicitly surveyed. Usually nothing whatever is asserted concerning the way in which the elementary events from the universe U do emerge by the procedure $\mathcal{P}$.

> The channel for the adduction of semantic substance from the "pool of reality" (in the sense of D2) into the considered probability space $[U, \tau, p(\tau)]$, is undefined and unexplored. It is only alluded to by mere words.

In each application of the abstract theory of probabilities, to some specific problem, the corresponding semantic substance is injected into the studied probability spaces in an intuitive unruly way. It might be argued that this is an intentional non-determination which endows the formalism

---

$A \cup B$ and $A - B$.

[44] A probability measure defined on τ consists of a set of real numbers $p(A)$, each one associated to an event A from τ, such that: $0 = p(A) = 1$, $p(U) = 1$ (normation), $p(\emptyset) = 0$, and $p(A \cup B) \leq p(A) + p(B)$ where the equality obtains iff A and B are "independent" in the sense of probabilities, i.e., iff they have no elementary event e_i in common ($A \cap B = \emptyset$). The number $p(A)$ yields the value of the limit – supposed to exist—toward which the relative frequency $n(A)/N$ converges when the number N of realizations of the involved repeatable procedure P is increased toward infinity ($n(A)$ being the number of outcomes of A when P is repeated N times).

with a maximal generality (interpretability). However the absence of any formal mould for the expression of a probabilistic concept as basic as the random phenomenon that generates the considered probability space, cannot be claimed to maximise the generality of the formalization. It clearly is just a lacuna entailing looseness.

Furthermore, from the standpoint of MRC the definitions of the elements from the probability space $[U, \tau, p(\tau)]$ are lacking precision. For instance: What is the descriptional status of the procedure $\mathcal{P}$? Is it an operation of generation of an object-entity? Is it an operation which only somehow involves an already previously generated object-entity? Or is it some *association* between an operation of creation or of only manipulation of a pre-existing object-entity, and an operation of examination of the result, by some view? It seems obvious that also some view is acting inside the procedure $\mathcal{P}$, since it is asserted that, notwithstanding the "identity" between all its realizations, the procedure brings forth "different" elementary events e_i. But "different" in what a sense? With respect to which view? In the absence of *any* view the elementary events cannot be perceived. They even cannot be imagined. So *a fortiori* they cannot be compared and mutually distinguished. So the content of the procedure $\mathcal{P}$ is obscure; it has to be elucidated.

Supposing now that indeed a view is found to be involved in what is called an elementary event, the unique index i for distinguishing between the elementary events e_i is not sufficient for cutting out a conceptual receptacle able to contain the full specification of the qualifications produced by this view. Even in the simplest case of a view with only one aspect, the definition D5.1 requires already two indexes, an aspect-index g and an index k of value of this aspect. The symbolic framework necessary for the expressibility of the qualifications of an object-entity, *via* the MRC concept of a view, is absent from Kolmogorov's representation. *The Kolmogorov concept of elementary event though it involves a view cannot be clearly referred to MRC-views; it even cannot be clearly referred to classical predicates.* The involved "properties" or "specificities" are just alluded to, but neither their logical status (or even only the grammatical or the descriptional one), nor their contents, are defined. Thereby *it is an a-logical concept.*

This circumstance becomes clearer by its consequences upon the events e from the algebra τ constructed on the universe U of elementary events. An event e is by definition "a subset of elementary events from the universe U". But—in general—this subset is not regarded as a class determined by some predicate. So it cannot be directly connected with syllogisms which are essentially tied to classes of predicates (all men are mortal; Socrates is a man; so Socrates is mortal). This is one of the main reasons why classical probabilities withstand the attempts at the specification of a general relation with classical logic: the elementary events and the events are introduced in set-theoretical *counting* terms, not in classical grammatical-logical terms (subject-predicate).

But the most fundamental question is this. Beyond its formal definition, what is the significance of the probability measure from a probability space? The semantic involved in the concept of probability measure remains very particularly cryptic. A remarkably complete study on this topic was made by Arthur Fine[30][45] in 1973. Karl Popper offered on this subject deep considerations that will be mentioned below, but these found little echo among mathematicians and physicists. So, like classical logic, the apparently so clear classical representation of probabilities achieved by Komogorov, when analyzed, reveals non-intelligibility.

5.2.3. MRC-reconstruction of the concept of probability generalities

In what follows we shall proceed under two conjugated sets of constraints: Kolmogorov's concept of a probability space, and the requirements of MRC.

Each *association* between a given random phenomenon $[\mathcal{P}, U]$ and a probability space $[U, \tau, p(\tau)]$ generated by it will be called *a probability chain* and will be symbolized by the writing

$$[\mathcal{P}, U] \rightsquigarrow [U, \tau, p(\tau)]$$

where the sign $\rightsquigarrow$ represents a connection of which the content and the structure have to be specified. According to the principle of separation P15 and the definition D16 of a metadescription, a probability chain involves explicitly a hierarchy of three connected but distinct descriptional levels. Indeed:

* the elementary events e_i are placed on a first descriptional level;
* the algebra τ of events is placed on a higher descriptional level, since it involves sets e of elementary events e_i from U;
* the probability measure $p(\tau)$ lies on a still higher descriptional level than τ since it qualifies numerically the relative frequencies $n(e)/N$ of the outcomes of the events e from the algebra of events τ.

And—again according to the principle of separation P15—the process of description achieved on *each* one of these three distinct levels involves its *own*

[45] I do not think that meanwhile the situation has evolved much. Kolmogorov himself, and Chaitin, have abandoned the concept of probability in the attempts at a mathematical characterization of complexity, because they could not identify the meaning of the concept or a conceptual guide for defining a probability measure in a given situation.

epistemic referential, which has to be specified. So, by confrontation with a Kolmogorov probability space, it appears now strikingly to what a degree the provisional definition of a probabilistic relative description contained in D14.1 was insufficient (cf. note 10),[46] and even from the point of view of MRC itself. When this provisional definition has been introduced, the principle of separation and the concept of relative metadescription were not yet defined, nor the concept of a genetic class, and the Kolmogorov concept of a probability space had not yet been introduced as a reference. But in the present stage of development of MRC it is obvious that the *unique epistemic referential* (G, V) *considered in D14.1, certainly cannot produce all the qualifications required by a probabilistic description able to include the whole—very complex—concept of probability introduced by Kolmogorov.* Other metareferentials certainly have to be brought in.

The following thorough elaboration of the content of a probability chain will suppress the initial lacunae. The results will permit to understand in a more concrete way the powers of systematic descriptional relativizations.

We shall proceed in three stages. In a first stage we shall develop the MRC concept of probability tree of a basic epistemic referential, inside which a unification between relativized logic and relativized probabilities will find

---

[46] We recall the definition of a *probabilistic relative description of a physical object-entity* contained in D14.1: Consider an epistemic referential (G, V) where G is a *physical* generator that generates a corresponding *physical* object-entity $œ_G$, and V is a *physical* view with respect to all the aspect-views V_g of which $œ_G$ does exist in the sense of D7 and which—as required by P8 and C9—contains a spacetime view V_{ET} introducing an *ordered* spacetime grating (D5.4). Furthermore consider, for *each* V_g from V, a big number N of realizations of the corresponding sequence $[G.V_g]$—in simultaneity or in succession—the time parameter being re-set at the same initial value t_0 for each realization of a sequence $[G \cdot V_g]$ .... Suppose now that, when the various successions $[G \cdot V_g]$ with $V_g \in V$ are realized N times, *not* all the successions $[G \cdot V_g]$ are found to reproduce identically one same configuration of $gk - E\vec{r} - Tt$-values; that at least for one $V_g \in V$ (not necessarily for all) the corresponding succession $[G \cdot V_g]$ produces a whole set $S_{gi} = \{c_{gi}\}$ of mutually distinct, dispersed configurations c_{gi} of $gk - E\vec{r} - Tt$-values (with $i \in I$ and I a finite index-set, to preserve the finitistic character of this approach); but that, for *any* succession $[G.V_g]$ which produces dispersed results, when N is increased toward infinity, the relative frequency $n(c_{gi})/N$ of occurrence of *each* configuration $c_{gi} \in S_{gi}$ converges toward a corresponding probability p_{gi}. In these conditions each configuration $c_{gi} \in S_{gi}$ will be called an *elementary-event-description* corresponding to the succession $[G \cdot V_g]$ with $V_g \in V$ and it will be denoted $D_{p(gi)}/G, œ_G, V_g/$. The epistemic referential $(G \cdot V)$ will be said to produce a *probabilistic relative description* of the physical object-entity $œ_G$, which will be denoted $D_p/G, œ_G, V/$.

So in D14.1 the concept of probability space was not explicated. For the algebra of events τ there is not even an implicit equivalent, while the distinction between the descriptional level where the elementary events are placed, and the level where the probability measure can be placed, remains obscure.

place. In a second stage, by intrinsic metaconceptualization, we shall obtain a minimal spacetime model for the random phenomenon which constitutes the physical ground of the probability tree of a basic epistemic referential; this model introduces a new sort of set called a genetic set (genset), that opens up the way toward a genetic relativized set-theory. In a last third stage we shall specify the MRC significance of a probability measure.

First stage: Probability tree of a basic epistemic referential

Elementary event from a basic probability chain. Consider a probabilistic description of a physical object-entity which moreover is a *basic* transferred description $D^{(o)}/G^{(o)}$,œ$^{(o)}$,$V^{(o)}/$.[47]

I shall now show that according to MRC a basic elementary event has the descriptional status of a relative description.

Let us concentrate upon the fact that in general the global basic view $V^{(o)}$ contains several basic aspect-views $V_g^{(o)} \in V^{(o)}$ that are not all mutually compatible in the sense of P10. So, in general, $V^{(o)}$ splits in a finite number n of subsets $b = 1, 2, \ldots, n$ of basic *branch-views* $V_b^{(o)}$ such that inside one given basic branch-view $V_b^{(o)}$ the aspect-views are all mutually compatible, while any two basic aspect-views from two different basic branch-views

[47] We recall also the definition D14.1.3 of *a basic transferred relative description* : The generator consists of a physical operation and it produces a physical object-entity that cannot be perceived directly by man. Such a generator will be called a *basic generator* and will be denoted $G^{(o)}$. The object-entity produced by a basic generator $G^{(o)}$ will be called a *basic object-entity* and will be denoted œ$^{(o)}$. The view able to draw phenomenal manifestations out of a basic object-entity is necessarily such that the phenomenological content of each *gk*-value of each involved aspect g, stems (by coding rules) from features of a material device for gk-registrations – biological, or not—but which always is *different from the studied object-entity*, these features emerging in consequence of interactions between the examination-and-registering-device and replicas of the considered basic object-entity. A view of the just specified kind will be called a *basic transfer-view* (in short a basic view) and will be denoted $V^{(o)}$. The aspect-views from $V^{(o)}$ will be called basic aspect-views and will denoted $V_g^{(o)}$. (The epistemic referential $(G^{(o)}, V^{(o)})$ will be called a *basic epistemic referential.*) A relative description in the sense of D14.1—individual or probabilistic—achieved with a basic generator and *one* basic transfer-aspect-view $V_g^{(o)}$, will be called a *basic transferred relative aspect-description* and it will be denoted $D^{(o)}/G^{(o)}$, œ$^{(o)}$, $V_g^{(o)}/$. A relative description in the sense of D14.1—individual or probabilistic—achieved with a basic generator $G^{(o)}$ and a basic transfer-view $V^{(o)}$ involving at least two *mutually incompatible* basic aspect-views $V_{g1}^{(o)}$ and $V_{g2}^{(o)}$, will be called *a basic transferred relative description* (also, in short, a basic description or a transferred description) and it will be denoted $D^{(o)}/G^{(o)}$, œ$^{(o)}$, $V^{(o)}/$ (in short $D^{(o)}$). A basic transferred description $D^{(o)}/G^{(o)}$, œ$^{(o)}$, $V^{(o)}/$ is posited to *characterize* observationally the involved object-entity œ$^{(o)}$, which means that it is posited that no other operation of generation $(G^{(o)})' \neq G^{(o)}$ can be found which, associated with the same basic view $V^{(o)}$, shall produce the same basic transferred description.

$V_e^{(o)} \neq V_f^{(o)}$, with e and f denoting two distinct values of b, are mutually incompatible (in the sense specified inside the principle P10 of individualizing mutual exclusion).

Consider now *one* given basic branch-view $V_b^{(o)}$. According to the preceding remarks its analyzed content can be conveniently symbolized by writing $V_b^{(o)} = \cup_j V_{bj}^{(o)}, j = 1, 2, \ldots, m,\ m \leq n$, where the $V_{bj}^{(o)}, j = 1, 2, \ldots, m$, are the m mutually *compatible* basic aspect-views $V_g^{(o)} \in V^{(o)}$ that belong to $V_b^{(o)}$.

Since by definition two different basic branch-views $V_e^{(o)} \neq V_f,^{(o)}$ tied with one same basic generator $G^{(o)}$ do *never* act simultaneously on *one* same *replica* of the involved basic object-entity œ$^{(o)}$, it follows that:

> An *elementary* event from the probabilistic relative description generated by the basic referential $(G^{(o)}, V^{(o)})$ is—always – a *branch-elementary-event* produced by just *one* realization of a *branch-sequence* $[G^{(o)}.V_b^{(o)}]$.

How does a branch-sequence $[G^{(o)}.V_b^{(o)}]$ work to bring forth an observable basic branch-elementary-event? This is not immediately obvious because $V_b^{(o)} = \cup_j V_{bj}^{(o)},\ j = 1, 2, \ldots, m$.

To explicate, let us note that *the m aspect-views* $V_{bj}^{(o)},\ j=1, 2, \ldots, m$ *from one same branch* $V_b^{(o)}$ *differ from one another* only conceptually.[48] Indeed, by their very definition introduced in the principle P10 of individual mutual exclusion, two or more "compatible" physical aspect-views can be all measured simultaneously on only *one* replica of the considered basic object-entity œ$^{(o)}$, if a convenient measurement-and-registering device is made use of. Let us call such a device a basic *branch-device*, in short a $b^{(o)}$-device. One act of $V_b^{(o)}$-examination by this unique $b^{(o)}$-device concerns only *one* replica of the considered basic object-entity œ$^{(o)}$, say œ$_r^{(o)}$, and it yields only *one*—factual–configuration of observable marks. Let us denote such

[48] We take immediately an example from quantum mechanics, in order to facilitate what follows. The momentum observable P and the observable $p^2/2m = T$ of kinetic energy, are compatible. So they can be measured by a same branch-device (by a method called "time of flight"). This device involves a screen. An examination of one replica of the studied microstate yields two data, namely a mark on this screen and the time when the mark occured, which constitutes a configuration of two factual "marks". From this unique configuration, one then calculates by rules specified in advance, on the one hand the vector-*eigenvalue* of the observable P, and on the other hand the scalar *eigenvalue* of the observable T. Each one of these two calculations "describes" the unique configuration of factual marks, in terms of an eigenvalue of one of the two involved compatible observables. Together, these two descriptions constitute one elementary-event-description $ee_{bi}^{(o)}$.

a configuration by $(bk)_f$ where b labels the branch, k labels the factual configuration, and f stresses the "factual" character (here $k \in K$ where K is an index set of which the cardinal is equal to the number of all the possible distinct configurations $(bk)_f$ permitted by the definition D5.1 of the basic branch-view $V_b^{(o)}$: a finite number, by construction). Now, by itself, a configuration $(bk)_f$ of factual marks is devoid of any significance in terms of values of the basic branch-views $V_{bj}^{(o)}, j = 1, 2, \ldots, m$, which compose $V_b^{(o)}$. Indeed the K factual configurations $(bk)_f$ are *not* themselves the values of $V_b^{(o)} \equiv \cup_j V_{bj}^{(o)}, j = 1, 2, \ldots, m$ in the sense of 5.1. While precisely a significance in terms of values of $V_b^{(o)} \equiv \cup_j V_{bj}^{(o)}, j = 1, 2, \ldots, m$ in the sense of 5.1 is researched when the basic branch-view $V_b^{(o)}$ is made use of. In order then to acquire a significance (an "interpretation") in the researched terms, the factual configuration $(bk)_f$ produced by one $V_b^{(o)}$-examination has to be furthermore coded in terms of "values" of the basic aspect-branch-views $V_{jb}^{(o)}, j = 1, 2, \ldots, m$. Now, each one of the m distinct but compatible basic branch-aspect-views $V_{bj}^{(o)} \in V_b^{(o)}, j = 1, 2, \ldots, m$ introduces its *own* values and its *own* coding rules in terms of these values. So the one registered factual configuration $(bk)_f$ produced by one $V_b^{(o)}$-examination admits of m distinct *conceptual* "interpretations" that can be *all* elaborated *via* the m conding rules involved by the m distinct but compatible aspect-views $V_{bj}^{(o)} \in V_b^{(o)}$, out of this—unique—configuration and concerning the—unique—replica $œ_r^{(o)}$ of $œ^{(o)}$. And the qualifications involved by $(bk)_f$ is exhaustively achieved only if it is achieved accordingly to *all* the basic aspect-views $V_{bj}^{(o)} \in V_b^{(o)}, j = 1, 2, \ldots, m$.

In short now. Each realization of one succession $[G^{(o)} \cdot V_b^{(o)}]$ produces *first* the observable registration of one configuration $(bk)_f$ of factual observable marks tied with the one involved replica $œ_r^{(o)}$ of the considered basic object-entity $œ^{(o)}$; and *then* this unique registred factual configuration is m-fold qualified in the m different but compatible aspect-values-languages introduced by the various aspects $V_{bj}^{(o)} \in V_b^{(o)}, j = 1, 2, \ldots, m$.

The above analysis permits to represent a basic branch-view $V_b^{(o)} \equiv \cup_j V_{bj}^{(o)}, j = 1, 2, \ldots, m$ as being *split* in two views which act in succession. Namely a *factual* basic branch-*aspect*-view, say $V_{bf}^{(o)}$ consisting of a unique aspect endowed with K factual "values" $(bk)_f$, $k \in K$ (in the sense of D5.1), which acts first, *via* only one *basic* $V_{bf}^{(o)}$-examination of only one replica $œ_r^{(o)}$ of $œ^{(o)}$, and a *conceptual coding-view*, say V_{bc}, containing m coding-*aspect*-views V_{bcj} – so $V_{bc} \equiv \cup_j V_{bcj}$, $j = 1, 2, \ldots, m$—which acts afterward on the registered factual configuration $(bk)_f$ and qualifies it *m-fold* by a "corresponding" *group* of m compatible "values" of the aspect-coding-views V_{bcj} (one such value for each aspect-coding-view V_{bcj}, $j = 1, 2, \ldots, m$). Now the referential being basic, another $V_{bf}^{(o)}$-examination of the involved basic object-entity $œ^{(o)}$ will necessarily introduce another replica of $œ^{(o)}$, say $œ_q^{(o)} \neq œ_r^{(o)}$. And since the final global description is probabilistic by hypothesis, another sequence $[G^{(o)} \cdot V_{bf}^{(o)}]$ will in general produce another configuration $(bk)'_f \neq (bk)_f$, i.e., there is *no* individual stability with respect to the repetitions of the sequence $[G^{(o)} \cdot V_{bf}^{(o)}]$. It follows that the answer which is relevant is that one realization of a sequence $[G^{(o)} \cdot V_{bf}^{(o)}]$ produces *a testimonial relative description* $\theta^{(o)}/G^{(o)}, œ_r^{(o)}, V_{bf}^{(o)}/$ in the sense of D14.2.2 (basic and factual, exactly as those involved in the basic testimonial relative propositions DL.4 studied in D5.1.2) yielding the qualification $(bk)_f$ of the involved replica $œ_r^{(o)}$ of the basic object-entity $œ^{(o)}$. This qualification $(bk)_f$ is than *meta*qualified by the branch-coding-view V_{bc} drawn from the branch aspect-views $V_{bj}^{(o)}, j = 1, 2, \ldots, m$, which here plays the role of a (conceptual) *meta*view $V_{bc}^{(2)}$. Then, accordingly to P15 and D16, the "legal" global MRC-expression of the realized sequence of results consists of saying that we are in presence of *a testimonial metadescription* $\theta^{(2)}/G^{(2)}, œ^{(2)}, V_{bc}^{(2)}/$ *with* $G^{(2)}$ *the conceptual metagenerator (a selector) of object-entity which selects as object-entity* $œ^{(2)} = \theta^{(o)}/G^{(o)}, œ_r^{(o)}, V_{bf}^{(o)}/$ *and qualifies it by the coding metaview* $V_{bc}^{(2)}$.

Since any relative testimonial description or metadescription is a relative *description*—a limiting case of relative description but *still* a description—this finally settles the question of the MRC descriptional status of a basic elementary event:

> According to MRC each elementary event from a basic probabilistic description generated by a basic referential $(G^{(o)}, V^{(o)})$, is produced by one basic *branch*-succession $[G^{(o)}.V_b^{(o)}]$ and it has the descriptional status of a *relative* `description`. So it involves a view (in classical terminology, predicates), which permits classifications.[49] This will appear just below to be crucial for the unification of the probabilistic approach, with the logical one.

[49]This is a detailed reconstruction of the content of the notation $c_{gi} \in S_{gi}$ from the preliminary definition of a probablistic relative description contained in D14.1 and quoted in the note 45.

Let us denote such an elementary event by

$$ee_{bi}^{(o)} \equiv \theta^{(2)}/G^{(2)}, œ^{(2)}, V_{bc}^{(2)}/$$

where $G^{(2)}$ is the conceptual metagenerator of object-entity—a selector—which selects as object-entity $œ^{(2)}$ the precedingly established basic relative testimonial description $\theta^{(0)}/G^{(o)}, œ_r^{(o)}, V_{bf}^{(o)}/$ defined above, and qualifies it by the conceptual coding metaview $V_{bc}^{(2)}$; the index i belongs to a (finite) index-set I and labels *globally* the m-fold qualification by $V_{bc}^{(2)}$ of the unique factual mark $(bk)_f$ that has emerged by the one considered basic examination $V_{bf}^{(o)}$ of the replica $œ_r^{(o)}$ of $œ^{(o)}$.

The branch-(random phenomenon) from a basic branch-(probability chain). By hypothesis, when a given succession $[G^{(o)}.V_b^{(o)}]$ is repeated a big number of times, the obtained factual results bk are dispersed. Then also the corresponding elementary events are dispersed: a whole branch-universe $U_b^{(o)}$ is produced by the repetitions. So to each branch-view $V_b^{(o)}$ from $V^{(o)}$ there corresponds a *branch-(random phenomenon)* that can be written as

$$[[G^{(o)}.V_b^{(o)}], U_b^{(o)}]$$

By identification of terms with the generic expression $[\mathcal{P}, U]$, it appears that in this case the *repeatable procedure* $\mathcal{P}$ *consists of the* succession *of epistemic operations* $[G^{(o)}.V_b^{(o)}]$. So we have:

$$\mathcal{P}_b^{(o)} \equiv [G^{(o)}.V_b^{(o)}], \quad U_b^{(o)} = \{ee_{bi}^{(o)}, i \in I\}, \quad [\mathcal{P}_b^{(o)}, U_b^{(o)}] = [[G^{(o)}.V_b^{(o)}], U_b^{(o)}].$$

This settles also the questions of the MRC-status of a branch-procedure $\mathcal{P}_b^{(o)}$ and of the content of a branch-random-phenomenon $[\mathcal{P}_b^{(o)}, U_b^{(o)}]$.

The meta[random-phenomenon] produced by a basic epistemic referential. Consider now the whole basic epistemic referential $(G^{(o)}, V^{(o)})$. It can be represented as a union

$$(G^{(o)}, V^{(o)}) = \cup_b (G^{(o)}, V_b^{(o)}), \quad b = 1, 2, \ldots, n,$$

of n mutually incompatible basic *branch-referentials* $(G^{(o)}, V_b^{(o)})$ containing all the same basic generator $G^{(o)}$ but with different basic views. These, because they are mutually incompatible, produce together a universe of basic elementary events U which is the union of n mutually exclusive *branch-universes* of basic elementary events, $U_b^{(o)}$, $b = 1, 2, \ldots, n$:

$$U^{(o)} = \cup_b U_b^{(o)} = \cup_b \{ee_{bi}^{(o)}, i \in I\}, \quad b = 1, 2, \ldots, n.$$

So the global random phenomenon produced by a basic epistemic referential $(G^{(o)}.V^{(o)})$ admits of the following sequence of equivalent but differently analyzed MRC representations:

$$[[G^{(o)}.V^{(o)}], U^{(o)}]] = [[\mathcal{P}^{(o)}, U^{(o)}]] = \cup_b[\mathcal{P}_b^{(o)}, U_b^{(o)}]$$
$$= \cup_b[[G^{(o)}.V_b^{(o)}], U_b^{(o)}] = \cup_b[[G^{(o)}.V_b^{(o)}], \{ee_{bi}^{(o)}, i \in I\}]].$$

> A basic referential $(G^{(o)},V^{(o)})$ generates a meta[random phenomenon], a whole family of *related* random phenomena, involving all *one same operation of generation of a basic object-entity,* but a (finite) set of distinct mutually exclusive branches brought forth by the mutually incompatible branch-views $V_b^{(o)}$ from $V^{(o)}$.

If in particular $V^{(o)}$ consists of only one branch-view $V_b^{(o)}]$, $b = 1$, this family reduces to only one random phenomenon $[\mathcal{P}_b^{(o)}, U_b^{(o)}] \equiv [[G^{(o)}.V_b^{(o)}], U_b^{(o)}]$ like in the classical Kolmogorov probabilities.

This, finally, is a complete, fully explicit and entirely relativized representation of the content of the random phenomena involved by a basic epistemic referential.

> *The channels for the adduction of semantic substance, from the pool of what is called "physical reality", into a basic probabilistic description, are now entirely* `represented`.

At the same time the powers of representation of the initial basic epistemic referential $(G^{(o)},V^{(o)})$ are now *exhausted.* This referential alone cannot produce the whole MRC equivalent of a Kolmogorov representation of a probabilistic description, nor only a probabilistic description in the more ancient sense, of von Mises, for instance. Indeed $(G^{(o)}, V^{(o)})$ does not contain the descriptional resources necessary for representing the generation of the object-entities and of the qualifications involved by an algebra of events constructed on the universe of basic elementary events $U_b^{(o)}$ produced by $(G^{(o)}, V_b^{(o)})$, nor those, still more complex, involved by a probability measure on this algebra. All that the initial epistemic referential $(G^{(o)}, V^{(o)})$ can produce, in fact, is only the basic transferred descriptions $ee_{bi}^{(o)}$ from $U_b^{(o)}$, $b = 1, 2, \ldots, m$, so also, at the limit, the m universes $U_b^{(o)}$ themselves.

The algebra of events on a branch-universe U_b. In order to redefine in MRC terms the algebra of events from a Kolmogorov probability space constructed on the branch-universe $U_b^{(o)}$, the principle of separation

P15 and the definition D16 of a metadescription require to pass now on a higher level of conceptualization (with respect to the initial one) and to form there a convenient new epistemic referential.

Consider first only *one* among the branch random phenomena that contribute to the meta[random phenomenon] $[[G^{(o)}, V^{(o)}], U^{(o)}]]$. Consider the branch-universe $U_b^{(o)}$ from this branch-random-phenomenon. The relativized elementary events $e_{bi}^{(o)}$ from $U_b^{(o)}$ have the MRC status of testimonial relative descriptions involving some definite branch-view $V_b^{(o)} \equiv \cup_j V_{bj}^{(o)}$. This entails the following consequences.

(a) The insertion into the representation of the MRC concept of *probability*, of the deep level of *logical* conceptualization brought forth in Sec. 5.1.2, namely the level tied with strict individuality. Indeed since each occurrence of an elementary event $e_{bi}^{(o)}$ possesses the descriptional status of a testimonial relative description $\theta^{(o)}/G^{(o)}, œ_r^{(o)}, V_b^{(o)}/$ in the sense of D14.2.2, involving a *given* replica $œ_r^{(o)}$ of the basic object-entity $œ^{(o)}$, its tentative assertion is a testimonial proposition $p[\theta_1(œ_j^{(o)})]$ in the sense of DL.4. We are in conditions which, in essence, coincide with those which in 5.1.2 have been found to restrict the applicability of the logical conjunction: two testimonial propositions $p[\theta_1(œ_r^{(o)})]$ and $p[\theta_2(œ_r^{(o)})]$ which assert two distinct occurrences of elementary events (descriptions),$ee_{bl}^{(o)} = \theta_1(œ_r^{(o)})$ and $ee_{bw}^{(o)} = \theta_2(œ_r^{(o)})$ with $l \neq w$, but which are asserted for a *same* replica $œ_r^{(o)}$ of the basic object-entity $œ^{(o)}$, cannot be composed by a logical conjunction, such a composition is meaningless because the resulting composed proposition cannot exist factually.

> This "explains (or justifies) logically" why in a Kolmogorov probability space no product is defined for two elementary events, while, if these elementary events are reconsidered inside the algebra from that space, as one-element sets, their intersection is systematically void.

It is satisfactory that this "logical explanation" is made available *inside* the concept of probability which is constructed here. This is a first manifestation of the intimate relation which arises inside MRC between probabilities and logic.

(b) A second consequence of the fact that the elementary events $ee_{bi}^{(o)}$ have the status of relative descriptions, is the definibility, on the branch-universe $U_b^{(o)} = \{ee_{bi}^{(o)}, i \in I\}$, of a *classifying* branch-algebra of events involving classes determined on $U_b^{(o)}$ by aspects and values of aspects from the acting branch-view $V_b^{(o)} \equiv \cup_j V_{bj}^{(o)}$, $j = 1, 2, \ldots, m$. We have already remarked that the Kolmogorov elementary events, introduced by a set-theoretic definition, do not directly offer themselves for classifications, so that classifications can be only super-imposed upon them *a posteriori* by an added, entirely exterior descriptional action. Whereas inside MRC the elementary events $ee_{bi}^{(o)}$, because they emerge as relative descriptions, *are* qualifications (predications, in classical terms), so they *incorporate* criteria for future classifications.

Let us take an example. Remember that each elementary event $ee_{bi}^{(o)}$ can be regarded as a description of the object-entity [one observable configuration of marks $(bk)_f$], *via* the coding-view V_{bc} extracted from $V_b^{(o)}$. Imagine now that the coding-view V_{bc} is such that a description $ee_{bi}^{(o)}$ produced by it consists of some spatial configuration of coloured forms. Suppose that we consider the maximal spatial dimension involved by each form, specified separately as a characteristic feature of $ee_{bi}^{(o)}$. Then, considering the class of all the $ee_{bi}^{(o)}$ of which the maximal spatial dimension of a form from it is less that 5 cm, amounts to making abstraction of any other specificity than this last one; while considering the class of all the red $ee_{bi}^{(o)}$ from $U_b^{(o)}$ amounts to making abstraction of any other specificity of an elementary event $ee_{bi}^{(o)}$ apart from being red; etc. So, by dropping this or that qualification involved by the coding view V_{bc} involved by $V_b^{(o)}$, one can define classes on $U_b^{(o)} = \{ee_{bi}^{(o)}\}$, "classifying" metadescriptions of sets of elementary events from $U_b^{(o)}$. In this way it is possible to define on $U_b^{(o)}$ algebras τ_b of classifying metadescriptions of sets of elementary events $ee_{bi}^{(o)}$, *via* metaviews extracted from the coding view V_{bc} involved by $V_b^{(o)}$. An algebra of such metadescriptions will be called a *classifying algebra.* When a classifying algebra on $U_b^{(o)}$[50] is posited to contain also all the elementary events $ee_{bi}^{(o)}$ themselves, then it becomes the total classifying algebra on $U_b^{(o)}$, which brings in, also, all the purely set-theoretic features of any set of basic elementary-event descriptions $ee_{bi}^{(o)}$.

Now, syllogisms being constructed with classes of predicates, the classifying algebras defined on $U_b^{(o)}$ permit an immediate embeddability of syllogisms into them (continuing the example given above: all the descriptions $ee_{bi}^{(o)}$ which involve the qualification of being coloured red constitute the class $\{ee_{bi}^{(o)}\}_r$; the description $ee_{bi}^{(o)}$ belongs to the class $\{ee_{bi}^{(o)}\}_r$; hence the description $e_{bi}^{(o)}$ involves the qualification red). So both levels of *logical* conceptualization become embeddable into the MRC concept of *probability,*

[50] The total algebra on a set S is the algebra on S (cf. note 43) which involves all the subsets of S, including the subsets of only one element from S.

not only the strictly individual level of logical conceptualization mentioned above—which so far remained entirely hidden outside MRC-logic – but also the usual statistical level of natural logic.

> *Inside MRC, the relativization of any elementary event, to a definite view, entails complete dissolution of the obstacle that stands in the way of an explicit definition of the relations between the classical logic and the classical probabilities.*

Together, the preceding points *(a)* and *(b)* indicate already in what a sense the MRC reconstruction of Kolmogorov's concept of probability, entails a deep and organic, as if spontaneous association between the logical conceptualization and the probabilistic one. This however will become much clearer in the sequel.

Consider now explicitly the question of the epistemic referential involved by an event e from a classifying algebra τ_b defined on $U_b^{(o)}$. Such an event (if it does not coincide with an elementary event $ee_{bi}^{(o)}$) is a metadescription with respect to the descriptions $ee_{bi}^{(o)}$, produced by a new, conceptual, non basic metareferential. This metareferential introduces a metagenerator of object-entity which acts on the zone of reality consisting of the universe $U_b^{(o)} = \{ee_{bi}^{(o)}, i \in I\}$ and it consists itself of just [the field of perceptibility of a metaview] extracted from the coding-view V_{bc} by some abstraction, by some dropping of values of aspects or of whole aspects from V_{bc} (consider the examples from the above point *(b)*). So this generator is the generator of a view. Let us denote it $G_{br}^{(1)}(V_{br}^{(1)})$ where $V_{br}^{(1)}$ is the view of abstraction that has been utilized; the lower index r labels the chosen classifying feature, while the upper index 1 stresses that we are now on a descriptional metalevel with respect to that one labeled by 0. The meta[object-entity] produced by this metagenerator is a class $\{ee_{bi}^{(o)}\}_r$ of elementary descriptions from $U_b^{(o)}$. So the involved epistemic referential is $(G_{br}^{(1)}(V_{br}^{(1)}), V_{br}^{(1)})$. The corresponding relative (meta)description is

$$\{ee_{bi}^{(o)}\}_r = e_{br}^{(1)} \equiv D_{br}^{(1)}/(G_{br}^{(1)}(V_{br}^{(1)}), \{ee_{bi}^{(o)}\}_r^{(1)}, V_{br}^{(1)}/.$$

So the event $e_{br}^{(1)}$ from a classifying algebra $\tau_b^{(1)}$ defined on $U_b^{(o)}$ is a *degenerate* metadescription because it involves the generator *of the acting view* $V_{br}^{(1)}$, like the implicitly achieved metadescriptions of which the "objects" from the classical logic consist (cf. Sec. 5.1.2)). From now on τ_b is renoted $\tau_b^{(1)}$.

Since $e_{br}^{(1)}$ depends on the metaview $V_{br}^{(1)}$ which in its turn depends on the sort of abstraction by which it is extracted from the coding-view V_{bc}, another abstraction will lead to another metaview and another metagenerator, so to another event-description characterized by another lower index r.

> *The algebra of events $\tau_b^{(1)}$ introduces a whole family of metareferentials of the type $(G_{br}^{(1)}(V_{br}^{(1)}), V_{br}^{(1)})$.*

The probability measure on a branch-algebra $\tau_b^{(1)}$. By definition the probability of an event $e_{br}^{(1)}$ from the algebra of events $\tau_b^{(1)}$ constructed on the universe of elementary events $U_b^{(o)}$, say $p(e_{br}^{(1)})$, is the limit—*supposed to exist*—toward which the relative frequency $n(e_{br}^{(1)})/N$ of the realizations of $e_{br}^{(1)}$ (of occurrences of any elementary event $e_{bi}^{(o)}$ from $e_{br}^{(1)}$) converges when N is increased toward infinity:

$$p(e_{br}^{(o)}) = \lim N \to \infty[n(e_{br}^{(o)})/N].$$

And the *probability measure on* $\tau_b^{(1)}$ is by definition the set $\{p(e_{br}^{(o)})\}$ of *all* the probabilities assigned to events from $\tau_b^{(1)}$

Let us specify the MRC descriptional *level* of the probabilistic estimations from a branch-probability space. On a level immediately successive to that of $\tau_b^{(1)}$—so here the level 2 with respect to the initial level 0 of the elementary elements—a convenient operational-conceptual generator of object-entity generates for each event $e_{br}^{(1)}$ from $\tau_b^{(1)}$ the corresponding relative frequency $[n(e_{br}^{(1)})/N]$ of occurrence of $e_{br}^{(1)}$ in a sequence of N iterations of the considered branch-random-phenomenon $[[G^{(o)}.V_b^{(o)}], U_b^{(o)}]$; and an *aspect-view of relative frequency* estimates the *numerical values* of the ratios $n(e_{br}^{(1)})/N$ from this sequence of N iterations, which are also the values *in the sense of D5.1* of the aspect-view of relative frequency. Afterward, on a subsequent level—so here the level 3 with respect to the initial level 0—a convenient operational-conceptual generator of object-entity, say $G_{br}^{(3)}$, selects as meta-meta-object-entity the whole sequence of ratios

$$\sigma_{br}^{(3)} = [n_1(e_{br}^{(1)})/N, n_2(e_{br}^{(1)})/N_2, \dots, n_q(e_{br}^{(1)})/N_q, \dots],$$

where $e_{br}^{(1)}$ is en event from the algebra of events $\tau_b^{(1)}$, and the number N of iterations of the involved random phenomenon is increased toward infinity *via* some sequence of increasing integers N_q, $q = 1, 2, \dots$. The meta-meta-object-entity $\sigma_{br}^{(3)}$ selected by $G_{br}^{(3)}$ is examined *via* an *aspect-view of probability* (convergence) $V_{b\,pr}^{(3)}$ say, which checks for the existence of a convergence in the sequence $\sigma_{br}^{(3)}$ and, if the convergence does exist, estimates the limiting numerical value

$$\left(p(e_{br}^{(1)})\right)^{(3)} = \lim .N \to \infty[n(e_{br}^{(1)})/N],$$

which is also a value *in the sense of D5.1* of the aspect-view $V_{b\ pr}^{(3)}$. So on this last descriptional level, of relative order 3, the acting epistemic referential is $(G_{br}^{(3)}, V_{b\ pr}^{(3)})$. The meta(metadescription) produced by it is

$$D^{(3)}/G_{br}^{(3)}, \sigma_{br}^{(3)}), V_{b\ pr}^{(3)}/ \equiv \left(p(e_{br}^{(1)})\right)^{(3)}.$$

So the probability measure on the whole algebra of events $\tau_b^{(1)}$, is

$$p_b^{(3)} \equiv p^{(3)}(\tau_b^{(1)}) \equiv \left\{\left(p(e_{br}^{(1)})\right)^{(3)}\right\},$$

where r runs over the whole index-set of events from $\tau_b^{(1)}$. Since $\tau_b^{(1)}$ is a `logical classifying organization` of the elementary events $ee_{bi}^{(o)}$ from $U_b^{(o)}$, the syllogistic constructions embedded in the algebra $\tau_b^{(1)}$ can be quite naturally associated with numerical probabilistic estimations. If furthermore $\tau_b^{(1)}$ is the total algebra on $U_b^{(o)}$, the probability measure $p(\tau_b^{(1)})$ defined on it concerns also the elementary events from $U_b^{(o)}$.

> *The MRC connection between classical logic and probabilities is fully achieved.*

This connection starts on the level of the elementary-event descriptions $ee_{bi}^{(o)}$ where *repetitions* of the involved random phenomenon are permitted. So—quite satisfactorily—it leaves out, *beneath* it, the strictly individual level of the non-classical MRC-logic, concerning testimonial propositions tied with *one replica* of an object-entity of a given sort: the MRC-logic begins at the same level as the MRC-probabilties, namely beneath the statistical level of classical logic. But the MRC-probabilities end above the MRC-logic and qualify numerically the statistical zone of the MRC-logic, by values of limits of convergent statistical sequences.

The branch-probability chain stemming from a branch basic epistemic referential. So a basic branch-probability-chain

$$[P_b, U_b] \rightsquigarrow [U_b, \tau_b, p(\tau_b)]$$

admits of the more specified MRC representation

$$[P_b^{(o)}, U_b^{(o)}] \rightsquigarrow [U_b^{(o)}, \tau_b^{(1)}, p^{(3)}(\tau_b^{(1)})],$$

which can also be written in various other more detailed forms. The elementary-event-descriptions $ee^{(o)}_{bi}$ are achieved inside the epistemic referential $(G^{(o)}, V^{(o)}_b)$, each event-descriptions $e^{(1)}_{br}$ from $\tau^{(1)}_b$ introduces its own epistemic metareferential $(G^{(1)}_{br}(V^{(1)}_{br}), V^{(1)}_{br})$, and the probabilistic description $p^{(3)}(\tau^{(1)}_b)$ of the algebra of events $\tau^{(1)}_b$ is achieved inside the epistemic meta-metareferential $(G^{(3)}_{br}, V^{(3)}_{b\ pr})$. On these writings one can read the whole essence of the genetic and hierarchical MRC structure of a branch-probability-chain.

A branch-probability-chain as represented above is the MRC equivalent of a classical Kolmogorov probability space for the case that a basic epistemic referential is at work. This equivalent transcends already a classical probability space. Each one of the elements introduced by it is explicitly relativized to the generator of object-entity and the view introduced by the epistemic referential involved in the generation of that element. The descriptional relativities cannot all be read directly on the final synthetic representation chosen above, but they are all explicitly available, and they can be made manifest in the symbolizations whenever this is wanted. The operational and the conceptual structure of the random phenomenon which founds the space, as well as the hierarchical structure of the space itself, become apparent. Each one of the involved descriptional entities (actions or results of actions) is endowed with an explicit definition and an own symbolization:

> *One disposes now of entirely specified moulds for* `expressing` *the whole genetic and hierarchical structure of a basic branch-probability space.*

A mathematician might perhaps hold that these specifications amputate the generality of Kolmogorov's purely set-theoretic-algebraic representation. But such a criticism would have to be dismissed. Indeed, as shown already, the mathematical generality of the classical concept of probability can also be regarded as a source of lacunae, and the MRC representation dissolves the lacunae without interdicting the use of more synthetic expressions and treatments.

We are now ready to introduce the major novelties produced by MRC inside the probabilistic conceptualization, namely the concept of probability tree and the correlative clarification and complexification on the meaning of what is called probabilistic independence or dependence.

The probability tree of a basic epistemic referential. It follows immediately that the integral probabilistic phenomenon which stems from a basic epistemic referential $(G^{(o)}, V^{(o)})$ where $V^{(o)} = \cup_b V_b^{(o)}$, $b = 1, 2, \ldots, n$, can be represented as follows:

$$[[P^{(o)}, U^{(o)}]] \rightsquigarrow \cup_b [U_b^{(o)}, t_b^{(1)}, p^{(3)}(\tau_b^{(1)})].$$

This representation points toward a new probabilistic metaconstruct. This metaconstruct constitutes a *probabilistic* unity, in this sense that in all the mutually exclusive branches involved by it, the same generator $G^{(o)}$ of object-entity acts, creating a common "trunk", namely one same sort of basic object-entity œ$^{(o)}$ which then plays a key role in the emergence of all the n distinct branch-probability-spaces $[U_b^{(o)}, \tau_b^{(1)}, p^{(3)}(\tau_{b(1)})]$, *connecting them genetically.* This new probabilistic metaconstruct will be called *the probability tree of the basic epistemic referential* $(G^{(o)}, V^{(o)})$. It will be symbolized by $\mathbf{T}(G^{(o)}, V^{(o)})$.

The classical theory of probabilities of Kolmogorov does not define such a construct.

But in quantum mechanics a particular instance of this very construct *does* manifest itself, though implicitly. One operation of micro-state-generation (playing the role of basic generator $G^{(o)}$) produces one microstate (holding the role of basic object-entity œ$^{(o)}$) and all the quantum mechanical probability measures defined for this unique microstate, but concerning the outcomes of all the mutually incompatible groups of commuting quantum mechanical observables (holding the role of basic branch-views $V_b^{(o)}$), are calculated from the *unique* state-function ψ associated with this microstate and the involved quantum observables. A given group of compatible quantum mechanical observables, produces a universe of factual elementary events (marks registered on a measurement-device)—each one codable in terms of this or that eigenvalue of an observable from that group—which has *no common element* with the universe of factual elementary events produced by another group that is incompatible with the first one; in this sense the mentioned universes of factual marks are mutually exclusive. So the algebras—*Boleean* algebras— constructed on each one among these mutually exclusive universes of elementary events, are equally mutually exclusive. Hence, by asserting probability measures on these mutually exclusive algebras, one finally obtains a whole *set* of distinct probability spaces, but *all associated with one* same *state-function* ψ: In MRC terms, one obtains a *quantum mechanical probability tree* [12-17]. This situation—but in the absence of an explicit concept of quantum mechanical probability tree—has been amply discussed (Mackey, Gudder, Suppes, Van Fraassen, and many others) because it is devoid of a corresponding general form in Kolmogorov's abstract theory of probabilities, so it does not yet possess a defined probabilistic status. In particular, various attempts have been made at defining *one* metaprobability measure corresponding to the unique involved state-function and involving somehow the branch-probability measures. But, as far as I know, no consensus has been reached as yet concerning a satisfactory solution. Therefore what is called "quantum probabilities" is still considered to constitute an unsolved problem of the probabilistic conceptualization.

Furthermore, the situation sketched out above has also induced attempts at the examination of the logico-algebraic nature of the *global* algebra consisting of the union of all the mutually exclusive branch-algebras of events $\tau_b^{(1)}, b = 1, 2, \ldots, n$, tied to one state function ψ.

> And this global algebra has been found *not to be Boolean*. Which constitutes the "problem of quantum logic".

Nowadays quantum-logicians seem to consider to have solved this problem by assigning a lattice-structure to this global algebra. But such a structure appears as *inadequate* as soon as one becomes aware that ***(a)*** the logical conjunction is not a universal logical connector (cf. [13] and 5.1.2) and ***(b)*** that logical complementation is a relative operation (cf. Refs. [24] and 5.1.2).

In this context, the interest of the general MRC-concept of probability tree, seems clear: it becomes possible to deal with the questions of quantum probabilities and of quantum logic inside a quite general and organized framework (it is in this way that meta[quantum mechanics] is developed (cf. the Introduction)).

But independently of this specific perspective involving quantum mechanics it is remarkable by itself that MRC, where exclusively the fundamental descriptional mould is drawn by generalization from the epistemic strategy practised in quantum mechanics, brings forth at the *top* of its elaboration the metaconstruct of a probability tree, of which the quantum probabilities appear *a posteriori* as a *particular* realization, and where a corresponding global algebra of events is contained that is by construction open to syllogistic-logical qualifications naturally tied with probabilistic qualifications.

On the logic obeyed by the global algebra of events from a probability tree. Consider the union $\cup_b \tau_b^{(1)}$, $b = 1, 2, \ldots, n$ of all the algebras of events from all the n distinct branches of a given probability tree $\mathbf{T}(G^{(o)}, V^{(o)})$; we denote it by $\tau_{\mathbf{T}}^{(1)}$ and call it the *global algebra from* $\mathbf{T}(G^{(o)}, V^{(o)})$. How can the logical specificities of $\tau_{\mathbf{T}}^{(1)}$ be pertinently represented? In the present context we make only the following remark.

A probability tree $\mathbf{T}(G^{(o)}, V^{(o)})$ is equivalent to the basic genetic class $C[G^{(o)}.V^{(o)}]$, supposed to end up with a probabilistic description of the involved basic object-entity œ$^{(o)}$. So the principles sketched out in Sec. 5.1.2 concerning a calculus with genetic classes, conjugated with the characterization of a probability tree achieved above, permit inside MRC a guided, a dominated specification of the logic of a global algebra $\tau_{\mathbf{T}}^{(1)}$, freed from arbitrary assumptions like an *a priori* posited lattice structure, and enriched by an explicit awareness of all the involved descriptional relativities (in particular the relativity of complementation) as well as of the logical consequences of the mutual exclusions that stem from strict factual individuality (lack of factual counterpart for logical conjunction (Sec. 5.1.2)).

Probability trees versus probabilistic dependence. Kolmogorow ([29]) wrote:

> "*... one of the most important problems in the philosophy of natural sciences is—in addition to the well-known one regarding the essence of the concept of probability itself—to make precise the premises which would make it possible to regard any given real events as independent.*"

But Kolmogorov's approach is purely mathematical. The criteria for probabilistic independence are researched exclusively and directly as formal criteria working on a directly posited abstract mathematical structure. The specificities of the involved *physical* phenomena are *never* taken into account. If only one probability space is considered, two events A and B from the algebra τ from this space are just posited to be independent if the numerical product $p(A)p(B)$ of their probabilities is equal to the probability $p(A \cap B)$ of the product-event $A \cap B$, in the set-theoretical sense. This same definition is generalized to also the case when A belongs to one algebra, and B to another one, but *presupposing* always that the conjoint outcome of A and B is possible, which, in MRC terms, amounts to embeddability of both algebras in one same space, by the definition of a conveniently enriched coding-view for the interpretation of a factual outcome. But the concept of probability tree of a basic transferred probabilistic description brings into evidence that

> *Kolmogorov's definition of probabilistic dependence or independence, is not a general definition.*

The winding line along which this definition fails when two distinct branches of a same probability tree are brought in, can be followed in detail.

Let b_1 and b_2 be two distinct branches of a probability tree $\mathbf{T}(G^{(o)}, V^{(o)})$. The product-event $A \cap B$ of two events A and B with A from b_1 and B from b_2, is systematically the null-event, because A and B, being produced by different random phenomena, cannot contain common elementary events: they belong to algebras $\tau_{b_1}^{(1)} \neq \tau_{b_2}^{(1)}$ constructed on two universes of elementary events $U_{b_1}^{(o)} \neq U_{b_2}^{(o)}$ which are produced by two distinct and *mutually incompatible* branch-examinations with $V_{b_1}^{(o)} \neq V_{b_2}^{(o)}$, and so contain no common elementary events. In these conditions $p(A \cap B)$ is always zero. Now, zero is different from the quantity $p(A)p(B)$ as soon as both A and B are possible, *so this might mean systematic dependence*. Therefore, at a first sight, one might think that finally Kolmogorov's definition works well, since for two events from two distinct branches of one tree, dependence can be explained by the common generator $G^{(o)}$. But $p(A \cap B)$ is always zero also for two events from two branches from two different probability trees, while in general $p(A)p(B)$ is not zero, again, and in this case why should there always be dependence? Obviously the seemingly satisfactory systematic nullity of the quantity $p(A \cap B)$ when it is calculated for events A and B from two different branches of a same tree, in fact is just an automatic, meaningless reaction of a formalism which in fact is exceeded by what is tried to be described by its use.

> Kolmogorov's formal definition of probabilistic dependence/independence simply is *alien* to the concept of probability tree. It stems from a classical experiential background where situations like those introduced by probability trees are not taken into account.[51]

Such situations have not even been conceived on the basis of the experiential background from which the classical theory of probabilities has been drawn.

On the other hand, according to the "theory of transformations" from the Hilbert-Dirac formulation of quantum mechanics, given two incompatible quantum mechanical observables X and Y and *one* state-function ψ, the probability $p_\psi(y_i)$ of the occurrence, for the microstate with state function ψ, of (any) one given elementary event consisting of an eigenvalue y_i of the observable Y, is a functional

$$p_\psi(y_i) = F[p_\psi(X)]$$

[51] In so far that it is always possible, for any set of correlated spaces, to construct by cartesian multiplication one space that contains all the spaces from this set, *the confinement inside one branch* holds also for the classical concept of probabilistic *correlation*, not only for that of dependence. Correlation, like dependence, is *exceeded* by the concept of probablity tree.

of the whole probability *measure* $p_\psi(X)$ concerning the same ψ and the observable X, the form of the functional F being specified by Dirac's calculus. Inside quantum mechanics this formal fact is regarded as just a calculational "rule" concerning the passage from the "representation" of the state-ket $|\psi>$ expressed in the basis (the Hilbert referential) corresponding to the observable X, to the representation of $|\psi>$ in the basis introduced by the observable Y: *no specifically* `probabilistic` *significance is assigned to the above-mentioned functional relation.* So *a fortiori* no physical significance either has been researched. But when it is reconsidered from the point of view of the MRC concept of a probability tree (cf. [12-17]), Dirac's transformation $p_\psi(y_i) = F[p_\psi(X)]$ acquires the significance of a relation of *probabilistic metadependence* which expresses the physical kinship, the `semantic kinship` *between the contents of all the various branches*: two distinct branches of a same probability tree refer indeed to two different and non commuting quantum mechanical observables, but they concern *one* `same` *microstate*, generated by a unique operation of state-generation and represented by a unique state-function. So it leaps to one's eyes that Dirac's transformations, apart from their formal calculational role, express also an effect of the uniqueness of the considered microstate, upon the nature of the contents from all the distinct branches. And they express it, not as a probabilistic dependence in the classical sense, but as a non-classical probabilistic *meta*dependence: *each* elementary probability $p_\psi(y_i)$ from the branch of the tree corresponding to the observable Y, depends, not *individually* on this or that other elementary probability $p_\psi(x_j)$ from the branch corresponding to X, but on the set $\{p_\psi(x_j)\}$ of *all* the elementary probabilities from the branch of X (on the whole probability measure $p_\psi(X)$).

In Sec. 3.2 it was shown how quantum mechanics has opened up the way toward the construction of MRC. Now it appears that MRC permits to identitfy a deep probabilistic meaning of Dirac's theory of transformations. We shall complete this process of spiraling double-way mutual influence as follows:

Suppose a probability tree $\mathbf{T}(G^{(o)}, V^{(o)})$ where, in every branch b, the total algebra on the involved universe $U_b^{(o)}$ is chosen (which contains also all the elementary events $ee_{bi}^{(o)}$). We make the natural—even inescapable—assumption that the unique generator of object-entity $G^{(o)}$ which contributes to the emergence of all the branches of $\mathbf{T}(G^{(o)}, V^{(o)})$, induces, *via* the corresponding object-entity $œ^{(o)}$, a semantic kinship between the contents of these branches, at all the three involved levels, the level of elementary events, the level of the algebras of events, and the level of the probability laws. We posit that there exists a "degree of similitude" between the contents placed at the same descriptional level of any two branches (on different levels there cannot be comparability), which is somehow determined by the "angle" between two mutually incompatible "directions" of $V_b^{(o)}$-examination (with $b = b_k$ and $b = b_q$) of the unique basic object-entity œ$^{(o)}$. So we expect observable manifestations of this kinship. Concerning these—on the basis of the fact that the quantum mechanical probability trees are particular instances of the general MRC concept of probability tree—we postulate what follows:

> The semantic kinship between the contents of the branches of a probability tree can be conveniently expressed mathematically on the probabilistic level, by admitting that each probability of an *elementary* event from a branch b_k of $\mathbf{T}(G^{(o)}, V^{(o)})$, depends on the *whole* probability *measure* from any other branch b_q, *via* a functional relation F of which the precise form has to be specified in each case by an experimental-theoretical approach appropriate to the particular nature of the involved phenomena.[52]

> *Then the set of all the distinct branch-probability measures from all the distinct branches of* $\mathbf{T}(G^{(o)}, V^{(o)})$, `interconnected by the above-posited functional relation`, *constitute together an* `observable` *metaconcept of probability measure that characterizes globally the probability tree* $\mathbf{T}(G^{(o)}, V^{(o)})$*: There is no need of a unique metaprobability-measure.*

Furthermore, since a probability tree $\mathbf{T}(G^{(o)}, V^{(o)})$ of the epistemic referential $(G^{(o)}, V^{(o)})$ is equivalent to the basic genetic class $C(G^{(o)}, V^{(o)})$ of the same referential, the possibility of a logical calculus with whole genetic classes draws attention upon the possibility of a corresponding probabilistic calculus with probability trees considered as *wholes* (for the particular case of quantum mechanincs, cf. [12-14]. General rules of composability of two or more different trees can be defined, involving specific sorts of probabilistic metadependence (or metacorrelation, this distinction will have to be redefined), namely between two distinct trees involving different generators of object-entity but the same branch-views, or *vice versa* the same generator and different branch views, etc. This completes the domain of probabilistic-logical research opened up by MRC.

---

[52] If the sort of basic object-entity that is involved has *not* a wavelike nature like in the case of a quantum mechanical microstate, there is no *a priori* reason for admitting a principle of superposition, though such a principle might be found to hold. If it comes out that the principle of superposition is semantically inadequate, the whole mathematical framework of a Hilbert *vector* space, would have to be conveniently modified. A blunt transcription of the quantum mechanical mathematics would be meaningless.

As for the probabilistic dependence between two events A and B belonging both to the algebra from one same branch b, Kolmogorov's definition holds, of course. But inside MRC it is furthermore "explained" semantically, namely again as a manifestation of a community of nature between any two elementary events or events from a same branch, induced *genetically* by the unique random phenomenon $[G^{(o)}.V_b^{(o)}], V_b^{(o)}]$ that produced them both. This last sort of dependence is certainly stronger than the metadependence postulated above, because it is induced by the conjugated actions of the involved basic generator $G^{(o)}$ *and* the involved basic branch-view $V_b^{(o)}$. Furthermore it is estimated in simple numerical terms, which is not the case for the probabilistic metadependence between distinct branches. These features explain why the classical probabilistic dependence has been remarked since a long time, while the metadependence brought forth by MRC has not been discerned. So:

> The classical concept of probabilistic dependence, considered so important by Kolmogorov, becomes *intelligible*, and it is included in a larger concept of probabilistic dependence which is organized in zones of dependences of different *natures* and of different *degrees*.

A last consequence of the MRC concept of probability tree, and not a minor one, concerns "causality".

> The hierarchical and probabilistic dependences brought forth by the concept of probability tree, and their connections as expressed by the to-be-elaborated calculus with whole probability trees, yield a new, very organized framework for the representation of the still so vague concept of causality. Inside this framework it might be possible to define in a precise way mutually distinct concepts of causality, dependence, and correlation, as well as the relations between them.

Note added in proof. C. G. Jung has introduced a famous concept of "synchronicity" (concerning phenomena of any nature, physical, psychical, or physical-psychical) concerning which he wrote in *Erinnerungen, Traüme, Gedanken* (registered by Aniela Jaffé):

> "My concern about the psychology of unconscious processes has obliged me, already since a long time, to research—beside causality – another principle of explanation, since the principle of causality seemed to me to be inappropriate for explaining certain surprising phenomena from the psychology of unconsciousness. Thus I found parallel unconscious phenomena which could not be causally connected to one another; but they had to be related otherwise by another development of the events. This connection between events seemed to me to be given essentially by their relative simultaneity which called forth the term "synchronicity. (My parentheses. Note the relaxing terms "essentially" and "relative", by parenthesis) .... So I make use here of the term of "synchronicity" in the specific sense of coincidence in time of two or more events devoid of causal relation and which possess the same content of

significance or a similar one, and this in opposition with "synchronism" which indicates simply the simultaneous emergence of two phenomena .... The coincidence between *events tied by significance* (my italics) can be thought as pure chance. But more they multiply and more the concordance is precise, more their probability diminishes and bigger becomes their unlikelihood, which amounts to saying that they cannot any more count as pure chance, but, given the absence of a causal explanation, they have to be regarded as significant arrangements. Their inexplicability does not stem from the fact that their cause is ignored, but from the fact that our intellect is unable to think it." (My own English translation from the French edition, Gallimard, 1966).

One has the feeling of watching a deep effort to express a class of circumstances of the same essential nature as those which—in MRC terms—bring forth elementary events or events from a probability tree of a probabilistic basic relative description, which are tied with one another *via* a same basic operation $G^{(o)}$ of generation of an unobservable basic object-entity $œ^{(o)}$ (facts "*devoid of causal relation and which possess the same content of significance or a similar one*"!) All this is about "the semantical kinship" between elementary events or events from different branches of the same probability tree. Of course, in the absence of the guidance drawn from quantum mechanics of which the construction of MRC did benefit, it was quasi impossible to reach a fully worked-out expression of circumstances of this kind. But it is quite remarkable, I think, that Wolfgang Pauli took Jung's concept of synchronicity very seriously. Indeed Pauli and Jung exchanged a number of letters on this subject (*Wolfgang Pauli und C. G. Jung: ein Briefwechsel*, Springer, 1992). For instance, in a letter to W. Pauli (30 November 1950), Jung wrote:

> Since it appears that physical discontinuities cannot be reduced by confining oneself to simple causality, they represent a "being", that is to say *a disposition or an "act of creation"* (my italics) like any case of synchronicity." (My own translation from the French edition, Albin Michel, 2000).

Again this intense though unorganized intuition of a common basic generator $G^{(o)}$ ("act of creation") of a deeply hidden basic object-entity $œ^{(o)}$ ("*being*", "*disposition*"!, see the comments in D14.3.1) which, in a way that transcends what is called causality, determines observable physical effects that appear as "*physical discontinuities*" (mutually *exclusive* elementary events $ee^{(o)}_{bi}$!). All this perceived like a strongly illuminated structure surrounded by thick mist.

In Pauli's answer (12 December 1950, *op. cit.*) one finds the following remarks concerning the "calculi of likelihood" (notice that he did *not* write "calculus of probability" because of the unsolved probabilistic status of what he called "the (mathematical) law of likelihood" from quantum mechanics:

> " ... in this moment "fundamental mathematics" are lost in a great confusion because they have tried to master these questions in a very ambitious, but unilateral and counter natural way. Inside the domain of "fundamental mathematics", the "foundations of the mathematical calculus of likelihood" are particularly touched by this crisis. While reading a special issue of a specialized review devoted to this problem, I felt completely consternated by the divergence of all the points of view and I have later learned that the specialists avoid discussing this question on the pretext that "as everybody knows" it is impossible to come to an agreement on this point! ... It seems to me indispensable that you expose clearly inside the chapter IV about *physical discontinuities* (my italics) the conceptual differences between nonpsychical acausal orders on the one hand and on the other hand the semi-psychical and psychical synchronicities."

It is striking that today the official status of "the (mathematical) law of likelihood" from quantum mechanics is still exactly as obscure as it was in 1950. It is still more striking to compare Pauli's formulations with the fact that the MRC general concept of probability tree of a basic probabilistic relative description, specifies the whole structure of the "non-psychical acausal orders" involved in the probabilistic conceptualization (see also the subchapter on *the significance of a probability measure*, forthcoming in the third stage of (5.2.3), offers a definite and deep-set framework for a unification of probabilities and logic *in terms of epistemological choices and actions, and of the observable physical results of these*, and thereby, contrary to what "everybody knows," settles the question of the status of the "foundations of the mathematical calculus of likelihood", that of the "quantum mechanical calculus of likelihood" included.

Second stage: A minimal spacetime model of the random phenomenon from a probability tree: genetic set (genset)

We have much stressed before that in consequence of the facts expressed by the frame-principle P8, we are unable to think about physical phenomena outside spacetime. Since a basic probability tree is founded upon a physical random phenomenon, as long as an explicit spacetime representation is not offered, in some way or other some inexplicit and unruled spacetime representation will nevertheless surreptitiously creep in. Which might produce confusion. So let us explicate accordingly to MRC a spacetime structure that can be assigned to the random phenomenon involved by a basic probability tree.

The minimal intrinsic metaconceptualization $[min.D_I^{(1)}/D^{(o)}]$ of a basic transferred description $D^{(o)}$ and the minimal model $[min.M(œ^{(o)}/V^{(o)}]$ extracted from it (D19.3), have "explained" the involved basic object-entity $œ^{(o)}$ in terms of a bulk of potentialities of future and relative observable manifestations located inside a spacetime domain $[\partial \mathbf{r}.t_o]$. They also posited that the basic processes of examination of $œ^{(o)}$ corresponding to the various successions $[G^{(o)}.V_g^{(o)}\ ]$ with $V_g^{(o)} \in V^{(o)}$, transpose the relative potentialities of observable manifestations confounded inside this bulk $œ^{(o)}$, into the *actualized* observable marks of which the transferred description $D^{(o)}$ consists. But the spacetime structure of these processes of actualization has been left inexplicit. In the definitions of the concepts $[min.D_I^{(1)}/D^{(o)}]$ and $[min.M(œ^{(o)}/V^{(o)}]$, the accent has been placed exlusively on the basic object-entity $œ^{(o)}$. In consequence of the frame-principle P8 and of the principle P10 of individual mutual spacetime exclusion, it seems clear that the posited processes of actualization possess a *tree-like* spacetime structure (as have anticipated the denominations of "trunk", "branches", and "probability tree"). Let us nevertheless establish this assertion.

Let us make use of the renotations in terms of branches b introduced at the beginning of 5.2.3. Consider the repeated realizations of the basic successions $[G^{(o)}. V_b^{(o)}]$ with $V_b^{(o)} \in V^{(o)}$ which generate the global random phenomenon $[P^{(o)}, U^{(o)}] = \cup_b [P_b^{(o)}, U_b^{(o)}]$ from a probability tree. *The processes from these successions are* **themselves** *physical entities.* So according to the frame-principle P8, each such process covers some spacetime domain. The process of generation by $G^{(o)}$ of a replica of œ$^{(o)}$, present in any realization of any sequence $[G^{(o)}. V_b^{(o)}]$, covers always a *same* spacetime domain with respect to an origin of times *renewed* each time that an operation $G^{(o)}$ is *started*, thus determining a common trunk of the spacetime representation of the random phenomenon. While P10 entails that the spacetime domains covered by incompatible $V_b^{(o)}$ examinations with $V_b^{(o)} \in V^{(o)}$—always started from an origin of times taken when the creation of a replica of œ$^{(o)}$ has just been *achieved*—cover *distinct* spacetime domains. So, together, all the successions $[G^{(o)}.V_b^{(o)}]$ with $V_b^{(o)} \in V^{(o)}$ cover indeed a tree-like spacetime domain.

What has been said so far concerns any basic transferred description, no matter whether individual or probabilistic. From now on we concentrate upon the probabilistic basic transferred descriptions.

We have shown that the initial definition denoted $D^{(o)}$ was insufficient in the case of a probabilistic basic description, and we have completed it, thereby obtaining the concept of probability tree $\mathbf{T}(G^{(o)}, V^{(o)})$. So, instead of $[min.D_I^{(1)}/D^{(o)}]$ we re-write now $[min.D_I^{(1)}/\mathbf{T}^{(o)}]$ where $\mathbf{T}^{(o)}$ is an abbreviation for $\mathbf{T}(G^{(o)}, V^{(o)})$.

Consider now the set $\{ee_{bi}^{(o)}, i \in I\}$ of elementary events (descriptions) produced in a given branch b from $\mathbf{T}(G^{(o)}, V^{(o)})$. From a logical point of view, these constitute a *class* – labelled by b and determined by the predicate "produced by sequences $[G^{(o)}.V_b^{(o)}]$" – of observable configurations of factual marks $(bk)_f$, each mark being coded in terms of only conceptually distinguished values of the various aspects bj from b, whereby an elementary description $ee_{bi}^{(o)}$ is obtained. But from a set-theoretic point of view, the elementary-event-descriptions $ee_{bi}^{(o)}$ constitute a set of such marks. Therefore we shall speak of the *class-set* b of elementary events $ee_{bi}^{(o)}$. Then the elementary events produced in all the distinct branches b from $\mathbf{T}(G^{(o)}, V^{(o)})$, constitute the set of class-sets $\cup_b \{ee_{bi}^{(o)}, i \in I\}$, $b = 1, 2, \ldots, n$. If now we associate to each element from $\{ee_{bi}^{(o)}, i \in I\}$, the spacetime representation of its *whole genesis* such as it is posited by the minimal intrinsic metaconceptualization $[min.D_I^{(1)}/\mathbf{T}^{(o)}]$, we obtain a *new* set of class-sets, with a new sort of elements, namely the geneses of all the elementary events œ$_{bi}^{(o)}$ from $\mathbf{T}^{(o)}$. We call it the *genetic* set of class-sets from $\mathbf{T}(G^{(o)}, V^{(o)})$, or the *genset* of the basic epistemic referential $(G^{(o)}, V^{(o)})$, and we symbolize it by

$Gen[(G^{(o)}, V^{(o)})]$. An element from a genset will be called a *genetic element*, in short a *genelement*. So the genset $Gen[(G^{(o)}, V^{(o)})]$ is the set of all the sets of genelements from $\mathbf{T}(G^{(o)}, V^{(o)})$. This endows us with the researched minimal spacetime representation of the physical random phenomenon involved by a probability tree $\mathbf{T}(G^{(o)}, V^{(o)})$ (the algebras of events $\tau_b^{(1)}$ and the probability measures $p(\tau_b^{(1)})^{(3)}$ on these, with $b = 1, 2, \ldots, n$, are *conceptual* metaconstructs of increasing order, *superposed* on the physical geneses of the elementary events which —alone— constitute the physical support of $\mathbf{T}(G^{(o)}, V^{(o)})$).

The concept of probability tree $\mathbf{T}(G^{(o)}, V^{(o)})$ leaves imprisoned in the only half-conceived, both the basic object-entity œ$^{(o)}$ and the geneses of the elementary events $\{ee_{bi}^{(o)}, i \in I\}$. The genset $Gen[(G^{(o)}, V^{(o)})]$ associated with $\mathbf{T}(G^{(o)}, V^{(o)})$ draws them into the clearly conceived and communicable. This is a pragmatic improvement.[53]

Consider now a genelement from a genset. It cannot be considered to *clearly* "belong" to the genset, because it does not entirely *pre*-exist. It possesses three mutually distinct modalities of existence that come into being successively. A genelement is first only abstractly and prospectively distinguishable inside the minimal intrinsic model $[min.M($œ$^{(o)}/V^{(o)}]$ assigned to the basic object-entity œ$^{(o)}$ alone, namely as "one" among the bulk of all the as yet non realized, mutually non individualized relative *potentialities* of which œ$^{(o)}$ is imagined to consist. This only mentally, prospectively individualized potentiality, undergoes then a *process of actualization*, whereby the previous potential whole labelled œ$^{(o)}$ is—entirely—*consumed*. And finally the observable end of this *process of actualization*, an elementary description $ee_{bi}^{(o)}$, is obtained as a stably *actualized* result, whereby the previous processual state of actualization also becomes in its turn entirely consumed. So there is a passage that leads from an undivided whole labelled œ$^{(o)}$, to this or that one among all the observable ends $ee_{bi}^{(o)}$, the material trace of the process that led to $ee_{bi}^{(o)}$ being each time effaced. This passage brings into play the whole depth of the Aristotelian and Kantian modal dimension which goes from potentiality to actuality, while it also shrinks down the potential undivided whole labelled œ$^{(o)}$, into this or that *individual*, actualized, phenomenal manifestation labelled $ee_{bi}^{(o)}$. So, saying that a genelement "belongs" to the genset $Gen[(G^{(o)}, V^{(o)})]$ to which it contributes, would amount

[53] The theory of elementary particles, more or less implicitly, works with gensets. It *associates* to purely predictional probability distributions of the type of those defined in fundamental quantum mechanics, *minimal models* constituting gensets in which the quantum mechanical sets of state-observables and of objectities (cf. notes 5 and 13) are enriched with other, sub-quantical state-observables and objectities and with hypotheses concerning their spacetime emergence.

to a brutal *a posteriori* simplification whereby the differences between the successively involved modalities of being are occulted, duration is eliminated, and instead, a fictitiously fully "present" whole is instated: a sort of surreptitious geometrization harboured by the totalizing word genesis. While in fact, as it will appear below, the temporal and modal characters *act* conceptually inside the genset $Gen[(G^{(o)}, V^{(o)})]$, they dictate there their own specific logical and probabilistic laws which are incompatible with co-presence. Indeed the fact that it is meaningless to write down the logical conjunction of two propositions concerning two testimonial descriptions that consist of two distinct elementary events asserted for one same replica of the object-entity $œ^{(o)}$, (5.1.2) is intimately tied with the modal dimension along which a genelement comes into observability; and thc same remark holds concerning the systematic nullity of the product of the probabilities of two distinct elementary events.

As far as I can see, the sort of set called here a genset has never before been conceived of and studied in general terms, neither in mathematics nor in logic (Peano's definition of the infinite set N of integers is also genetic, but in another sense). The concept of genset stems from the necessity, at the limit of an exhaustive representation of the very first phase of a chain of conceptualization, to accomplish separately *two*, and mutually *independent* epistemic operations, first an operation of basic generation of an unknown object-entity, and then a subsequent basic operation of qualification of this basic object-entity. Only physics, only modern microphysics in fact, has been able to reach this limit and thus to bring forth the explicit recognition of the necessity specified above. In mathematics the connections with pure factuality are much too remote to bring into evidence a so highly counterintuitive necessity. Moreover, though instated inside microphysics as an implicit practice, this two-steps cognitive strategy has furthermore had to be *recognized* to bear the germ of an innovating general descriptional method.[54] And this recognition has then had to be worked out into a fully explicit and general concept of transferred basic description, explicitly connectable with classical logic and probabilities *via* the general concept of intrinsic metaconceptualization of a basic transferred description. A quite peculiar and long way to be gone through. So it is not surprising that the sort of set called here a genset has not yet been considered so far.

The mathematical theory of the gensets—like the calculus with genetic classes, and in relation with it—remains to be elaborated under the guidance of the nucleus of MRC. What operations can be defined between the genelements from one given genset (internal calculus)? What a sort of calculus do obey two or more gensets, considered *globally* (external calcu-

[54] The present author is still rather isolated in this recognition.

lus) ? What are the relations between the classical set theory and the genset theory (what are the *specific* conceptual consequences of the genset theory)? From the start, on the basis of the results already brought forth in Secs. 5.1.2 and Sec. 5.2.3, one can assert what follows:

> *The to-be-achieved theory of the gensets is tied with a deep non-classical unification between the epistemological foundations of modern microphysics, set-theory (so mathematics), logic, and probabilities.*

Indeed in so far that mathematics as a whole can be derived from the concept of set, the unification between logic and probabilities achieved by the concept of probability tree, should in its turn be embeddable into a still wider unification, namely between logic and mathematics, as founded on genetic sets.

Inside MRC the classical concept of set can be regarded as a sort of projection of the concept of genset, onto a vault, onto a covering metasurface. A projection that imprints all the as yet mutually non-individualized potentialities from $œ^{(o)}$, simultaneously and directly, onto the final level of the already individualized-and-actualized, thus smuggling away the peculiarities of strict individuality, the initial status of mere prospective and relative potentiality, and the subsequent processes of actualization with their non removable *relativity* to views. Time is thus eliminated, and an absolutizing totalisation is performed, a "geometrisation" on a surreptitiously introduced metalevel of description. The "problem" of actualized infinities might be intimately related with this kind of hidden conceptual leap. As G. Longo put it:[55] "the classical concept of set is newtonian, a hypostatic concept chained to the thin upper stratum where only technicalities of the superficiality are at work". But the concept of genset might lead to a calculus with sets of *processes* that start at the local, purely factual and strictly individual origin of this or that chain of conceptualization, and then involve the whole modal dimension that leads from potentiality to phenomena. This, probably, would achieve, for the definition of a set, the *maximal* liberation of *a priori* constraints. Indeed the primitive sets were introduced by pointing toward the elements, one by one. This confined to a finite number of pre-existing and directly perceivable elements. Then Cantor and Frege introduced sets defined by predicates P. This enlarged the concept of set to the case of also an infinite number of elements, material or conceptual, but restricted by the requirement of a *pre-decided* common property. The physical operational definition of the geneses from a genset frees now of also this last restriction:

[55] During a session of the CeSEF.

it produces a set of "long elements" where the final observable structure of qualifications appears as the result of a succession $[G^{(o)}.V_b^{(o)}]$ of two operations, so the choice of the succession can be closed *after* the realization of the first fragment, the operation $G^{(o)}$; so the *a priori* constraints on the production of this or that configuration of observable qualifications, are left open as long as possible.

At the root of the chains of conceptualization, the MRC-concept of genset knits together physical factuality, and communicable knowledge, by spacetime representations of physical operations and processes. Thereby it stabilizes and amplifies the mental perception of the local, strictly individual zero-points of the chains of conceptualization, and incorporates explicitly their unifying consequences. Here, like in the basic transferred descriptions, the seminal action is the generation, out of the depths of pure factuality, of as yet unknown object-entities, each one of which is conceivable as a factually specified bulk of non-conceptualized *being*. J. B. Grize, in a private comment on this work called this "une motte de *quid*, sémantisable mais encore non- sémantisée". And then, in the spacetime representation offered by a genset, one can clearly follow how, out of this initial bulk, *via* appropriate operations of examination and codings of the observable results of these, are drawn phenomenal manifestations that can be incorporated into language-and-knowledge. It becomes clear that the evolutions of this sort, though mute and ignored, can be conceived to *proceed incessantly*, defeating the impossibility, with mere words, to genuinely grasp being (Aristotle's ens, Spinoza's substance, Kant's *thing-in-itself*, Heidegger's triad *Seiende-Dasein-Sein*, Wittgenstein's *unspeakable*), or even to only insure that the surface of being is touched, that we do not float far above it in the fluid conceptual substance that surrounds the nets of words. Reference, explicitly rooted into physical factuality, *beneath* language, is tied inside a fully intelligible spacetime model with a structure of communicable words and symbols which point toward it thus showing it to the mind.

Third stage: On the significance of a probability measure

Throughout the preceding development it has been supposed that in each branch from a probability tree, the relative frequencies of the outcomes of the events from the algebra *do* converge toward a corresponding probability law. What is the meaning of this hypothesis? And what does a probability measure *represent*, when it exists?

The answer to the first question has become rather obvious in the course of the elaboration of the nucleus of MRC. Given an epistemic referential (G, V), basic or not, if the generator of object-entity G and the view V do mutually exist in the sense of D7, then they can be usefully conserved only if *furthermore* many repetitions of all the successions $[G.V_g]$ with $V_g \in V$ produce some stable global structure of gk-values, a structure that offers a support for being named, communicated, for being used as a basis for intersubjective knowledge and for action. The existence of a probability amounts to just the existence of a such a relative stability, namely a "feeble" non-individual one.

The answer to the second question is less straightforward. I introduce it by an example. Imagine a puzzle consisting of 100 small squares; each square is covered by a small coloured form and bears on it a tiny inscription of the values of two space-coordinates $x = 1, 2, \ldots, 10$ and $y = 1, 2, \ldots, 10$. The available forms can be labelled by $j = 1, 2, \ldots, m$, with m much smaller than 100 so that the same form can occur on several different squares. If the squares are arranged in the spatial order indicated by the xy-values, a certain rough picture is obtained, say of a landscape. But let us ignore the xy-values, mix well the 100 squares, and put them in a bag. We then play the following "probability game". We draw a square from the bag, we note in j-terms (j-value) what image we see on it, we put the square back into the bag, and we mix well the squares. We repeat this procedure a big number of times N, say 1000. What will happen? In general, *all* the $m < 1000$ distinct "values" j of coloured form will come out, and each one of these will appear with a certain relative frequency $n(j)/1000$. If we then increase N more and more, for instance by choosing first $N = 10^6$ and afterward $N = 10^7$, etc., what will happen? In the first place, more and more "certainly" *all* the m "values" j of colored form will come out; and, in the second place, most among the relative frequencies corresponding to the various notations j, will manifest a convergence toward the total number $n_L(j)$ (L: landscape) of j-images from the picture of a landscape on which the puzzle is founded. And if N continues to be progressively increased, this convergence will progressively appear for *all* the m distinct notations j, thus determining a probability law $\{p(j), j = 1, 2, \ldots, m\}$ with $p(j) = n_L(j)$ for any j. It seems clear, I suppose, that this will happen. We are convinced of this. But *why*? Because, we think, the picture of a landscape *is in the bag*, parcelled and mixed up, extracted out of its ordering spatial support, but nevertheless constantly the same before each new trial as for its content of small-coloured-forms-in-a-square. So even though we do not take into account the xy spatial coordinates to effectively reconstruct the form, this global form will nevertheless finally manifest its stable presence inside the bag, when N is increased toward infinity. Namely *via* precisely the convergence of the relative frequencies $\{n(j)/N,\ j = 1, 2, \ldots, m\}$ toward the limiting probability law $\{p(j),\ j = 1, 2, \ldots, m\}$: this "law", for each value

j, connects the relative frequency $n(j)/N$, to the number $n_L(j)$ which is a characteristic of the puzzled landscape. So in this case we believe in the existence of a probability law $\{p(j), j = 1, 2, \ldots, m\}$ *as an expression of the global picture of a landscape, coded in the parcelling language of relative frequencies of values j of coloured form by which we have access to this global picture.*

The above example is extremely simplifying. In general, when we perceive events obeying a probability law we have no *a priori* knowledge of a global form of *gk*-space values associated with the studied random phenomenon. Furthermore the situations similar to the puzzle are far from being the rule. Indeed the coloured forms on squares, like the global picture itself, are just intrinsic *models* extracted implicitly from spontaneously accomplished intrinsic metaconceptualizations. But we do not always perceive directly results of spontaneously accomplished intrinsic metaconceptualizations while the corresponding basic transferred descriptions are achieved by reflex processes genetically wired in our automatic neuro-physiological functioning. Often we are *exclusively* in presence of transferred data, as it happens systematically in microphysics and also quite often in biology, medicine, cosmology, etc. Moreover usually *time* comes in also, like in meteorology, in the study of the accidents on highways, and so on. Nevertheless the example provides us with essential clues which permit to integrate the following general conclusion.

Everything which in the physical world can produce communicable knowledge, can produce it only insofar it can be conceived as a form of spacetime-aspect values endowed with some stability in the sense of D14.1, i.e., as a description which obeys the frame principle P8. Sometimes, that which in the description plays the role of object-entity is such —with *respect* to what plays the role of view—that the description comes out to be probabilistic, not individual in the sense of D14.1.

> *But then, in the obtained probability law—systematically—* `certain organizing spacetime features get lost`. The existence of the probability law, however, by itself is a sign that these spacetime features exist, that also other representations involving the semantic content that is brought in by the considered probability law are possible, inside other epistemic referentials which bring into play meta-aspects that we have not perceived, and which are essentially *tied to spacetime qualifications* (distances, angles, etc.) and therefore lead to a spacetime-(aspect values *gk*)-*metaform*] which "makes a global sense." While the particular *structure* of the probability law characterizes cryptically this [spacetime-(aspect values *gk*)-metaform].

Which means *only* that, if it were known, this metaform would "explain" the observed probability measure, in the following sense. According to the above interpretation, the relative frequencies $n(gk)/N$ that characterize the outcomes of events $e \equiv gk$ from an algebra τ from a probability space, can be regarded as coded "messages" stemming from an unknown meta*form* of $g'k'$-spacetime-values, where $g'k'$ are meta-values of one or several meta-aspects with respect to g, say g', *which are* **indelibly** *tied with spacetime values* (spacetime distances separating gk-values, inside the metaform, spacetime "directions" (space directions or successivities, or both) of gk-qualifications, etc.). By their convergence, the relative frequencies $n(gk)/N$ construct progressively, by parcelled random touches, a gk-coded purely numerical representation of this unknown metaform. A sort of random and approximate but asymptotic "reading" of this unknown metaform, which offers only cryptic reflections of the global structure of spacetime-metavalues-$g'k'$; reflections which are *impoverished and* **pulverized** *by the extraction from this structure of the spacetime specifications (so also of the $g'k'$ qualifications which disappear when spacetime is abstracted away)*. The elementary probabilities $p^{(3)}(gk)$ are the ideal point-like limits toward which the reading of these coded messages tends when N is progressively increased toward infinity (the descriptional level where the gk-values are placed is here conventionally taken as the first level 1; the relative frequencies $n(gk)/N$ are on the relative level 2; so the probability law is on the level 3). And the whole probability measure $\{p^{(3)}(gk)\}$, considered globally, is a precise, numerized—but only ideal—expression of this entire cryptic, impoverished, pulverized, randomized reading in terms of relative frequencies $n(gk)/N$, of the unknown metaform of spacetime metavalues-$g'k'$.

> The *intelligible* referent—in the sense of the frame principle P8—of a probabilistic description, is a corresponding metaform of **spacetime**-metavalues-$g'k'$ involving "globalizing" meta-aspects g' which are different from the aspects g involved in the countings of relative frequencies $n(gk)/N$ from the considered probabilistic description, and are essentially tied with space, or time, or both space and time.

As long as the spacetime integral metaform that got lost in the pulverizing extraction of gk-values which led to the considered probability law $p^{(3)}(gk)$, is not reconstructed, this referent stays unknown. Then, in consequence of the frame principle P8, the probabilistic description involving the probability law $p^{(3)}(gk)$ floats in a no-man's-land between communicable knowledge and absence of knowledge, because it is a description without an intelligible referent. (Until some 20 years ago, only probabilistic meteorological descriptions were available, expressed *exclusively* in terms of relative frequences of gk-events (rain, wind, sun, etc.) and tied with empirical probability laws. Nowadays these events are *explained*, they are *understood* as "messages" which "code" for a definite integral metaform of spacetime-g'-values (g': zones of depression, etc.). We are shown every day on our television screens such a metaform of spacetime-g'-values). This is so even if the description is not a *basic*, a transferred description, like in the class of situations examined before in the sequence of definitions D19. Such is the force of the frame principle P8, its irrepressible demand of intelligibility in terms of spacetime models. And here lies the mystery of the probabilistic descriptions.

It is striking to note to what an extent the MRC significance of a probability law specified above, is consonant with the Popperian concept of "propensities".

> "Take for example an ordinary symmetrical pin board, so constructed that if we let a number of little balls roll down, they will (ideally) form a normal distribution curve. This curve will represent the *probability distribution* for each single experiment, with each single ball, of reaching a possible resting place. Now let us "kick" this board, say, slightly lifting its left side. Then we also kick the propensity, and the probability distribution, ... Or let us, instead, remove *one pin*. This will alter the probability for every single experiment with every single ball, *whether or not the ball actually comes near the place from which we removed the pin. ...* we may ask: "How can the ball 'know' that a pin has been removed if it never comes near the place?" The answer is: the ball does not "know"; but the board as a whole "knows,", and changes the probability distribution, or the *propensity*, for *every* ball; a fact that can be tested by statistical tests."

Furthermore, it is also striking to note that an "information source" in Shannon's sense consists by definition of an "alphabet of signs" (these can be denoted gk) on which a probability measure is posited ($p^{(3)}(gk)$, with this notation), and the theorems of the theory of information concern the source as a whole, not this or that individual message expressed in terms of the signs from the alphabet; which is the major queerness about the theory of information. Thereby:

> The metaform of some *abstract*-space-and-metavalues-$g'k'$, cryptically expressed by $p^{(3)}(gk)$ in a pulverized way, is omnipresent throughout the information theory.

But even then it still is a form, i.e., it admits of a model in an abstract space where a certain topology organizes distances in simultaneity or in succession, and also abstract directions, of which the probability laws $p^{(3)}(gk)$ yield only pulverized reflections.[56]

5.2.4. Conclusion on the MRC probabilities

With reference to Komogorov's theory of probabilities and to quantum mechanics, the method of relativized conceptualization produces a deepened and enlarged theory of probabilities which is intimately tied with the MRC logic, with the theory of sets, and with the information theory. A genetic class that leads to a probabilistic description, and the corresponding probability tree, appear as two faces of one same logico-probabilistic concept, of which a genset offers a minimal spacetime intrinsic model, while the theory of information associates to it algorithms and theorems concerning communications coded in terms of "signs" extracted from the metaform of abstract or physical space-time $g'k'$-values tied with the involved probability laws.

Thereby MRC endows with the outline of a deep-rooted and strong unification between probabilities, logic, set-theory and the theory of information. As soon as the deepening unification between probabilities, logic and information is perceived, it becomes clear that the most specifically adequate, the most efficient mathematization of the method of relativized conceptualization will not be that one achieved here in terms of the theory of categories: it will be a mathematization in terms of multivectors, including a Hilbert-space formulation as a particular case applicable only when a principle of superposition can be asserted concerning the involved object-entities, which in general is not the case.

6. MRC *VERSUS* OBJECTIVITY IN THE SENSE OF THE RELATIVISTIC APPROACHES FROM MODERN PHYSICS

Einstein's theories, which marked the whole thinking of this century, are called theories of relativity. The present exposition of a method of "relativized" conceptualization, cannot be closed without specifying briefly the relations of this method, with the relativistic approaches which, since Einstein's work, keep being so intensively developed in modern physics.

The relativistic approaches are developed quasi exclusively under constraints of formal consistency in the sense of *classical* logic, imposed upon

[56]For instance, in English, the word "mother" is nearer to the word "love" than is the word "war". So, after having received in an on-going message the word (the sign) mother, the conditional probability for the next word to be "love"—as calculated from the relative frequency for the successions of two words in English—is bigger than the conditional probability that the next word be "war".

the *mathematical representations* of the objects of study and—above all—of quantities employed for qualifying these objects (the views, in MRC terms). The major aim is to construct representations of the physical reality insuring a maximized degree of inter-subjective consensus. Notwithstanding that Einstein's analyses of the way in which measurements of spacetime coordinates or distances are achieved, have played such a basic role for the formulation of the theory of special relativity, *factuality possesses no explicit importance in the relativistic approaches.* The object-entities are supposed to pre-exist "out there", exactly like in classical logic, while the views are constructed *formally* according to aims of inter-subjective consensus. Thereby the relativistic approaches escape the peculiar sort of semantic control insured by the syntactical structure of fundamental quantum mechanics, where the accent lies upon the *factual* production of the object-entities and of the qualifications of these. From this point of view there subsists a scission inside modern physics as it now stands.

I shall indicate very briefly the main stages of the development of this alternative way of making use of descriptional relativities, referring them explicitly to MRC in order to facilitate the comparisons.

Limiting conditions and laws. Let us go back to the fact that only descriptions can be known in a communicable way. Now, it is obvious that it would be nonsense to wish to describe "all" that "exists": at any given time the possible object-entities constitute an open and evolving infinity of which the cardinal is bigger than that of the continuum. So the idea of a choice to be made has naturally imposed itself as a non transcendable constraint. It has been tacitly agreed that only "regularities" can be regarded as an object for scientific description, only relations endowed with a certain stability, concerning which it is possible to insure a certain consensus, and which permit predictions. Relations of this type were called natural laws.

But according to what criteria, exactly, can one identify what can be object of a natural law? Up to this day the answer to this question has never ceased reorganizing itself. The main stages of this process can be regarded as fundamental features of the development of scientific thinking. The beginning of the process is relatively recent. It emerged during the epoch that separates Kepler from Newton: Kepler still tried to find, concerning the geometrical dimensions of the planets, laws of the same kind as those that he had formulated concerning the trajectories of the planets. While Newton considered already that the geometrical dimensions of the planets were "inessential" so that one had to isolate them from the researched laws and, if wanted, to introduce them afterward in connection with limiting conditions (spacetime values on the frontier of the spacetime domain covered by a given physical phenomenon) in order to specify and predict this or that particular manifestation of a law. So, by definition, what is called law is categorial, regular, and generates predictions; while limiting conditions are singular, accidental, non predictable, just a set of data that have to be registered or supposed, and have to be used in order to explicate the individual predictions that one wants to draw from a law.

Notice that in this first stage the distinction between law and limiting conditions is introduced as absolute, as intrinsic: this *is* essential, regular, that *is* non-essential, accidental. Just obvious "facts". No criterion is given for distinguishing what is essential and what is not.

"Physical spacetime" and spacetime referentials versus *the frame principle*

All the representations of physics presuppose spacetime. So, if one wants to construct mathematical representations, it is necessary to specify in mathematical terms how spacetime features have to be taken into account. This essential question runs straight into metaphysics, whereby the specific competence of a physicist is exceeded. So it is not surprising that the treatment of this question brings in a mist of ambiguous ways of speaking that hinder an acceptable connection of physics, with epistemology and philosophy.

In classical physics it was currently asserted that void "physical space" (without any mass) admits of an *absolute* mathematical representation consisting of a continuous 3-dimensional variety that is indefinitely differentiable, homogeneous (all the points are equivalent), and isotropic (all the directions are equivalent). It was furthermore admitted that "physical time"[57] can be represented by a continuous 1-dimensional variety that is indefinitely differentiable, homogeneous, and endowed with an arrow (a direction). According to classical physics these two varieties can pertinently be *juxtaposed* in a unique 4-dimensional variety representing the "physical spacetime". But according to the theory of special relativity, the 4-

[57] I have achieved a detailed MRC reconstruction of the concept of time (to be published soon), which leads to the conclusion that, in a certain sense, "physical time" is just a verbal label for a remarkably complex conceptual construct. By this construct, certain basic features from the inner universes of all the (normal) observer-conceptors, corresponding to what is called "inner psychological time", are connected in a definite way with descriptions of physical object-entities, which leads to a family of descriptions called "relative physical changes". In particular, a relative physical change can concern physical object-entities of a category called "clocks". Thereby it becomes possible *to export time-qualifications from inside the inner psychological universes, into the exterior physical world,* and *to import* **measure** *from the exterior physical world, into the inner psychological universes.* This permits to define a conventional inter-subjective concept of time: "the" time, or "the physical time".

dimensional variety representing the "physical spacetime" cannot be separated in a mere juxtaposition of two representations, as mentioned above: these two representations merge to form an organic whole. This whole, however, is characterized in *integral* mathematical terms, namely by a Eucledian metric. While inside the general theory of relativity, "physical spacetime" is represented by a non separable spacetime variety which is characterized in *differential* mathematical terms, namely by a Riemannian metric.

In order to give a *communicable* (conventional) mathematical form to the descriptions of physical entities (rigid bodies, fields, physical phenomena[58] in general), a spacetime referential (a system of 4 reference-axes, endowed with a centre and with units of space and of time) is immersed in the 4-dimensional variety that represents "physical" spacetime. This permits to associate communicable numerical labels—spacetime-coordinates—with each point of the spacetime variety. These labels can be explicitly combined with the qualifications of the studied physical phenomenon *via* gk-aspect-values where $g \neq ET$ (according to the frame principle P8, spacetime qualifications alone cannot describe a physical phenomenon, but they do irrepressibly emerge in any description of a physical phenomenon, even if *a posteriori*, if convenient, some or all the spacetime qualifications can be eliminated by projection). A spacetime referential is distinct from the spacetime variety itself in which it is immersed.

We are now ready to specify the ambiguous ways of speaking mentioned above. They concern the expression "physical spacetime", and the assertion that what is indicated by this expression, "possesses" a metric. I hold that from a philosophical point of view such formulations have to be dismissed. Indeed, as posited by Kant, as accepted in modern philosophy, and as re-expressed in the frame principle P8, spacetime itself is not a physical entity; it is an "*a priori* form of the intuition" which (a) *pre-conditions* any description of a physical entity; (b) *contributes* to any description of a physical entity, namely in the role of a spacetime frame-view V_{ED} associated with at least one other aspect-view $V_g \neq V_{ED}$; (c) *alone*, in the absence of, rigorously, any other sort of effectively perceived or at least imagined aspect $g \neq ET$, cannot generate impressions, nor only conceived impressions.

> `Spacetime is not a physical phenomenon.` "Physical spacetime" —as such—does not "exist" physically, it is just an *intrinsic model* (in the sense of D19.2) associated to an *a priori* form of the human intuition. We can call this *a frame-model*.

So—rigorously—one cannot speak of physical spacetime, nor, *a fortiori*, of the metric of physical spacetime. One can only speak of a metric chosen for

[58] Here the word phenomenon is used in the current sense, not the philosophical one.

the representation of spacetime by a 4-dimensional variety (a mathematical frame-model) endowed with a spacetime referential (a view V_{ET} which, in a relative description of some physical entity $œ_G \neq$ [spacetime], is associated with one or more aspect-views $V_g \neq V_{ET}$ in the construction of a representation-space). The expression "the structure of physical spacetime" points in fact toward *structures of results of measurements on object-entities* $œ_G$, measurements of lengths of some aspect $g \neq ET$ with respect to which this or that $œ_G$ exists in the sense of D7 (or distances, or surfaces, or volumes of some aspect $g \neq ET$) and durations of such aspects $g \neq ET$.

Finally notice also that the adequacy of the conditions of continuity and of indefinite differentiability of the 4-dimensional variety by which spacetime itself is represented, certainly is not universal (Laurent Nottale [32] has well brought this into evidence). Indeed according to MRC all the relativities involved in descriptions of physical entities have to be systematically taken into account. So in particular one has to take into account also the relativities to *a view of order of magnitude of the presupposed spacetime units.* Such a view is always involved in a description of physical phenomena, and it is always discrete and even finite, whereby it entails exclusions by mutual inexistence in the sense of D7 (anything that introduces dimensions of a smaller order of magnitude than the units, is not perceived by the view which acts in the description).

Principles of Symmetry and Translational Invariants. Conservation Laws

It is admitted that "spacetime is homogeneous" i.e., that all the spacetime points are "equivalent", and this is called *the principle of homogeneity of spacetime.* This principle amounts to the requirement that, in the descriptions of physical phenomena, what is "essential" be independent of translations of the spacetime referential (i.e., changes of exclusively the position of the centre of the referential); in other terms, the requirement that what is "essential" shall stay invariant when a translation of the spacetime referential is performed. According to this requirement, the spacetime coordinates (positions) are not essential, while the differences of the coordinates (distances) are essential. Consequently any velocity is essential because, as a ratio of two differences of coordinates, a difference of space-coordinates and a difference of time-coordinates, it is globally invariant with respect to translations of the spacetime referential.

So there appears now a formal criterion that permits to distinguish between what is essential and what is not. This criterion brings into evidence a pair of connected concepts. On the one hand, a concept of homogeneity—a "symmetry"—assigned abusively, in current speaking, to "physical spacetime" itself, but which in fact designates only an invariance of certain features from descriptions of physical object-entities (cf. the preceding discussion of metrics "of spacetime"); and on the other hand, a correlative class (a group in the mathematical sense) of changes of the state of observation, expressed by changes of the spacetime referential, namely by the group of "geometrical" or "static" translations of the referential (called so because exclusively the positions of the centre of the referential are changed, in the absence of any rotation and any movement of the referential). So the invariants tied with the principle of homogeneity of spacetime, are essential in *this* sense that, when changes of only the position of the centre of the spacetime referential are operated, they manifest a descriptional independence with respect to these changes, an indifference, a recurrence of a descriptional form, an in-variance, a *conservation law.* While the coordinates of the physical events, because they do change when the centre of the spacetime referential is translated, are regarded as inessential; this qualification of non-essentially being asserted notwithstanding that it is absolutely *necessary* to know the coordinates of the involved events in every particular case in which one wants to be able to make predictions concerning this case.

Analogous considerations are valid concerning the posited equivalence of all the spatial directions, i.e., *the principle of isotropy of space.* In this case other invariants or conservation laws are involved, tied with the group of spatial rotations of the spacetime referential, in the absence of motion.

Note now that velocity, which is by construction fully invariant with respect to translations—the direction as well as the norm – is not invariant in direction with respect to rotations also. As for the coordinates of the involved events, again they are inessential in this *new* sense that in general they change by a rotation of the referential. So the concept of "essentiality" is now explicitly regarded as relative to the considered group of transformations of the spacetime referential, i.e., as relative to the corresponding set of observers.

But why are these distinctions and ways of speaking been introduced? Are they imposed by factuality? It is quite clear that they are not, that another sort of reason founds them.

For the observer tied with *any* given referential, the time-coordinate of an act of observation of an event keeps changing *irrepressibly.* As for the space-coordinates, by the very definition of a space-referential they necessarily change by passage from one space referential to another one, so from one observer to another one. These are indeed psycho-conceptual-physical *facts*, not mere free conceptual constructions. So, if one wants to elaborate descriptions endowed with *stability* and able to insure a certain consensus among distinct observers, then one *has* indeed to find ways of organizing a conceptualization that shall bring forth invariants with respect to the universal and unavoidable changes mentioned above, of the time coordinate and, in another way, of the space-coordinates. While these themselves *have* to be regarded as non-essential, accidental features: an opposite attitude would be hopeless. So what is obviously impossible from the start, is renounced. Now the aim might have come out to be impossible nevertheless. It could have appeared that no sort of descriptional stability whatsoever can be found, no matter what strategy is adopted. Then there would have been neither "natural" laws, nor science. In fact however the aim has been found to be possible, but only *relatively* to this or that group of transformations of the state of observation (of the spacetime referential), which then selects a corresponding set of invariant descriptional features. This restricted possibility is already very remarkable. But it should be quite clearly understood that such an invariance is never a "physical fact". It is just an abstract artifact involving a whole adequate conceptual network: invention or choices of "convenient" spacetime varieties and referentials (spacetime-frame-views V_{ET}), Cartesian, curb, etc.); deliberate construction of "convenient quantities" (aspect-views V_g), (velocities, accelerations, angles, total-energies); delimitation of convenient systems (object-entities $œ_G$), "rigid bodies", "material points", "fields", etc. All these descriptional elements being conceived in such a way that when the network formed with them is superposed to physical factuality[59] it leads to descriptions of which certain features stay invariant under this or that corresponding group of transformations of this or that aspect of the states of observation, thereby insuring a certain corresponding potential of inter-subjective consensus. In order to realize to what a degree this is so, it suffices to consider that the equivalence of all the spacetime-points from the 4-dimensional variety representing spacetime, where one immerses the spacetime referentials, is by no means a physical fact. It is just a posited idealization, an abstraction, a useful strategic abstraction. The water does not boil at the same temperature here or on the Himalaya, and the astronomers know well that the laws evolve throughout the history of the universe. As for the directions from our life-space, they "are" not at all always equivalent either, since a stone falls downward, not *vice-versa* nor from left to right The physicist just posits abstractions by which he obtains the concept of spacetime that permits best to construct relative consensuses and corresponding predictions. And a velocity, an energy, even a distance, even a position, are not "facts", they are *constructs* concerning the representation of certain phenomenal perceptions. Think of the position. Inside a

[59] I say factuality, not phenomena, in order to include basic generators of basic object-entities, as well as basic views, which act on as yet non-phenomenal (non-perceived) zones of the physical reality.

4-dimensional mathematical variety that *represents* spacetime, there simply are no "positions", there are only "points": *Position is a concept that is definable only if also a referential has been immersed in the variety.* And one cannot even assert that inside what is pointed toward by the verbal label "physical spacetime" there "is" what we call "place"; there is only what *we* have in our mind when we utter this word and when we point toward this or that source assigned to a perception *via* some intrinsic model, thus using approximately our own body as a space-referential. Science is just a cognitive strategy in which factuality and phenomenal perceptions are dealt with under constraints of stable representatability, of intelligibility, of consensus and of predictability.

Let us now go further in the examination of the aims of intersubjective consensuses with respect to which certain descriptional choices are convenient, and others are not.

Principles of Relativity and Dynamical Laws

We have considered above groups of geometrical, static transformations of the state of observation. The different referentials from such groups are considered to be at rest with respect to one another. One can imagine the whole group as immersed in *one* big reference-receptacle containing replicas of itself, with shifted centers, or with axes displaced by rotations: an observer could circulate freely from one of these replicas to any other one. Such a view entails no conceptual difficulties.

But one can also imagine referentials that are moving with respect to one another. It is tacitly admitted that in this case each observer is tied to its own referential, even if he can communicate with the others by signals. This is a rule of the conceptualization game which physicists play with one another. What does this rule involve? Does it still permit to insure a certain inter-subjective consensus? The answer is given by the positing of *principles of relativity*, the principle of restricted (or special) relativity, and the principle of general relativity.

* *The principle of special relativity* posits that all the observers tied to *inertial referentials* (moving with respect to one another with constant velocities), perceive identically all the *dynamical laws* of physical phenomena, i.e., all the relations between measurable quantities involving accelerations (changes of velocity), this being indelibly connected with the assertion that, when one passes from the description of a phenomenon achieved in a given referential, to the description of this same phenomenon but achieved in another referential, all the involved spacetime coordinates have to be changed

accordingly to a definite "law" for the transformation of the coordinates:[60]

> According to the principle of restricted relativity, inside the set of all the observers from a set of mutually inertial referentials, there exists an inter-subjective consensus tied with a definite group of transformations of the spacetime coordinates, the corresponding new invariants being this time the dynamical laws.

It is noteworthy that the geometrical invariants are not invariants with respect to also the new group of transformations specified above. Though the dynamical laws are expressed by making use of also the quantities precedingly constructed such as to insure geometrical invariance, these quantities in general change their numerical characterizations when the inertial referential changes and the asserted law of transformation of the spacetime coordinates is applied (such is the case for distance, velocity, mass, energy, etc.). So again, what is regarded as essential changes with the considered group of transformations of the referential. Once more the relativity of essentiality, to the type of the researched consensus, manifests itself. We are now far indeed from the initial notion of an intrinsic essentiality or accidentality of the qualifications.

** *The principle of general relativity* goes still much farther on the direction of the increasing degrees of constructional freedom practised by the modern physicist. According to this principle the dynamical laws "are" invariant[61] with respect to any change of the spacetime referential, expressed by any transformation of the spacetime coordinates.

The basic motive that caused Einstein to posit this principle—very striking indeed—is the fact that there is no way for deciding whether yes or not a given referential is "really-inertial-by-itself". One can only find out whether yes or not a given referential is inertial with respect to another given referential. The qualification of inertiality cannot be assigned a "final" significance, it involves a sort of indefinite regression, of undecidability.

> In such conditions Einstein considered that—for philosophical reasons—it was imperative to transcend the limitation to inertial referentials involved in the principal of restricted relativity.

[60] The admitted law of transformation of the spacetime coordinates has first been that proposed by Galileus. In 1905 Einstein has proposed a modified law (the Lorentz-Einstein transformations) that reduces to that of Galilei for velocities that are small with respect to the velocity of light.

[61] Such a way of speaking, though current, is deeply inconsistent with the very essence of the relativistic approaches from modern physics, which are *constructive*: the dynamical laws, like any sort of laws, are *built* under deliberately chosen constraints of invariance which then entail the ways in which the qualifying quantities (the views which are made use of) are defined.

And he realized this transcendence, but only for the case of gravitational macroscopic interactions. The method elaborated by Einstein in order to achieve this descriptional aim is very impressive by the demiurgic degree of liberty taken with respect to the concept of "physical facts". However, paradoxically, it involves the way of speaking in terms of metric "of physical spacetime" that was criticized before.

In this context it would be as inappropriate to try to expose Einstein's method in only several lines, as to try to expose it thoroughly. So I shall just remark that in this new step, again, the change of the set of observers among which consensus is researched, entails a change of also the object of consensus. Einstein's description is constructed in such a way that the object of consensus, the invariant, becomes the *geometrical form*, in a convenientnt representation-space, of the trajectory of the studied body: this geometrical form is always a geodesic of the—Riemannian, *differentially* characterized—metric assigned to the variety that represents "physical" spacetime (filled with fields and masses). Whereas the invariants relative to the inertial group of transformations of the spacetime coordinates cease to be invariants in connection with Einstein's general principle of relativity: the principle is "general" in the sense that it concerns all the conceivable *observers*, but the corresponding invariant is the form of the dynamical law (the energy also is represented so as to be endowed with a conservation law). And Einstein's general form-invariant is so abstract that its factual semantical content nearly vanishes out of the realm of what can be genuinely imagined. One has the feeling that a sort of law of compensation operates inside the processes of conceptualization of the physical reality, according to which when the extension of the class of consensual observers is increased, the factual semantical content of the object of consensus is correspondingly diminished.

A fundamental question raised by the principle of special relativity, is the status of what is usually called the Einstein-Lorentz transformation "laws" for the coordinates, but also sometimes the transformation *rules*. Indeed, considered in the perspective of the principle of general relativity, the status of the principle of special relativity becomes uncertain. According to the principle of general relativity any transformation of the coordinates has to be posited to insure the form-invariance of the dynamical laws, in which case there is no "law" of transformation any more. Then what relevance, exactly, the "factual truth" of the special "law" of transformation, would possess? By the passage from the principle of special relativity, to the principle of general relativity, the conceptual status of what is called principle of relativity seems to have surreptitiously undergone a mutation from an assertion believed to express an *empirical truth*, to an expression treated as a methodological condition for constructing a representation of the physical reality which will be "acceptable" from a philosophical point of view.

Of course, one would like to be assured on explicit grounds of something `else`, namely that there exists a possibility to know with certainty whether yes or not two observers which are not at rest with respect to one another and are each one imprisoned inside his own referential, are indeed considering the `same` event, phenomenon, the same physical situation, whatever this be. It is not obvious that form-invariance of the dynamical laws, alone, entails with necessity a definite answer.

The foundations of both principles of relativity are still hidden in an as yet insufficiently analyzed coalescence of scientific descriptional strategies and of feebly elaborated philosophical decisions, but which entail major descriptional consequences. Indeed, on the level of conceptualization where the modern relativistic approaches are placed, the question of descriptional *method*—so also of descriptional `aims`—which, more or less implicitly, has triggered the beginnings of physics as an independent science, again draws attention upon it, this time with a new, imperative power. How *should we want* to represent the physical reality? What structure of pragmatic-philosophical criteria should be adopted, and *why*, on the basis of what *reasons*, whether pragmatic or metaphysical? The relativistic approaches from modern physics have entered a zone of such degree of abstraction of our conceptualization of physical reality, of such vertical distance from physical factuality, that *it becomes now vital for physics to construct explicitly and systematically its own philosophy and its own epistemology*, if it wants to stay deeply true to its own modern aim of maximizing consensus: indeed also a philosophical consensus should be constructed explicitly, not only this or that new particular sort of observational consensus, fabricated accordingly to an inertially followed fashion that has developed didactic roots.

Summarizing Considerations

In the relativistic approaches, the search for objectivity has explicitly transmuted into methods for deliberate construction of classes of inter-subjective consensus, each one relative to a definite group of transformations of the state of observation. When the group of transformations changes, the objects of consensus in general change also. For each group, what is qualified as essential is that what is invariant inside that group: *essentiality relative to consensus*. The aim to construct consensus, and inside a class of observers as rich as possible, is given absolute priority, on grounds which first were pragmatical but later evolved into philosophical requirements.

The search of observational invariants concentrates the attention upon the mathematical representation of qualifiers, of views. The whole approach is mainly marked by requirements of logico-mathematical coherence concerning the construction of pairs [(group of transformations), (corresponding views yielding invariant descriptions)]. When the construction is achieved, its experimentally testable consequences—sometimes very rare—in general pledge the theory only globally and, whether for confirmation or falsification, in a way that is more cumulative and diffuse than sudden and definite. The way of producing the involved object-entities, factually, independently, is left wholly in the dark, so an explicit tie with basic transferred descriptions in the sense of D14.3.1 is very rare if not inexistent. Like in classical logic, the object-entities are simply supposed to pre-exist. This is so even when the formal representation of the object-entities is thoroughly reconstructed for reasons of logico-mathematical coherence with previously constructed representations, like in the case of the methods of gauge-invariance, or similarly, like in general relativity: *The mathematical representations of the considered object-entities are constructed via the views*, whereby their factual generation and content are surreptitiously abandoned to arbitrary and uncontrollable restrictions. Though the results of measurements of the spatial lengths or of the durations involved in phenomena are relativized to the state of observation (which is a revolution with respect to the current way of thinking about space and time), the relativistic approaches from modern physics operate wholly inside the realm of classical logic which starts from the spontaneous intrinsic models offered by the current languages. This is a consequence of the fact that the relativistic approaches appeared first inside macroscopic physics, where an enormously thick layer of preceding theoretical conceptualizations of directly observable physical phenomena underlie them, while the generation of the object-entities that one wanted to study, seemed not to raise problems. So the canonical structure of a basic transferred description simply did not appear. *A fortiori* the peculiar characters stemming from the generation of object-entities independently of any subsequent qualification, remained wholly hidden. The contrast with the case of quantum mechanics, where the main innovation is the role played by the generation of object-entities independently of any subsequent qualification, is striking. A genuine polarity. So, since the beginning of this century the construction of objectivity in physics has advanced on two front-lines, in two opposite directions. The front-line created in quantum mechanics roots the construction of objectivity into physical factuality, down to an unprecedented depth. Thereby it permitted us to explicate in full detail how the conceptualizations incorporate and carry Being, which is the hard core of any observational objectivity. On the other hand the front-line created by the relativistic approaches erects with a vertiginous degree of descriptional freedom, rigorous abstract representations constrained by requirements of a type of consensus founded upon the expressional form of views, a sort of linguistic constraints on just views. The connection and unification between these two distinct progressions is not yet worked out inside modern physics. There subsists a scission. MRC—since it incorporates the epistemic specificities of quantum mechanics, in a *general* epistemological method—should permit to develop a unification.

7. FINAL REMARKS

The method of conceptualization exposed in this work is founded upon a descriptional mould that has been drawn from fundamental quantum mechanics and has been generalized, because, by the systematic relativizations to the involved epistemic actions which it has been found to incorporate, this mould seemed to us able to hinder any descriptional ambiguity, false problem or paradox.

The approach practiced inside the method exposed in this work breaks with a tradition. Indeed up to now the processes by which knowledge is created have always been studied from a psychological or a neurobiological point of view. And the studies have always been worked out in a spirit of "neutral submission to the natural facts", so also of "neutral objectivity", the human conceptor being (implicitly) involved only in the non-subjective, general sense of the term. Methodological applications have always been left for a later stage, and so far the aims of such applications have been mainly pedagogical, commercial, etc., never to optimize the process of conceptualization itself. The modern cognitive sciences continue this tradition. While MRC, on the contrary, is quite essentially a methodological, normative representation of the processes of conceptualization, though the conceptor's fully subjective epistemic curiosities and decisions are explicitly included and play a key-role. So it might seem that there is a radical divergence between the method of relativized conceptualization and the cognitivistic approaches. Therefore I want to remark what follows.

The bio-psychological studies of the ways in which knowledge emerges, are *themselves* processes of conceptualization: no science can escape the emprisonment inside conceptualization. And, as pieces of conceptualization, the bio-psychological studies fall inside the domain of entities which the method of relativized conceptualization claims to rule. This establishes a zone of necessary superposition. A very strange zone of superposition, indeed, reminiscent of Escher's Print Gallery where the picture incorporates the street. The methodological suggestions which could be extracted from this zone might appear to be particularly fertile because in both the cognitivistic approaches and MRC, what is tried to be represented is the same phenomenon, namely generation of knowledge. But while MRC does not deal with the bio-psychological processes which take place inside the body (of which only the final effects, the "phenomena," are taken into account), concentrating exclusively upon an exterior deliberate cognitive strategy, the cognitive sciences, on the contrary, put the accent quasi exclusively upon the neurobiological processes from inside the body. The *encounter* of the results obtained in these two different ways, if well characterized inside MRC, might be illuminating. Wittgenstein[32] remarks:

> "There exists a tendency to speak of "the effect of a piece of art", the feelings, the images, etc. It is then natural to ask: "Why listen to this minuet?", and one tends to answer "in order to obtain that or that effect". And the minuet by itself does not count? (Smythies adds in his notes: Does this mean that everything would be all right if, leaving apart the piece of painting, you would produce upon someone exclusively its effects? It is certain that what comes first is that you see a picture or that you recite the words of a poem. A syringe that would produce the effects, would do that exactly as the painting does?"

(my own retro-translation into English from the French edition, Gallimard 1992). It might bring forth reflexions, upon the exterior epistemic strategy from MRC, of known neurobiological processes: what inside MRC is introduced as deliberate methodological choices insuring certain "desired" pragmatic optimalities, might appear to be related with certain inescapable neurophysiological characters of the epistemic actions. Such is certainly the case concerning the frame principle and the intrinsic metaconceptualizations, but possibly also concerning many other methodological choices, like the principle of separation, the cellular and hierarchical organization of the descriptions, and even the fundamental relativities of any description to a delimited object-entity and a filter for qualification. This, *vice versa*, could then suggest pertinent research on neurobilogical features of the processes of conceptualization. *Which eventually might yield certain rules of translation.* Thereby much intelligibility would be gained.

ACKNOWLEDGMENTS

The views exposed in this work have, in the very first place, benefited by long and repeated exchanges with my former collaborators, Daniel Evrard and François Thieffine, who have patiently encouraged my attempt, spending on it much attention and time.

During the last five years, my view has integrated organising influences from the part of all my colleagues and friends from the Center for the Synthesis of a Formalized Epistemology (CeSEF). I warmly thank them all.

Vincent Schächter credited MRC enough for trying to apply it to qualifications of complexity, which led to very stimulating exchanges.

Awareness of the fact that MRC can be regarded as a "normation" of the processes of description was specifically triggered by two lectures on Kantian epistemology delivered by Jean Petitot in the framework of the CeSEF.

The initial version of MRC contained non-mathematical, ideographic symbolizations. Since a long time already Michel Paty tried to convince me to drop the symbols, which, he felt, obscured the meanings, and to attempt instead a formulation in the terms of usual language. Jean Petitot supported this same opinion and furthermore he conveyed to me strong arguments for attempting, separately, a genuinely mathematical formalization. I tried here to follow both these injunctions.

Michel Bitbol, with whom I usually am in remarkable consensus, formulated a strong reservation concerning the necessity of the realist postulate. Jean Louis Le Moigne and Hervé Barreau formulated—in essence—the same reservation. These reactions, while they led to precious explicitations in my mind, finally stabilized my own choices concerning the presentation of the concept of realism.

Jean Louis Le Moigne and Evelyne Andeewsky insisted on the role, in the generation of basic object-entities, of pre-existing conceptualization. This led to quite important clarifications.

Giuseppe Longo, throughout our CeSEF-sessions, has drawn attention as often as possible upon the decisive role, in any formalization, of intuitive perceptions and of "pre-formal" knowledge and reasoning. This has notably strengthened my own positions in this respect, as expressed in MRC.

Francis Bailly, Hervé Barreau, Michel Bitbol, and Giuseppe Longo accepted to read successive forms of MRC and to express very useful questions, informations and remarks. Their various competences have provided me with useful clues. My exchanges with Hervé Barreau have been particularly rich. I am also indebted to Sylvain Fautrat for a thorough reading of an exposition of MRC and for suggestive questions.

Not least, J.B. Grize read two previous expositions of MRC and communicated me precious remarks and agreements.

Georges Yves Kervern manifested a constant interest and agreement with the successive forms of MRC, as well as a deep intuitive understanding of it, and he tries to apply it to the "science of danger"—which gave me moral support for which I am grateful.

I am thankful to Doina Gayet for a precious support concerning bibliographic questions, and also for her constant faith in the importance of this research.

The sketch of a formalization of MRC in terms of the theory of categories has benefited from remarks by Matthieux Amiguet.

REFERENCES

1. E. P. Wigner, *Symmetries and Reflections* (Indiana University Press, 1967), p. 37.
2. K. R. Popper and J. C. Eccles, *The Self and its Brain* (Springer, 1977).
3. *Non-Existence and Predication*, Rudolf Haller, ed. (1985).
4. J. Petitot, "Objectivité faible et Philosophie transcendantale," in *Physique et réalité, un débat avec Bernard d'Espagnat"* (Frontières, 1997); "Debate with Jean Petitot on mathematical physics and a formalized epistemology," in this volume.
5. M. Bitbol, "Some steps towards a transcendental deduction of quantum mechanics," *Philosophia Naturalis* **35**, 253-280 (1998).
6. J.-P. Changeux, *L'Homme neuronal* (Fayard, 1983).
7. G. M. Edelman, *Bright Air, Brilliant Fire: On the Matter of Mind* (Basic Books, 1992).
8. A. Damasio, *The Feeling of what Happens. Body and Emotion in the Making of Consciousness* (Harcourt Brace, 1999).
9. N. Goodman, makes the same point in *Ways of Worldmaking* (Hackett, 1978).
10. W. V. O. Quine, *Ontological Relativity and Other Essays* (Columbia University Press, 1969); *Le mot et la chose* (Flammarion, 1977).
11. H. Putnam, *Reason, Truth and History* (Cambridge University Press, 1981; and reprinted regularly, the last time in 1997).
12. M. Mugur-Schächter, "Spacetime quantum probabilities I," *Found. Phys.* **21**, 1387-1449 (1991).
13. M. Mugur-Schächter, "Toward a factually induced spacetime quantum logic," *Found. Phys.* **22** (1992).
14. M. Mugur-Schächter, "From quantum mechanics to universal structure of conceptualization and feedback on quantum mechanics," *Found. Phys.* **23**, 37-122 (1993).
15. M. Mugur-Schächter, "Les leçons de la mécanique quantique (vers une épistémologie formelle)," *Le Débat* **94**, 169-192 (1997).
16. M. Mugur-Schächter, "Mécanique quantique, réel et sens," in *Physique et réalité, un débat avec Bernard d'Espagnat* (Frontières, 1997).
17. M. Mugur-Schächter, A: "Esquisse d'une représentation générale et formalisée des descriptions et le statut descriptionnel de la mécanique

quantique," *Institut de la Méthode, Lausanne*; B: "Une méthode de conceptualization relativisée: vers une épistémologie formelle apte à faire face aux complexités," *Rev. Internat. Systémique* **9**, 269-303 (1995).
18. M. Mugur-Schächter, "Spacetime quantum probabilities II: Relativized descriptions and Popperian propensities," *Found. Phys.* **22**, 235-312 (1992).
19. U. Eco, *Kant et l'Ornithorynque* (Grasset, 1999), p. 29; my translation is from the French edition.
20. R. Firth, "Reply to Sellars," *Monist* **64**, 91-101 (1981); the quotation is from p. 100.
21. J.-B. Grize, "Pensée logico-mathématique et sémiologie du langage," in *Pensée logico-mathématique ... Nouveaux objets interdisciplinaires*, Olivier Houdé et Denis Melville (P.U.F., 1993). The "natural logic" developed by J. B. Grize is the sort of logic that seems the nearest to that which acts throughout the elaboration of the nucleus of MRC.
22. Matthieu Amiguet, in a private communication, has made interesting suggestions in this respect.
23. J-B. Grize, "Logique des classes et des propositions, dans Logique et Connaissance Scientifique," in *Encyclopédie de la Pléiade* (Gallimard, 1967).
24. N. Hadjisavvas, F. Thieffine, and M. Mugur-Schächter, "Study of Piron's system of questions and propositions," *Found. Phys.* **10** (9/10) (1980).
25. L. Wittgenstein, *Remarks on Logical Form* (Aristotelian Society, 1929).
26. B. Russell, *Principles of Mathematics* (Allen & Unwin, 1948).
27. K. Gödel, "Über formal unentsheidbare Sätze der Principia Matematica und verwandter Systeme," *Monatsh. Math. Phys.* **38**, 173-198 (1931).
28. A. Whitehead, and B. Russell, *Principia Matematica*, 2nd edn. (Cambridge University Press, 1925).
29. A. N. Kolmogorov, *Foundations of the Theory of Probabilities* (Chelsea, 1950), translation of the original German monograph, *Grundbegriffe der Wahrscheinlichkeitrechnung* (1933).
30. T. L. Fine, *Theories of Probability* (Academic, 1973).
31. K. Popper, "Quantum mechanics without the observer," in *Quantum Theory and Reality*, Mario Bunge, ed. (Springer, 1967); *A World of Propensities* (Thoemmes, 1980).
32. L. Nottale, *La relativité dans tous ses états* (Hachette, 1998).
33. L. Wittgenstein, *Lectures and Conversations* (Blackwell, 1966).

8

MATHEMATICAL AND FORMALIZED EPISTEMOLOGIES*

Robert Vallée

Université Paris-Nord
Paris, France

A *mathematical epistemology* at the macroscopic level is proposed, based on the process of perception represented by an *observation operator*. The linear case introduces a Volterra composition with the two extreme cases of multiplication and convolution. Presented in terms of observation operators, are introduced the concepts of *epistemological indiscernibility* and of *epistemological inverse transfer*. The case of *perception of duration* is considered, as well as *time-space selection* and *time-space filtering*, which give rise to rather general modelings of familiar observation devices. If the observing system also has the ability to decide, a *pragmatic operator*, the product of observation and decision operators, may be introduced. It generates *pragmatic indiscernibility* and *pragmatic inverse transfers*. The resulting actions modify the evolution of the *supersystem* composed of the system and its environment, thereby creating a feedback loop allowing the construction of a *mathematical epistemo-praxiology*, which may be seen as a step toward other formal epistemologies not restricted to the macroscopic domain.

Key words: algebra, epistemo-praxiology, formal epistemology, inverse transfer, mathematical epistemology, observation operator, perception of duration, pragmatic operator, pragmatic inverse transfer, supersystem.

1. A MATHEMATICAL EPISTEMOLOGY

1.1. Observation Operators

As we have said in our introductory article ("On the possibility of a formal epistemology"), a first step in the representation of the acquisition of knowledge at the macroscopic level, can consist of a representation of the process of perception, be it biological or metaphorically physical.

*See "Important Note" on p. xviii.

So we consider a macroscopic dynamical system S able to perceive the evolution of its environment E and of itself. We describe the process of perception by a mapping $\mathcal{O}$ of the set X of all the possible pairs $x = (s, e)$ of functions representing the factual evolution s of the state of system S and the factual evolution e of the state of environment E, on the set Y of the corresponding pairs y of perceived evolutions. We write $y = \mathcal{O}(x)$ where $\mathcal{O}$ is called an "*observation operator*" [11]. This can be seen as a first step toward what we proposed to consider as a "*mathematical epistemology*" [16, 18]. Indeed x, since it contains as a pair both the evolution of system S and that of its environment E, is a representation of the evolution of the whole universe U; while y is a representation of the evolution of the perception of S concerning U, a perception which is a *description* of x. Since the operator $\mathcal{O}$ necessarily acts, at any given instant, only on the past and the present of x, it is classically said to be *hereditary or causal.*

It is very natural to introduce algebraic considerations concerning observation operators. If observation operator $\mathcal{O}_1$ is a mapping of the set X on the set Y and observation operator $\mathcal{O}_2$ a mapping of Y on Z, by concatenation, or composition in *series*, of the corresponding compatible observation processes, we have

$$y = \mathcal{O}_1(x), z = \mathcal{O}_2(y), z = \mathcal{O}_2\mathcal{O}_1(x),$$

so $\mathcal{O}_2\mathcal{O}_1$, which is obviously an observation operator defined on X with values in Z, is the *product* to the left of $\mathcal{O}_1$ by $\mathcal{O}_2$. Of course the most simple case is obtained when all the observation operators considered are mappings of X on X. Then generally $\mathcal{O}_2\mathcal{O}_1 \neq \mathcal{O}_1\mathcal{O}_2$. If moreover all the considered operators possess inverses and if one of them is the identity operator, we have a *group* noncommutative in general. When observation operator $\mathcal{O}$ is a mapping of X on X, we may look for evolution functions x for which $\mathcal{O}(x) = x$. Such an evolution function is a fixed point of $\mathcal{O}$ and we may say that the observation device represented by $\mathcal{O}$ is *transparent* to x.

1.2. The Linear Case

In the *linear* case at instant t, the state $x(t)$ of the considered universe U belongs to a linear space, for example $\mathbb{R}^m$ or $\mathbb{C}^m$, or even to an infinite dimensional linear, real or complex, space, and $y(t)$ which is the state of U as perceived by S, belongs also to a linear space such as $\mathbb{R}^n$ or $\mathbb{C}^n$, or to an infinite dimensional space[1]. So the sets X of functions x and Y of functions y are linear spaces and it may happen that an observation operator be linear,

[1]We shall see later that $x(t)$ may depend also of a point M of physical space. The state, at instant t, is then $x(t, M)$ so a field like the electromagnetic one. The action of an observation operator involves then an action in physical space $\mathbb{R}^3$ as well as in time $\mathbb{R}$.

provided it is hereditary. In such a case, if we use Schwartz's distributions, this operator is *an hereditary Volterra composition* [11, 13]. So we have

$$y(t) = \int_{\mathbb{R}} L(t,s)x(s)ds,$$

where $x(s)$ and $y(t)$ are column matrices and $L(t,s)$ a rectangular matrix, reduced to a null matrix for $s > t$ and so rendering hereditary the operation of composition (no influence of the values of $x(s)$, after instant t, on $y(t)$).

If $L(t,s) = A(t)\delta(t-s)$, where δ is the Dirac distribution (the unit impulse) and $A(t)$ a generally rectangular matrix, we have, according to an obvious property of δ, a *multiplicative* operation

$$y(t) = A(t)x(t).$$

In this first extreme case we shall say that the observation operator is of *type* $\mathcal{A}$:it acts only on the *present* of x. But there are other instantaneous linear operators which are not of type $\mathcal{A}$, as it will appear later. When $A(t)$ is a positive or null scalar $a(t)$, it may be considered as a *factor of attention* paid, at instant t, to $x(t)$ by observing system S [19]. So if $a(t)$ is equal to one on interval $[t_1, t_2]$ and to zero outside it, the observation device plays the role of a *temporal window* permitting perfect observation from t_1 to t_2 and excluding any observation before t_1 and after t_2. If $a(t)$ is equal to one at regularly distributed instants and to zero everywhere else, the observation device acts as a *stroboscopic apparatus* realizing a temporal sampling.

But as announced above, the observation operators of type $\mathcal{A}$ do not represent all the linear operators acting on the mere present of x as announced above. We may have [21]

$$L(t,s) = \sum_{k=0,n} A_k(t)\delta^{(k)}(t-s)x(s),$$

$\delta^{(k)}$ being the kth derivative of δ and $\delta^{(O)}$ distribution δ itself. According to a classical property of $\delta^{(k)}$, we have then

$$y(t) = \sum_{k=0,n} A_k(t)x^{(k)}(t).$$

Then if the A_k are null, except A_0, A_1 and A_2, reduced to scalars a_0, a_1 and a_2, we have

$$y(t) = a_0 x(t) + a_1 x'(t) + a_2 x''(t).$$

Then if a_0 and a_2 are small compared to a_1, the observation operator offers an acceptable representation of a *tachymeter*, the measurement of $x'(t)$ being slightly blurred by $x(t)$ and $x''(t)$. The same kind of remark can be made concerning an *accelerometer* if a_0 and a_1 are small compared to a_2.

Let us consider the case where $L(t, s) = Z(t - s)$, the kernel $Z(t - s)$ being a null matrix for $s > t$. We have an hereditary Volterra matrix convolution and

$$y(t) = \int_{\mathbb{R}} Z(t - s)x(s)\, ds.$$

For example, if $Z(t - s) = \delta(t - s - a)$, with $a > 0$, the observation operator gives the values of x with a time lag equal to a. Coming back to the general hereditary Volterra matrix convolution, we say, in this second extreme case, that the observation operator is of *type* $\mathcal{Z}$. Its matricial effects being set appart it acts mainly as a *filter of frequencies* of the Fourier transform of x. More precisely, if $\mathcal{F}_x$,$\mathcal{F}_y$ and $\mathcal{F}_z$ are the Fourier transforms of x, y and z respectively, we have a multiplicative process, ω being a frequency,

$$(\mathcal{F}_y)(\omega) = (\mathcal{F}_z)(\omega)(\mathcal{F}_x)(\omega).$$

1.3. Algebraic Considerations

The algebraic point of view presented above in the general case can be considerably extended in the *linear case* [13, 14]. By concatenation of two compatible linear observation devices we get another linear one represented by the *product* $\mathcal{O}_2\mathcal{O}_1$ of the two linear observation operators $\mathcal{O}_1$ and $\mathcal{O}_2$. But apart from concatenation, or composition in series, it is possible, in the linear case, to realize also a composition in *parallel*: if both $\mathcal{O}_1$ and $\mathcal{O}_2$ are linear observation operators, mapping the linear space X on the linear space Y, their composition in parallel gives the sum $\mathcal{O}_1 + \mathcal{O}_2$.

If $\mathcal{O}$ is a linear observation operator mapping X on X, we can look for an evolution function x, different from the null one, for which

$$\mathcal{O}(x) = \lambda x,$$

λ being a real or complex scalar. For this x, which is an eigenfunction of $\mathcal{O}$, the observation device acts as a mere *multiplicator* by λ, an eigenvalue of $\mathcal{O}$, so there is no distorsion.

Coming back to composition of two linear observation devices in series or in parallel, we can consider more precisely the two extreme cases of observation operators of types $\mathcal{A}$ and $\mathcal{Z}$. It is obvious that, for two operators of type $\mathcal{A}$, composition in parallel and in series involves, respectively, the product A_2A_1 of matrix functions of compatible formats and the sum $A_1 + A_2$ of matrix functions of the same format, corresponding to operators of type $\mathcal{A}$. Generally the product of square matrix functions A_1 and A_2 do not commute, which means that the two corresponding observation devices cannot generally be exchanged without modification of their effect. We have rather similar results for two observation operators of type $\mathcal{Z}$. If the kernels attached to these operators are matrix functions Z_1 and Z_2, composition in series and in parallel, for matrices of compatible formats, corresponds respectively to $Z_1 + Z_2$ and $Z_1 \otimes Z_2$, $\otimes$ representing a convolution. The product of operators of types $\mathcal{A}$ and $\mathcal{Z}$ gives of course a linear observation operator which is not of type $\mathcal{A}$ nor of type $\mathcal{Z}$. If $A(t)$ and $Z(t)$ are square matrices of the same format, we have, for products of, respectively, types $\mathcal{AZ}$ and $\mathcal{ZA}$, the integrals

$$y(t) = \int_{\mathbb{R}} A(t)Z(t-s)x(s)ds, \quad y(t) = \int_{\mathbb{R}} Z(t-s)A(s)ds,$$

with no commutation. Observation operators, such as those of $\mathcal{AZ}$ or $\mathcal{ZA}$ types, are quite able to describe rather realistic observation processes. Besides the matricial effect, an operator of type $\mathcal{A}$ can act as a *selector* of the instants where the process occurs really (for example a time window), a type $\mathcal{Z}$ one can help to see x without its meaningless irregularities by a *smoothing* induced by the attenuation of high frequencies.

1.4. Indiscernibility and "Inverse Transfers of Structures"[2]

The use of observation operators is helpful to study *indiscernibility* [16, 17]. If observation operator $\mathcal{O}$ is a one-to-one mapping of the set X of all the possible evolutions of the considered universe, (the system and its environment) on the set Y of all corresponding perceived evolutions, the process $x \to \mathcal{O}(x) = y$ involves a deformation but no loss of information. When there is such a loss, observation operator $\mathcal{O}$ takes into account the limitations of cognitive capacities of the observing system which play a part in its bounded rationality, to use the terms of Herbert Simon [10]. Even if $\mathcal{O}$ has no inverse, which seems to be the general case due to "Nature's horror of inversibility" as we may call it, the reciprocal image mapping $\mathcal{O}^{-1}$ exists, $\mathcal{O}^{-1}(y)$ being by definition the set x^* of all x such that $\mathcal{O}(x) = y$. So

$$\mathcal{O}^{-1}(y) = x^*, \quad \mathcal{O}(x^*) = y.$$

If $\mathcal{O}$ is not inversible it may happen that $\mathcal{O}(x') = \mathcal{O}(x)$ with $x' \neq x$. Then these two evolutions of the universe are indiscernible for observing system S. In other terms x and x' are *identical* for S using the observation device represented by $\mathcal{O}$. We are here in presence of an equivalence relation J which we may call a *relation of epistemological indiscernibility* representing

[2] Many of the results presented in this paragraph, concerning indiscernibility and "inverse transfers of structures" [16, 17], have also been proposed, later, in a slightly different way, by Robert Rosen [9]."

an aspect of the *epistemological subjectivity* of observing system S. We write classically $x'Jx$ or x' equivalent to x according to J. It means also that x and x' belong to x^*. The set of all the equivalence classes such as x^*, or quotient X/J, is a partition of X which defines the *epistemological resolving power* of the observation process, each class corresponding to a given y belonging to Y.

Let $\mathcal{O}_1$ and $\mathcal{O}_2$ be two observation operators mapping X on Y. If the partition defined on X by the relation of indiscernibility attached to $\mathcal{O}_1$ is a subpartition of the partition corresponding to $\mathcal{O}_2$, we may say that *the resolving power of $\mathcal{O}_1$ is greater than that of $\mathcal{O}_2$*. If we consider the case where the observing system possesses two observations devices represented by their observation operators $\mathcal{O}_1$ and $\mathcal{O}_2$ mapping X on, respectively Y_1 and Y_2, we may consider the pair $(\mathcal{O}_1, \mathcal{O}_2)$ as one composite observation operator mapping X on $Y_1 \times Y_2$. Then we can write

$$(\mathcal{O}_1, \mathcal{O}_2)(x) = (\mathcal{O}_1(x), \mathcal{O}_2(x)) = (y_1, y_2) \in Y_1 \times Y_2.$$

An equivalence class of the relation of indiscernibility involving $\mathcal{O}_1$ and $\mathcal{O}_2$ is defined by all the x' such that

$$(\mathcal{O}_1, \mathcal{O}_2)(x') = (\mathcal{O}_1, \mathcal{O}_2)(x) \quad \text{or} \quad \mathcal{O}_1(x') = \mathcal{O}_1(x), \mathcal{O}_2(x') = \mathcal{O}_2(x).$$

This equivalence class containing x is the intersection of the equivalence classes of $\mathcal{O}_1$ and $\mathcal{O}_2$ containing x, so the *resolving power of $(\mathcal{O}_1, \mathcal{O}_2)$ is greater than that of $\mathcal{O}_1$ and $\mathcal{O}_2$ separately*, a result we may call a *stereoscopic epistemological effect* [27].

Indiscernibility is not the only aspect of the epistemological subjectivity of an observing system. We consider an observation operator $\mathcal{O}$ devoid of an inverse. If X is the set of all the possible evolutions of the universe and Y is the set of all the corresponding perceived evolutions, we have $\mathcal{O}(X) = Y$ and $X = \mathcal{O}^{-1}(Y)$. Metaphorically we can say that Y is a kind of screen on which appears the image $y = \mathcal{O}(x)$ of the elements of X. Y is not amorphous, it possesses intrinsic structures of which system S is aware, even if "unconsciously", and which it is tempted, "unwillingly", to attribute to X about which *a priori* it knows nothing.

More generally when we have a mapping h of a set A on the set B, and a structure on B, there is a classical process to obtain on A a naturally induced structure involving the reciprocal image mapping h^{-1}. Léon Motchane gave of this process an epistemological interpretation [3] which we integrated in the framework of our dynamical formalism of "observation operator" [17]. For example if we have a binary relation $\mathcal{R}$ (equivalence, order, similarity) on $Y = \mathcal{O}(X)$, a corresponding binary relation $\mathcal{T}$ is induced on X and defined by

$$xTx' \Longleftrightarrow \mathcal{O}(x)\mathcal{R}\mathcal{O}(x').$$

So x and x' are subjectively perceived, by the observing system S, to be related by $\mathcal{T}$, when their images through $\mathcal{O}$ are related by $\mathcal{R}$. This is very natural because the only relations the system can know are those occurring in $Y = \mathcal{O}(X)$. There is a kind of propagation back to X of structure $\mathcal{R}$, intimately incorporated in the observing system, a propagation which we call *epistemological inverse transfer of structure* [17]. If Y posseses a topology, defined by its open sets, it is transferred to X by changing each open set W of Y into an open set $\mathcal{O}^{-1}(W)$ of X. This subjective choice of a topology in X is the only one that system S can do, because it knows only the open sets of Y. If there is a measure m, for example attached to a probability density, naturally defined on Y, it is transferred subjectively to X as a measure n such that, for any m-measurable set E of Y, $n(E) = m(\mathcal{O}(E))$.

Concepts such as inverse transfer of structure and relation of indiscernibility, remind us of *Plato's metaphor of the cave* presented in *The Republic* (*Book VII*). In this metaphor men see only the rear of a cave where they are enchained, behind them is a fire and between the fire and themselves is a wall along which procede statues and objects born by hidden people. The enchained men see only the moving shadows projected on the screen constituted by the rear of the cave and when they hear voices or sounds they believe that they come from the shadows. We have here a poetical presentation of the imperfections of perception: the very structures of the rear of the cave (two dimensions, irregularities, distances) are transferred to the objects observed, two distincts objects may be considered as identical if they have the same shadow [22,23].

All the above considerations involving observation operators, their algebra, indiscernibility and inverse transfers of structures with their subjective aspects, are elements of what we propose to consider as a *mathematical epistemology* at the macroscopic level.

2. TIME AND SPACE

The process of *perception of duration* may be considered from the point of view of observation operators. Parameter t, belonging to $\mathbb{R}$, is considered by definition as a *reference time*, all instants have the same interest, t is a "uniform" time. So the importance of interval $[t, t + dt]$ is measured by dt, the density $e(t)$ of importance of instant t of reference time is constant and equal to one. Let $a(t)$ be the *density of attention* paid by an observer to instant t, the subjective importance of interval $[t, t + dt]$ is then $a(t)e(t)dt$ and the *density of subjective importance* of instant t is $a(t)e(t)$ with of course

$a(t) \geq 0$. Then the *subjective duration* of an interval $[t_0, t]$ is given by the integral over $[t_0, t]$:

$$\theta(t) = \int_{[t_0,t]} a(s)e(s)ds = \int_{[t_0,t]} a(s)ds.$$

In other words,

$$\theta(t) = A(e),$$

A being an observation operator of Volterra composition type, with a kernel independent of t, which we may call *integral attention operator* [19]. Of course, if $a(t) = 1$, we have $\theta(t) = t - t_0$ and the subjective duration is equal to the *reference duration.*

But if the perception of duration involves *memorization,* we may consider that the perceived duration $a(s)\,e(s)\,ds$ of interval $[s, s + ds]$, when remembered at instant t posterior to s, is equal to $\rho(t,s)a(s)e(s)ds$, where $\rho(t,s)$ is a *factor of memorization,* inferior to one, which must tend to zero when s tends to $-\infty$. Moreover for the sake of coherence, we must have

$$\rho(t,s)\rho(s,r) = \rho(t,r).$$

So, after some elementary calculations,

$$\rho(t,s) = exp\left(- \int_{[s,t]} b(u)du \right),$$

where $b(u) \geq 0$. We must add that the integral must be divergent for $s = -\infty$ since $\rho(t,s)$ must, as we have said, tend to zero when s tends to minus infinity. So *the subjective duration with memorization* of the interval $[t_0, t]$ is

$$\theta(t) = \int_{[t_0,t]} exp\left(-\int_{[s,t]} b(u)du\right)a(s)e(s)ds = \int_{[t_0,t]} exp\left(-\int_{[s,t]} b(u)du\right)a(s)ds$$

or

$$\theta(t) = A_m(e),$$

A_m being an observation operator of the same type as A, which may be called *integral attention-memorization operator* [19].

The density of attention paid by the observer to each instant may depend upon the sensorial channel used. So, if there are two sensorial channels, there are two observation operators for duration and, as we have seen above, there is a "stereoscopic" effect. Moreover these two observation operators and, more generally, n such observation operators open the way to a *n-dimensional subjective time* [24,28].

Up to now we have considered that the state $x(t)$ of the universe, constituted by system S and its environment E, belongs to $\mathbb{R}^n$ (or $\mathbb{C}^n$). It could also belong to a linear space of infinite dimension such as a functional space. For example $x(t)$ may be a physical field defined on the usual 3-dimensional space. More explicitly we could have

$$(t, M) \to x(t, M),$$

with $t \in \mathbb{R}$ and $M \in \mathbb{R}^3$. We can then consider *observation operators* acting on functions x of both time and space [11,13]. Of course these operators must be causal, which means that, at any given instant t, they must act only on the past and present of x (and, in a *relativistic context*, at any given (t, M) on the anterior part of the light cone of summit (t, M)).

We may consider now the *linear case* which involves a *causal Volterra matrix composition*, implying instant t and point M. We have, with obvious notations,

$$y(t, N) = \int_{\mathbb{R}^4} L(t, N, s, M) x(s, M) ds dv,$$

where dv is the elementary infinitesimal volume in $\mathbb{R}^3$ and $L(t, N, s, M)$ vanishes for $s > t$ (or outside the anterior part of the light cone of summit (t, M) in relativity). If

$$L(t, N, s, M) = A(t, M)\delta(t - s, N - M),$$

where δ is the Dirac distribution, we get

$$y(t, N) = A(t, N)x(t, N).$$

This is an extreme case and we shall say again that the observation operator is of *type* $\mathcal{A}$. If $A(t, M)$ is a scalar $a(t, M)$, positive or equal to zero, we have a *factor of attention* in time and space. If $a(t, M)$ is equal to one on a set and to zero outside, the observation operator acts as a *frame* in time and space or as a frame in space changing with time. It *selects* what is interesting, for example by a sampling in time and space. If $L(t, N, s, M) = Z(t-s, N-M)$ we have a *causal Volterra matrix convolution*, involving time and space,

$$y(t, N) = \int_{\mathbb{R}^4} Z(t - s, N - M) x(s, M) ds dv.$$

In this second extreme case we say that the observation operator is of type $\mathcal{Z}$, as we did in the purely temporal case. If we consider the quadrimensional Fourier transform of x, y and z, we have again a multiplicative process

$$(\mathcal{F}_y)(\omega, w) = (\mathcal{F}_z)(\omega, w)(\mathcal{F}_x)(\omega, w),$$

ω and w being respectively time and space frequencies. The observation operator acts mainly as a *filter of time and space frequencies*, introducing a *subjective point of view*.

The use of both types $\mathcal{A}$ and $\mathcal{Z}$ operators, when A and Z are square matrix functions of the same format, introduces the products of types $\mathcal{AZ}$ and $\mathcal{ZA}$ which correspond to very general modellings. For example we may suppose, for the sake of simplicity, that the points M and N belong to given planes and that we consider an observation operator of type $\mathcal{A}$, defined by a scalar $a(t, M)$ taking only value one and zero, and an observation operator of type $\mathcal{Z}$ defined by the scalar kernel $z(t - s, N - M)$. The action of the type $\mathcal{A}$ operator may be that of *framing* a spatial two-dimensional scene and of *selecting* very short intervals of time, outside which no attention is given to what is going on, as it happens with movies. The action of an operator of type $\mathcal{Z}$, which is a *frequency filter*, may be an attenuation of high time frequencies and also a suppression of high space frequencies. The attenuation of high time frequencies generates a phenomenon like *remanence* on a screen. The suppression of high space frequencies gives effects of *diffraction*, for example in optics with monochromatic light, when the observed field is the luminous field, not its intensity, as shown, after Ernst Abbe, by Michel Duffieux [1]. This diffraction effect introduces a *graining* in the constitution of the two-dimensional image. If space frequencies are suppressed outside a given rectangle centered on zero frequencies, we can use Claude Shannon's theorem (due in fact to Charles de La Vallée Poussin) famous in the purely temporal signal theory. The image is perfectly defined by the value of y on the nodes of a rectangular lattice and we have here a simple example of the *structuration* of space[3] due to the observing device itself [25]. In the case of static optics, with the luminous field, the action of the type $\mathcal{A}$ operator may be to select the view to be observed, in which case the type $\mathcal{Z}$ operator yields the optical image and we have a product of type $\mathcal{ZA}$. But the action of the type $\mathcal{A}$ operator may occur after the obtention of the optical image and then we have a product of type $\mathcal{AZ}$.

3. INTRODUCTION TO A "MATHEMATICAL EPISTEMO-PRAXIOLOGY"

We suppose now that our system S is not only able to perceive the evolution of its environment and of itself but that it is also able to *decide* and to *act* upon its environment and itself, even if only in a metaphoric sense. After perception, represented by an observation operator $\mathcal{O}$, we must have

[3] The case of the graining of time and of the "width of an instant" may also be considered [24] but in a rather different way.

also some process of decision described by a *decision operator* $\mathcal{D}$. So x, representing the evolution of the universe U, the evolution of the perception of system is $y = (x)$. Then, in this rather simple modelling, we assume that y generates $z = \mathcal{D}(y)$representing the evolution of the decisions taken by S. We have

$$x \rightarrow \mathcal{DO}(x) = \mathcal{P}(x) = z,$$

where operator $\mathcal{DO} = \mathcal{P}$ is what we call a "*pragmatic operator*" [17] and z a *pragmatic image* of x, a description of the evolution of the universe in the *language of decisions*. Besides being *hereditary*, or *causal*, operator $\mathcal{P}$ has the same formal properties as $\mathcal{O}$. For example we may consider a *pragmatic indiscernibility*, which means that x and x' are pragmatically indiscernible if they generate the same decision function

$$\mathcal{P}(x) = \mathcal{P}(x'),$$

a circumstance which occurs, with $x \neq x'$, when $\mathcal{P}$ has no inverse. In a similar way we may introduce the concept of *pragmatic inverse transfer* [17, 23]. Operator $\mathcal{P}$ is a mapping of the set X of all the possible evolutions x of the universe, on the set $Z = \mathcal{P}(X)$ of all the corresponding decision functions z. As well as structures, inherent to the "screen" $Y = \mathcal{O}(X)$, generate structures on X through epistemological inverse transfers, the structures of Z induce also structurations of X.

As announced, we also admitted that system S is able to act. So we introduce an action process which, acted upon by the decision function z and also influenced directly by x, generates the very evolution function s of S itself. Let G_S represent the corresponding *effect operator*, generally hereditary. We have

$$G_S(z, s) = s.$$

We assume that the environment E is passive. It does not perceive nor decide or act, but it is influenced by s, so if G_E is its operator of evolution, we can write

$$G_E(s) = e,$$

e being the evolution function of E. The two above equations may be written synthetically

$$G(x) = x,$$

if we remember that $x = (s, e)$ and $z = P(s, e)$. So, initial conditions being implicitly taken into account, the evolution of the *supersystem* U constituted by both S and E, is given by a *fixed point*[4] of the synthetic operator G.

[4]This fixed point may be compared to Heinz von Förster's eigen-behaviors [2].

We have a dynamical modelling of the retroaction loop *perception-decision-action* upon what is to be perceived. This is a step towards an *epistemo-praxiology* [20, 23, 28] which introduces an *epistemo-praxiological cognition.* Universe U is both observed and acted upon through a part S of itself, so an apparence of some *pre-established harmony* must appear at the levels of intelligibility and feasibility. The inverse transfers, both epistemological and pragmatic, linked to the imperfections of perception and decision, of intimate structures of S on the set of all the possible evolutions of U, make universe U look more *familiar* to S. All that cannot be perceived or decided is eliminated from epistemo-praxiological cognition, making the universe appear more understandable, manageable and secure, even if in an illusory way.

4. TOWARDS OTHER FORMALIZATIONS

Up to what extent could the above epistemological considerations concerning the macroscopic world, be transferred to quantum physics, thereby giving rise to more general approaches of the problem of a *formal epistemology* [6,7,8]. Classical physics is said to separate completely the object from the subject, but this is only an idealization, if not a caricature. Except in classical astronomy, the object has always had some connection with the subject, feeble as it may be. As we have seen, conceptual simplicity appears, when we consider the *supersystem* constituted by the subject and the object (the system and its environment). Here is a point of view which can be transposed from the macroscopic to the microscopic on level of observation where it is accepted that an interesting entity consists of a pair *observer-object*, even if the relation between subject and object is not classical.

We have seen the role, at the macroscopic level, of what we have called *inverse transfers of structures.* Has this concept any impact in quantum theory ? Through well chosen experimental processes, we ask questions to Nature in our macroscopic language embodied in macroscopic devices. We give no choice to the answer, it must be expressed in the same style. It seems that Nature does its best to satisfy us: if we ask the question in corpuscular language we get an answer in corpuscular terms, if we use a wave language we obtain an answer in wave terms. Nature tells us in a Pirandellian way: I am "as you desire me". Can we not say that there is an inverse *transfer of the macroscopic structures, or categories, of the macroscopic observational device to the quantum world*, making it apparently more intelligible but in fact betraying it?

It is well known how important *Fourier transform* is in signal theory, in both the classical cases of a purely temporal signal $t \rightarrow x(t)$ or of a time and space one $(t, M) \to x(t, M)$, function x taking, in all cases, its values in $\mathbb{R}^n$ (or $\mathbb{C}^n$) and M being a point of physical space $\mathbb{R}^3$. For the sake of simplicity we consider the case where point M, represented now by m, belongs to $\mathbb{R}$. At a given instant t, it is possible to define a central value for $x(t, m)$ if we make use of $|x(t, m)|^2$ and a dispersion $D_m(t)$ after we have normalized $|x(t, m)|^2$, all the involved integrals being supposed convergent. Let $(\mathcal{F}_m x)(t, w)$ be the Fourier transform of $x(t, m)$ with respect to m given by the integral over $\mathbb{R}$

$$(\mathcal{F}_m x)(t, w) = \int_{\mathbb{R}} exp(-2i\pi wm) x(t, m) dm,$$

and $D_w(t)$ the dispersion corresponding to $|(\mathcal{F}_m x)(t, w)|^2$. Classically we then have $D_m(t) D_w(t) \geq \pi/4$. This inequality is formally similar to Heisenberg's inequality $D_m(t) D_p(t) \geq h\pi/4$ where $D_m(t)$ and $D_p(t)$ are the dispersions corresponding respectively to $|\Psi(t, m)|^2$, $\Psi(t, m)$ satisfying Schrödinger's equation with one space variable m, and to $|\Phi(t, p)|^2$, Φ being given by

$$\Phi(t, p) = h^{-\frac{3}{2}} \int_{\mathbb{R}} exp(-2i\pi pm/h) \Psi(t, m) dm.$$

In both cases the inequalities result from the properties of the Fourier transform. For the first inequality we may say that the space and frequency dispersions of the signal cannot be arbitrarily small at a given instant, the frequency being here a space frequency. For the second one, concerning position m and momentum p, which are not commutable observables, the position and momentum dispersions cannot be arbitrarily small and, in contradistinction with the first inequality, appears the quantum of action h. Anyway we may think that the Fourier transform, or one of its possible extensions, has to play a part as a factor of unification between the macroscopic and the microscopic worlds [12,15,26].

The concept of operator, more precisely of *observation operator*, plays an important part in our proposition of a "mathematical epistemology" (and even "epistemo-praxiology"). But these operators are rather far from those of quantum mechanics. When we write, for an observation operator, $\mathcal{O}(x) = \lambda x$, we look for the evolutions x that are perceived with no more deformation than a multiplication by λ and, if $\mathcal{O}$ is linear, the possible x and λ are the eigen-functions and eigen-values of $\mathcal{O}$. But when Q is a quantum mechanical operator, corresponding to some observable, necessarily linear and even Hermitian, if we write $Q(\rho) = \lambda \rho$, the eigen-values λ are the possible values of the observable and the eigen-functions ρ give, by a well known process, the corresponding probabilities. So, if we believe that the role of operators is of great importance in the construction of a formal epistemology, analogies must examined with circumspection.

Still dealing with operators let us compare the *observation operators of types $\mathcal{A}$ and $\mathcal{Z}$* [11,13] and the *delimitators and views* [4,5]. They participate, respectively, to the construction of a "mathematical epistemology" (even an "epistemo-praxiology") and of a "formal epistemology". As we have seen, the role of an $\mathcal{A}$ operator is, in the most simple cases, to select a part of space-time, in other words to choose a frame. A delimitator acts in a rather comparable way, cutting out from reality $\mathcal{R}$ an entity for examination. A $\mathcal{Z}$ operator acts, for example, after an $\mathcal{A}$ operator, as a filter of space-time frequencies. A view, for example with only one aspect, acts on the result of the action of a delimitator. For $\mathcal{A}$ and $\mathcal{Z}$ on one side and for a delimitator and for a review on the other, we get $\mathcal{Z}\mathcal{A}(x)$ and the product of a view and a delimitator acting on $\mathcal{R}$. These formulations, in fact different, have some common features, they both give the elements of a representation. Indiscernibility in the first kind of formulation and identity in the second, are defined respectively with reference to an observation operator and to a view. An observation operator, of $\mathcal{Z}\mathcal{A}$ type for example, if we take into account an effect operator, following a possibly unconscious decision as explained above, induces a modification of what is observed and so introduces the consideration of iterations through an epistemo-praxiological loop [17,22,28], as well as for the product of a view and a delimitator creating a kind of circular consciousness process ([4]). We have here very general mathematical and formal attempts towards a new epistemology deserving the name of *formal epistemology.*

REFERENCES

1. M. Duffieux, *L'intégrale de Fourier et ses applications à l'optique* (Rennes, 1946).
2. H. von Förster, "Formalisation de certains aspects de l'équilibration des structures cognitives," in B. Inhelder, R. Garcias, and J. Voneche, eds., *Epistémologie génétique et équilibration* (Delachaux & Niestlé, Neuchatel, 1976), pp. 76-89.
3. L. Motchane, "Structures formelles du monde réel," in F. Le Lionnais, ed., *La méthode dans les sciences modernes* (Science et Industrie, Paris, 1958), pp. 105-110.
4. M. Mugur-Schächter, "The general relativity of descriptions," *Analyse de Systèmes* **11**, 40-82 (1985).
5. M. Mugur-Schächter, "Space-time quantum probabilities, relativized descriptions, and Popperian propensities I," *Found. Phys.* **21**, 1387-1449 (1991).
6. M. Mugur-Schächter "From quantum mechanics to universal struc-

tures of conceptualization and feedback on quantum mechanics," *Found. Phys.* **23**, 37-122 (1993).
7. M. Mugur-Schächter, "Une méthode de conceptualisation généralisée: vers une épistémologie formelle apte à faire face aux complexités," *Revue Internationale de Systémique* **9**, 963- 994 (1995).
8. M. Mugur-Schächter, "Les leçons de la mécanique quantique, vers une épistémologie formelle," *Le Débat* **94**, 169-192 (1997).
9. R. Rosen, *Fundamentals of Measurement and Representation of Natural Systems* (North-Holland, Amsterdam, 1978).
10. H. Simon, *The Science of the Artificial* (MIT Press, Cambridge, MA, 1969).
11. R. Vallée, "Sur deux classes d''opérateurs d'observation'," *C.R.A.S (Paris)* **233**, 1350-1351 (1951).
12. R. Vallée, "Aspect informationnel de certaines relations d'incertitude," *C.R.A.S (Paris)* **233**, 1580-1581 (1951).
13. R. Vallée, "Un point de vue algébrique en théorie macroscopique de l'observation," *C.R.A.S. (Paris)* **241**, 1979-1980 (1955).
14. R. Vallée, "A note on algebra and macroscopic observation," *Information and Control* **1**, 82-84 (1957).
15. R. Vallée, "Expression asymptotique, pour les grandes valeurs du temps, de l'information associée à la fonction d'onde dans le cas d'un corpuscule libre," *C.R.A.S. (Paris) B* **267**, 529-532 (1968).
16. R. Vallée, "Sur la formalisation mathématique en théorie de l'observation," in *Actes du 7ème Congrès International de Cybernétique* (Namur, 1993) (Association Internationale de Cybernétique, 1974, Namur), *Vol. 1*, pp. 225-232.
17. R. Vallée, "Observation, decision and structure transfers in systems theory," 2nd European Meeting on Systems Research, Vienna 1974, in R. Trappl and F. Pichler, eds., *Progress in Cybernetics and Systems Research*, Vol. 1, pp. 15-20 (Hemisphere Publishing, Washington, 1975).
18. R. Vallée, "Sur le problème de la perception en théorie des systèmes," *Cybernetica* **19**, 39-44 (1976).
19. R. Vallée, "Memorization in systems theory and perception of time," in E. G. Lasker, ed., *Applied Systems and Systems Research, Vol. 2* (International Congress on Applied Systems Research and Cybernetics, Acapulco, 1980) (Pergamon, New York, 1981), pp. 697-700.
20. R. Vallée, "About Heinz von Förster's 'eigen-elements'" (Systems Science VIII, Wroclaw 1983), *Systems Science* **10**, 25-28 (1984).
21. R. Vallée, "Sur la modélisation de la perception," in H. B. Lück, resp. *Biologie Théorique, Solignac 1985* (Editions du CNRS, Paris, 1987),

pp. 277-280.
22. R. Vallée, "La caverne de Platon revisitée," in E. Bernard-Weil and J.-C. Tabar, eds., *Praxis et cognition* (Perspectives Systémiques 2, Cerisy-la-Salle, 1988) (L'Interdisciplinaire, Limonest, 1992), pp. 25-32.
23. R. Vallée, "Plato's cave revisited," *Kybernetes* **19**, 37-42 (1990).
24. R. Vallée, "Perception, memorisation and multidimensional time," *Kybernetes* **20**, 15-28 (1991).
25. R. Vallée, "Observation et structuration de l'espace," in *Actes du 13ème Congrès International de Cybernétique* (Namur, 1992) (Association Internationale de Cybernétique, Namur 1993), pp. 701-705.
26. R. Vallée, "Aspects of complementary," in M. E. Carvallo, ed., *Nature, Cognition and System II* (Kluwer Academic, Dordrecht, 1992), pp. 7-85.
27. R. Vallée, "Perception, decision and action," *J. Biol. Systems* **2**, 43-53 (1994).
28. R. Vallée, *Cognition et système, essai d'épistémo-praxéologie* (L'Interdisciplinaire, Limonest, 1995).

9

AGO-ANTAGONISTIC SYSTEMS

Élie Bernard-Weil

Foundation A. de Rothschild
Departement of Neurosurgery I
26-29, rue Manin, F-75940 Paris Cedex 19, France

Today, bio-medical sciences and human sciences in general are demanding some new epistemological paradigms, in the same manner that quantum physics began to proceed to a renewal of this kind eighteen years ago. Such paradigms seem to be connected with systems science, and especially a special branch of it, called agonistic-antagonistic systemics (AAS), combining co-operativity and conflict between two poles. AAS is under the necessity of considering, at the same time, both sides of whatever phenomenon—which may appear as contradictory, opposite or only different—and, finally, of taking into account the unity to which both sides belong. The dynamics study of the behavior of these couples, or of the so-called agonistic-antagonistic networks, allows to better understand the occurrence of amazing phenomena, as well as to consider special types of control, when agonistic antagonistic unbalances have occurred.

Key words: systems sciences, general or transdisciplinary models, complexity, self-organization, chaos, mathematical models.

> "They are all wrong, so much more dangerously so that each one of them follows a truth. Their fault is not to follow a false truth but to not follow also another truth .... And usually it so happns that, being unable to conceive of a relation between two opposite truths and believing that the confession of one entails the exclusion of the other one, they cling to one, they exclude the other one, and they think that we do the opposite."—*Blaise Pascal.*

1. INTRODUCTION

It looks as though systemics—the science of systems—were failing to obtain the place which it seemed to have a right to claim in the unraveling of the paradigms which define the general orientations of the various modern sciences, so in the constitution of the nowadays *episteme*. An element of explanation might be that, though the works developed inside the framework of systemics have produced concepts which, here and there, have been seminal, other such concepts aroused resistance.

Systemics, as is well known for now, has been developed by reaction to a tendency in modern science, toward a reductionist, immanentist view, according to which knowledge concerning all the parts of a whole would *additively* entail a complete knowledge of also that whole. To this tendency systemics opposed a science of—specifically–the whole. It brought into evidence the henceforth undisputed awareness that in general, from any given point of view, the whole appears as either *more* than its parts or less than its parts, because certain manifestations exhibited by the whole emerge from only the interactions or the relations between the parts, while manifestations which the parts do exhibit when observed isolately might be occulted in the behavior of the whole. In short, systemics drew methodic attention upon contextuality, making clear the necessity to consider the functioning of an element from a studied system, only in relation to the functionings of the other elements from that system, and furthermore in relation also with the systems from the environment of that system, especially the environment consisting of the "observing system". It correlatively brought into evidence that, when no criteria can be produced for specifying a dominant system, such contextual relations can take on aspects of "entangled hierarchy".

However systemics did not stay entirely true to itself. So, its endeavor toward influence inside the scientific community induced a surreptious yielding to certain prejudices or latent ideologies that are at work inside this community. One of these is an open hostility against *transcendence*. Of course the immanentism of most of nowadays researchers can be fully understood when one counts the blows stricken upon scientists by religious authorities, since six or seven centuries, in the name of "revealed truths". But not any transcendence amounts to a revealed truth, and we have now reached a stage where strict immanentism—in a quasi-sacred form—becomes, in its turn, an obstacle to the further development of science. Despite very numerous and valid modern notions due to system sciences, they still seem, in fact, to be dominated by strict immanentism, and this is not, as far as I know, clearly dismissed inside systemics. In this sense systemics indulges itself into silent inner contradictions.

In what follows one will find a reaction to this state of things. The objections which I oppose to the concept of self-organization, followed by brief considerations concerning the concept of complexity, will introduce to the exposition of *an ago-antagonistic method for the control of open homeostatic homeorhetic systems* which—quite explicitly and essentially—involves transcendence.

2. INSIDE, OUTSIDE SELF-ORGANIZATION, HETERO-ORGANIZATION

That toward which the notion of self-organization points requires principles of which it cannot be asserted that they are wholly *derivable* from experience or experiment. It requires a certain *posited* transcendence, at least in the minimal sense entailed by the unavoidable necessity of a recourse to the abductive method. Let me explain.

The view labeled by the term self-organization applies only to *open* systems, able of states of *dynamical* equilibrium characterized by continual fluxes of dynamical interaction with their *outsides*. One never speaks of self-organization to designate the evolution of a closed system (devoid of dynamical interaction with its outside) toward a state of static equilibrium. In such a case, quite on the contrary, the second principle of thermodynamics asserts a type of process that is usually characterized as a trend toward the maximal *disorder*; but—mind this—the maximal disorder that is *compatible with the global constraints*. So even in the case of closed systems, i.e., in the posited absence of any dynamical interaction between the studied system and its outside, a rigorous formulation of the inner evolution of the system cannot omit the decisive role of the outside of the system : indeed the "global constraints" are nothing else than the quintessential extract—with respect to the studied system—of the mechanically static, the only energetic "pressure" upon the whole of that system, of precisely its outside. Then so much more in the case of an open system which by definition does dynamically interact with its outside exchanging with it both energy and *matter*, it simply is a contradiction-in-term to require a strict immanentism throughout the construction of models. And nevertheless it is precisely this that is required for the models of self-organization, as the vocable "self" explicitly states : any assertion of an action stemming from the outside of the studied system is banished from the accepted models. By "outside" we mean not so much the environmental field around a considered system than the presence of this quintessential extract or this type of logos we will call "hetero-organization".

Under these conditions, the modelization of a process of inner organization of an open system accordingly to the concept of self-organization, is usually developed as follows. The studied system is considered to consist of elements that interact *only locally* and thereby bring forth "emergences" of new properties. These elements can be termites "in front of" their future termitary, or "inhabitants" just preparing to built a city, or cells or a mass of cells preparing to built an organism (caricaturing, one should say that a stomach, a brain and muscles, etc., meet by chance thereby subsequently entailing the emergence of an organism). *But why are these elements already there together, at half-way between a supposedly random meeting place and what will now "emerge" in consequence of their local interactions?* A taboo—the term is not too strong—is acting here, according to which it is forbidden to produce the faintest explanation of this meeting and of the consecutive emergences, by transcendent considerations suggesting that exterior factors involving a hetero-organization would also play some part, via some constraints acting directly upon the whole frontier of the considered system (wherefrom they then would spread into the inner elements, by "influences" on these and on their mutual interactions).

This scheme of modelization expresses the essence of the immanentist notion of "pure" *self*-organization. It stays valid, I think, for the totality of the self-organization models, which are a host.

The explanations provided by the models of self-organization for the emergence of new properties inside the studied system, remain vague. This, of course, does not come as a surprise. So much more that certain particularly purist authors seem reluctant to admit of even an origin identifiable in the micro-structure of the system *itself* before the emergence of new properties observable on the level of its macro-structure. At this extreme limit of orthodoxy the word "emergence" is used as if—just by itself—it were already a sufficient explanation: a gliding into sheer incantation. Other authors, however, are less orthodox. Some transgress the canonical modelization fully unaware of their transgression. Others present the transgression as an innovation without always clearly realizing that in fact this innovation breaks the fundamental rules, amounting in fact to the assertion of a transcendence [2,3].

The first category mentioned above can be illustrated by a model proposed for a beehive [4] where the concentric disposition of the honey, the pollen, and the brooder is obtained by simulation. The modelization is explicitly asserted to have strictly excluded any other presupposition besides *local* interactions: none among the insects was allowed to play a hierarchical role, not even the queen, so none was assumed to possess, explicitly or even implicitly, "a representation of the task to be accomplished". However the proposed simulation is ruled by a *system of partial differential equations.* So the behaviour of "local" interactions is submitted to some rules (moreover any definite solution requires the specification of limiting conditions). Thereby the authors, without realizing the fact, have posited a hetero-organization and they have immersed their representation into it.

The second category mentioned above—that of the "innovators"—is well illustrated by the quotation of the legend of a photograph of an embryo which, according to the author, can be regarded to "result from a programmed self-assemblage" [5]. This principle, the author writes "is promising, but the research engineers only begin to understand the type of structures which they might obtain and the type of tasks that could be assigned to these structures. The programmed self-assemblage is a concept of the XXIth century". The author seems to be rather unaware of the inner inconsistency of a concept that associates in one whole the idea of *program* and that of *autonomy* of the system which *executes* the program! This comes very near to the "paradoxical injunctions" of Gregory Bateson (be spontaneous, my son). It seems clear that the authors of such "contributions" to the doctrine of self-organization, though they are still unprepared to raise an explicitly critical debate, nevertheless—via the *language* they make use of—do already point toward conceptual feelings or even toward an inexplicit theory underlying inside their own minds the experiments communicated in their works, which in fact are *illustrations* of the unavoidable character of the transgression of the concept of self-organization; and they show *already* specific ways in which also a hetero-organization can be taken into account.

The practical consequences that could be drawn from the half-truths generated by the concept of self-organization are worrying. According to this concept, a social system, for instance, would reach an optimal equilibration between ruling powers and developmental trends, if a self-organizational process were introduced in its functioning. This, however, is far from being certain. The specialists of management are quite right to claim that "strategies of emergence", though they might well contribute to generate new types of organization, will not necessarily bring forth solutions to the dysfunctions inside the firms, if they are not combined with deliberate strategies.

We now sum up. The way in which the concept of self-organization fails, once perceived, strongly draws attention upon the essential impossibility of "monistic systems". The limiting concept of "The Universe" set apart, *any system has its outside*, and this, though obvious, has to be claimed as long as long as concepts which contradict still have course and enjoy credit. Now this entails the necessity of a systematic introduction of *pairs* [(a system with its organization), (the *outside* of that system with its *hetero* organization)]. Indeed it seems clear that if the system is considered monistically only truncated or ill-grounded conclusions can be obtained, at least in the case of an open system that interacts dynamically with its outside, whereby the organization of the system, at any time, depends explicitly on the hetero-organization in question.

It might seem difficult to specify in general terms the laws according to which the two elements of such a pair are linked to one another. It will however appear, I hope, that the ago-antagonistic method announced in the introduction offers an explicit example of rules for achieving such a linkage. But for the moment let us continue with a second set of brief critical remarks, namely concerning the concept of complexity.

3. COMPLEXITY, SIMPLICITY, THE OUTSIDE OF A QUALIFICATION

Together with the concept of self-organization, systemics—among other approaches—has furthermore popularized another concept also, namely that of *complexity*. In the present context this concept is interesting because, drawing more specifically attention upon the *relations* between the parts of a studied system and between these and the parts of its outside, it displaces the accent from *systems on qualifications*, on condition that "qualification" be associated, as in Mugur Schächter's method of relativized conceptualization, to another pole, of *"delimitation* " (let us remark that by emphasizing this point it becomes possible to connect agonistic-antagonistic systems to other springs of the so-called "formal epistemology"). Thereby another sort of monistic tendency comes into evidence, a tendency toward also a certain "*qualificatory* monism". This, like the position of monistic *systems*, leads to unacceptably absolute views that are contradicted by obvious transcendences.

Two main measures of *algorithmic* complexity have been defined so far. One of these, the Kolmogorov-Chaitin measure, consists of the minimal *length* of a program able to yield an informatical definition of the system, a "plan" of it in the form of the whole information necessary to a computing machine for representing the system. Then, it is considered as a measure the random complexity. Since any regularity discerned in the structure of the studied system permits to *shorten* this program, the value of the Kolmogorov-Chaitin measure is sometimes regarded as a numerical estimation of the degree of chaos involved by the system, though such a conclusion is not always valid (the same algorithm may give rise to periodic as well as chaotic dynamics; cf. later).

The other measure has been distinguished by C.H. Bennett and it consists of the computation-*time* necessary for the realization of the aforementioned program of minimal length. This time can be long even if the program is short. For instance a fractal structure, because of the big *number* of distinct space-time levels brought into play, can require a very long computation time, notwithstanding that the program can be shortly expressed by very simple equations. So Bennett's definition can be regarded to better yield a numerical estimation of the total "quantity of organization" involved by the system. In *this* sense it is sometimes called the *organized complexity* of the system. It seems clear however that *none of these two sorts of complexity is absolute.* In both cases the value of the measure is crucially relative to the way in which have been chosen the "elements", the aspects that qualify these elements, and the (meta)-aspects that relate either the considered qualifications, or the "meta-elements" consisting of bulks of elements. The organized complexity of a program for the representation of a cat, for instance, can be made arbitrarily big by increasing the number of required specifications of regular features (molecular constitution of the hairs, the skin, or some characteristics depending on the genus and the species the cat belongs to, etc.), while the random complexity can be largely varied as a function of the required specifications of random or chaotic aspects (a very precise specification of the length of each hair, etc.). So also the ratio between the two sorts of complexities can be varied with much freedom.

But the main point—in the present context—is that *both* sorts of complexity-measures can be defined for any *one* given system, and their ratio can have *pragmatic* significance. I give a simple example.

For a normal subject the cardiac rhythm seems—at a first sight and at least during a period of rest—to be regular, so organized. But more precise estimations show that "in fact" (i.e., with respect to smaller unities of space and of time) the rhythm is chaotic. This leads to a language in terms of a chaos of feeble or of high "dimension", of course *with respect to a norm* established by statistical data and referred to the danger involved for the subject's life. Furthermore, one can also construct the "topology" of the corresponding strange attractor, namely the mean value of the relative frequency of the heart beats, which is an *invariant* qualification of the whole set of performed registrations. This is a maximal, a limiting value of organized complexity associable with the studied system. Now, nothing hinders a simultaneous consideration of, *both*, the chaotic complexity of the cardiac rhythm (to be measured) *and* the corresponding invariant mean (to be constructed): these two qualifications, though distinct *and* of a same nature, are not mutually exclusive, since the second one is constructible from the first one. It is always possible to specify them both "at the same time", for one given set of registrations (this set plays here the role of the "system", i.e., of *what* is qualified) [7].

Nevertheless this double qualifying aim is seldom pursued. Why? Because of the words used for coding the involved pair of qualifications : these obscurely suggest a violation of the well known logical rule called "the excluded third".

> *Current logic has thus structured the nowadays scientific mentality that certain scientists are reluctant to qualify one system as—at the same time—both "chaotic" and "invariant".*

This reluctance induced by current language-and-logic is so strong that it has led to claim homeostasis—the action which arises inside a system when it deviates too far from the norm and which draws it back toward the norm—to be an obsolete concept doomed to be *replaced* by the concept of chaotic complexity [8].

The main point, however, is that, given a system, since there is a huge liberty in the choice of the elements and the qualifications in terms of which this system is studied, one can *always* deliberately try to distill conjugated *pairs* [(a chaotic complexity A of the system), (a corresponding *organized* complexity A' of the system)] where the organized complexity A' is an invariant associable with the chaotic complexity A, a "simplicity" connectable with the "complexity". This, according to a well known formulation, amounts to the possibility to extract some "order" out of a given "chaos". Thereby the monistic, absolutizing tendency in the qualifications of complexity—this system "is" complex, this system "is" simple—is *transcended.* There is then no reason whatever to think that "complexity is stronger than us" (as certain publicities suggest) and to surrender. On the contrary it is quite possible to make a deliberate and dominated use of the concept of complexity and to increase by this the expressivity of the representations of systems and of their control. The hopes of those who first introduced the concept of complexity as a new dimension of liberty, can be fulfilled.

The preceding remarks were intended as an illustration of false mutual exclusions suggested by current language-and-logic, in consequence of illusory absolutizations.

4. AGO-ANTAGONISTIC [9] CONTROL OF OPEN HOMEOSTATIC HOMEORHETIC SYSTEMS

So far I only tried to despond certain illusory monistic states of mind which, I think, stand in the way of an adequate representation of processes of which a correct understanding might entail important effects. Thereby I have explicated the ground where, inside systemics, I rooted the theory of ago-antagonistic systems: *explicit refusal of any sort of monism, systematic assertion of transcendence*—a transcendence that has to be epistemologically coupled with an immanence.

Now, by itself, naked so to say, the idea of transcendence cannot lead to essential innovations. Only the identification of its consequences in this or that specific circumstance can bring forth the radical changes that it operates in the realm of rationality [10] and the sometimes disconcerting techniques which it permits to define and to put to work. In what follows I provide an example. I shall show the following:

> Consider an open and homeostatic homeorhetic system the dynamical equilibrium of which involves (in particular) a given pair of "opposite" features, forces or fluxes. If the system runs stably out of the equilibrium concerning this pair of features, it is possible to compensate for the failing homeostatic and/or homeorhetic reaction by instating a control involving the simultaneous application upon the system of a pair of actions of the *same nature* as the initially distinguished pair of system-features (forces or fluxes, respectively) but stemming from the *outside* of the system, and combined in—both—an antagonistic *and* an agonistic manner. This class of phenomena—a rich class—admits of a non trivial *ago-antagonistic model* entailing counter-intuitive effects that cannot by guessed a priori.

I shall first describe the seminal experiment that generated the mentioned results. I shall then sketch out the general mathematical ago-antagonistic model. I shall then close the characterization by adding a paradigmatic metaphor that endows the model with an intuitive relief. The announced characterization will be followed by a specification of the *limits* of the domain of validity of the ago-antagonistic model. Since the homeostatic homeorhetic systems—*if* they are physical or bio-physical—belong to the wider class of material systems in dynamical equilibrium that is described by the thermodynamics of irreversible processes, I shall first mention the *specificities*, inside this more general class, of the sub-class of material systems obeying the ago-antagonistic model (of the intersection between the class of material dynamical systems and the class of systems admitting of the ago-antagonistic model).

Then, taking also into consideration social, or psychological, or economical systems, etc., obeying the ago-antagonistic model, I shall try to distinguish the own contours of the *whole* ago-antagonistic class, with respect to an informal conceptual environment where many extensions might come out to be valid but where one is also exposed to many surreptitious glidings into *false* generalizations.

Finally I shall point out the utilities that I assign to the ago-antagonistic model.

4.1. The Example

My medical studies ended with a period of practical work in a hospital. During this period I was impressed by the importance which the Head of the Department assigned to the problem of what he called the *adrenal-post-pituitary antagonisms.* Schematically, what had been observed was the following. The destruction of the neuro-post-pituitary (deliberately provoked for experimental aims, or due to a pathology) entailed the suppression of a hormone—the antidiuretic hormone—and this produced an insipidus diabetes (emission of 5 to 10 liters of urine per day). If then the anterior pituitary was also destroyed, the insipidus diabetes was observed to *vanish.* This unexpected effect had been shown to be due to the vanishing of another hormone, a diuretic hormone—the cortisone—which the adreno-cortical glands fabricate under the control of the anterior-pituitary. The conclusion drawn from this was that a definite symptom, polyuria, *depended upon a lack of* balance *between two agents and that it was not possible to make it to depend upon only one of these agents.*

Such a reasoning, still unusual at the time, seemed to me worth beeing completed so as to explain also other pathologies and to orient their control. In particular, I was interested by the case of unbalance with *excess* of cortisone *and a still bigger excess of antidiuretic hormone* : an unbalance of this sort appeared *not* to be sensitive to administration of only cortisone. Then, since the treatment dictated by "common sense"—to give *more* (diuretic) cortisone because there was a dominating excess of antidiuretic hormone—usually produced no effect, we administered *also* (I was working with Haguenau) a certain studied amount of antidiuretic hormone, notwithstanding that the organism contained already this in dominating *excess.*

> In this way I was led to take into consideration also an *"agonistic"* dimension of control, i.e., also two simultaneous actions of the two significant factors going both in the *same* sense (of augmentation, in this case).

The model that we constructed entailed progress for both the understanding of the involved pathologic phenomena and the corresponding therapeutic strategy. I quote only the discovery in 1968 of the effects of the antidiuretic hormone as a stimulus for the growth of cultures of malignant cells [11] and the general definition of *bipolar treatments.*

Later I produced other proofs of the efficiency of a bipolar strategy, in particular by explicit comparisons between the effects produced by cortisone alone and those produced by a combination of cortisone plus antidiuretic hormone [12].

These results finally led to a general characterization of what, for the sake of simplicity, I called *ago-antagonistic "systems"*.

4.2. General Mathematical Representation

The essence of the general characterization of ago-antagonistic systems is contained in a mathematical model.

Let S designate an open homeostatic homeorhetic dynamical system. Let (x, y) (two variables) represent a pair of ago-antagonistc features—forces or fluxes—used for the representation of the *inner* organization of S (in the example given above x can represent the effect of the diuretic factors and y the effect of the antidiuretic factors). By *definition*, if the functioning of the system is "physiological" or "normal" the two following conditions, maintained by the homeostatic homeorhetic regulation, are both satisfied:

$$x - y = c_1, \tag{1}$$

with c_1 a constant (usually equal to zero) which characterizes what is called the *antagonistic equilibrium of* x and y, and

$$x + y = c_2, \tag{2}$$

with c_2 another constant (or a variable in relation to time) which characterizes what is called the *agonistic equilibrium of* x and y. The two constants c_1 and c_2, are established as a *norm* extracted from experimental measurements as the mean of the values found for "well-functioning" replicas of S. So, by construction, if these values are satisfied, we observe a stability of the singular point x, y, whatever be the perturbations brought to the system.

Only together can the two conditions (1) and (2) characterize what is called the *normal dynamical equilibrium of* x *and* y maintained by regulation. (For instance, if S represents a budget, x the gains and y the expenses, the antagonistic equilibrium $x-y=c_1$ alone could not permit to distinguish between a rich man and a poor one, while the value c_2 of the agonistic equilibrium $x+y=c_2$ alone contains no information concerning the separation of gains from expenses).

If S exhibits a deviation from the norm, due for instance to a stress, that is found to be bigger than the tolerated mean fluctuation, S might nevertheless, by some altered regulation, *stabilize* in a state with either $x-y \neq c_1$, or $x+y \neq c_2$, or both. In such a case S is said to have reached a *pathological dynamical equilibrium* between x and y: such stabilized disfunctionings are at the origin of the concept of "pathological autonomy".

Consider now a deviation of S from its normal dynamical equilibrium, either in the form of a new, a pathological dynamical equilibrium between x and y, or in a steadily progressive form doomed to destroy the system if it were not stopped. The clinical example reported above shows that a *control* of (x, y) can be achieved by the application upon S—from the outside of the system S—of a conveniently dimensioned ago-antagonistic pair of two actions X and Y of the *same nature* as, respectively, x and y. (One can choose to built a representation where the control pair (X, Y) obeys equations of the same form as those which command the equilibrium of the pair (x, y), but involving different parameter values; this, however, is not a logical necessity). The control is required to either restore to S its normal equilibrium, or at least to insure for S a *stable* new—physiological—critical point with $x + X - y - Y = c_1$, $x + X + y + Y = c_2$ i.e., a new homeostatic homeorhetic regime. The therapeutic evolution of S can be conveniently represented by a system of equations involving the state variables x, y, the control variables X, Y, the first time derivatives of all these variables, the constants characterizing the equilibrium, other constants, and other functions insuring synchronization, not necessary for the existence of limit-cycles, but playing a role in determining their patterns. The *possibility* of the conditions for a solution of this system (their compatibility with other vital conditions) define the prognostic.

More detailed information can be found in the Appendix. Here we add only the following important remark.

Basically the ago-antagonistic approach concerns only *one* couple (x, y) of state-variables and the corresponding couple (X, Y) of control-variables. But a realistic representation of concrete situations usually requires a *network* of such pairs of state-and-control-couples (necessarily involved in the prognostic). And if such more complex representations are achieved it appears that: The whole unbalanced system can be equilibrated at *all* its "levels" by acting with ago-antagonistic control upon only *one* among these; see the Appendix. This seems noteworthy for at least two reasons:

(1) It relativizes the concepts of "hierarchy" and of "autonomy", since any change of "level" in the consideration of an open homeostatic homeorhetic system brings one in presence—again—of the same "fractally" recurring ago-antagonistic logos (in the case of biological systems, for instance, the processes of stimulation-inhibition run like a red thread across all the "levels" and all the sub-systems of an organism).

(2) It explains how a global unbalance can be neutralized by a local intervention: the mathematical model of a network (indicated in the Appendix) represents well the process of transmission of a local control, into the whole system.

4.3. A Metaphor

I close this characterization by a metaphor which, though simplifying, nevertheless yields an immediate intuitive understanding of the mathematical model.

Consider a "flying scaffold" sustained by two ropes the extremities of which are fixed on two motors placed on the roof of the building, the whole being controled by a computer that insures for the scaffold a permanent horizontality as well as localizability at any convenient working-height. This system is obviously constrained by two conditions. An antagonistic condition of horizontality that requires: x (the length of the rope from the right hand-side)$=$ y (the length of the rope from the left hand-side), with $x - y = 0$, and, on the other hand, an agonistic condition also that regulates the localization at a convenient working-height, namely $x + y = 2x = m$, where $m/2$ measures the desired height. This system, however, is not well-controlled if the computer fails. In this respect the metaphor is incomplete, though it could help for a better understanding of the control principles, by pointing out the impossibility to check an antagonistic unbalance by pressing on the higher end of the scaffold, and the possibility to check it by pressing on both ends (the higher and the lower ones!). Nonetheless, being very concrete and well-known, it might be found useful. (I mention that it has inspired a successful work in the domain of the theory of signals).

4.4. On the Limits of the Ago-Antagonistic Model

A physical or bio-physical (biological) ago-antagonistic system falls inside the class of material systems in dynamical equilibrium which, so far, satisfy the general laws formulated in the thermodynamics of irreversible processes. But since a physical or biological *ago-antagonistic* system is by definition homeostatic homeorhetic, it can assume, as mentioned already in the mathematical model, *various* pathological equilibria belonging to a sort of spectrum of possibilities (continuous or discrete). These have also been called states of *pathological homeostasis*, as mentioned already, or *pathological autonomy* also. The choice of these alternative denominations stresses the tendency of an abnormal stable state, to oppose *resistance* to the "therapeutic" attempts at bringing it back to some equilibrium : in a certain sense the system is mimicking its own behaviour in the state of normal dynamical equilibrium where it opposed homeostatic resistance to perturbations stemming from the environment. Now, such additional possibilities of equilibrium are not common to *all* the physical systems in dynamical equilibrium. So, inside the general class of systems concerned by the thermodynamics of irreversible processes, they draw a certain line of contour around the sub-class of material ago-antagonistic systems. This sub-class possesses a high specific interest, from several points of view. I mention in particular the possibility of a central role of the biological ago-antagonistic systems in the *adaptive* selective transitions that lead to the evolution of the species.

But the thermodynamics of irreversible processes cannot be directly applied to systems that transgress the domain of physics-and-biology, as for instance social, or economical, or psychological, or conceptual systems, etc. However many such systems exhibit ago-antagonistic features. But not *all* of them, of course. So it is necessary to specify the criteria which distinguish an ago-antagonistic system of *any* nature among the *quasi general* category of open homeostatic systems. An immediate general answer consists of the requirement of *applicability of the mathematical model* of an ago-antagonistic system. This entails immediately a radical restriction, since obviously not any pair of features of which the verbal designation suggests an opposition, can form a pair satisfying this mathematical model. For instance the pairs designated by the verbally opposite couples (balance, unbalance), or (life, death), or (good, evil), etc., quite obviously stay outside the domain of validity of the mathematical model of ago-antagonism.

> *This is satisfactory: The concept of ago-antagonistic system is not a hold-all.*

In particular I want to stress, even though this is obvious, that the concept of an ago-antagonistic control does not inertially pretend to englobe *any* transcendence, under cover of the fact that in each case it presupposes transcendences, namely both a transcendence of the studied system (since the control acts from the outside of the system) and a qualificatory transcendence (since the qualifications that define one of the two involved features are achieved by transcending the qualifications defining the other one). Transcendence is a *basic* universal feature of conceptualization which ... transcends the concept of ago-antagonistic system, as it also transcends any other given concept or theory.

Nevertheless, let us evoke a special logical configuration: in the case where a model would claim a qualification of "universal", we can say that, even if universal, it will require yet a meta-model [14]—and the agonistic antagonistic model itself would not escape from such a condition. Let us add that the "universal" model cannot give information about its meta-model, except that this last one would be the place for what cannot be modelled by the "universal" model, i.e., creation, innovation and freedom.

4.5. Utilities

I hold that an *explicit and specific* characterization of the ago-antagonistic systems entails quite non trivial and important utilities, of various natures.

I begin with the general conceptual consequences because these spread to all the other ones. When a feature taken into consideration for characterizing the organization of a system is found to run away from the norm, the first and quasi irrepressible intuitive tendency, if one wants to restore the norm, is, for supposedly maximal efficiency, to act "at the point", i.e., directly and *exclusively* on that feature, and namely in the direction that is opposite to the deviation. Since in fact this tendency is in general inefficient, it is clearly important to combat it. But by purely intellectual arguments it is very difficult, if not impossible, to convince of its inadequacy. One needs material and specific proofs in order to impose counter-intuitive ideas. Therefore the clinical examples, their mathematical representation, and the associated metaphor, carry with them an important conceptual weight. They can help to *educate less simplistic intuitions*, able to respond with real efficiency to the so numerous undesired deviations from the norm, of organized systems.

Passing now to the class of biological systems, on the study of which the mathematical ago-antagonistic model has been directly founded, the importance of the model for the general mentality that acts when therapies are devised, seems obvious. This remark can be extended for the search of various *techniques* (not only therapies) concerning complex organized systems, biological *or* physical techniques.

Furthermore the model has *explanatory* power also, in particular, as already remarked, *for theories of evolution*—by pointing out the persistence of agonistic antagonistic couples (catalysis vs. information, gene expression vs. repression, stimulation vs. inhibition...) quite along its successive steps—and namely by the possible transition role of abnormal equilibria; for theories of biological evolution but *also* of psychological or conceptual evolution, and in particular for *learning*. This seems important, possibly fertile in consequences.

But probably by far the most spectacular consequences concern the social and economical systems. There the *paradoxical strategies* entailed by the bi-polar ago-antagonistic model address to a domain of thought still so feebly structured, so sparse, that the ago-antagonostic model—at the same time so largely valid, so definite and so counter-intuitive—might lead to numerous *essential* improvements which otherwise could not be hoped for. I only quote titles, the rest can be immediately inferred: mondialization versus local identities, European institutions versus European nations, socio-economical reglementation versus liberalism, education versus repression, etc. But, of course, for any such application one should first succeed to define *measures* of the involved ago-antagonistic features, if more than only a qualitative orientation is researched: for each genuinely technical application a specific research has to be conducted first. This suggests a whole vast class of non trivial investigations.

5. RETROSPECT

Since I began—twenty five years ago—to examine, for open systems, the specific effects of these combinations of conflict and cooperativity which I called ago-antagonistic pairs, the notion and the study of such pairs has been much developed. It has even suffered generalizations as well as banalizations, in general without scrutinizing the underlying dynamics, for instance via terms like "organizational antagonism" (E. Morin) or simply via slogans, like "for an open, viable, productive conflictuality" (the movement, attractive besides, called "confrontations").

Of course, more or less vaguely, the general contours of the notion have been apprehended and made use of before by many authors. Roger Caillois, for example, has introduced in his anthropological analyses which I would retrospectively call ago-antagonistic considerations, and François Perroux made use of such considerations in his theoretical representations of economic processes. And, as it happens for any new view, for ours also roots can be found still much longer before, as far in fact as one chooses to go back in history. A whole well known current of thought can be now regarded as an anticipation of the ago-antagonistic style of thought. Inside a halo of sufficient vagueness, and avoiding antiquity, one can evoke names like Hegel, Kierkegaard, Nietzsche, Freud, Lacan, Wittgenstein, Lupasco, and Bateson. But the thinker to whom personally I feel very close indeed, as the motto chosen to introduce this work testifies, is Blaise Pascal. So I close with another quotation from his work, intended by him mainly for theologists, but which nowadays can adequately be addressed to any scientist tempted by a "unique" way of thinking, as well as to any other scientist who criticizes the former one by claiming an opposite "unique" view:

> "When one wants to admonish usefully and to show to somebody else that he is wrong, one has to find out from which side he considers the thing, for usually from that side it is true, but the side from which it is wrong must be pointed out to him. He then will accept, because he will see that he was not wrong but only missed seeing the thing from all the sides : one does not take offense at not having seen all, but having been wrong is refused ... "

6. APPENDIX: MATHEMATICAL VERSION OF THE MODEL FOR THE REGULATION OF AGONISTIC-ANTAGONISTIC COUPLES

Then, the following mathematical model is able to modelling the functioning of the numerous agonistic antagonistic couples reported up to now, by simulating alternation between each ot both poles. Control method itself might find applyings in the case of an attempt to check imbalanced epistemological couples. Chaotic dynamics study will allow to illustrate several previously evoked points. Finally, using this model to only epistemological ends seems possible in the special case of ambiguity breaking.

The mathematical model for the regulation of agonistic antagonistic couples (MMRAAC). It is supposedly known [16] Most of our researches on the MMRAAC is firstly concerned with the behaviour of one couple of agents. However a super-MMRAAC (now called an *AA network*) has been built in order to better understand the effects of a combination of couples (balanced or imbalanced)—as it is always the case in the real systems, and also to study some types of dynamics not observable with the elementary model, i.e., a possible chaotic dynamics.

It is necessary to understand the epistemological background, the principles of the model functioning, its meanings and the intended objectives on which the MMRAAC must appear. Let us say here only that it is a *general model*, common to various systems, or the *model of a function* [the functions of balance and growth (or decrease)]. Mathematically, it is formed by non-linear differential equations, with phenomenological and not physical parameters. When functioning is *physiological*, it enables the antagonistic balance ($x = y$ for instance) and/or the agonistic balance ($x + y = m$ for instance) of both variables to be restored after a perturbation. But, in case of a pathological functioning, another different critical point becomes stable ($x \neq y$, $x + y \neq m$) and we have to propose a *control* (cf. later).

The mathematical formalization of the elementary MMRAAC is the following:

$$\begin{aligned}
\dot{x} &= \sum_i k_i(u+r)^i + \sum_i c_i(\nu+s)^i, \\
\dot{y} &= \sum_i k_i'(u+r)^i + \sum_i c_i'(\nu+s)^i, \\
\dot{X} &= \sum_i k_{i+4}(u+r)^i + \sum_i c_{i+4}(\nu+s)^i + \Lambda_1, \qquad (1)\\
\dot{Y} &= \sum_i k_{i+4}'(u+r)^i + \sum_i c_{i+4}'(\nu+s)^i + \Lambda_2,
\end{aligned}$$

$$\Lambda_1 = \sum_j \lambda_j (X - \bar{X})^j, \quad \Lambda_2 \sum_j \lambda'_j (Y - \bar{Y})^j;$$
$$i = 1, 2, 3, \quad j = 1, 2, 3.$$

In these equations, Λ_1 and Λ_2 are some "penalty functions" designed to avoid the "drift" of the limit-cycle of dimension 4 (x, y, X, Y). Furthermore,

$$u(t) = x(t) - y(t) + n + p(t), \quad r(t) = X(t) - Y(t),$$
$$\nu(t) = x(t) + y(t) - m + g(t) + q(t), \quad s(t) = X(t) + Y(t),$$

where $x(t)$ and $y(t)$ are state variables, while $X(t)$ and $Y(t)$ are control variables of the same nature as $x(t)$ and $y(t)$; k_i, c_i, m, n denote constant parameters (usually $n = 0$) or sometimes variable in relation to time; $g(t)$ represents a synchronizer such as $g(t) = A \sin(\omega t + \phi)$, where, in the simulation of circadian rythms in biomedicine, the synchronizer simulates the day-night alternation; finally $p(t)$ simulates an antagonistic input and $q(t)$ an agonistic input.

Principles of agonistic-antagonistic control. Since 1970, we have suggested controlling such a model in the cases where it simulates an imbalance between x and y (antagonistic and/or agonistic) by *adding control equations* (X, Y) *similar, except for the parameter values, to state equations.* The trajectories of the variables $(x + X)$ and $(y + Y)$ may thus be again balanced. Therefore, *an imbalance between two endogenous AA variables has to be controlled by the administration of the same but exogenous variables* (bilateral strategies or bipolar therapies).

An optimization process in relation to a performance index (Davidon-Fletcher-Powell's method) allowed to determine the control parameter values.

Figure 1 shows, on the left, physiological circadian rhythms of cortisol (x) and vasopressin (y) [c.u. = common units; 0.4 c.u. = 7.7 mcg/100 ml of plasmatic cortisol = 1.1 pcg/ml of plasmatic arginine-vasopressin (daily mean values)]. On the right, after a change in parameter field, appearance

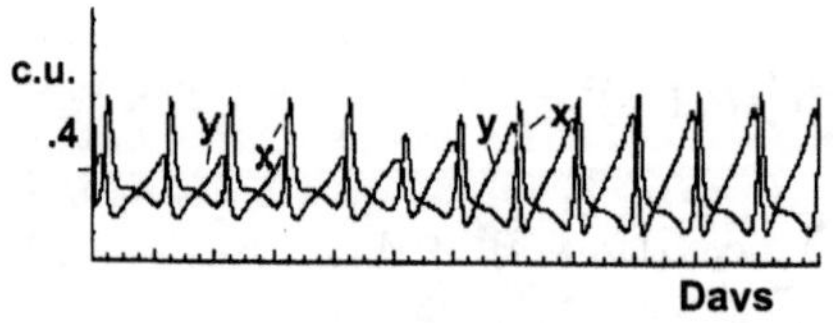

Fig. 9.1.

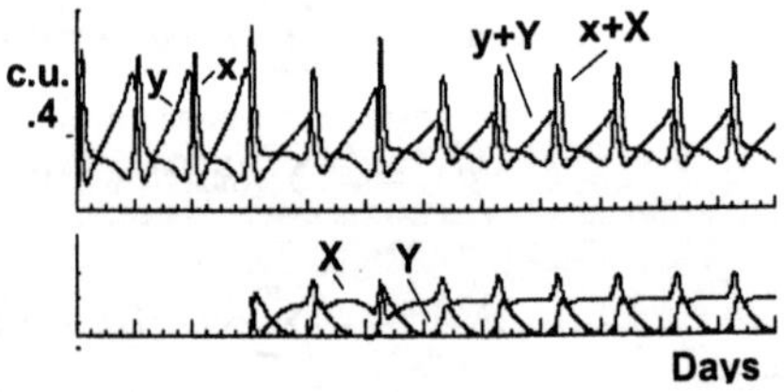

Fig. 9.2.

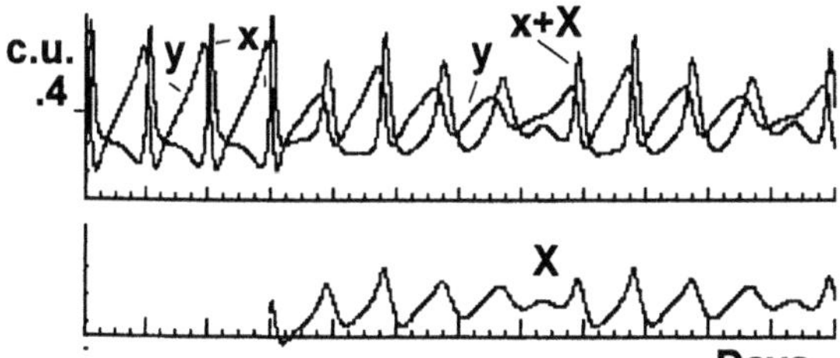

Fig. 9.3.

of pathological rhythms (y sum $>$ x sum). Figure 2 shows a fairly effective control by the four Eqs. (1), due to both hormones $X$and Y administration. Figure 3 shows that not effective control was performed by means of only the first three equations of (1) (with X, similar to the deficient hormone, and without Y). If the administration of the deficient variable alone was *ineffective*, on the contrary, *the administration of the sole variable already in excess may reestablish the balance, leading to the notion of unipolar "paradoxical" therapy* (the agent already in excess has to be added with a convenient dynamics, but the doses are not infinitesimal, in a physiological range; cf. simulation in [16,1992]).

Applications to the study of chaotic dynamics. In the first instance, we proposed a combination of elementary agonistic antagonistic models, the so-called agonistic antagonistic networks (AAN), in order to take into account the effects of the interconnection of balanced and/or unbalanced AA elementary models. Secondly, by using control equations according to the strategy exposed, some simulations [17] allowed us to give an answer (positive) to the following question: *Is it possible to equilibrate the different "levels" of an imbalanced biological system by acting upon one of them only?* [18]. Finally, given that we disposed of at least four state variables, the same formalizing will allow the simulation of chaotic dynamics:

$$\begin{aligned}
\dot{x}_i &= \sum_j k_{ij}(u_i + r_i)^j + \sum_j c_{ij}(\nu_i + s_i)^j \\
&+ \sum_j \bar{k}_{ij} \sum_i (u_i + r_i)^j + \sum_j \bar{c}_{ij} \sum_i (\nu_i + s_i)^j, \\
\dot{y}_i &= \sum_j k'_{ij}(u_i + r_i)^j + \sum_j c'_{ij}(\nu_i + s_i)^j \\
&+ \sum_j \bar{k}'_{ij} \sum_i (u_i + r_i)^j + \sum_j \bar{c}'_{ij} \sum_i (\nu_i + s_i)^j,
\end{aligned}$$

$$
\begin{aligned}
\dot{X}_i &= \sum_j \hat{k}_{ij}(u_i + r_i)^j + \sum_j \hat{c}_{ij}(\nu_i + s_i)^j \\
&+ \sum_j \tilde{k}_{ij} \sum_i (u_i + r_i)^j + \sum_j \tilde{c}_{ij} \sum_i (\nu_i + s_i)^j + \Lambda_1, \\
\dot{Y}_i &= \sum_j \hat{k}'_{ij}(u_i + r_i)^j + \sum_j \hat{c}'_{ij}(\nu_i + s_i)^j \\
&+ \sum_j \tilde{k}'_{ij} \sum_i (u_i + r_i)^j + \sum_j \tilde{c}'_{ij} \sum_i (\nu_i + s_i)^j + \Lambda_2,
\end{aligned} \tag{2}
$$

wherein

$$\Lambda_1 = \sum_j \lambda_{ij}(X_i - \bar{X}_i)^j, \quad \Lambda_2 = \sum_j \lambda'_{ij}(Y_i - \bar{Y}_i)^j, \quad i = 1, 2 \ldots m, j = 1, 2 \ldots n,$$

$$u_i = \sum_i (x_i - y_i), \quad \nu_i = \sum_i (x_i + y_i - m_i), \quad \mathbf{r_i} = \mathbf{X_p} - \mathbf{Y_p}, \quad \mathbf{s_i} = \mathbf{X_p} + \mathbf{Y_p}.$$

By means of these last expressions concerning r_i and s_i, we resorted to a simplified version of AAN models, due to the fact that only *one single* pair of variables (X_p, Y_p), was used for the entire system in an attempt to re-establish global balance. The 2M-dimensional vector (x_i, y_i) and the vector (X_p, Y_p) are defined to be composed of (a) a sum directly corresponding to its counterpart in (1) and (b) an interconnection modulus combining all the balances (or imbalances). Synchronizers g_i were also added in ν_i, their use may be important for the appearance of strange attractors.

Firstly, in the case of a complex imbalanced system including very numerous imbalanced elementary couples, then it has been demonstrated that the use of *only one* couple of control variables could be capable to restore the balance of the *whole* system. Secondly, if the true dynamics of the biological couples would be close to a chaotic dynamics (strange attractors) and not to a periodic or quasi-periodic one, the possibilities of a control seem to remain valid. Indeed, by considering the place of the attractor in the phase-space (and the ergodic theory hypothesizes that this settlement has a meaning), *the control of a pathological strange attractor (SA) would consist in allowing to displace the strange attractor towards a more physiological localization in the phase space (without necessary asking to the disappearance of the CD).* These facts are illustrated by Figs. 4 and 5 simulations: an AAN was formed by two couples, the first corresponded to our data about the pathological pattern (sum of $y_1 >$ sum of x_1) and the second one to an hypothetized couple of agents with the same actions; this last one is also unbalanced (sum of $y_2 >$ sum of x_2); by some changes in the parametric field of this AAN, it

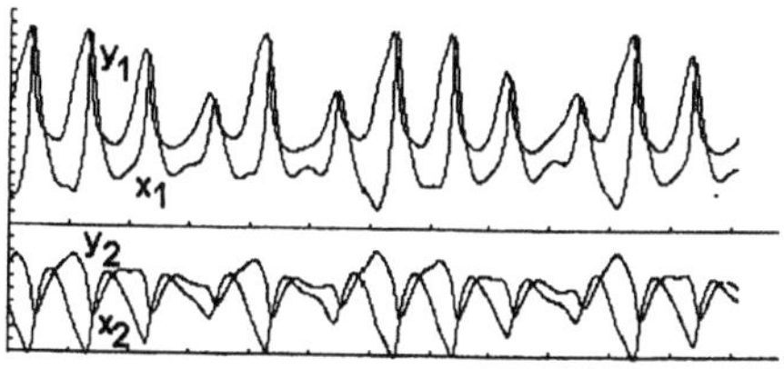

Figure 9.4.

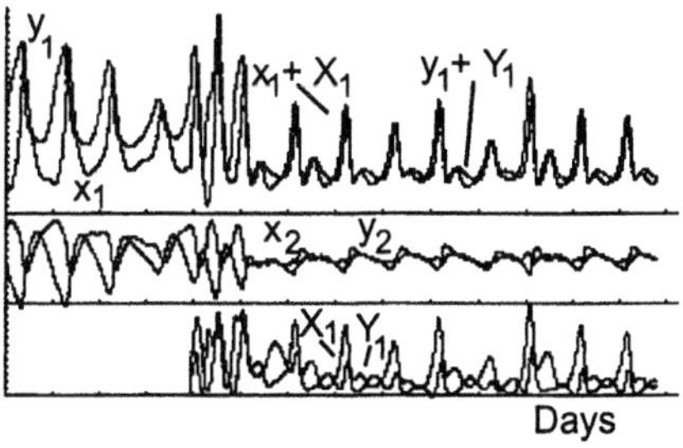

Figure 9.5.

was possible to obtain chaotic dynamics appearing (Fig. 4). Then (Fig. 5), the computer simulation showed that a couple of control agents (X_1, Y_1) was theoretically able to fairly check the unbalance of (x_1, y_1) and *also of this of* (x_2, y_2),*without making system chaotic dynamics disappear* (higher positive Lyapounov's exponent, correlation dimension equal to 4, and other measures demonstrating the chaotic nature of these rhythms). Besides that, such a way to approach chaotic dynamics control might help on a better understanding of the couple *causality* vs. *random* (or topological invariant vs. trajectories impredictability; cf. above).

Ambiguity breaking. Wave/corpuscules duality (*coherence*) may be understood as a virtuality—ghost-like for an outsider in physics—which is "waiting for" a *decoherence* due to some state-preparations. Even if the following remarks cannot concern quantum physics, this type of status seems to be convenient in order to caracterize oscillations between two opposite directions, before a *choice*, i.e., by actualizing one of these virtualities. In Fig. 6, we find at the top a simulation of Epimenides Paradox by means of the MRAAC: its indecidability is well-visualized. At the middle, we see a (virtual) oscillation between two possible decisions ; by using some particularities of the MMRCAA—here inputs $p(t)$ and $q(t)$—finally one of the alternatives is actualized. In the bottom, opposite decision has been reached. It would be interesting to better know the neurophysiological mechanisms which could correspond to cognitive virtual oscillations ending in this type of decision.

Such a 'formalized' metaphor seems to deny some principles of agonistic antagonistic science, i.e., the fact that we must not favour an element of an agonistic antagonistic couple in relation to the other. We find again this science principles if one considers that not any definitive choice has been assumed (the same holidays problem will be set later), as it is also the case,

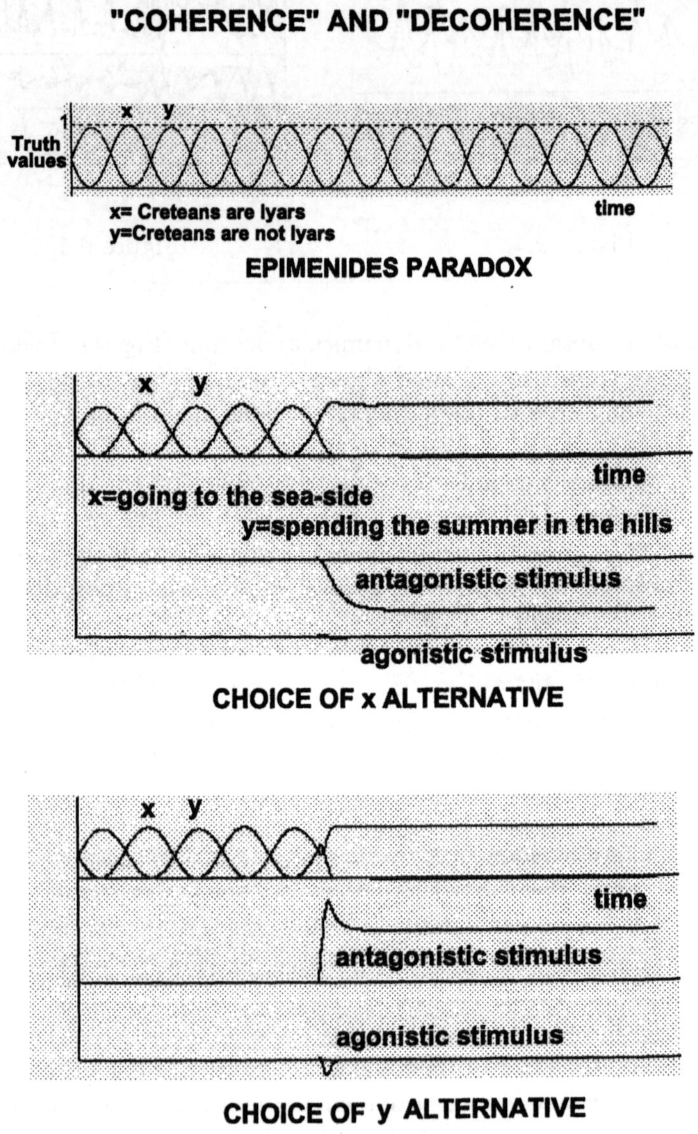
"COHERENCE" AND "DECOHERENCE"
x
y
1
Truth values
x= Creteans are lyars
y=Creteans are not lyars
time
EPIMENIDES PARADOX
x
y
time
x=going to the sea-side
y=spending the summer in the hills
antagonistic stimulus
agonistic stimulus
CHOICE OF x ALTERNATIVE
x
y
time
antagonistic stimulus
agonistic stimulus
CHOICE OF y ALTERNATIVE

Fig. 9.6.

with another time scale, for a sailor which beats up to windward. Therefore, alternation in the "decoherence" orientations would correspond to another type of AA modelling: it would explicitely integrate too the agonism which corresponds to the common goal of opposite managings, actions in general, cognitive processes or behaviours.

REFERENCES AND NOTES

1. "Homeostasis" is convenient for defining a balance control between two agents, but some people do not accept that this term could define the level where such a balance occurs, because *normal* values of this level may vary in relation to some changes in the environment. Then "homeorhesis" corresponds to this level control.
2. H. Haken, *Information and Self-Organization* (Springer, New York, 1988).
3. E. Schwarz, "The labyrinth of the world: Looking for the key, or a meta-model for the emergence of order in nature and the evolution of self-organizing systems toward complexity and autonomy," in *Proceedings, Second European Congress on Systems Science (Prague), Vol. 2,* I. Vajda, ed. (UES, Prague, 1993), pp. 476-485.
4. S. Camazine, J. Sneyd, J. Jenkins, and J. Murray, "A mathematical model of self-organizing pattern formation on the combs of honeybee colonies," *J. Theoret. Biol.* **147**, 553-571 (1990).
5. G. Whitesides, "L'auto-assemblage des matériaux," *Pour la science* **217**, 114-117 (1995).
6. A. C. Martinet, "Planification stratégique," in *Encyclopédie de la Gestion,* P. Joffre and Y. Simon, eds. (Economica, Paris, 1996).
7. P. Bergé, Y. Pomeau, and C. Vidal, *L'ordre dans le chaos* (Paris, Hermann, 1984).
8. A. L. Goldberger, "Cardiac chaos," *Science* **243**, 1419 (1989).
9. In an agonistic antagonistic pair, "agonism" corresponds to some actions in the same sense exterted by every pole of the pair, while "antagonism" corresponds to some actions in opposite senses; cf. Sec. 4.2.
10. The transcending systemic rationality does not preconise the suppression of the nowadays dominant, reductionist, monistic, rationality. Here, these two types of rationality are claimed to constitute together "the" rationality, a much more fertile, double-sided rationality.
11. E. Bernard-Weil and C. Da Lage, "Inhibition by cortisol of the favorable effect of lysine-vasopressin on the growth of HeLa Cell cultures," *Experientia* **24**, 1001 (1968).
12. E. Bernard-Weil, "Evaluation of the addition to corticoids of a growth

factor (vasopressin) in the palliative therapy of malignant brain tumors," *Neurol. Res.* **13**, 94-101 (1991).

13. Such an unbalance may be in relation to a *bipolar stability* in the space phase of the model equations, when the variable's trajectory crosses the separatrix between both attractors, or, seemingly more often, to a change in the parameter field values of these equations, *making unstable the physiological singular point* and provoking the appearance of*a stable pathological singular point.*
14. E. Bernard-Weil, "(Theo)logique et logique moderne," in *Savoir innover en Droit*, D. Bourcier, ed. (Documentation Française, Paris, 1999), pp. 253-287.
15. E. Bernard-Weil, "Does help to decision in biology help to decision in human sciences? An approach to bipolar strategies," *Acta Biotheoretica* **48**, 243-257 (2000).
16. E. Bernard-Weil and J. Mulletin, "A mathematical model for the study of adrenal postpituitary interrelationships: Its use in the correction of an antagonistic imbalance," *Math. Biosc.* **8**, 181-190 (1970). E. Bernard-Weil, "A general model for the simulation of balance, imbalance and control by agonistic antagonistic biological couples," *Mathemat. Modelling* **7**, 1587-1600 (1987); "Agonistic antagonistic systemics: An introduction to bilateral—and paradoxically unilateral—strategies," *Kybernetes* **21**, 47-66 (1992).
17. E. Bernard-Weil, "Homeostasic control of imbalanced strange attractors without asking to a change in chaotic dynamics," in *Advances in Intelligent Computing-IPMU'95*, B. Bouchon-Meunier and R. R. Yager, eds. (Springer, Berlin, 1995), pp. 483-492.
18. P. Nelson and E. Bernard-Weil, "Justification d'un modèle agoantagoniste," *Int. J. Biomed. Comput.* **11**, 145-162 (1980).

Part Three

Further Explorations

10

COMPLEXITY OF THE "BASIC UNIT" OF LANGUAGE: SOME PARALLELS IN PHYSICS AND BIOLOGY[1]

Evelyne Andreewsky

Inserm, Paris, France
andreews@ext.jussieu.fr

It is first brought into evidence that such fundamentally different domains of research as physics, biology, and language sciences, present—with remarkable unity—the same tendency to base their theoretical constructs upon "bricks," from where "information" is carried upwards by a one-way flow of causal determinations. It is then shown that in all these three domains, such theoretical constructs are far to be supported by the present data and conceptions. This brings into evidence the interest of a formalized epistemology, enabling the general mould in which are cast our representations of "reality" to be explicitly reconsidered in keeping with these data and conceptions.

Key words: basic elements, "bricks" of information, dynamical meaning, interdisciplinarity, neurolinguistics, formal transdisciplinar analogies.

1. INTRODUCTION

Many theoretical scientific *approaches* are relying on the concept of "basic elements." In the theoretical constructions of classical physics, "particles" were considered to constitute individualized and stable elements, playing the role of *bricks* of the structure of any material object from the physical world. In present biology, the genes are assigned a similar role; and in language sciences, the meaning of words are treated as essentially stable and definite fundamental entities. Correlatively, in both biology and in language sciences, the genes as well as the words are supposed to contain all the "information" providing the basic data for respectively phylogenetic processes or processes of understanding. In classical physics, assumptions of the same

[1]This work has been supported in part by a DGA grant No. 96070.

nature worked, more or less implicitly, but with a considerable force.

In this paper, we call specific attention to phenomena which invite us to thoroughly revisit this type of view. Indeed, "basic elementary" entities such as particles, genes or words-meaning appear in fact to be involved in a continuous *interplay of complex determinations.* They are therefore far too changing and unstable to be taken as proper "bricks" for any construction. It will in particular appear that in the case of language, a *complex dynamical determination of meaning acts at each word occurrence.* This dynamics obviously entails an egg and chicken problem for the classical view, insofar as in the framework of this view, the meaning of sentences is a function of that of the involved words—which are in turn, given their dynamical determination, function, among other, of the meaning of the sentence at hand...

Similar problems appear in a more or less similar way in physics and in biology – therefore, the need for a very basic epistemological approach, underlying all these disciplines, and able to endow us with explicit methodological rules for theoretical achievement. Such an interdisciplinary framework, emerging from the following consideration, provides a step on the way toward a `formalized epistemology`.

2. BIOLOGY AND THE LANGUAGE SCIENCES

We may illustrated as follows, in the framework of both *biology* and the *language sciences*, how complex phenomena are currently simplistically reduced to a simple one-way "transfer of information" (cf. Stewart and Andreewsky [18]):

In both genetics and psycholinguistics, reference is made to two principal levels of organization: an "elementary" level 1, and a "terminal" level n. The relationship between these two levels is considered to be virtually direct on the ground that, *if all other factors are held constant*, a difference at level 1 corresponds to a difference at level n, which can be expressed by a one-to-one relation between these two levels. Any intermediate level being thus eliminated, the approach claims that level 1 *determines* level n. Given that in both classical genetics and in psycholinguistics, one of these two levels is *not observable*, such a scheme involves audacious jumps and arbitrary simplifications.

Since the genes were not directly observable before the advent of molecular biology, they were just, for Mendelian genetics, theoretical constructs, operationally defined as being the elements, the differences of which cause the observed difference in a given character. Geneticists have even traditionally designated a given gene by a label derived from the observed character. For example, in Drosophila genetics, w (white), v (vermilion), vg (vestigial), etc. These labels obviously reinforce the notion of a *direct one-to-one correspondence* between genes and characters, which induce confusion between the two levels...

The situation is remarkably analogous in the study of language where the level n – that of the "psychological meaning" of words – is not directly observable. Here, the assumption is that each word "possesses" one (or a small set of) definite and well individualized "literal meaning" designated by the lexical item uttered at the observable level 1. This again induce the postulation of a one-one correspondence. In this context also, it is usual to consider that, *if all other factors are held constant*, differences at the observed level 1 reflect differences at the unobserved level n. As in genetics, this reduces *a priori* and arbitrarily the complex relations between these levels, that is, the activity of the dynamic cognitive system which interprets the linguistic stimuli.

The weakness of the approach will be very shortly illustrated here (cf. reference above for more details).

Genetics. Let us consider an example, the notion of "gene for intelligence." Individuals, homozygous for a certain mutant gene, suffer from a condition known as phenylketonuria. In such individuals, a basic enzyme deficiency renders them unable to correctly metabolize the amino-acid tyrosine. A consequence of this is a generalized disturbance in amino acid composition, which in turn perturbs normal brain development, and therefore these individuals are mentally retarded. In short, the gene which codes for the enzyme in question conforms impeccably to the operational criterion for being a "gene for intelligence": *if all other factors are held constant*, a difference in this gene causes a difference in intelligence. Yet it is obvious that *knowledge* of this gene, however detailed, contributes practically nothing to our understanding of the phenomenon of "intelligence." Genes are certainly indispensable components of living organisms, but the biological organization does not simply proceed from a mere transfer of information; it is rather epigenetically structured in a hierarchical succession of *emergent levels* [20].

Language. For *language*, the traditional approaches – based on the notion of an objective, definite informational *content* of words, and on the notion of word understanding as merely an information retrieval process – present the same type of *weakness* brought into evidence above. The meaning of words is traditionally viewed as represented by *data* stored in the mental lexicon,[2] the address of which being provided by the lexical item at hand.

[2]The "mental lexicon" is traditionally viewed as an inner Webster, supposed to contain

Here again, a *one-to-one correspondence* is postulated between lexical stimuli and meanings. This assumption is driven by a naïvely "constructive" view of the meaning of sentences, relying on the raw material to be used – the bricks of the construction, i.e., the meanings of the words occurring in the sentence. These bricks must be available, that is, "retrieved," to undertake the construction. However this framework, though quite widespread, is of no help whatever in understanding essential phenomena such as homographs, polysemy, metaphors, and so on, which cannot be derived from any *content* of individual words. Lakoff and Johnson [9] and Shanon [17], among others, have pointed out the difficulties that these phenomena entail for the classical approaches. For example, consider the following statements:

Dupont[3] studies English.
Durand studies English.
François studies English.

It is clear that these statements are very similar. This naturally brings about a formalization such as:

For $\forall x \subset$ (set of persons able to learn a language), *x studies English,*

means something like: *x learns English,*

However, this is not accurate for the statement:

Chomsky `studies` *English,*

which, at first sight, seems to perfectly fit the formalization. In this case, the statement means something like:

Chomseky `does research` *on English,*

since the famous American scientist is known for his research in linguistics (with English as a specimen of human language).

In this framework, how is one to embed unstable bricks such as "study" into the process of "constructing" meanings? The construction metaphor seems largely inadequate to handle this type of phenomena, which are recurrent.[4]

the psychological "meaning" of lexical items. Such "objective meanings" are internal representations reflecting the objective outside elements (the world being cut up into elements). Let us recall that we are unable to directly observe neither these asserted psychological meanings, nor the mental lexicon. In this framework, as for the quantum mechanical microstates [12], in order to gain any knowledge, we must rely on experiments which, by producing interactions with these supposed objects, lead to observable "marks" that we are able to perceive and to translate in terms of measurable quantities.

[3]Dupont and Durand are typical French family names.

[4]Many other metaphors of the understanding process are available, such as to *understand is to compute*; see Andreewsky E. [2] for some alternative metaphors, for example: *to understand is to sculpt.*

3. PHYSICS AND THE LANGUAGE SCIENCES

In classical physics, as in the language sciences, many researches are rooted (cf. M. Bitbol [4]) in the double presupposition of the permanence of elementary constituents and of a minimal degree of stability of their determination. In the domain of atomic physics, the atoms as well as the electrons, protons, neutrons, do satisfactorily meet these presuppositions, being individualized and stable enough to play indeed the role of "bricks" for the construction of what is called the "physical world".

It is however well known that inside the framework of present-day micro- physics, the way in which physicists represent matter and the world has undertaken a deep metamorphosis. What is studied ceases to be, as it was in atomic physics, the properties of particles, these "*independent existing entities*". In quantum mechanics the main object of study is the *states* of particles, and these transmute into pure configurations [16]. Thereby, bascially, one no longer only finds models of entities that are conceived of as existing *independently* of human cognitive actions. Rather, one finds abstract sets of relations, expressing mathematically the interactions with physical entities produced by human observers, in order to obtain knowledge concerning these entities.

Indeed, as Niels Bohr (cf. M. Bitbol, supra) said concerning quantum physics: "Experimental results are not enough invariant by modifying the experimental sequences, to be free from these *experimental* `contexts`; we cannot deal with them as if they reflected a determination `belonging` to the elementary object on *its own*". Within this framework, the relativity of phenomena to the experimental context may be well understood, as asserted by M. Bitbol [5], not by viewing this context as *adding* some perturbation to the phenomenon (as is considered in the classical approaches), but rather, in a `reverse perspective`, by regarding it as the root, the heart of the whole phenomenon. Neither the states of any elementary object can be conceived to be independent of the experimental situation, nor the properties of this object.

A similar reverse perspective taken for `language` by a few authors (cf. B. Shanon [17]), may be defined in the same terms as in the case of physics, *without changing an iota*, as is shown below.

4. ON "WORD'S PSYCHOLOGICAL MEANING"—SOME EXAMPLES FROM EVERYDAY LIFE

Our theses is that *the process of assigning meaning to words – traditionally viewed as a mere* `information retrieval` *process – is a far cry from classical computer based models. It is, on the contrary, a highly* complex *process, embedding our personal experiences and our skills in finding and testing* hypothesis, *which, in fact,* are *the "*meaning*" that we assign to a particular word, produced in a given space and under given circumstances.* As it is, given its complexity, we do not know how to fully analyze this process...

According to Piaget and Garcia [15], there exists a "functional continuity between everyday cognitive elaboration and scientific ones", because we have to deal with the circumstances under hand, both in everyday life and in scientific research. Cognitive everyday *interpretation* somehow recalls *reasoning* featuring scientific research: both deal with the production and testing (falsifying) of hypothesis and theories, both aim at an explanation. This characterizes abduction, the hermeneutic ability to produce proper hypotheses for the interpretation of circumstances.

Abduction, according to Peirce [14], is an insight to suggesting a theory able to explain a given *unexpected* phenomenon. Such theories "are new suggestions, even if all their elements were already in mind, since we never dream to put these elements together". Returning to language, we may observe that a given utterance – except for a few exceptions – is not likely to provide much information if both its occurrence and its meaning are expected phenomena. This explains why interpreting non-trivial utterances may present some similarities with suggesting theories, as far as it leads (like for theories) to the emergence of something new, i.e., some relevant hypotheses on the meaning of what is being uttered (or written), within given circumstances.

An abductive framework of the sort, emphasizing *hypotheses fitting circumstances* as a *main* cognitive process to interpret language, is obviously required when the meaning of a given sentence, such as the one which follows, is strongly driven by the circumstances under which the sentence is uttered:

It's better to give than to get.

Here, "the" meaning could be one thing... or exactly the opposite, depending on whether the utterance has been made, for instance, by a boxer during a match or by a priest during a sermon! This clearly demonstrates that sentence meaning (at least for this example and for similar cases) is *driven* by circumstances – here, by our models of both the person which is speaking, and the place of action. The meaning of the statement, far from resulting from any function of "objective" components,[5] derives sooner from recurrent abductive hypotheses (triggered by *domains of experience* rooted in our

[5]trying to build meanings out of such "objective" components seems therefore hopeless!

culture, history and experience) on what the speaker is likely to say. Such hypotheses are involved in most processes of understanding, as we will try to point out.

The abductive framework best fits everyday life utterances where "the" meaning of a given word, as in the above example, is strongly driven by the circumstances in which this word occurs. Word-meaning may indeed entirely change from one occurrence in a given circumstance to another one, tied with a different circumstance, and may even be doted of a radically new meaning, differing from *any* earlier one. This is obviously the case of a word occurring in a new "metaphor" [9]. But from one circumstance to another one, a given word in a given utterance may undergo a genuine *metamorphosis* of its meaning. This will be shown hereafter, out of psycholinguistic experimental data. It is also self-evident in examples of everyday life such as the following one, built on a common word such as *water*:

there is no more water!

This statement, read under a title like "Supermarket", might mean that there are no more bottles of mineral water left on the shelves; the lexical item *water* therefore signifies "bottles containing mineral water". Under a title such as "Lost in the Desert" the same item comes to mean something like "drink vital to survival", and the claim *there is no more water* might be a cry of despair. And this would turn into a cry of joy in context with the assertion "he relentlessly baled out his frail boat" in a paper about a Pacific Ocean crossing in a row boat—the meaning of the item *water* becomes "salty ballast, undesirable, slowing down the boat." In a scientific paper commenting on spin-offs from studies on the memory of water, likely to bring about changes in our conceptualization of the liquid, it would become something like "our traditional concept of H_2O". Or it could signify also "heavy water" if used in reference to a dramatic incident in an atomic factory; it could even mean "vegetable soup", if the incident is at home, and the soup about to burn... We are obviously strongly "updating" the meaning of the word "water" for each new considered circumstance, by developing appropriate *hypotheses* fitting the factual and conceptual context.

So the complexity of word-understanding withstands reliance on over-simplified models in terms of storage and retrieval of ready-to-hand, *static* word meanings. Rather, it suggests a *dynamic framework for the continual and recurrent co- production of word and sentence meanings.*

In order to assess the relevance of this framework, we shall now present some **behavioral experiences** falsifying the attempts at considering word understanding in terms of retrieval processes. Indeed, quite a number of experimental data challenge this traditional approach.

5. ON "WORD'S PSYCHOLOGICAL MEANINGS": SOME EXPERIMENTAL DATA

Niels Bohr's statement quoted above concerning quantum physics[6] can be applied such as it stands, without the slightest change, to also experimental data concerning the meaning of words.

5.1. The "Structure" of the Mental Lexicon

The strong contribution of the experimental context will be illustrated by the following experience, founded on the supposition of a `mental lexicon`,[2] taken *to house* the meaning of words, or, in other words, the *bricks of signification.*

In experimental psycholinguistics, many experiments have explored the so- called *semantic distance* between words "stored in the mental lexicon", which is granted to define the lexicon's "structure". In these experiments, called `lexical decision tasks`, subjects are required to "decide", for each item displayed on a screen whether it is a word (this is for instance the case for *apple*)—or not (as for instance *mirpe*—which is a called a "non-word"), that is whether or not the item belongs to the language. This task, like in the quantum mechanical experiments, if it is repeated on enough subjects, provides the observer with an observable "mark" that can measure a quantity, namely the "reaction time," defined as the *mean-time* taken by the considered set of subjects for "deciding" if a given item displayed on a screen does or not belong to the language. This quantity, in its turn, is taken to be correlated with a non-observable quantity, that is the "semantic distance" between the words stored in the lexicon. At this point comes in a remarkable fact: The subject's "reaction time" for a word **b**, *displayed on the screen just after* **a**, is shorter if **a** and **b** are *semantically* related (if **b** = *table*, for instance, this time will be shorter when **a** = *chair* than when **a** = *zoo*. Here, *chair* is said to be "priming" *table*). So the procedure offers indeed a non-trivial way for measuring a "semantic distance" between words.

Unfortunately however, such objective and easy-to-measure data are not at all stable; they depend upon context – namely, upon the list of words displayed for subjects before ***a*** and ***b*** (cf. Parisse and Andreewsky [13]). For instance, indeed, *the type of relations* between successive words prevalent in the list (a parameter not usually noticed by experimenters) has a great impact on the subject's reaction time. In the framework of an experiment where these relations are, for instance, mostly *syntagmatic* (such as in: apple-

[6] "experimental results are not enough invariant by modifying the experimental sequences, to be free from these `experimental contexts`; we cannot deal with them as if they reflected a determination `belonging` to the elementary object on `its own`".

orange-grapefruit), the "reaction time" for **b**, displayed just after **a**, will be short *if, and only if,* **a** and **b** are also *syntagmatically* related (for example **a** = table, **b** = chair). If they are paradigmatically related (such as for **a** = table, **b** = dinner), it will be much longer – and *vice versa.* The semantic "distance" between a given pair of words, as measured by the reaction time, is therefore driven by all the other words the experiment makes use of! The so-called "structure" of the lexicon, a function of these distances, resembles a kaleidoscope: it undergoes an abrupt metamorphosis at each passage from a given experimental sequence to another one!

It turns out from the above experiment that – like in quantum physics – behavioral data such as the reaction time are *far from being invariant when the experimental sequences are modified.* Neither `lexical items` nor microstates seem to be endowed with predefined determinations *of their own.*

5.2. "Subliminal" Experiments

Let us recall once more that the classical approach to *sentence understanding* is worked out in terms of a *construction* achieved with basic bricks: the meaning of each word of the sentence. It requires to first *identify* the "bricks" for the construction. This identification is assumed to be an *all-or-none process.* Lexical stimuli are not supposed to "transport" their meanings (despite the frequent use of a metaphor such as *the idea "conveyed" in this sentence*). The role assigned to the lexical stimuli in the process of comprehension is, classically, to define the "address" in the mental *lexicon* where the "representation" of the corresponding meaning is to be found. Within this theoretical framework, the understanding of a lexical stimulus (a word) is supposed to be achieve as follows: The stimulus is understood if one *accedes* to the representation of its meaning in the lexicon; otherwise, it is *not* understood.

We will return to the lexical decision tasks and the "priming" phenomenon, i.e., the speed up of the reaction time for a given word, when it is displayed after a word semantically related to it. This phenomenon manifests the remarkable feature of occurring even when the presentation of the "priming" item is *subliminal* (so short that subjects cannot identify this prime, or even notice it). This means that a lexical item presented in this way, though *not identified*, can nevertheless determine an *effect linked to its meaning.* This phenomenon, if it is analyzed accordingly to the classical framework, in terms of a mental lexicon and an "all-or-none" accessing to lexical meanings, obviously leads to a paradox: *one has to admit that the subjects* `have-and-have-not` *had access to the meaning of the prime...*

5.3. "Odd-Word-out-Test"

"Aphasia" is the generic name for neuropsychological language disturbances. It does not refer to an impairment of all language performances. On the contrary, one may observe specific behavioral features, suggesting certain properties of the normal psycholinguistic mechanisms.[7] *Alexia* is a specific form of this disease, where the main language problems occur in reading.

The following behavioral experiment with alexic patients give rise to contradictions similar to the subliminal one. "Global" alexia concerns patients who seem totally incapable of processing any written material. This is assessed by ensuring that patients are unable either to recognize any isolated letter or to identify a written word. Written lists of words were presented to such patients, with one odd-word such as "hat" in:

cat, dog, pig, hat, cow.

Patients were required to detect the odd word. Their first claim was that such a task was impossible, given the drastic difficulties the patients had with written material. Nevertheless, most patients were perfectly able to point to the odd-word, i.e., *hat*, but without being able to also *explain* their choice. Like in the case of subliminal experiments, such a phenomenon, in the framework of an "all-or-none" *access* to lexical meanings, implies that alexic patients `have-and-have-not` *"accessed" the meanings of the words in the list.*

6. WORD UNDERSTANDING PROCESSES: EMERGENCE VS. INFORMATION RETRIEVAL?

Neither the above contradictions nor our former everyday examples are in agreement with the traditional models of word-understanding in terms of information retrieval processes. The experimental data suggest, rather, a *creative* process involving successively *emergent levels* (as in the case of genetics).

Indeed, in the first place, this alternative view provides room for the *dynamics* of a recurrent elaboration of meaning, coping with the contexts and with the involved domains of experience. *This dynamics simply erases*

[7] The general conceptual framework allowing to take aphasic behaviors as data to study normal mechanisms relies on the assumption that these pathological behaviors reflect the action of spared sub- mechanisms. Aphasia provides, therefore, a kind of "pseudo-experimentation" on the brain which may result in behavioral data accounting for mechanisms usually embedded in the redundant psycholinguistic system. Such data, lacking with normal subjects, are essential pieces to complete the puzzle of language understanding mechanisms.

the problem of the "precedence" of word meaning vs. sentence meaning, given the joint emergence of both.

It also provides the ability to explain the above contradictions without *ad hoc* hypotheses. The development of an emergence requires a *starting* point, and the emergence is likely, in specific cases, to be restricted to this point (that is to some `first steps`). According to F. Gonseth [7], and his "Open Methodology", the production of abductive hypotheses (either in scientific domains or elsewhere), requires four main steps. The first (the one we will make use of here) is the `delimitation of a referential` consisting of the fuzzy space of knowledge and experience relevant for the considered case. As far as:

(i) in certain conditions, written word understanding turns out to be *restricted* to exclusively such a delimitation;
(ii) subliminal presentation and alexic problems belong to these conditions;

the mere delimitation of a referential—that is the first step of word understanding—is directly able to explain the above contradictions. Indeed:

The alexic behavior for the "odd word out test" (cf. supra) could be accounted for with such a delimitation. If the understanding processes of the patients are indeed restricted to this initial cognitive delimitation, this, on the one hand, suffices for lead to the detection of the odd word (defined as the item of the list that is endowed with an odd referential), and on the other hand, this is not sufficient for also understanding written items. The fuzzy referential of a given item somehow plays, with respect to the meaning of this item, a role similar to that played by a block of marble (see note 4) chosen by Michelangelo, with respect to the masterpiece which he has drawn out of that block! In short, if we assume that global alexic patients can only identify the referential of written words, we become able to explain that these patients are altogether *totally unable to understand written words, and very smart at detecting the odd-word* in a list of written items.

The same type of explanation may be provided for subliminal experiments. If it is assumed that a subliminal presentation of a given lexical item entails an understanding of this item restricted to the implicit cognitive delimitation of its fuzzy referential, then, given the closeness of the referential of "semantically close" words, such as a prime and its target, a mere "remanence" of the first referential corresponding to the prime is directly providing a large part of the referential corresponding to the target. This explains the observed *speeding up* of the reaction time for "deciding" whether the target is, or not, a word from language, no matter whether yes or not the subjects have noticed the prime.

We shall now present similar functional hypotheses for the case of `sentence understanding` enabling to cope with other specific experimental data.

7. A PUZZLING BEHAVIOR: DEEP DYSLEXIC PATIENTS (*clues concerning the "emergence of sentence's meaning"*)

The meaning of a sentence is anything but a set of *isolated concepts.* As argued above, it is rooted in:

(i) the *entire cognitive domain of experience* triggered for a given subject by the words belonging to the sentence,
(ii) the whole *context* of this sentence, and
(iii) its *structure.*

The following experimental data may provide some insight into the relations between these three parameters.

Let us first consider a very simple example. In the framework of *natural language processing* in *Artificial Intelligence* (A.I.), with attempts at applications which are somehow *connected* with meaning, it appears quite clearly that certain aspects of the `structure` of the sentence have to be taken into account even for the most elementary attempt to deal with meaning. For example, consider the two following sentences, made up of identical items:

I can leave a will vs. *I will leave a can*

It is obviously impossible for an A.I. system to *decide* whether the two sentences have the same meaning, or different meanings, without—at least—taking into account the *grammatical class* of the ambiguous items "can" and "will" at each occurrence. In other words, given the huge number of grammatical ambiguities currently occurring in natural language, *it is impossible for a system, be it automatic or cognitive, to perform even such a merely "semantic decision" as to compare two strings of identical lexical items, without, at the very least, a* `syntactical disambiguation` *of these items.*

With this remark in mind, the behavior of a particular kind of alexic patient, the *deep dyslexics*, becomes very puzzling. If we take for granted the traditional description of deep dyslexia[8] [6], where patients demonstrate altogether a seemingly acceptable understanding of every day current written language, but on the other hand, are known as having *lost their syntax*, we are faced with an obviously strong contradiction with the evidence reported in the note 8: indeed, there is here a system (the deep dyslexic patients reading system) which, in spite of the fact that it is unable to deal with syntax – as assessed by several psychological tests – nevertheless demonstrates a reasonable understanding of written language...

Puzzled by such a contradiction, one may further investigate the syntactical abilities of deep dyslexic (or agrammatics) patients [3]. A specific test was designed for these patients who were neither able to read aloud nor understand function words, be they isolated or in a sentence. This test is rooted on *grammatical ambiguities* such as, for instance, the French word "car" ("*bus*", a noun—vs. "*because*", a conjunction), and the word "or" ("*gold*", a noun—vs. "*now*", a conjunction).

---

[8]The reading behavior of deep-dyslexic patients seems rather difficult to interpret in terms of the classical reading mechanisms approaches ($b, a \Rightarrow /ba/$). These patients are indeed neither able to identify any *isolated letter*, nor to read *non-words* or *function words* aloud (they cannot use grapheme to phoneme conversion rules). Nevertheless, they are able to read out loud certain words – as far as they belong to such specific*grammatical categories* as nouns or verbs – but often produce errors that are quite peculiar. They utter words that bear little morphological resemblance to the target but are semantically close to it (e.g., they utter *church* for the written word *cathedral*), thus demonstrating some *understanding* of these items. But sentence understanding is granted to be agrammatical, since deep dyslexic patients do not rely on syntactical cues to interpret sentences: they are well known as having "lost" their syntax. Evidence for such a "syntactical loss" is rooted on a set of well described behavioral data. Namely, patients always fail to properly understand sentences the meaning of which relies on function words, as specifically assessed by a number of different tests. For instance, matching the sentence:

the diamond is below the club

with either Fig. 1 or Fig. 2 (a match which primarily relies on the function word "below") is achieved by deep dyslexic patients at only chance level.

Fig. 1	Fig. 2
◇	♣
♣	◇

On the other hand, deep dyslexic patients demonstrate a rather acceptable understanding of sentences and texts (as assessed out of a set of neuropsychological classical tests). This understanding appears to be mainly founded in a certain peculiar sort of *expectation and knowledge* which is only a part of the whole understanding processes. For instance the deep dyslexic patients are always "normalizing" contra-intuitive sentences (such as "the patient treats the doctor").

French sentences such as:

le car ralentit car le moteur chauffe
(the bus slows down because the engine overheats)

including an item switching from content word to function word, depending on its `position` in the sentence (here, such ambiguity is provided by the item "*car*") were presented to patients. And it comes out that *the item "car" is always read aloud by the patients when occurring as a noun (i.e., "bus") and never when the same item occurred as a conjunction (i.e., "because").* Such a clear cut behavior establishes beyond any doubt that patients which nowadays are labeled "agrammatics", far from having lost their syntax, do systematically perform an (implicit) *syntactical disambiguation... The puzzling contradiction involved in deep dyslexic understanding, supposedly without syntactical processing, can therefore be over-ruled.*

We might consider this experiment as a kind of `pseudo-experimentation` on our normal understanding processes, enabling us to point to some usually *hidden* properties of these processes, such as an *isolated* syntactical processing devoted to—specifically—grammatical disambiguation (which is not in keeping with the classical approaches.).

We will try to suggest the role of such a specific process in the very starting of (normal) language understanding:

In the framework of the traditional approaches for explaining language understanding, deep dyslexic behavior (cf. note 8) appears as a set of disparate phenomena, and no issue has been raised in order to explain why these phenomena do occur *together* for deep dyslexic patients.

Yet if we assume that sentence understanding—like single words understanding—is providing by the development of successively `emergent levels`,[9] a self-consistent structure comes into light, which organizes in one whole the seemingly disparate set of phenomena manifested by deep dyslexic patients. According to this assumption, all these phenomena reflect the very first steps of understanding, namely *the* `delimitation` *of the sentence referential.*

There is a formal procedure, overused in the field of automatic documentation, which may shed more light on the assumption of such a delimitation [1], namely the `indexing procedure` which points to the key-words of a given text in a set of texts (that is a data base). These key-words enable to provide a very first "glance" into what this text is about (which amounts

[9]Models of understanding in terms of emergence have been developed at large in the framework of A.I. (i.e., Winograd and Flores [21]). These models lead to thoroughly revisit the classical approaches according to which the meaning of a given sentence is a mere combination (function) of the meaning of the words it includes.

to delimiting a "referential" of this text), and key-words are essential for automatic documentation precisely in consequence of this.

The deep dyslexics' reading behavior shares striking similarities with the computerized indexing procedures. Indeed:

(i) "Indexing" specifically points to `content words`. *Deep dyslexic patients understand and read aloud only content words*;

(ii) Indexing requires a `syntactical disambiguation` to detect these content words (as suggested earlier out of the grammatical ambiguities "can" and "will"). *Deep dyslexic patients, traditionally considered as having lost their syntax, are nevertheless able to operate syntactical disambiguation (e.g., the above experiment with the French grammatical ambiguity "car").*

Such a disambiguation should occur at the first steps of understanding, given that:

- On the one hand, a given written sentence directly provide, before starting understanding processes, the *positional* data enabling to point to its content-words.
- On the other, the delimitation of the referential of a given sentence, on the basis of the referential of the content words it includes, should trigger the abductive processing [7] of which we assume the sentence-understanding to consist.

(iii) Indexing must deal with `synonymy` in order to link each content-word with a key-word itemized in a "thesaurus". *Deep dyslexic patients make errors such as producing a synonym instead of the item to be read, when reading aloud.*

(iv) Indexing must "`normalize`" words in order to fit dictionary or thesaurus entries: if a word is a verb, it must be infinitive; if it is an adjective, it must be masculine, singular, etc. *Deep dyslexic patients do achieve such normalization: they read aloud* `verbs without their flexion`, *and in the case of a language such as the Russian one [11],* `nouns without declensions`, *etc.*

So as asserted in the above framework, the behavior of deep dyslexics no longer appears as a set of disparate phenomena; the co-occurrence of all these phenomena can be viewed as reflecting a pertinent coherence, namely that of the *indexing procedures*, which is also that of the *modeling* of the `first steps` of understanding as the identification of a referential.

All in all, the theoretical approach of understanding as a process rooted in a succession of `emergent levels` [21], enables to avoid many crucial theoretical problems which are puzzling classical theories. It furthermore provides room for abduction, the cognitive ability to generate hypotheses on meaning, in the framework of the emergence. Finally, as illustrated here, it quite satisfactorily explains the striking syndrome of deep dyslexia. This is more than sufficient, we think, for seriously taking this approach into consideration.

8. CONCLUSION

Complex phenomena as those evoked in the present paper prevent from relying on *ready-made concepts* such as "basic bricks" for the intelligibility of our cognitive abilities, and of our world. This intelligibility should sooner be conceived to emerge from cross-disciplinary approaches to *complexity*, weaving together the specific attempts and hypotheses into a broad *formalized epistemology.*

To conclude, let us stress in another way the complexity of meaning, with a quotation of the great psychologist L. S. Vygotsky, expressing this complexity with strength and poetry: *A word endowed with meaning is a dewdrop reflecting the sun, a microcosm of human consciousness.*

REFERENCES

1. E. Andreewsky, ed., *Systémique et Cognition* (Paris, Dunod, 1991).
2. E. Andreewsky, "Quelques questions inhérentes à la compréhension du langage," in J.-L. Le Moigne, ed., *Intelligence des Mécanismes, Mécanismes de l'Intelligence* (Fayard, Paris, 1986).
3. E. Andreewsky and X. Seron, "Implicit processing of grammatical rules in a classical case of agrammatism," *Cortex* **XI**, 379-390 (1975).
4. M. Bitbol, *Mécanique Quantique; Une Introduction Philosophique* (Flammarion, Paris, 1996).
5. M. Bitbol, *Physique et Philosophie de l'Esprit* (Flammarion, Paris, 2000).
6. M. Coltheart, K. Patterson, and J. Marshall, eds., *Deep Dyslexia* (Routledge & Kegan Paul, London, 1980).
7. F. Gonseth, "Le référentiel, univers obligé de médiatisation," *Dialectica* (1975).
8. L. Henderson, "Word recognition: A tutorial review," in M. Caltheart, ed., *Attention and Performance XII: The Psychology of Reading* (1987), pp. 171-200.

9. G. Lakoff and M. Johnson, *Metaphors We Live By* (University of Chicago Press, Chicago, 1980).
10. J.-L. Le Moigne, "Sur les fondements épistémologiques de la science de la Cognition," in E. Andreewsky, ed., *Systémique et Cognition* (Dunod, Paris, 1991).
11. A. R. Luria, *Basic Problems of Neurolinguistics* (Mouton, The Hague, 1976).
12. M. Mugur-Schächter, "Les leçons de la mécanique quantique – vers une épistémologie formelle," *Débat* **94**, 169-192 (1997).
13. C. Parisse and E. Andreewsky, "Dynamique lexicale et cognition," *MCX/APC, la Dialectique du Savoir et du Dire* (Poitiers, 1997).
14. Ch. S. Peirce, *Le Raisonnement et la Logique des Choses* (Lectures of 1898; Cerf, Paris, 1995).
15. J. Piaget and R. Garcia, *Psychogenèse et Histoire des Sciences* (Flammarion, Paris, 1983).
16. E. Schrödinger, *L'esprit et la Matière* (Seuil, Paris, 1990).
17. B. Shanon, *The Presentational and the Representational* (Harvester Wheatsheaf, New-York, 1993).
18. J. Stewart and E. Andreewsky, "From information to autonomy: Analogies between biology and the language sciences," *Kybernetes* **21** (5), 15-32 (1992).
19. L. S. Vygotsky, *Pensée et Langage*, trans. by F. Sève (Paris, Messidor, 1985).
20. C. H. Waddington, *Principles of Embryology* (Allen & Unwin, London, 1956).
21. T. Winograd and F. Flores, *Understanding Computers and Cognition: A New Foundation for Design* (Ablex, Norwood, NJ, 1986).

11

ABOUT THE EMERGENCE OF INVARIANCES IN PHYSICS: FROM "SUBSTANTIAL" CONSERVATION TO FORMAL INVARIANCE

Francis Bailly

Laboratoire de Physique du Solide de Bellevue
CNRS, 1 Place Aristide Briand
92195 Meudon Cedex, France

The role and place of relativity principles and related symmetries in physics are emphasized. It is proposed to organize part of a formalized epistemology around such invariance principles.

Key words: invariances, symmetry, formal structures, Noether's theorem.

1. INTRODUCTION

The question of the problematic relationships between the empirical variability of the phenomena and the invariance of some of their properties has deeply influenced cultures and civilizations. Some of them have organized their own signification systems through this very contrast, like many Far Eastern *Weltanschauungen*—particularly in China—which fundamentally associate reality and permanence, illusion and change [1,2]. In Western countries the arising of History as a discipline, with Herodote, at the same time as Philosophy and Mathematics, broke off with such radical views. Nevertheless, considering either the role of Platonic forms as invariant models for further realizations or Aristotelian contrasts between a sublunary world (corruptible and changeable) and a celestial permanence, this duality between variability and invariance remains a central theme in western culture, playing a structural role in the explanation and the comprehension of the world and even in governing the action of man in the world. This fact entails one of the main conditions for science to rise and develop, i.e., the distinction between subject and object for constructing knowledge. It seems thus that this very duality constitutes a deep cognitive invariant of human categorization itself, which simultaneously needs keeping stability as a reference and treating effectively all the empirical modifications it encounters.

Medieval theology itself took into account such categories (for instance through the opposition between the spiritual and the temporal or the eternal and the perishable), but modern science at its beginning did not reject this concern, although it progressively freed itself from this conceptual-discursive way of theorization to the benefit of a mathematical objectivation of phenomena. In turn, it deeply modified its problematics by ceasing to oppose eternal supranatural essences to natural effective changes and, instead, extracting from the observed changes what had nevertheless to be conserved. In a first stage it mainly concerned substantial quantities (like mass or electrical charge), but later on – and quite rapidly – it extended to much more abstract structures, which only mathematical formalization was able to reveal and to treat.

From that time, physics and chemistry appeared (except the particular studies they ensured by providing new concepts and discovering new scientific laws) as more and more abstract enterprises in constructing theoretical invariant quantities supposed to be able to explain in a very general way the properties experimentally shown by the phenomena. Even more: as soon as these invariant structures were constructed and organized in fundamental understanding schemes, they appeared as prescriptive tools for the physical order, to such an extent as to acquire predictive forces and become guides for experimental as well as for theoretical research (cf. the recent discovering of the "top" quark, predicted by the standard model). Nowadays these formal invariant structures seem to constitute the very cognitive skeleton of physical theories, and they induce not only disciplinary research but also epistemological reflection. It follows that the way these invariant structures have been elaborated and developed, the manner they operate in constructing scientific knowledge, appear as essential elements to build up a formal plausible epistemology (in the sense in which J. Petitot—to whom I have been indebted for many years in his Seminar on these topics—uses this term [3,4]).

In the following, we briefly recall some historical and theoretical aspects of the rise and of the role of some fundamental invariant structures in physics and chemistry. Starting with the discovering of many "substantial" conservation laws related to matter, energy, electrical charge, etc., we consider more and more abstract properties and more and more formal structures, up to those encountered in quantum physics or in physics of critical phenomena (part 1). In part 2 we are mainly interested in the generalization and rigorous mathematical formalization of these empirical and theoretical features. To this end, we describe the theoretical and conceptual situation resulting from Noether's theorem (for the case of continuous symmetries) and from the TCP theorem (for the case of discrete symmetries in the frame of quantum fields theories). Along the way, we try to distinguish carefully what concerns purely mathematical properties and what depends on physical principles. To conclude (part 3), we recall the essential role played in this domain by mathematical group theory and we discuss the status of the physical principles associated with these symmetry groups. Finally, we make use of these examples in an attempt to determine some fundamental features, which can reveal themselves as important for constructing a formalized epistemology.

2. ON THE APPEARANCE AND TREATMENT OF SOME FUNDAMENTAL INVARIANT QUANTITIES. "SUBSTANTIAL" CONSERVATION

From the very beginning of the mathematization of physics, determining conserved quantities constituted an important theoretical goal: remember, for instance, the controversies between Descartes and Leibnitz about the conservation of kinetic momentum and of kinetic energy in mechanics. But chemistry and the theory of chemical reactions proved the fundamental basis for considering the conservation of some quantities as scientific principles: the earliest conservation rules first regarded the very nature of the substances (despite the various reactions in which they were involved – which would be explained later thanks to the atomic theory), but also the quantity of matter (nothing gets lost, nothing is created, all things are mutually transformed!). Such principles were then used as reliable guides for research, instead of being considered as "accidental" features or, on the contrary, as metaphysical rules.

This conceptual and theoretical frame was considerably enlarged with the theory of thermodynamics and the role played by energy (first principle). At the same time, and from a complementary point of view, so to speak, there appeared the question of finding pertinent criteria regarding the evolution of thermodynamic systems, i.e., expressing and quantifying what, on the contrary, is not conserved (second principle: spontaneous increase of entropy in isolated systems, minimization of the free energy). Such a duality between conserved quantity and evolution criteria appeared again in mechanics itself: conservation of the Hamiltonian on one side, extremalization of the Lagrangian on the other side. But the meaning of this latter situation is very different since the terms of the corresponding duality reveal themselves theoretically equivalent: they lead to the same equations of motion and even if the conceptual contents are different, the relevant facts and the governing laws are the same. Moreover, on another level of interpretation, we may note that this Lagrangian minimum principle denotes some underlying conservation: the actual trajectory has to be the critical one under weak perturbations (minimal length in the relevant space, expressing the geodesic principle).

The Hamiltonian invariance formalizes the principle of energy conservation in a dynamic way. On a more abstract level, the resulting commutativity of Poisson's brackets translates mathematically the structural effects of this physical principle. Reciprocally, the autonomy and generativity of these mathematical structures tend to prescribe – and not only to describe – the physically possible phenomena. In this way the scientific intelligibility goes a new step further: the ruling formal invariance replaces the substantial conservation (which remained bound to the object under study) and the basis is established for further progress, where symmetry principles will govern the possible physics and determine its actualization (cf. below).

But first, let us stress the fact that conservation principles appeared also in many other disciplinary sectors besides chemistry, thermodynamics, or mechanics. Electricity is partially based on the principle of the electrical charge conservation and its elementary unit, the electronic charge is a fundamental invariant. At its turn, quantum physics, through quantum numbers, introduces many other types of invariance, much more abstract and less intuitive. Their treatment resorts to abstract group theory and refers to very general frames corresponding to relativity principles associated with them. In the case of continuous symmetries, these treatments are principally based on the generalization and application of Noether's theorem. But other invariances (like TCP invariance considered below) are associated with discrete symmetries.

From a conceptual point of view, we have to emphasize that the essential game is henceforth played through the relationships between the normative formal mathematical structures and the corresponding physical principles: from the formalisms ("syntax"), physics, in a first step, conserves only what can receive a specific semantic content (a physical meaning); reciprocally it can be assumed that to each fundamental physical conservation principle there corresponds some abstract symmetries. In a second step what prevails is the proper generativity of these abstract symmetries regarding the physical situations they organize as well as for the research they suggest and guide.

This trend is well illustrated by the methods used in quantum field theories, and also by the scale invariance analyses (dilatation symmetry), in the most discontinuous cases of change considered by physics, i.e., critical phenomena. Indeed, renormalization group analyses of these phenomena lead to the characterization of universality classes which free the models from their specific features and from the details of their properties: they are transformed into generic representatives of classes, determined by such general aspects as the dimensions of the relevant spaces where they are defined, or of their order parameters, or even as the topological properties (critical points, singularities) of their characteristic functions. In dynamic systems, these analyses include and explain the appearing of numerical invariant quantities associated to scaling properties and specifying the nature of the mappings (for instance Feigenbaum's invariant values for discrete unimodal mappings).

Thus, it seems clear that a formalized epistemology has to take into account such theoretical situations and their conceptual and cognitive implications. Let us provide some more details in order to pursue in the same manner.

3. RELATIVITY PRINCIPLES AND SYMMETRY GROUPS. "FORMAL" INVARIANT STRUCTURES

It is well known that the main invariant quantities of physics are tightly linked with relativity principles as connected with underlying symmetries in the relevant spaces (usual spacetime, but also various functional spaces). These symmetries are themselves formalized through group theory which, for these reasons, has acquired a fundamental status in the epistemology of physics [5]. In this part we study more precisely the two main cases mentioned above: Noether's theorem (involving continuous symmetry) and the TCP theorem (involving discrete symmetries).

3.1. Continuous Symmetries. Noether's Theorem and its Applications

Noether's theorem [6] was first established as an important mathematical result inside the general frame of research of invariant structures (and conserved quantities) related to variational problems. Of course, in the perspective of physical applications, the paradigm of such problems was the expression of Hamilton's principle (last action principle) for dynamical systems, enabling us to derive the equations of movement.

In the Appendix, we present an outline of the demonstration of the theorem (extended by Klein [7] to the case of the Einsteinian gravitation theory) following Hill [8]. In this part we only try to determine its presuppositions and to extract its most significant elements from a technical as well as from a conceptual point of view.

What does this theorem deal with, and what are the constraints to wich it responds? As stressed above, the original aim was clearly to exhibit and investigate invariant mathematical quantities and structures related to symmetry transformations associated with dynamical systems defined in the frame of variational approaches. Historically, at the beginning of the twentieth century, these investigations were inscribed in a general attempt to study the most general conserved quantities and invariant mathematical structures belonging to systems undergoing various transformations. Of course, from a physical point of view, the theorem corresponded to the aim of determining what does not change when many things may change, and in particular to the aim of characterizing systems through more abstract and universal features than the usual empirical or "law-theoretical" ones (as, for instance, the Newtonian laws of movement).

Let us describe in some detail the main entanglements between mathematical determinations and physical principles that the theorem implies and obeys.

(i) *Mathematical substrate* (initial data of the representation): There exists a sufficiently regular (continuous, derivable as far as necessary) functional, depending on space and time variables and on field variables (which are themselves continuous and derivable on space and time) as well as on their derivatives. This functional is the Lagrangian action and its spacetime density is the fundamental function of the problem. In some cases it may be interpreted as the metric of an abstract space.

(ii) *Physical principle*: The equations of motion (the behavior of the physical system) are given by applying Hamilton's principle, which means that the actual trajectory is defined by the vanishing of the perturbative variation (at constant time) of the Lagrangian action. Mathematically speaking, a principle of variations concerns a variation calculus; physically, it is the expression of a principle of extremality which may be interpreted as expressing a geodesic principle on a relevant manifold (cf. Eisenhart's theorem [9, 10]). Let us stress that considering such a geodesic is the only "natural" way to distinguish a specific trajectory among all the generic ones. Indeed the geodesic corresponds to a stable critical trajectory.

(iii) *Definition of the infinitesimal symmetry transformations*: these operate a gauge transformation on the Lagrangian density. They have to leave invariant the equations of motion obtained above. Let us stress the fact that the variations associated with these transformations differ from those used in the former calculus (minimization of the Lagrangian action) because time itself is now subject to variation.

(iv) *Determination of the invariant quantities* (conservation laws): one uses together the equations of motion, the variations of the Lagrangian action under the symmetry transformations and the possible gauge variation of the Lagrangian, in order to deduce a conservation equation. In this equation the important quantities (kinetic moments, energy, etc.) appear to be correlated with certain types of principles of relativity – symmetry transformations (respectively, space translations, time translation, etc.).

Beyond the own interest of such a result, let us recall the importance of this theorem in the frame of the gauge theories for quantum field theory. It provides a basis for the possibility to classify elementary particles and analyze their mutual transformations, due to their reciprocal interactions. For instance, this theorem, used in the frame of group theory, allowed the construction of the quark theory (QCD, for quantum chromodynamics) and nowadays serves as a foundation in the research for the theoretical unifying of the various interactions (electroweak, strong and even, possibly, gravitational, in a supersymetric universe).

Finally, from a more general conceptual and cognitive point of view, let us emphasize the fact that Noether's theorem plays both an explanatory and a heuristic role. Indeed, it makes explicit the physical-formal source of the invariant structures related to a given problem and to this extent it appears as a main factor of scientific intelligibility. Reciprocally, it appears as a guide for research itself (for instance in looking for phenomenological Lagrangians able to take into account evolving of systems the full determinations of which are not completely known). It thus plays an essential role in the so-called abductive approach. Moreover, we will see in the subsequent discussion that this theorem may serve as a basis for clearly determining what depends on the choice of the reference frames and their mathematical representations on the one hand, and, on the other hand, on the physical characteristics of the systems themselves (their physical "identities").

From a complementary point of view, the theorem enables us to interpret geometrical – and particularly symmetry – properties not only as constraints acting on a system, but even as some formal causality principle determining the general behaviour of the system. From an epistemological viewpoint, it tends to substitute, at a more abstract level of cognition, formal structural determinations to usual efficient causality linked to "forces" or even to fields and potentials (as it becomes clear in gauge theories where local gauge invariance may give rise, or suppress, such fields and forces) (see also [15] for a more detailed discussion).

3.2. Discrete Symmetries. TCP Invariance

Noether's theorem deals with invariant structures accounting for continuous symmetries (and even for infinitesimal transformations), but it remains mute about possible discrete symmetries which may be associated to a system and about the invariant quantities possibly related to them. Actually such symmetries, as for instance time reversal ($t \to -t$), are often included in the equations of motion (such is the case for classical mechanics where the presence of a second-order derivative in time expressing the acceleration – and the absence of a first-order derivative in the conservative situation – makes this invariance evident; of course, such a symmetry is broken in the case of dissipative processes). The most usual discrete symmetries are related to time variables (time reversal, just mentioned, and represented by the operator T) and space variables (parity, i.e., space variable reversal ($x \to -x$, etc.) represented by the operator P). But quantum physics has lead to considering many other types of symmetries: for instance charge conjugation C, already mentioned, relating in some way matter and antimatter (electron-positron conjugation, for instance in Dirac's equation) and also the symmetry or antisymmetry of wave functions under the permutations of the particles labeling. It is well known that these last transformations give rise to quantum statistics (regarding bosons for symmetry and fermions for antisymmetry) and to observable properties associated with them (Bose condensation for bosons, Pauli's exclusion principle for fermions). In such cases too, it appears that the invariance rules strongly determine the effective behaviors and select the actually realizable events among all the *a priori* available possibilities. Let us stress the fact that such a distinction between fermions and bosons leads to distinguish fundamentally (at least in a first approximation) between fields of matter and fields of interactions respectively. Among all these cases of discrete symmetries, let us further detail this TCP invariance which is particularly interesting and cannot be considered, at the present time, as fully understood.

First we have to notice that the question of time reversal is strongly related with that of an intrinsic reversibility or irreversibility of the physical phenomena. It is also related to the question of their mutual relativisation as linked to the question of redefining the concepts of space and matter [11]. From this point of view, irreversibility as tied to time reversal enables us to hope [12] that its treatment could lead to explain, or at least to clarify, the question of the "time arrow" [13], i.e., one of the deepest structures of our sensorial experiences and cognitive processes. Simultaneously, and strongly coupled with time reversal, we find in this theorem a reference to the other fundamental structure of our experience and cognition, i.e., space. It was one of the most surprising results of the 50's and 60's that in some weak interactions the space invariance was locally violated with respect to the parity operation, thus revealing an intrinsic chirality that enabled us to define physically a right side and a left side. This intrinsic spatial dissymetry of the universe would then be coupled with a time dissymetry.

But actually these partial symmetry breakings are mutually coupled and compensated by a more global invariance (the invariance of the product of the three operators T, C, and P) which fully justifies the opinion that it is no more possible to consider separately the three main components of the physical universe: space, time, and matter (in the broad sense, including interactions and antimatter). This physical "external" inseparability (for distinguishing it from the quantum "internal" non-separability which involves a quite different approach) between energy distribution and spatio-temporal geometry, goes even further than the one introduced by the general relativity theory. Indeed, we are concerned here with the evidence of some global invariance of the physical world, which may reveal features of its very nature as a theoretical and empirical whole (even if more "regional" invariance laws keep their importance and fecundity in their proper domains of relevance).

4. DISCUSSION. CONCLUSION

Through all the historical and theoretical examples that we have presented here, it is clear that the physico-mathematical invariant structures play at least three essential roles. They contribute to determining stable and conserved *scientific objects* despite the variability of the phenomena in which they are involved (charge conservation, for instance). They specify the *ruling of the interactions and exchanges* between these objects (energy conservation and Hamiltonian invariance, for instance). They enable us to characterize the *evolution rules* that the the physical systems obeys (Lagrangian variation and Hamilton's principle, for instance). Thereby they sketch and mark out a cognitive landscape that responds to the scientific norms of intelligibility and categorizes this scientific reality.

Their importance and their increasing formalization together with the theoretical developments also indicate a rise towards abstraction in characterizing the physical reality. After being restricted to the quasi-intuitive recognition of conserved quantities endowed with a strong substantial connotation (matter, energy, nature of chemical elements, etc.), the physical invariant structures have become correlated with formal prescriptions regarding entities as abstract as transformation groups (in particular, the Lie groups) operating on fibers of internal spaces. To such an extent that, from now on, it has become "natural" to search for the the "causes" of the main physical features in mathematical structures expressing symmetries, as in applying principles regarding purely physical ontology. At the same time, as mentioned in discussing the possible use of Noether's theorem, the emergence of physical invariant structures has brought to light the importance of abductive reasoning in physics. This feature, emphasized by Peirce [14], is now well recognized in the epistemology of physics, where it takes its full place besides the inductive and deductive approaches. Thus, it seems necessary to consider that, while elaborating a formalized epistemology, one has to take into account not only the theoretical concepts, their relationships and their structures of intelligibility, but also the main types of reasoning which guide the research itself and its heuristics (including the elaboration and achievement of experiments), beyond the technical demonstrations.

To go further, we have to stress the extraordinary feature, made evident through Noether's theorem, that the loss of some information (due to the symmetry properties) entails a gain of determination for the physical system itself by exhibiting the corresponding invariant quantities which contribute strongly to its identity. To pursue the analysis though, we have to carefully distinguish between the nature of what is lost in comparison to what is more and more determined. Indeed, we have to remark that the symmetry transformations generally act on the *reference systems* (the various space, time, fiber bundles, etc.) supporting the mathematical representation of the phenomena, i.e., the *frame-universes* which enable us to describe them objectively. On the contrary, the subsequent invariant structures (energy, electric charge, etc.) belong to the physical world itself and to its "empirical" properties. The articulation between these two levels (mathematical reference spaces and physical systems) is ensured by Noether's theorem which links the indetermination rules (symmetry properties) of the first to the determination of the second (conserved quantities).

Thus, by taking an increasing part in constructing the objectivity of physical reality (and so its intelligibility) and by taking an essential place in physical reasoning, the physical invariant structures appear as irreplaceable elements in building a plausible formalized epistemology. Indeed, they endow with some objectivity such fuzzy notions as the one of "identity" (for they prove able to extract some stable features from highly variable phenomena) or of "interaction rules" (for they prove able to determine them strongly through the conservation of a little number of quantities) and they enable us to clearly distinguish what depends on the mathematical reference frame of description from what depends on the "identity" of the physical system.

5. APPENDIX. DISCUSSION OF A DEMONSTRATION OF NOETHER'S THEOREM

Although we provide each step of its demonstration, we will not exhaustively detail Noether's theorem. Nevertheless, we propose, following Hill [8], a very explicit outline and some comment in order to characterize the role and the importance of these steps, with the aim of determining their conceptual contents and replacing them in the context of the theoretical whole to which they belong and which expresses their significance.

In the following, we apply Einstein's convention for the summation on repeated indices.

5.1. Variation of the Lagrangian Action Functional. The Mathematical Problem

First, it is postulated that there exists a Lagrangian density $L(x^k, \psi^\alpha, \psi_l^\beta)$, continuous and differentiable as much as needed, where the x^k are the independent (space and time) variables and ψ^α the field variables. Let us note that using only first-order derivatives of the field variables in expressing the Lagrangian density is an assumption (which corresponds to many real cases) that simplifies the calculus, but our demonstration does not depend on this limitation.

We are principally interested in the functional integral of the Lagrangian action

$$J = \int L(x^k, \psi^\alpha, \psi_l^\beta) dv,$$

and we will study its change under the variation of the independent variables:

$$x'^k = x^k + \delta x^k.$$

These variations entail variations of the field variables and of their derivatives:

$$\begin{aligned} \psi'^\alpha(x') &= \psi^\alpha(x) + \delta\psi^\alpha(x), \\ \psi_k'^\alpha(x') &= \psi_k^\alpha(x) + \delta\psi_k^\alpha(x). \end{aligned}$$

We have to take care that, in these expressions, the field variables are not all taken at the same point. We will have to take this condition into account in the following. Nevertheless, to first order in δ, we have $\delta\psi^\alpha(x') = \delta\psi^\alpha(x)$. The variation of J is expressed as

$$\delta J = \int_V L(x'^k, \psi'^\alpha, \psi_l'^\beta) dv' - \int_V L(x^k, \psi^\alpha, \psi_l^\beta) dv,$$

wherein we have, to first order,

$$L(x^k + \delta x^k, \psi^\alpha + \delta\psi^\alpha, \psi_k^\alpha + \delta\psi_k^\alpha)$$
$$= L(x^k, \psi^\alpha, \psi_k^\alpha) + \frac{\partial L}{\partial x^k}\delta x^k + \frac{\partial L}{\partial \psi^\alpha}\delta\psi^\alpha + \frac{\partial L}{\partial \psi_k^\alpha}\delta\psi_k^\alpha.$$

Moreover, the change from v to v' for the differential elements corresponds to

$$dv' = \frac{\partial v'}{\partial v}dv = \left(1 + \frac{\partial \delta x^k}{\partial x^k}\right)dv,$$

where $\partial v'/\partial v$ is the Jacobian of the transformation. Thus we can write

$$\delta J = \int_V \left[L\frac{\partial \delta x^k}{\partial x^k} + \frac{\partial L}{\partial x^k}\delta x^k + \frac{\partial L}{\partial \psi^\alpha}\delta\psi^\alpha + \frac{\partial L}{\partial \psi_k^\alpha}\delta\psi_k^\alpha\right] dv.$$

As the field variables are not taken at the same point in defining their variations, we will generally have $\delta\psi_k^\alpha \neq \partial\delta\psi^\alpha/\partial x^k$. In order to get an equality, we have to consider another variation which avoids this disadvantage by defining the variation $\delta*$ such that

$$\begin{aligned} \psi'^\alpha(x') &= \psi^\alpha(x) + \delta * \psi^\alpha(x), \\ \psi_k'^\alpha(x') &= \psi_k^\alpha(x) + \delta * \psi_k^\alpha(x); \end{aligned}$$

and this time we have indeed, to first order: $\delta * \psi_k^\alpha = \partial\delta * \psi^\alpha/\partial x^k$. The link between the two types of variations δ and $\delta*$ is given by

$$\begin{aligned} \delta\psi^\alpha &= \delta * \psi^\alpha + \psi_k^\alpha \delta x^k, \\ \delta\psi_k^\alpha &= \delta * \psi_k^\alpha + \psi_{k,l}^\alpha \delta x^l. \end{aligned}$$

The variation of J then reads

$$\begin{aligned} \delta J &= \int_V \left[L\frac{\partial \delta x^k}{\partial x^k} + \frac{\partial L}{\partial x^k}\delta x^k + \frac{\partial L}{\partial \psi^\alpha}\delta * \psi^\alpha \right. \\ &+ \left. \frac{\partial L}{\partial \psi^\alpha}\psi_k^\alpha \delta x^k + \frac{\partial L}{\partial \psi_k^\alpha}\delta * \psi_k^\alpha + \frac{\partial L}{\partial \psi_l^\alpha}\psi_{k,l}^\alpha \delta x^k\right] dv. \end{aligned}$$

Let us now introduce the differentiation operator D (corresponding to the total derivative of implicit functions):

$$\frac{D}{Dx^k} = \frac{\partial}{\partial x^k} + \psi_k^\alpha \frac{\partial}{\partial \psi^\alpha} + \psi_{k,l}^\alpha \frac{\partial}{\partial \psi_l^\alpha}.$$

We stress that, if functions of the coordinates only are concerned, the action of this operator simplifies, since

$$\frac{D\delta x^k}{Dx^k} = \frac{\partial \delta x^k}{\partial x^k}.$$

The expression for δJ then reduces to

$$\delta J = \int_V \left[\frac{D(L\delta x^k)}{Dx^k} + \frac{\partial L}{\partial \psi^\alpha}\delta * \psi^\alpha + \frac{\partial L}{\partial \psi_k^\alpha}\frac{\partial \delta * \psi^\alpha}{\partial x^k} \right] dv.$$

By integrating the last term by parts, we arrive at

$$\delta J = \int_V \left[\frac{D}{Dx^k}\left(L\delta x^k + \frac{\partial L}{\partial \psi_k^\alpha}\delta * \psi^a \right) + [L]_\alpha \delta * \psi^\alpha \right] dv,$$

where

$$[L]_\alpha := \frac{\partial L}{\partial \psi^\alpha} - \frac{D}{Dx^k}\left(\frac{\partial L}{\partial \psi_k^\alpha} \right)$$

is the Lagrangian derivative of L with respect to ψ^α. Reverting to the variations in δ, we may finally write, on using the familiar Kronecker symbol δ_{kl}^K,

$$\delta J = \int_V \left\{ \frac{D}{Dx^k}\left[\left(L\delta_{kl}^K - \frac{\partial L}{\partial \psi_k^\alpha}\psi_l^\alpha \right)\delta x^l + \frac{\partial L}{\partial \psi_k^\alpha}\delta\psi^\alpha \right] + [L]_\alpha \left(\delta\psi^\alpha - \psi_m^\alpha \delta x^m \right) \right\} dv.$$

The foregoing approach is purely mathematical; up to now, no physics has been involved.

5.2. Equations of Motion and Hamilton's Principle. Physical Postulates

Applying now Hamilton's principle, we suppose that two supplementary conditions are fulfilled:

(i) the integration domain V remains unchanged under the variations;
(ii) the variations of the state functions (fields ψ^α) vanish identically at the boundaries of the domain V, i.e.,

$$\delta x^k|_V = \delta\psi^\alpha|_V = 0.$$

The variation of the Lagrangian action then reduces to

$$\delta J = \int [L]_\alpha \delta\psi^\alpha dv.$$

Under these conditions, Hamilton's principle states that

$$\delta J = 0, \quad \forall \delta\psi^\alpha, \quad \text{i.e.,} \quad [L]_\alpha = 0,$$

which expresses Euler-Lagrange's equations of motion.

It is essential to note that the choice of the Lagrangian density which leads to these equations is not unique. Every divergence transformation such as

$$L \to L + \frac{DG^k}{Dx^k},$$

where the G^k are any functions of the x^k and ψ^α but not of the ψ_k^α, leads to the same result; it is easy to show that then $\left(DG^k/Dx^k\right)_\alpha = 0$. Let us add that if second-order derivatives, for instance, are allowed in the Lagrangian density, then G^k may depend on the first derivatives of the fields, and so on.

The Euler-Lagrange equations solve the problem of the mathematically describing, and thus in principle discovering, trajectories and other features of particle motion.

5.3. Symmetry Transformations. Mathematical Determination of the Physically Invariant Quantities

We are now interested in the transformations that leave invariant the equations of motion (but not necessarily the Lagrangian density itself). Towards this end, we define a Lagrangian density L' such that (with obvious notations):

$$L'(x', \psi', \psi'_k)dv' = L(x, \psi, \psi_k)dv.$$

We have seen that, to leave unchanged the equations of motion (and it is shown in Sec. 5.1. that this is necessary), it is sufficient to require

$$L'(x', \psi', \psi' k) = L(x', \psi', \psi'_k) + \frac{DG^k}{Dx'^k}.$$

Let us then consider an infinitesimal symmetry transformation with an associated variation Δ such that

$$x'^k = x^k + \Delta x^k,$$

$$\psi'^\alpha(x') = \psi^\alpha(x) + \Delta\psi^\alpha(x), \qquad (1)$$

$$\psi_k'^\alpha(x') = \psi_k^\alpha(x) + \Delta\psi_k^\alpha(x).$$

We then have

$$L'(x + \Delta x, \psi + \Delta\psi, \psi_k + \Delta\psi_k)dv' = L(x, \psi, \psi_k)dv$$

and, to first order,

$$L'(x + \Delta x, \psi + \Delta\psi, \psi_k + \Delta\psi_k)dv'$$

$$= L(x + \Delta x, \psi + \Delta\psi, \psi_k + \Delta\psi_k)dv' + \left(\frac{D\Delta G^k}{Dx^k}\right) dv;$$

hence, to the same order,

$$L(x, \psi, \psi_k)\, dv = L(x + \Delta x, \psi + \Delta\psi, \psi_k + \Delta\psi_k)dv' + (\frac{D\Delta G^k}{Dx^k})dv. \qquad (2)$$

The Jacobian of the transformation is

$$\frac{\partial x'}{\partial x} = 1 + \frac{\partial \Delta x^k}{\partial x^k},$$

so that

$$L(x + \Delta x, \psi + \Delta\psi, \psi_k + \Delta\psi_k) = \left[L(x, \psi, \psi_k) - \frac{D\Delta G^k}{Dx^k}\right]\left(1 - \frac{\partial \Delta x^k}{\partial x^k}\right).$$

To first order, we also have

$$-\frac{DG^k}{Dx^k} = \left[\Delta x^k \frac{\partial}{\partial x^k} + \Delta\psi^\alpha \frac{\partial}{\partial \psi^\alpha} + \Delta\psi_k^\alpha \frac{\partial}{\partial \psi_k^\alpha} + \frac{\partial \Delta x^k}{\partial x^k}\right] L.$$

This last equation enables us to verify that a given transformation is actually a symmetry transformation (its right-hand side has to be written under the form of a divergence, which permits the identification of the function ΔG^k). We remark that, if this term vanishes and if the Jacobian equals unity ($\partial\Delta x^k/\partial x^k = 0$), then the Lagrangian density itself remains invariant under the symmetry transformation (which usually is the case in the field theories).

5.4. The Conservation Theorem. Physical Interpretation of the Mathematical Result

It suffices now to integrate Eq. (2) (on the left over V' and on the right over V) by taking into account the preceding mathematical determinations:

$$\int_{V'} L(x + \Delta x, \psi + \Delta\psi, \psi_k + \Delta\psi_k)\, dv' = \int_V \left[L(x, \psi, \psi_k) - \frac{D\Delta G^k}{Dx^k}\right] dv.$$

Per definition of the functional variation of J under Δ,

$$\Delta J := \int_{V'} L(x + \Delta x, \psi + \Delta\psi, \psi_k + \Delta\psi_k)dv' - \int L(x, \psi, \psi_k)]dv$$

and thus

$$\Delta J + \int_V \frac{D\Delta Gk}{Dx^k} dv = 0.$$

By using the formal correspondence $\Delta \leftrightarrow \delta$ (where δ is the variation used in Sec. 5.1.), we find

$$\int_V \left\{ \frac{D}{Dx^k} \left[\left(L\delta^K_{kl} - \frac{\partial L}{\partial \psi^\alpha_k}\psi^\alpha_l \right) \Delta x^l + \frac{\partial L}{\partial \psi^\alpha_k}\Delta\psi^\alpha + \Delta G^k \right] \right.$$
$$\left. +[L]_\alpha(\Delta\psi^\alpha - \psi^\alpha_l \Delta x^l) \right\} dv = 0,$$

which has to be true whatever the integration domain; hence

$$\frac{D}{Dx^k} \left[\left(L\delta^K_{kl} - \frac{\partial L}{\partial \psi^\alpha_k}\psi^\alpha_l \right) \Delta x^l + \frac{\partial L}{\partial \psi^\alpha_k}\Delta\psi^\alpha + \Delta G^k \right]$$
$$+[L]_\alpha(\Delta\psi^\alpha - \psi^\alpha_l \Delta x^l) = 0.$$

This expression is to be viewed as a simple consequence of the symmetry transformation. Now, if we take into account the equations of motion $[L]_\alpha = 0$, we obtain the wanted conservation relation

$$\frac{D}{Dx^k} \left[\left(L\delta^K_{k,l} - \frac{\partial L}{\partial \psi^\alpha_k}\psi^\alpha_l \right) \Delta x^l + \frac{\partial L}{\partial \psi^\alpha_k}\Delta\psi^\alpha + \Delta G^k \right] = 0. \qquad (3)$$

This general expression corresponds to different conservation relations depending on each infinitesimal symmetry transformation imposed upon the system. Moreover, since the full set of such symmetries form a group, we have here the conservation theorems associated with this symmetry group.

Remark. In classical mechanics of the moving point ("fields" concentrated on the position variables q^j and with $L = \int L d^3x$, which indicates an integration over the space variables), the fundamental equations transform into the equivalent of Eq. 2 after integration over space:

$$\left(\Delta t \frac{\partial}{\partial t} + \Delta q^j \frac{\partial}{\partial q^j} + \Delta q'^j \frac{\partial}{\partial q'^j} + \frac{d\Delta t}{dt} \right) L = -\frac{d\Delta G}{dt},$$

where t is the time, $q'^j = dq^j/dt$ are the velocities, and G is the function replacing the G^k in the case of the densities and expressing the so-called gauge variation of the Lagrangian; and, regarding the conservation, we have

$$\frac{d}{dt} \left[\left(L - \frac{\partial L}{\partial q'^j} q'^j \right) \Delta t + \frac{\partial L}{\partial q'^j} \Delta q'^j + \Delta G \right] = 0,$$

the equivalent of Eq. 3 integrated over space.

This last equation is the expression of *Noether's theorem* for classical mechanics. The invariant quantity is easily identifiable depending on symmetry transformations over space or time. Let us stress the fact that here the Δ variations are symmetry variations involving the time too (while the corresponding δ variations of Hamilton's principle applied to this classical case and leading to the corresponding Euler-Lagrange equations leave the time unchanged, so that, for any function f, one has $\Delta f = \delta f + (df/dt)dt$.

The conservation equation may be brought into the dual form

$$\frac{d}{dt}(-H\Delta t + p_j \Delta q^j + \Delta G) = 0,$$

(with $H = p_j q'^j - L$ denoting the Hamiltonian of the system, $p_j = \partial L/\partial q'^j$ the momenta conjugate to the q^j, and G the gauge variation function for the Lagrangian), wherein the energy/time and momentum/position dualities are rendered manifest. To conclude, let us note that, by coupling this approach with Eisenhart's theorem, it is possible to give a simple geometrical interpretation of Noether's theorem in the case of classical mechanics as considered here. We will now elaborate on this point.

5.5. On a Geometrical Interpretation of Noether's Theorem in Classical Mechanics

Let us first recall that Eisenhart's theorem states that the equations of motion are the projection in an $(n+1)$-dimensional manifold (n spatial dimensions plus one dimension of time), along the supplementary variable, of such a geodesic in an $(n+2)$-dimensional manifold, the metric of which is mainly determined by the Lagrangian function [the $(n+2)$th variable being itself defined from the Lagrangian action]. Let us begin with a brief exposition of the theorem, following Lichnerowicz [9,10].

Because of the fact that the equations of motion are obtained through a variational principle applied to the Lagrangian action, it is tempting to construct some metric in the tangent manifold, the coefficient of which could be related to the Lagrangian itself, in order to obtain the equations of motion as geodesics of this metric and to provide thus a particularly simple geometrical interpretation of these equations and of the principle governing them. For instance, we might write $ds^2 = 2Ldt^2 = g_{\alpha\beta}dq^\alpha dq^\beta$ and calculate the equations of the corresponding geodesics. But operating in this direct way does not lead to the right result. Indeed, if the accelerations

$$\gamma_i := \frac{d}{dt}\left(\frac{\partial L}{\partial q'^i}\right) - \frac{\partial L}{\partial q^i}$$

are effectively zero, the last one, γ_0, is not zero but equal to dL/dt, which rules out any simple geometrical interpretation of the desired type.

In order to overcome this difficulty, we introduce a supplementary variable $u := q^{n+1}$ and consider the new $(n+2)$-dimensional manifold with the metric

$$d\sigma^2 = 2Ldt^2 + 2dtdu = g_{AB}dq^A dq^B,$$

so that $g_{0,n+1} = 1$ and $g_{i,n+1} = g_{n+1,n+1} = 0$.

Calculation of the new geodesics now leads to the expected results. Indeed, for $A = i, \gamma_i = 0$, while for $A = n+1, d^2t/d\sigma^2 = 0$. From this it is easy to determine u via $du/dt = k^2 - L$, with k a constant such that $k^2 = u'(0) + L(0)$. Integration gives:

$$u = k^2 t - \int_0^t L dt + a = k^2 t - S + a, \quad a := u(o),$$

which shows the link between u and the Lagrangian action S. For $A = 0$, $\gamma_0 + d^2u/dt^2 = 0$, i.e., $\gamma_0 = dL/dt$ (from the preceding definition of u).

Thus, the trajectories emerge clearly as the projection along the supplementary variable u of the geodesics of the $(n+2)$-dimensional manifold with the convenient metric; this result is *Eisenhart's theorem.*

Note, at this stage, that the new $(n+2)$-dimensional metric $d\sigma^2$ is derived from the old $(n+1)$-dimensional ds^2 by changing L into $L + du/dt$, which has just the form of the "gauge variation" considered above (with $G = u$), and that the new equation for u causes just this quantity $L + du/dt$ to be a constant k^2.

This remark indicates that it must now be quite easy to show that Eisenhart's theorems may provide an interpretation of Noether's theorem in terms of the supplementary coordinate introduced above. Indeed, from the equation determining u, it follows that $u - k^2 t = a - S$, and we obtain directly

$$-\frac{d}{dt}\Delta(u - k^2 t) = \frac{d\Delta S}{dt} = \Delta L + L\frac{d\Delta t}{dt}$$

$$= \left[\Delta t\frac{\partial}{\partial t} + \Delta q^i \frac{\partial}{\partial q^i} + \Delta q'^i \frac{\partial}{\partial q'^i} + \frac{d\Delta t}{dt}\right] L, \tag{4}$$

which reduces to the usual expression on simply setting $\Delta G = \Delta(u - k^2 t)$.

As a consequence, proceeding either directly or taking into account together the equations of motion, the expression of u, and the time behaviour of the symetry transformation, it is straightforward to derive the result

$$\frac{d}{dt}\left[\left(L - q'^i \frac{\partial L}{\partial q'^i}\right)\Delta t + \Delta q^i \frac{\partial L}{\partial q^i} + \Delta(u - k^2 t)\right] = 0, \tag{5}$$

which, of course under the same condition $\Delta G = \Delta(u - k^2 t) = \Delta(a - S)$, reproduces the conservation equation. Thus a simple geometrical interpretation of Noether's theorem becomes available in relationship with the behavior of the geodesics in the $(n+2)$-dimensional manifold (which are of length $\sqrt{2}kt$): the symmetry transformation of the arbitrary function has to be related to the symmetry transformation associated with the new variable $u := q^{n+1}$. Let us make some remarks:

(i) We do not need to suppose that $G := u - k^2t$ in order to fulfill the correspondence; it suffices for the Δ variations of these quantities to be equal.
(ii) In the case $G = u - k^2t = a - S$, the "gauge variant" Lagrangian vanishes along the trajectories $L + dG/dt = 0$.
(iii) We emphasize that the expression $u - k^2t$, must be treated as a function of the time and the position variables alone when it is varied in order to correspond to the same limitation on G; this condition is related to the fact that the Lagrangian is classically supposed not to depend on derivatives of the position variables higher than the first.

Now, let us consider the formal new "gauge Lagrangian" Λ, referred to the $(n+2)$-dimensional manifold, i.e., such as $\Lambda = L + q'^{n+1}$ (recall that $u := q^{n+1}$), with the corresponding "Lagrangian action"

$$\sum = \int \Lambda dt = k^2 t.$$

The associated new momenta then are $\pi_j = \partial\Lambda/\partial q'^j$ $(j = 1, 2, \ldots, n+1)$; thus, $\pi_j = p_i$ for $j = i$ and $\pi_{n+1} = 1$ for $j = n+1$. Note then that the new Hamiltonian η corresponds to the old one, for $\eta = \pi_j q'^j - \Lambda = H$. Of course, the Euler-Lagrange equations corresponding to the Lagrangian Λ are fulfilled: for $j = i$ they correspond to the known equations for L in the $(n+1)$-dimensional manifold, and for $j = n+1$ one gets the identity $0 = 0$, since $d\pi_{n+1}/dt = 0$, and Λ does not depend on the "position" q^{n+1}.

Let us now rewrite both equations expressing Noether's theorem for the system under consideration in the new formalism. After some manipulations, the first equation reduces to

$$\Delta\Lambda = \Delta(k^2), \tag{6}$$

a trivial result, remembering that $\Lambda = k^2$. The second member of Eq. (6) is not necessarily zero, although k is a constant as a function of t; for, as defined above, k^2 equals the initial value $u'(0) + L(0)$, which can be varied.

Finally, the second expression takes also a reduced form:

$$\begin{aligned}\frac{d}{dt}\left(\pi_j \Delta q^j - \eta\Delta t - \Delta k^2 t\right) &= \frac{d}{dt}\left(\pi_j \Delta q^j - \eta\Delta t - \Delta\Sigma\right)\\ &= \frac{d}{dt}\left(\pi_j \Delta q^j - H\Delta t - \Delta\Sigma\right) = 0,\end{aligned} \tag{7}$$

reproducing the corresponding form of the conservation theorem in the $(n+1)$-dimensional manifold, i.e.,

$$\frac{d}{dt}(p_i \Delta q^i - H\Delta t - \Delta S) = 0, \quad \text{with } \frac{d\Delta G}{dt} = -\frac{d\Delta S}{dt}.$$

REFERENCES

1. M. Granet, *La pensée chinoise* (Albin Michel, 1968).
2. J. Needham, *La science chinoise et l'Occident* (Seuil, 1973).
3. J. Petitot, "Idéalités mathématiques et réalité objective. Approche transcendantale," in *Hommage à Jean-Toussaint Desanti* (TER, 1991).
4. J. Petitot-Cocorda, *Physique du Sens* (Editions du CNRS, 1992).
5. J. Ullmo, *La pensée scientifique moderne* (Flammarion, 1969).
6. E. Noether, "Invariante Variationsprobleme," *Kgl. Ges. Wiss. Nachr. Math.-phys. Kl.* **2**, 235 (1918).
7. F. Klein, "Über die Differentialgesetze für die Erhaltung von Impuls und Energie in der Einsteinschen Gravitationstheorie," *Kgl. Ges. Wiss. Nachr. Math.-phys. Kl.* **2**, 171 (1918).
8. E. L. Hill, "Hamilton's principle and the conservation theorems of mathematical physics," *Rev. Mod. Phys.* **23** (3), 253 (1951).
9. A. Lichnerowicz, *Théories relativistes de la gravitation et de l'électromagnétisme* (Masson, 1955).
10. A. Lichnerowicz, *Eléments de calcul tensoriel* (Armand Colin, 1962).
11. G. Cohen-Tannoudji and M. Spiro, *La matière-espace-temps* (Fayard, 1986).
12. H. D. Zeh, *The Physical Basis of Direction of Time* (Springer, 1989).
13. E. Klein and M. Spiro, eds., *Le temps et sa flêche* (Frontières, 1994).
14. C. S. Peirce, *Ecrits sur le signe* (Seuil, 1978).
15. B. C. van Fraassen, *Laws and Symmetry* (Oxford University Press, 1989).
16. R. Courant and D. Hilbert, *Methods of Mathematical Physics* (Interscience, 1962).
17. A. Sommerfeld, *Mechanics* (Academic, 1964).
18. H. Goldstein, *Classical Mechanics* (Addison-Wesley, 1950).

12

FORM AND ACTUALITY

Michel Bitbol

Centre de Recherche en Epistemologie Appliquee du CNRS
1 rue Descartes
75005 Paris, France

A basic choice underlies physics. It consists of banishing actual situations from theoretical descriptions, in order to reach a universal formal construct. Actualities are then thought of as mere local appearances of a transcendent reality supposedly described by the formal construct. Despite its impressive success, this method has left major loopholes in the foundations of science. In this paper, I document two of these loopholes. One is the problem of time asymmetry in statistical thermodynamics, and the other is the measurement problem of quantum mechanics. Then, adopting a broader philosophical standpoint, I try to turn the whole picture upside down. Here, full priority is given to actuality (construed as a mode of the immanent reality self-reflectively *being* itself) over formal constructs. The characteristic *aporias* of this variety of "Copernican revolution" are discussed.

Key words: quantum mechanics, statistical physics, Gödel, transcendental illusion.

> *Think about form, but do not construct theories about form.*—
> Nâgârjuna, Mûlamadhyamakakârikâ, IV, 5; translation by J. Garfield.

1. INTRODUCTION

According to Hegel, the now is just *what is not any longer when it is* [2]. By contrast with this expression of extreme instability, physics has inherited from Parmenides a very strong tendency towards the formulation of formal

[1] A few paragraphs of this paper were borrowed from my short preface to [1].

invariants, independent from any personal, spatial and temporal point of view. The very notion of state, which is so fundamental for physics, concentrates in it an ambition to convey the primacy of Being over the process of Change [3]. This quest of immutability culminated with Minkowski's geometry of space-time, and with Einstein's Parmenidean characterization of the flow of time as an illusion [4]. Thus, at first sight, "now" is exactly the kind of term which is to be banished from the language of physics. One should not be surprised that "physics is missing any concept of the now" [5].

In fact, the lack of any concept of the now is only one instance of a momentous choice which underlies physics as a whole. This choice consists of banishing actual situations and indexical elements of speech from theoretical descriptions, in order to reach a universal formal construct relative to which actualities and indexical terms can be considered as reflecting mere particular standpoints. Physics could be defined, *inter alia*, as a systematic attempt at pushing *actuality* aside and bringing *form* to the fore. True, the formal descriptions which are the theoretical end-products of physics have to connect somewhere with actuality. But one should notice that even this connection is dealt with in such a way that any direct reference to actuality is avoided. Indeed, in physics, applying a general law to a particular case, namely connecting the theoretical description to the practice of the experimentalists, involves two steps which both carefully avoid strict adherence to actuality. First, one refers to mesoscopic individual objects, namely to preparative devices and measurement apparatuses that can be handled within the laboratory. Second, by ordering the individuatable events which may arise from experimental configurations, and by defining a metric on this ordered set, one generates a set of numbers taken to be the measured values of the variables which enter into a certain law. Now, mesoscopic individual objects and individuatable events are related to actuality, but only *sketchily*, by means of one of their aspects or profiles, at a certain time and in a certain perceptual context. As for the measured values, they retreat even farther from actuality; for they tend to substitute abstract locations on a numeric scale for actual events, just as stating a property P means substituting an abstract location on the logical binary scale [P, not-P] for an actual perception.

To sum up, even though a physical formalism has to connect repeatedly with actuality, it usually does so through the mediation of the halfway concepts of object and event. Invariant structures are defined in such a way that they are not *directly* instanciated by actual perceptions, but rather by individual objects and particular processes involving changes in properties or in measured values. Physical laws connect only *indirectly* with actuality, usually by providing a mapping of a set of individual *objects* and measured *values* onto another set made of the same individuals and modified values [6]. The previous remarks can be understood as follows: physics articulates a second- level objectivation (the level of laws and structures); and at the same time, it presupposes a first-level objectivation which underlies language, everyday activity, and ordered experimental activity as well. The first-level objectivation amounts to defining tacitly (by acting and by using language) mesoscopic permanent objects (or 'things'), enduring properties, and causal processes, of which singular actual percepts are supposed to be mere special appearances. This first-level is definitely out of the scope of physics, for the contents of physics could not even be stated without taking it for granted. We shall say that physical laws are related to actuality through the intermediary of pre-objectivized *delegates* of actuality (namely the individual 'things' and measured values).

Such a shift from actuality to its delegates has recently been documented by Hintikka. According to Hintikka [7], Kant's theory of knowledge can only apply to modern science if one replaces *passive* sensorial reception with *activities* of seeking and finding, and also if one replaces the intuitive mode of selection of particulars, which was advocated by Kant [8], with logical instantiation. But there is an important difference between intuition and logical instanciation, which bears on their respective relation to actuality. Kant's intuition is directly connected to sensorial actuality. By contrast, logical instantiations are once again *halfway* between forms (or concepts) and actuality. Indeed, on the one hand, logical instanciations are 'particular representations of concepts' according to Hintikka; and on the other hand we have seen that they are able to operate as *delegates of actuality.* We can understand this momentous difference of focus between Kant and modern advocates of transcendental philosophy as follows. Kant's reference to *intuition*, and to conceptual organization of the material afforded by intuition, is typical of a situation where science is so young that it needs some kind of (transcendental) foundation of the preliminary (first-level) objectivation of individuals, properties, and events, which it presupposes. By contrast, Hintikka's focus on logic is typical of a situation where science is sufficiently mature; so much so that one may consider that the pre-objectivation of individuals and properties has been given enough *intra-scientific* justifications by its successful use in the past, especially in *classical* physics, to be in no need of further extra-scientific (metaphysical or transcendental) foundations. No wonder that Bohr relied on *classical physics* as a necessary ground of the elaboration of quantum physics.

Now, this flight farther and farther from actuality, towards more and more universal hierarchies of forms, raises some problems. It is of course legitimate in so far as it is aimed at extending ever increasingly the scope of our inter- subjective discourse and our common mastery of every aspect of our environment. But exclusive fascination for the supposed target of the flight, to the detriment of the original actuality, may also have some serious drawbacks. I shall list some of them in this paper. To begin with, in paragraphs 2 and 3, I'll discuss those drawbacks which concern physics itself. Then, in paragraphs 4 and 5, I shall give a hint about some other drawbacks which concern our civilization as a whole in so far as it takes physics as a cultural paradigm, and objectivation as a quasi-exclusive value. Finally, in a short conclusion, I shall draw the teachings of this rehabilitation of actuality for the general project of a *formal epistemology.*

2. FROM THE LAWS OF PHYSICS TO ACTUALITY I: STATISTICAL MECHANICS

The hypothetico-deductive method, which is so widespread in physics, has had a great impact on the philosophical conceptions of the physicists. It is essential to this method that precise predictions bearing on phenomena can generally be derived from both the formalism and the initial conditions, but not the other way around. As a consequence, it is very common among physicists to reify the formal skeleton of the model; namely to behave as if the model were a faithful (though possibly approximate) representation of some absolute reality, and as if the phenomena themselves were only partial and relative appearances involving both the represented "absolute" reality and our special mode of experimental investigation. The most traditional way of expressing such a hierarchy between the model and the phenomena is the Descartes-Locke distinction between primary qualities (which belong to a geometrical representation) and secondary qualities (whose significance is relative to the receptive structures of our senses). Such a distinction is still popular today among physicists, provided one accepts to shift the focus of primary structures from the ordinary space to extended abstract spaces, and to replace the senses with experimental devices.

Unfortunately, this conception is not devoid of difficulties. To begin with, one may wonder, in terms borrowed from Schrödinger [9], whether the (hypothetico-deductive) method is based on the good reasons one has to think that a stable, unified and universal model will eventually be able to represent reality as such, or whether, conversely, this realist belief is based on the (provisional) success of the method. Another difficulty is that the more a formalism becomes autonomous, the more it incorporates symmetries which enable it to deal with a great variety of (experimental) situations, and the less it is able to deal with the essentially asymmetric features of particular situations without *ad hoc* assumptions. Two (possibly interrelated) instances[2] of this loss of relevance to immediate experience are:

(i) Statistical mechanics and its difficult connection with irreversibility,
(ii) Quantum mechanics and its purely probabilistic connection between the continuous unitary evolution of state vectors and the discontinuity of experimental outcomes.

Let us begin whith statistical mechanics (quantum mechanics will be dealt with in §3). It has been accepted, since the debate between Boltzmann and Loschmidt, that any purely mechanical description of a system made of a great number of molecules is bound to be time-symmetric and reversible. Gibbs' method has made this point even clearer [10]. Therefore, in order to account for the second law of thermodynamics in mechanical terms, it appears to be necessary to impose some extrinsic rules, or approximations, which have the effect of breaking the formal time-symmetry. But such rules or approximations (for instance Gibbs' coarse graining, Boltzmann's assumption of molecular chaos in a probabilistic framework, Jaynes' minimal information, or the comparison between the time of observation and the Poincaré's cycle), clearly refer to the spatial or temporal scale of the experimenters. This latter point has either been taken as a proof that the second law has some irreducible "subjective" aspects in it, or as an incentive to go beyond the Boltzmann-Gibbs version of statistical mechanics in order to look for a purely "objective" account of the second law. In both cases one considers that, for a law to be "objective", it should arise directly from the model, namely from the second-level of objectivation which is typical of physics, and not from our interest-relative way of dealing with our environment. The requirement is that the model must be capable of generating its own symmetry-breaking processes.

Many attempts at reaching the "objectivity" of the second law of thermodynamics *in this sense* have been carried out. But when carefully studied, such attempts all exhibit the features they aimed at disparaging, though under highly elaborated aspects. One recent example is Prigogine's theoretical description of how time-symmetry breaking occurs by means of what he himself calls the "laws of chaos" [11]. At first sight, the introduction of the high sensitivity to changes in initial conditions, which is typical of chaotic regimes, is all that was needed in order to show that the time-symmetry breaking of the dynamics of molecular systems is "intrinsic" (namely self-generated by the formalism). However, as I. Stengers rightly pointed out [12], Prigogine's result does not so much demonstrate the possi-

[2]Another celebrated instance is the difficult connection between the Einstein-Minkowski four-dimensional block-universe, the concept of 'now', and the past-future asymmetry.

bility of deriving the second law of thermodynamics from a proper consideration of the traditional object(s) of statistical mechanics, as the necessity of considering a completely new kind of object. In the so-called *baker's transformation*, for instance, fibers must be substituted for material points; for it is precisely this substitution which allows one to avoid any reference to coarse graining. Thus, in order for the model to generate its own symmetry-breaking process, one must modify the very mode of objectivation which yields it. Whereas the old mode of theoretical objectivation (which extrapolates the spontaneous mode of objectivation of our everyday speech and activity one step farther, namely from 'things' to 'material bodies' or 'material points') had the consequence that some aspects of the phenomena were to be ascribed to the relation between the object and the experimenters, the new mode of objectivation *incorporates* the relational aspect [13], and thus exempts the physicist from explicitly mentioning it. Accordingly, the leading question of the physicists no longer bears on the properties of an autonomous pre-given object of nature, but on how it is possible to frame a new type of object in such a way that it can be treated *as if* it were autonomous, and yet able to encompass the typical asymmetry of the situations with respect to which it has been provided this "as if" autonomy.

This circumstance has been taken by I. Stengers as a proof that, nowadays, physicists are still able to fulfil their good old dream of a series of models construed as increasingly faithful representations of reality, in spite of the fact that they have inadvertently made clear the interest-relative components of their models by the very attempt of hiding them. However, from a critical viewpoint, the same circumstance can be seen in a very different light. In this perspective, Prigogine's move shows how a great specialist of statistical thermodynamics is eventually forced to recognize, by his having recourse to a new kind of (purposedly constructed) hybrid objects, that he uses *constitutive procedures* (in Kant's sense) during the initial phase of his work. True, the modern statistical physicist has rehearsed successfully the fascinating game of projection of the normative rules of experimental practices onto a model and a set of objects. But the new version of this *projection* is so thoroughly modified that it misses almost completely its original target. For the projected object now manifestly encompasses the "descriptional relativity" [14] which was to be eliminated from the model. It has become obvious, from the very process of construction of the new object of statistical physics, that it cannot pretend to represent a reality construed *in the absolute*, irrespective of any relation with the particular experimental situations in which it manifests itself.

No wonder that another lineage of physicists and philosophers of physics [15] have tried to reverse completely the problem. I. Stengers herself accepts, at least in principle, the appropriateness of this kind of reversal, which gives priority to the asymmetric presuppositions of the experimental practices over the internal symmetries of the model. As she notices, "(...) the well-known physical laws which assert the equivalence between 'before' and 'after' have been made possible by measurement operations; (but) the least measurement apparatus denies this equivalence" [16]. However, as a philosopher of science, she does not wish to hold on to that simple remark. According to her, this would make difficult any communication between philosophers and physicists, since the latter are traditionally more fascinated by their intentionally aimed at objects and models, than by their own practices. I personally think that, on this point, her attitude is exceedingly ambiguous. Being respectful of the internal aims and historical choices of the community of physicists should not prevent a philosopher from throwing strong light on the background which underlies their experimental activity, and from insisting that this background *cannot* be completely wiped out from the theoretical end-product of their investigation without major inconvenients. After all, some physicists are not unable to understand what is at stake in the critical approach of their science, and this may provide their practice with additional self-consciousness. One should not forget that even though the usefulness of philosophical lucidity is admittedly doubtful during the periods of 'normal science', it has proved crucial during the past major scientific revolutions.

So, let us turn to the arguments of those thinkers who advocated an equivalent of Kant's "Copernican revolution" in the domain of statistical physics. To begin with, according to Bohr, the concept of observation already implies a fundamental irreversibility. In thermodynamics, the reason for this is quite obvious: "(...) the very concept of temperature stands in an exclusive relation to a detailed description of the behaviour of the atoms in the bodies concerned" [17]. In other terms, the operational definition of the temperature variable and the mechanical description of microscopic behaviour are "complementary". It is this extrinsic (though fundamental) methodological point, rather than any intrinsic feature of the mechanical model which, says Bohr, "(...) allows us to solve the apparent contradiction between the law of increase of entropy and the general reversibility of the individual mechanical processes" [18]. In the same way, Görnitz, Ruhnau and Von Weizsäcker [19] recently emphasized that temporal asymmetry is already at work in a very basic presupposition of experimental science construed as a process of information gathering: namely the pre-requisite of a difference between possibility and fact. Their problem is thus to explain the time-symmetry of the fundamental laws of physics given the time-asymmetry which is the precondition of experience, and *not* to explain the time asymmetry of the most familiar processes by taking the time-symmetry of the laws for granted. According to them, one possible answer is that this artificial time-symmetry is due to an abstraction leading one to transform semi-groups into groups whenever

(i) clock-time is represented by a real-valued continuum, and
(ii) laws of motion are formulated by means of differential equations.

Of course, one may then wonder why symmetric laws such as that of mechanics are not flatly falsified by experiments which are supposed to involve an all- pervasive asymmetry. I think that the reason of this absence of straightforward falsification is that no experiment is ever compared *directly* to symmetric laws. Actually, experiments are compared to *altered* symmetric laws, namely to symmetric laws implicitly modified by an additional *ad hoc* asymmetric assumption. They are compared to the laws of mechanics modified by the hypothesis that motion takes place from earlier times to later times, or to the laws of electromagnetism modified by the selection of retarded solutions, or to the laws of relativity modified by an extrinsic distinction between the future cone of light and the past cone, and so on and so forth. It is in this way that, in the everyday work of physicists since the seventeenth century, the urge for universality and symmetry has been made compatible for all practical purposes with the fundamental asymmetry of experience.

Let me add at this point a little qualification of what I have written in the previous paragraphs. By advocating the logical priority of the actuality over the formalism, of the asymmetric presuppositions of the experimental work over the symmetric form of the laws, I do not wish to deny the value of the work of those physicists, such as Prigogine and many others [20], who attempted to show how symmetry-breaking can be generated by the model itself provided some additional assumptions are made or some new choice of objects is performed. But I take this value to be quite different from what was usually indicated by the physicists themselves. Demonstrating the possibility for a model to generate its own symmetry-breaking does not mean that one has eventually disclosed the way an essentially symmetric reality manifests itself by asymmetrical appearances. It only means that physics is mature enough to be able to provide in its own terms a proof of self-consistency of the overall list of basic assumptions on which it relies. The axioms of the theory, which tend to reach maximal objectivity by means of generalized symmetry, and the asymmetric presuppositions of experimental practices, are thus shown to be mutually compatible. To paraphrase Quine [21], the crucial point is that the physicist is not confronting a challenge from some external reality whose basic symmetry is to be connected with the obvious asymmetry of experience. He is confronting a challenge that arises from *within* his science. This challenge runs as follows: *if* the models of theoretical physics were to be taken *litterally*, how could we make sense of the practice of experimentation? The problem of the physicist is that of finding ways, in keeping with his models, whereby human beings (and especially experimenters) can live in an asymmetric environment. To summarize, it is a problem of logical closure of science, not of ontology.

3. FROM THE LAWS OF PHYSICS TO ACTUALITY II: QUANTUM MECHANICS

In quantum mechanics, the issue of a connection between the formalism and actuality is even more stringent than in statistical mechanics. For, in this case, what is apparently missing in the model is not only an isolated feature of actuality, such as the asymmetry between fact and expectation, but it is a proper equivalent of *actuality itself in its univocity.* Accordingly, the most pervasive problem of the interpretation of quantum mechanics is that of "actualization" (or "transition to actuality"): namely that of the compatibility between

(i) the continuous evolution of a state vector construed as a description of the manifold *potentialities* of some experimental situation involving a putative microscopic object, and
(ii) the very circumstance that a single outcome is *actually* obtained at the end of the experiment. The first move in order to cope with this problem of compatibility has been to project the uniqueness of the outcome onto the formalism by a *fiat*, that is by means of Von Neumann's projection postulate. But many questions then remain to be answered: the questions about *where, when and how* the so-called "reduction of the wave packet" occurs.

In fact, things are even more intricate (and more interesting) than what this short presentation of the problem tends to show. Indeed, one additional distinctive feature of quantum mechanics is that its formalism puts the concept of pre-objectivized *delegates* of actuality, such as individual objects endowed with properties, or objective events, under strong pressure. Individuality lacks criteria within the quantum paradigm, and it must be bracketed in quantum (Bose-Einstein or Fermi-Dirac) statistics; furthermore, intrinsic properties are generally[3] replaced by contextual observables. As a consequence, the very notion of mutually exclusive past objective events, which is grounded on the idea that previous properties of objects have been permanently modified in a well-defined way, does *not* possess any formal

[3]The case of superselective observables must be taken apart.

equivalent within the framework of the quantum theories (if we put aside the artificial projection postulate). In B. d'Espagnat's terms, "within standard quantum mechanics, (there are) no 'really existing' facts" [22]. Not surprisingly, in view of this two-step analysis, the proposed solutions to the so-called measurement problem of quantum mechanics essentially fall under two categories.[4] There are solutions which tend to cross directly the gap between the quantum formalism and each *unique* actual event. And there are other solutions which only aim at showing how the *notion* of pre-objectivized delegates of actuality (the 'properties' and the 'objective events') can be recovered at the macroscopic scale.

The first kind of solution consists in modifying pure unitary quantum mechanics in such a way that it acquires its own mechanism of transition, from a state vector to one eigenstate of the relevant observable. This approach, which consists in adding to the Schrödinger equation a small term of random discontinuous jump which adds up when macroscopic bodies are involved, has been initially developped by Ghirardi, Rimini, and Weber [23], and it has then been advocated by John Bell [24]. It perfectly fits the general requirement that a model should be able to generate its own processes of symmetry-breaking, since the symmetry of a state vector written in terms of a linear superposition of eigenstates is broken so that only one eigenstate remains. But it also has many defects. One defect is that the term added to the usual Schrödinger equation is completely *ad hoc*, and that several physicists are now at great pains to provide it with convincing justifications. A second defect [25] is that one does not see how it is possible to account in this framework for the case of macroscopic superpositions (instantiated by superfluidity or superconductivity). A third defect is that one may wonder why and how a certain basis of eigenstates should be priviledged for the spontaneous collapse. This is the well-known "preferred basis problem" which is common to all the interpretations of quantum mechanics which tend to provide the Hilbert-space model with a sufficient autonomy [26]. Solving that problem by just mentioning that the choice of a basis may depend on some extrinsic criterion (such as the correspondence principle) would be tantamount to giving up the project of identifying a completely *intrinsic* mechanism of symmetry-breaking in the quantum-mechanical model.

But this is not all. A much more fundamental criticism which can be directed against the spontaneous collapse interpretation is that it is flatly irrelevant; that it aims at solving within the framework of physics a problem

[4] I shall not discuss here the Hidden variable theories. Indeed, they do not offer any solution to the well circumscribed problem of the connection between the quantum formalism as it stands and actuality: they rather substitute a new formalism for the quantum formalism.

which is in principle out of the scope of physics (and of science in general), namely the problem of the uniqueness of actuality. As R. Omnès wrote, "(...) the actuality of facts is something that needs not be explained by a theory" [27]. Accordingly, the aim of a good theory of quantum measurements is not to account in each case for a transition from a state vector representing potentialities to *one* particular eigenstate representing the actual actuality; it is only to show how the quantum model may be made compatible with the very idea that a measurement process leads to one *or* another well-defined outcome, embodied in permanent properties of pointers and recorders. In other terms, the project here is to show how one may recover the general notion of pre-objectivized delegates of actuality within the quantum paradigm. The most efficient strategy which has been followed in order to do so can be described in two steps. The first step consists in encompassing not only the measurement apparatus, but also an environment with a great (possibly infinite) number of degrees of freedom, within the account of the measurement process by pure unitary quantum mechanics. The second step is to show that the interference terms of the corresponding density matrix tend to decay very fastly, so that one witnesses a transition from a pure state to an approximate statistical mixture of mutually exclusive alternatives. In other terms, one shows that there is a transition from the 'and' of a superposition to the 'or' of a mixture. This is the essential claim of the *decoherence theories.* Another, more recent, claim of the decoherence theories is that they can also account somehow for the choice of a basis of eigenstates [28].

The main difference between spontaneous collapse and decoherence can now be seen very clearly. One cannot say that they both perform the same job, though in two different ways. They rather arise from two radically different conceptions of the job to be performed. In the spontaneous collapse strategy, a mechanism of symmetry breaking is offered. But in the decoherence strategy, what is asked to the model is not to break its internal symmetries by its own means, but only to have the capacity of transforming those symmetries in such a way that they offer a natural point of contact with an asymmetric well-defined actuality. This natural point of contact is the notion of mutually exclusive *events.* The necessity of reaching such an intermediate step between the probabilistic model and actuality is *typical of quantum mechanics*; it has no equivalent in classical physics. Indeed, the said intermediate step is straightaway available in classical stochastic theories, but not in quantum mechanics.

All the problems are not solved at this stage, however. Decoherence theories, which claim to be able to make the Hilbert-space model generate a transition from 'and' (superpositions) to 'or' (mixtures) by their own means, are *also* pervaded by interest-relative (or anthropomorphic) postulates [29]. They all involve some statements or assumptions which presuppose that the processes described by the model must eventually result in an acceptable macroscopic world for anthropoid creatures to speak about and to live in. The most important among these statements is Zurek's basic hypothesis that the overall state vector can be analyzed into three parts: one part for the object, one for the apparatus, and one for the environment. But admittedly, this partitioning only makes sense relative to a cognitive and experimental process involving mesoscopic instruments. In the same way, Gell-mann's theory of decoherent histories involves a superimposed coarse graining of the set of consistent histories [30]; and this coarse-graining is clearly relative to the characteristics of the so-called IGUSes ("Information Gathering and Utilizing Systems"), whose anthropomorphic flavour is unmistakable. This being granted, it is clear that, except in the remote perspective of a completely convincing strategy of "closing the epistemological circle" [31] of object and subject within the framework of the Hilbert-space model, the decoherence theories cannot pretend to have made this model able to generate a disjunctive structure by its own means. They has not succeeded to show convincingly how a classical world may emerge *by itself* out of a *completely self-sufficient* Hilbert-space world.

In view of this partial failure of the attempts at making the model able to self-generate structures which are isomorphic enough with actuality, one may be tempted by a renewed gesture of reversal of the problem. After all, if one transposes I. Stengers previous remarks from statistical mechanics to quantum mechanics, one lends into the following statement:*the (unitary quantum- mechanical) model which yields statements such as the cat paradox has been made possible by measurements; but the least single outcome of a measurement process flatly denies that the measurement chain is not in a well-defined state.*

Giving a logical priority to phenomena over the Hilbert-space model was the attitude Bohr recommended in the interpretation of quantum mechanics: "Strictly speaking, the mathematical formalism of quantum mechanics and electrodynamics merely offers rules of calculation for the deduction of expectations about observations obtained under well-defined experimental conditions specified by classical concepts" [32]. According to Bohr, the measurement problem thus arises from two essential mistakes. The first one bears on the status of the Hilbert-space model which is usually taken at face-value by physicists as describing "states" of "systems", whereas it only represents a purely mathematical tool for calculating "expectations" (namely probabilities) in an overall experimental situation. The second one is that, in the quantum theory of measurement, we improperly "(...) treat the instrument as an object" [33] to which a quantum "state" is ascribed. But one should not forget that, according to Bohr, instruments *must* be left in the (classical) background, rather than treated as (quantum) objects; for the instrument *must* fall under classical concepts in order that unambiguous communication between experimenters be possible at all. This is a kind of transcendental condition for experimental knowledge, and it cannot thus be ignored. Such a position is well-known, but it was soon discarded by physicists who hoped that the quantum theory of measurement would be able to self-generate its own classical level. When decoherence theories were formulated, Bohr's position appeared all the more superseded since the hoped-for result seemed close at hand. However, the conceptual loopholes of the decoherence theories (or rather the discrepancies between their ambitious aims and their methods) led to a recent renewal of Bohr-like arguments.

One very striking example is M. Mugur-Schächter [34], who both emphasizes that "In a probabilistic interpretation of quantum mechanics, there is *no* measurement problem", and that "the quantum mode of description *presupposes* the instrument as *a primary non-represented given*".

Another interesting example is S. Saunders, who started with an examination of Everett's interpretation, who then put this interpretation in the light of the decoherence theories, and who finally recognized his affinities with a very sober statement of Bohr's views. Let us try to understand these three successive moves, in the framework of the present study.

(1) In the decoherence strategies, one tries to make the model compatible with the idea that some event has occurred *an sich*, but that we do not know which one; then one considers that the "actual actuality" just reveals *which* objective process, leading to a certain event, was taking place. In other terms, the decoherence theory aims at displaying a formal equivalent of a list of alternative pre-objectivized delegates of actuality (the 'events').

(2) But is such an attempt at showing how the intermediate-level concept of delegate of actuality may be made compatible with the Hilbert-space model, really indispensible? After all, one can perfectly dispense with this concept of *delegate* of actuality, provided one accepts to deal directly with the connection between the formalism and *actuality*. This is exactly what Everett attempted to do by means of his "relative state" interpretation of quantum mechanics (which has to be carefully distinguished from later many- worlds interpretations). In the "relative state" interpretation, the connection between *actuality* and the various possible experimental outcomes exhibited by the formalism is direct and purely *indexical*, in the same way as the connection between *now* and a set of tensed proposition. As S. Saunders writes, "Whilst 'Event

E is past; Event E is future' are *prima facie* contradictory, introducing new events T, T^* we obtain: 'E is past relative to T; E is future relative to T^*' and there is no longer a difficulty. Likewise: 'Observable X has value r; Observable X has value r^*' are inconsistent. But introducing a new observable Y we may say instead: 'X has r relative to u of Y; X has r^* relative to v of Y' and there is no longer a contradiction" [35]. Thus, if one does not try to reconstitute the distance between actuality and objectified (absolutized) processes or events too rapidly, one may have the chance to realize that quantum mechanics has the structure of *two-level relativized description*. The first level is well-known: each set of observable values is relative to some given type of apparatus. The second-level is typical of the indexical reading of Everett's interpretation: each single value ascription for an observable is relative to a value ascription of another (apparatus) observable. In this scheme where no actual object or event is defined in the absolute, actuality can only arises relatively, for somebody who *partakes of* the chain of relations. But, this being accepted, shouldn't we adopt directly *our own* standpoint, namely the standpoint of someone who is caught into the network of relations supposedly constitutive of the world? Does it make sense for us to assert (from a position in "cosmic exile", so to speak) that our standpoint within the network of relation is "only" a *local* standpoint?

(3) S. Saunders, as some other philosophers, takes the latter remarks very seriously into account. So seriously that, at the end of his highly non-bohrian itinerary of thought, he fully recognizes the value of Bohr's (strictly anthropocentric and local) approach of the measurement problem: "What is the solution of the measurement problem? I say it is this: on measurement of X with eigenstates ϕ_1 outcome x_1 is observed with probability $|<\psi|\phi_1>|^2$, where ψ is the initial state. This is what we return to, so it will do for a beginning as well" [36]. At this point, the "copernican revolution" of our appraisal of the measurement problem has been completed: the unicity of each experimental result comes first, and the probabilistic formalism of quantum mechanics is subordinated to it.

Of course, here as in the case of statistical mecanics, we must add an important qualification to what has just been said. Advocating the logical priority of actuality over formalism, of the fundamental presuppositions of any cognitive process over the form of the model, does not mean denying the value of the work of the specialists (of decoherence) who attempted to show how the model may generate by itself the structure of objectivized delegates of actuality which any experimental work takes for granted. But this value is quite at variance from what is usually indicated by the physicists. Demonstrating the possibility for the Hilbert-space model to generate its own structure of mutually exclusive events does not mean that one has eventually disclosed how an essentially wave- like interferring reality may have emergent classical features. It only means that quantum mechanics is mature enough to be able to provide in *its own terms* a proof of consistency of the overall list of basic assumptions on which both its formalism and the experimental procedures used to test it, are based [37]. Here, as in the case of statistical mechanics, we are confronted with a problem of logical closure, not of ontology. The *ontological* problem would only arise if the Hilbert-space model were taken at face value, the state vectors being either considered as the basic constituants of the world or as expressing intrinsic determinations of the basic constituants of the world. By contrast, the problem of *logical* closure arises even if we consider the Hilbert-space formalism together with, say, the Born rule, as a mere instrument of generalized probability assessment.

Indeed, the problem of this instrument of probability assessment is that on the one hand it claims to be able to afford probabilistic valuations for any univocally defined experimental phenomenon, and that on the other hand, whenever it is extended to second-order experiments (measurements bearing on the first- order measuring instruments) it becomes *prima facie* incompatible with the simple statement that the first-order instrument has recorded a univocally defined phenomenon that we may happen to ignore. In other terms, if applied universally, this generalized probability theory *appears* to leave no room for the elementary notion of pre-objectivized delegate of actuality (the 'event' or the 'property of a pointer') which it itself presupposes. Decoherence shows that, actually, the Hilbert-space-Born's-rule mode of estimating probabilities can be made *approximately* compatible with the presuppositions of experimentation. Provided decoherence theories are given this very restricted significance, the interest-relative assumptions which are indispensible to them in order to be worked out are no longer embarassing. For in this case, one *only* needs to show that the interest-relative assumptions which are injected at one end are not *necessarily* inconsistent, given the model, with the interest-relative presuppositions which are to be respected at the other end. In more precise terms, one only needs to demonstrate (and one has indeed demonstrated by means of the decoherence theories) that when applied to a preliminary anthropocentered division of the world into objects, apparatuses, IGUSes, and environment, the quantum probability theory is not unable to give us back the mutually exclusive event-structure which human experimenters need to posit as a basic methodological assumption.

4. BEYOND PHYSICS: FORM AND ACTUALITY IN LIFE AND PHILOSOPHY

What is at stake in this problem of the relations between form and actuality goes well beyond the respective status of the postulates of theoretical physics and the presuppositions of experimental practice. It is also a basic issue for the Western culture, and especially for its ability at circumventing the blind spot which was generated by its characteristic tendency to emphasize the exclusive value of objectivity.

The priority given to the formal model over actuality had both a minor consequence and a major consequence in the basic attitudes of the West. The minor consequence is what I shall call the Golem complex. Namely the mixture of hope and fear that, in the end, the creature of man will exceed the power of its creator. A purely intellectual variety of the Golem complex is the tacit conviction that theoreticians somehow *think in order to avoid thinking any longer* [38]. Indeed, among other things, the use of mathematics is aimed at replacing the adventurous manipulation of fluctuating concepts by fixed definitions and mechanized derivations. This being granted, the hope and fear is that the mathematics 'knows more than the theoretician', and that it thus acquires a kind of autonomy with respect to the intellectual power of the scientist. Of course, at the present stage of science this situation is met only within the restricted domain of validity of some theories of physics. It is still necessary to think and to mould concepts near the margins of this domain, when the issue of the relation of one theory with another one (or with its successor) is at stake. But the very urge towards unification, the very dream of a 'theory of everything' which be the 'final theory' at the same time, shows that many of us consider that this is a provisional situation, which should ideally be replaced by one in which the theory is absolutely universal, self- sufficient, and thus able to dispense anyone from the obligation of further thinking. It is interesting to notice that this software variety of the Golem complex is strongly coupled with a hardware version, that we may call 'the Deep Blue complex'. Here, it is the autonomy of a material embodyment of our cognitive operations which is both hoped-for and feared. The sought result of this process consists in reaching a mastery of the mental aspects of 'all that is the case' [39] in the same terms as the physical aspects, and thus, finally, obtaining a complete 'closure of the epistemic circle' (*including actual appearances*), within the methodological framework of the physical sciences. If, moreover, computers that work out some sort of 'theory of everything' by themselves could be conceived, the merging of the software and hardware variety of the Golem complex would be close at hand.

Of course, this ambitious program could well prove to be an utopia, as a matter of principle. Let us then discuss this possibility. Many thinkers, the most prominent of whom are Gödel [40], Lucas [41] and Penrose [42], have provided some reasonings tending to show that complete closure of the epistemic circle by means of a mechanistic model (or, more generally, within the framework of a *computable physics*) is impossible. All these arguments rely heavily on self-reference and related incompleteness theorems. According to R. Penrose, for instance, "Gödel's theorem has the clear implication that mathematical understanding [and other kinds of human understanding as well] cannot be reduced to a set of known and fully believed computational rules" [43]. As for Lucas' classical argument, it runs thus: "We (...) construct a Gödelian formula (such as 'this formula is unprovable in the system') in [a given] formal system. This formula cannot be proved-in-the-system. Therefore, the machine cannot produce the corresponding formula as being true. But *we* can see that the Gödelian formula is true: any rational being could follow Gödel's argument and convince himself that the Gödelian formula, although unprovable in the given-system, was nonetheless - in fact for this very reason - true." [44]. Let us examine carefully the implications of these sentences. What the "mind" is supposed to do in order to *see* that the Gödelian sentence of the machine is true, is to formulate a meta-description of the relationship between the machine and its Gödelian sentence. But, after all, one may notice that this can be done by a second-order machine as well. Lucas therefore demonstrates that introducing higher-order machines does not help solving the difficulty in a purely mechanistical way, since this only leads to an infinite regress (the second order machine generates a Gödelian formula referring to its own formal system, etc.). Why is it then that the mind is not confronted with the same difficulty as any higher order machine? Lucas' answer is the following: "We are trying to produce a model of mind which is mechanical - which is essentially 'dead'. - But the mind, being in fact 'alive' can always go one step better than any formal, ossified, dead system can". Leaving aside the purely biological aspects of "life", one can reformulate this remark as Gödel himself did in his own argument against a mechanical model of mind: "(...)mind, in its use, is not static, but constantly developing" [45]. Mind is no well-defined higher order procedure; it is the *ability* to produce an arbitrarily high order reasoning, ever adapted to the stage reached by the problem at stake. In more general terms, avoiding a too narrow focus on minds and machines, one might say that what Gödel and Lucas are trying to convey in their reasonings, is that *actuality* gets always ahead of any attempt at encompassing its features within a formal model. Their reasoning thus challenges the software variety of the 'Golem complex', just as much as the hardware variety.

Now, what are we to think of this family of arguments? At first sight, they are quite convincing. But many sound counter-arguments have also been provided, for instance by P. Benacerraf [46] and J.C. Webb [47]. One of these counter-arguments is especially striking, because it uses the very existence of the argument against it. Let us quote J.C. Webb: "Such is the basic dilemma confronting anti-mechanism: just when the constructions used in its arguments become effective enough to be sure of, (...) a machine can simulate them. In particular, it implies that our very behavior of applying Gödel's argument to arbitrary machines - in order to conclude that we cannot be modelled by a machine - *can indeed be modelled by a machine.* Hence any such conclusion must fail, or else we will have to conclude that certain *machines* cannot be modelled by any machine! In short, anti-mechanist arguments must either be ineffective, or else unable to show that their executor is not a machine" [48]. This is perfectly right, but what does it show exactly? If taken at face value, it shows that *any effective argument*, be it an argument trying to appropriate 'life' or 'constant development', can be simulated by a machine. More generally, arguments which attempt to involve directly or indirectly actuality, at one step or another of their development, are somehow self-defeating. This is so because they pathetically tend to capture "what makes itself manifest" within the field of "logic" in the sense of Wittgenstein's *Tractatus.*

The consequence of these remarks is that the Gödel-Lucas family of arguments are invalid. Their invalidity however does not entail that they are useless. It only shows that they must be restricted to the status of a Tractarian "ladder" which *has* to be thrown away after one has climbed on it. Admittedly, part of the present article has itself this kind of status. But this should not be taken as a symptom of failure either. Only as a sign that whenever one tries to display the flaws of mechanistic or formalist positions by accepting the rules of the mechanist-formalist language-game, the well-foundedness of the whole move is undermined. Moreover, this kind of defect is not strictly specific of the criticism of mechanism and formalism. To be fair, one should also notice that those mechanist-formalist reasonings which go against the choice of giving actuality a priority over formal models are undermined for converse reasons. Indeed, they do not content themselves with an internally consistent chain of derivations; they try to promote their position against their opponents by relying on the *actual understanding* of their interlocutors, thus taking the explicitly denied primacy of actuality as an implicit basic premise.

To summarize, actuality should not intervene in the discourse of either its supporters or its opponents, lest they accept to be caught into inextricable performative contradictions. Its opponents should content themselves with pursuing their regulative ideal of ever-increasingly comprehensive models tending to close the epistemic circle. And its supporters should content themselves with displaying the lacunae which are left in one's description of the world by the successive realizations of the regulative ideal of the opponents (paragraphs 2 and 3 of this article illustrate this latter attitude). Talking of actuality can nevertheless become indispensible, as an auxiliary trick, whenever the upholders of the mechanist-formalist trend of thought become so fascinated by the faster and faster run in the direction indicated by their regulative ideal that they become deaf to the remarks of those who display the recurring lacunae. This is why I decided to give "actuality" such a prominent role in this article, though I was not unaware of the insuperable (and well- known) difficulties this would raise.

As I said formerly, giving priority to the formal model over actuality does not have only the "Golem complex" among its consequences; it also has another consequence that I described as *major*. This consequence is that it promotes and keeps very efficiently alive what Kant called "the transcendental illusion". But what is exactly the transcendental illusion? It consists in reifying the ideally completed aim of a rational investigation, so that one views it as an adequate representation of some absolute reality. For theoretical enquirers, it consists in taking at ontological face value every formal element which provides them with a precise orientation in the attainment of knowledge. In other words, the transcendental illusion is a natural tendency to *forget* that the reason why one is committed to formal regulative ideals of research is essentially *practical*, and accordingly to interpret the corresponding forms as retaining something of the nature of the independently real. Along with such a perspective, it appears that even though modern science has grown out of a radical criticism of scholastic and aristotelician ontology (especially the ontology of natural place) [49], *in fine* it has promoted this kind of forgetfulness more powerfully than ever. This is the case because of the very success of the scientific method. Indeed, in its highest achievements, it manages to incorporate all the normative aspects of a class of efficient experimental practices within a formal model. So much so that the model itself tends to be hypostasized, to the detriment of a lucid recognition of the practical component in it. No wonder that the discourse of so many scientists of our time is flatly pre-critical in Kant's sense: as Kant himself before the *Critique of Pure Reason*, they take for granted that actual perceptions or experimental outcomes represent things *as they appear*, whereas the theories and formalized models elaborated by our intelligence tend asymptotically to represent things *as they are* [50].

A very serious question must be raised at this point. Kant explained at length that, according to him, even if it is disclosed, the transcendental illusion is persistent and unavoidable [51]. This assertion is best justified by the commitment of any practice to its internally presupposed target. A transcendental illusion is likely to arise imperatively from within the practice whose interests are embodied by it. The man-in-the-street is committed to the targets of his action and discourse, and this commitment gives rise to what A. Fine named the Natural Ontological Attitude. As for the scientist, he/she is committed to the targets of his/her experimental practice, as well as to the heuristic guides of this practice. Extrapolating the *Natural Ontological Attitude* to the objects and models of science is then the normal expression of the *seriousness* with which the scientist undertakes the research at stake [52]. As a consequence, many philosophers of science consider that, as Putnam [53] would have it, "science taken at face value implies realism", or that "realism is so to speak science's philosophy of science".

But then, by taking the exact counterpart of this internal committment of scientists, namely by adopting an external view on science, we can also imagine a very different situation. Provided one *stands back* from the practices which generate a favourite intentionally aimed at picture of the world, one may have an opportunity to see the loopholes of this picture (and thus to be freed of the transcendental illusion associated with it). In order to submit Kant's view that the transcendental illusion is unavoidable to a moderate criticism, I shall therefore proceed in two steps. To begin with, I shall briefly evoke the loopholes of the current pictures of the world from a viewpoint poorly defined as 'that of somebody who has decided to step back from the practices associated to the pictures of the man-in-the-street and of the scientist'. Later on, in paragraph 5, I shall give some precisions about the various ways of stepping back, and about the various depths of the move.

Typically, the loophole left in the above-mentioned pictures of the world can be described as follows: exclusive interest for what Thomas Nagel calls 'The view from nowhere', and complete inability to account for any 'view from somewhere' aspect of 'all that is the case'. An old example of this, in the moral science, is the inability of scientists to find an agreement between the so-called "freedom of the will" and a deterministic picture of the world, and also their tendency to think (incorrectly [54]) that the solution of this riddle is to be found in some indeterministic features of the natural processes. Such a disarray is not surprising if, as L.W. Beck [55] points out, the scientific description comes about within a disengaged *view from nowhere*, whereas freedom is the necessary presupposition of any actor *engaged somewhere.* The major mistake here amounts to trying desperately to fit what pertains to the standpoint of the actually engaged actor into a disengaged and timeless picture.

Another example is the extreme reluctance of specialists, especially during the first half of the 20th century, to recognize *contextual* aspects in semantics or in the physical science. Nowadays, contextuality has virtually pervaded every field of knowledge, but there are also enduring symptoms that some consequences of it have not been fully accepted. Indeed, the predominant tendency is to look for a way of encompassing the low-order contexts within the field of a higher-order non-contextual discourse or description. But such a regress from a lower level of contextual description to a meta-level of non- contextual characterization of the contexts must have an end. Part of the contexts must be left in the background (see the case of quantum mechanics). Absence of recognition of this necessity has had unfortunate consequences in philosophy.

One of these consequences is the poor analysis provided by philosophers of language about the *indexical* components of everyday speech. According to the current view, indexical terms such as *here, now, I, this, etc.*, are all to be considered as token-reflexive devices. 'Here' is supposed to be used to refer to the *place* from which it is uttered; 'now' to the *time* of utterance; 'I' to the *person* who utters it, and 'this' to the *item* pointed towards by the person who utters it. But this simple token-reflexive analysis leaves aside a very important aspect of the use of indexical terms. That this is so is especially obvious for 'now' and for 'I'. One of the most striking components of the meaning of 'Now' is what we could call its self- elusiveness: namely the fact (already pointed out by Hegel in the introductory sentence of this paper) that as soon as Now is taken as an object of awareness, it is no longer *now*. Similarly, it was recently emphasized [56] that, in performative sentences, 'I' does much more than merely referring to the person who utters it. It conveys personal *commitment*. In other words, it is clear that 'I' has not only the function a *pronoun*; for replacing it by a *noun*, say in a promise, often fail to convey the same meaning.

It is not so difficult to overcome these difficulties provided one makes a clear distinction between the presuppositive and the denotative function of an indexical. The denotative function of indexicals enable them to partake of the expression of a formalizable 'view from nowhere'; but their presuppositive function is definitely irreducible to this view, and it indirectly points towards the too obvious and hence forgotten *actuality*. In the case of 'now', one should for instance establish a pragmatic distinction between the presupposed presence and the denoted instants. In the case of 'I', the model for a distinction between the presupposed and the denoted is already available in G.H. Mead's work about the difference between 'I' and 'me', and in recent commentaries on G.H. Mead by J. Habermas [57]. True, Mead [58] starts his analysis by endorsing the traditional opposition between the transcendental and the empirical, when he writes: "The 'I' is the transcendental self of Kant. The self-conscious, actual self in social intercourse is the objective 'me' (...)". But he then clearly promotes the pragmatic way of thinking when he points out that, during a conversation, *"(...) 'I' is a presupposition, but never a presentation of conscious experience"*, whereas the objective 'me' can be presented. It is interesting to notice at this point that there is an obvious twofold parallel:

(i) between 'I' and the actual now, and
(ii) between 'me' and the referred to instant of vocal utterance of the sound 'now'.

Making full use of this parallel would lead to the following paraphrase of Mead's statement about 'I': *The real Now is a presupposition of speech, even though it cannot be spoken about. The token-reflexive 'now' can be spoken about, but it does not deserve to be called "Now".*

That this is more than a mere analogy can be guessed from the detailed *temporal* analysis of 'I' and 'me' as given by Mead and Habermas. According to these authors, 'I' can but be given to me by means of memory; 'I' is always a historical figure, if it is to be a figure at all; the 'I' is either what you were *one second ago*, or it completely eludes thematization. Similarly, 'now' either receives *ex post facto* characterization or it eludes any characterization. Such an overlapping irresistibly suggests the idea of a common origin of the plurality of particular indexical terms such as 'now' and 'I'. It makes likely that they have all been derived, in some remote (and possibly mythical) prehistory of language, from a single general indexical term "Aha!", that may be said to stem from "absolute actuality" (in phenomenological terms) or from "act force" [59] (in pragmatic terms). Whereas each particular indexical (I, here, now, this) presupposes only a particular aspect of the context of speech, the general indexical "Aha!" would presuppose the whole actual context.

This primeval all-encompassing indexical would have some affinities with the "inarticulate sound" with which, says Wittgenstein [60], some philosophers would like to start their investigation. Wittgenstein is perfectly right to emphasize that this *inarticulate sound* cannot really be taken as the explicit departure point of philosophy, because "(...) one cannot begin before the beginning". But I also think that the all-pervasive implicit role of what is expressed by this sound should underpin each single word of the work of a philosopher, if he is to avoid improprieties and dissonances with respect to *what it is like to be a sentient being.*

5. THREE REMEDIES AGAINST THE TRANSCENDENTAL ILLUSION

Freeing oneself from the transcendental illusion would mean being able to broaden one's awareness so as to encompass the whole of actuality (including the immanently operating regulative ideals), rather than letting oneself be carried away by exclusive fascination for the interest-relative objects of thought construed as transcendent. As I mentioned previously, the preliminary condition for this liberation consists in *stepping back* from the practice whose orientations are embodied by a set of objects of thought. But of course, such a move is not easy to perform. The more one gets close to the basic practices of life, and the more it becomes difficult. In view of this difficulty, I shall adopt a progressive approach. I shall discuss successively three strategies aiming at freeing oneself from deeper and deeper layers of the transcendental illusion. The first strategy pertains to (Kant's) critical philosophy; the second stragegy to Wittgensteinian 'therapy'; and the third strategy to Indian (Hinduist and Buddhist) soteriology.[5]

The standpoint of critical philosophy has been adopted repeatedly in this paper; and the very concept of a transcendental illusion has been borrowed from it. That it consists in stepping back from the main scientific practices can easily be appreciated from the two-sidedness of Kant's discussion on the implications of transcendental philosophy for scientists. On the one hand, in the *Transcendental aesthetic* section of his *Critique of Pure Reason* Kant states that space is not a concept abstracted from our outer experiences, but rather the a priori form of all outer intuitions. It is only this way that one can understand how it is possible to have a knowledge of the *necessary* propositions of geometry. But on the other hand, in par. 13 of his *Prolegomena*, Kant also accepts that, with respect to any possible experience and to any possible geometrical practice, everything remains exactly *as if* ("als ob") space were an intrinsic feature of things and of their relations. The critical attitude thus stems from the meta-standpoint of the philosopher, and it proves mostly irrelevant from the ordinary standpoint of the man-in-the- street or the scientist who are immersed into their more or less sophisticated practices.

But once one has stepped back from the ordinary standpoint of practitioners, once one has adopted the philosopher's meta-standpoint in Kant's sense, some consequences become unavoidable. One crucial consequence is complete disconnection between objectivity and ontological reality, between the intentional objects and the putative 'thing-in-itself'. This distinction is usually found very difficult to understand by scientists. Indeed, most of them

[5]'Soteriology' means 'doctrine of salvation'.

take for granted that framing objective entities is tantamount to grasping reality; they accept without discussion that the striving for invariance is at the same time a striving towards reality in the absolute; they cannot figure out that universally valid relations, as expressed by a formalism, do not tend unavoidably to be identical with (ontologically) real relations. Their main argument is that, *by definition*, the sought absolute reality has to be independent of any particular perspective and of any special mode of experimental investigation. *Therefore*, they say, increasing the range of perspectives and modes of experimental investigations with respect to which our formal models are independent, can but bring us nearer and nearer from reality in the absolute. But this reasoning is manifestly flawed. First, the fact that invariance with respect to any generalized standpoint is a *necessary* condition for defining "absolute reality" does not entail that it is a *sufficient* condition. The absoluteness of this reality has invariance as a consequence, but the converse has yet to be proved. Second, there is a gap between defining a concept of "absolute reality" *in abstracto* and trying to characterize it. This is so because the very acceptance of the definition of an "absolute reality" makes the project of finding out its determinations self- defeating. Indeed, if this definition is taken at face value, the project has to assume that it makes sense to seek what is reality independently of any activity of seeking; or to characterize reality relative to no procedure of characterization at all [61]. Characterizing something, even in such a way that part of the characterization becomes invariant with respect to contexts and perspectives, involves two steps, not one. It involves one step of defining determinations *relative* to a large (but not arbitrary) class of contexts, and then another step of abductive[6] extraction of a *stable* element among these determinations. Asserting that this invariant tends to represent something of an "absolute reality" disconnected from *any* contextual background, is only possible if one has forgotten the initial step of the procedure by means of which the invariant was extracted. In order to avoid this inaccuracy, one should not lose sight of the fact that the kind of universality and invariance science is able to reach only holds for a wide class of perspectives, of methodological approaches, and of interests within the world; it does *not* hold for some utopic "nowhere" having nothing to do with perspectives, situations, methods and interests. As F. Klein would have it, every invariant must be referred to its group of symmetry; it is only the invariant *of this group*. Here again, objectivity implies independence with respect to situations belonging to a certain comprehensive class; it does not imply absolute lack of relevance of the concept of situation. The most obvious reason why many scientists (and also philosophers of the analytic tradition) are so prone to forget it, is that they just

[6]in the sense of C.S. Peirce.

happen to be immersed in these situations, to adopt these approaches, and to share these interests. But there is also another, more subtle, reason for this forgetfulness. It is the philosophical circumstance that if the emergence of invariants of a wider and wider class of modes of investigation is not to be ascribed to some convergence towards some pre-structured independent reality, then one usually does not know how to explain it.[7] This kind of remark can be found, for instance, in B. Williams' *Ethics and the limits of philosophy*: "In a scientific inquiry there should ideally be convergence on an answer, where the best explanation of the convergence involves the idea that the answer represents how things are" [63]. The problem is that the strong version of this belief, according to which realism "(...) is the only philosophy that does not make the success of science a miracle" [64], might well arise from the same family of prejudices as that of somebody who identifies himself so strongly to a certain set of perspectives (here a familiar set of foundational perspectives) that he loses sight of the fact that they are nevertheless only perspectives. Here, freeing oneself from the prejudice would mean remaining open to a variety of interpretations of the evolution and of the success of scientific theories. After all, the so-called "convergent realism" is not necessarily the best, and by no means the only, explanation of the growing generality and success of the invariants of scientific investigation. On the one hand the soundness of this explanation has recently been challenged with a series of strong arguments coming from the neo-rationalist and the neo- empiricist philosophies of sciences as well [65]; and on the other hand alternative explanations are not out of reach. Among these alternatives, let me emphasize the pragmatic-transcendental explanation of the success of quantum mechanics I have myself suggested [66]. It consists in showing that it is perfectly possible to regard the Hilbert-space structure of quantum mechanics, and the general form of its equations of evolution, as an embodiment of the necessary pre-conditions of a wide class of activities of seeking and predicting. This being granted, the quantum theory no longer appears as a reflection of some (exhaustive or non-exhaustive) aspect of a pre-given nature, but as the structural expression of the *co-emergence* of a new type of experimental activity and of the 'factual' elements which constrain it. Nothing then prevents one from extending tentatively the latter conclusion to other branches of physics, and to cognitive activities in general.[8] This would involve recognition that the major invariants of scientific theories are neither to be taken as a direct expression of some independent reality, nor as the mere projection of the structure of our thought, but as a formal expression of the conditions for the *co-stabilization* of a class of objects and

[7]On this point, see the long and thorough discussion of B. d'Espagnat, in [62].

[8]This view of cognitive science was developed by [67].

its (bodily or instrumental) modes of investigation. In terms borrowed from F. Varela et al., "cognition in its broadest acceptation consists of *enaction*, that is to say making a world emerge through a viable history of structural coupling" [68].

In view of this alternative orientation of the philosophy of science, there is no reason left to give any *metaphysical* priority to form over actuality, or to invariants over the flux of appearances, beyond the *epistemological* priority it understantably has for scientists.

I am aware that it sounds paradoxical to advocate renouncement to any attempt at catching some absolute reality beyond phenomena by invoking Kant's dissociation between ontological reality and objectivity. After all, the elementary concept of a 'thing-in-itself' underpinning the immanent appearances is likely to stimulate, rather than to inhibit, the project of looking for something immutable and true below the changing and sometimes deceptive actuality. This is so because the dualism of phenomenon and thing-in-itself unavoidably generates a representation of transcendence.

But one should not forget that the concept of 'thing-in-itself' has undergone a momentous evolution in the work of Kant, and then in the analysis of the successive generations of neo-Kantian philosophers. The key-process of this evolution was that of a progressive merging of the 'thing-in-itself' into the flux of immanence. According to L. Ferry, for instance, "The thing-in-itself should no longer be construed as a cause of the representations, *but as the very fact of representation*" [69]. Along with this move, the concept of 'thing-in-itself' has completely lost the power of suggesting that there is something *out there* which causes the appearances, and that scientists tend to grasp it asymptotically by identifying more and more comprehensive formal invariants. *Accordingly, reality is no longer construed as something very deep, very abstract, very general, far beyond the narrowly located actualities, but as essentially akin to actuality in general.*

It is interesting to notice that this immanent conception of reality, which was advocated long ago by neo-kantian or pragmatist philosophers, is also pervading the views of some contemporary realist and materialist philosophers. T. Nagel, who defines his own position as a variety of realism, emphasized repeatedly that objectivity and invariance do not exhaust reality: "The way the world is *includes (local) appearances*, and there is no single point of view from which they can all be fully grasped" [70]. As for M. Lockwood, who presented an interesting materialist view of mental processes in his *Mind, brain and the quantum*, he insisted that objective knowledge of the brain events by means of perception and elaboration of formal models is only *one* possible way of access to these events. Another way in which the same events might be known is "(...) self-awareness: knowing certain brain events (...) 'from the inside', by living them, or one might almost say, by self-reflectively *being* them" [71]. The actual complex of experienced *qualia*, of perceptive identification, and of intentional directedness, precisely represents this kind of apprehension of reality from within, according to M. Lockwood. If one makes an exception of the inaccuracy which consists in conflating the internal and external standpoints, namely asserting that one has knowledge 'from the *inside*' of a series of events (the brain events) which were initially defined relative to the *external* mode of access, this remark contains an important insight. It consists in pointing out that actuality is not to be considered, dualistically, as a pure local appearance-for-us of some transcendent reality which formal models tend to describe. Actuality should rather be thought of, non-dualistically, as an admittedly bound and partial *mode* of the immanent reality self-reflectively *being* itself.

As I mentioned previously, the second strategy tending to undermine the transcendental illusion is Wittgenstein's "therapy". As Wittgenstein writes, "The philosopher's treatment of a question is like treatment of an illness" [72]. Now, the etiology of this illness is not very difficult to elucidate: it is the powerful spell of language. As a consequence, "Philosophy is a battle against the bewitchment of our intelligence by means of language" [73]. The Wittgensteinian therapy is then primarily directed against the philosophical disease which consists in reifying the presuppositions of everyday life and speech, and elaborating a metaphysics out of this. Accordingly, the job of a Wittgensteinian philosopher consists in undoing the complex metaphysical architectures inherited from past philosophy, and pointing out its roots in the use of language. Each possible locus of bewitchment by language has to be explored in turn. A reasonable list includes the use of substantives, predicates, and (grammatical or mathematical) rules.

To begin with, one of the greatest source of "philosophical bewilderment" is that "a substantive makes us look for a thing which correspond to it" [74]. Substantives like 'meaning' or 'truth' seem to force us to point to something, and our incapacity to do so produces a "mental cramp". It is only if one transforms the question of correspondance into a question of use, that the mental cramp is cured at its source. Wittgenstein's criticism of our fascination for substantives also extends to what one might call the urge for Substances, namely for a metaphysical ground of the division of 'all that is the case' into individualized intrinsically existent objects. Firstly, we have no need of such a ground: "Children do not learn that books exist, that armchairs exist, etc. - they learn to fetch books, sit in armchairs, etc." [75]. Secondly, attempting to identify a metaphysical ground unavoidably generates sceptical reactions which are almost impossible to overcome by arguments. Our certainties do not arise from any firm ontological knowledge;

they only express the interplay of our linguistic and gestural practices. "My life shews that I know or am certain that there is a chair over there" [76]. "The end (...) is an ungrounded way of acting" [77]. One teaching quantum physicists should draw from this analysis of certainties concerns the emergence of a macroscopic quasi-classical world from the so-called quantum world. Instead of trying desperately to make the macro-world come out of the internal functioning of the Hilbert space model construed as a reasonably faithful description of reality, they should realize that this macro-world partakes of the ungrounded knowing-how of experimenters; and that their tentative theoretical knowing-that being based on this initial *knowing-how*, it can pride itself on no logical or metaphysical priority whatsoever. My reading of the decoherence theories in paragraph 3 was in good agreement with this anti- foundationalist stance. It said that decoherence does *not* show that the appearance of a classical world can literally be grounded on a real quantum world. Decoherence only displays the possibility of a reasonable quantitative agreement between the initial knowing-how of the quantum physicist and his/her theoretical end-product; it is part of a demonstration that the overall epistemic process can be made self-consistent, in spite of its being ungrounded. A metaphor used by Wittgenstein nicely expresses this substitution of a feed- back loop for the traditional foundationalist stratified scheme: "(...) one might almost say that these foundation-walls are carried by the whole house" [78].

Another aspect of the philosophical illness to be cured is fascination with concepts. Concepts seem to require rigid limits, and therefore possibility to locate unambiguously an object on one side or the other of the limit. This is a prerequisite for the extensional definition of concepts, and this appears to be indispensible if one is to grasp a true 'natural kind' by means of a concept. But, says Wittgenstein, the situation in which we may define the strict limits of our concepts is exceptional. The ideal of such a situation has a purely regulative function, and the meta-concept of 'natural kind' is to be construed as a way of hypostasizing this ideal. The usual case is that of a fuzzy definition of the domain covered by a concept, by means of some 'family resemblance' [79]. Of course, one could argue against Wittgenstein that 'family resemblance' is only useful in everyday language, and that science has nothing to do with it because it provides strict definition of its concepts. But even here, things are not so clear-cut. Enactment of a concept, i.e. *making use of it*, in actual experimental science as in actual life, supposes a sufficient plasticity of its form. An interesting example is provided by modern physics, whose persistent talk of "particles" has only been made possible by a remarkable capacity of extending the range of this concept well beyond what would have been acceptable in the context of classical science, and by acceptance of a certain amount of extensibility of its limits. "Particle" is what H. Putnam [79] would call a "broad spectrum notion".

The third and last element of the philosophical disease in Wittgenstein's sense, is the belief that, when we perform an ordered activity, we follow an inner rail called a 'rule'. But this way of putting things is misleading. For saying that somebody's actions are in accordance to a rule is not tantamount to saying that the person is explicitly guided by the rule. According to S. Kripke's reading [80] of Wittgenstein's analysis of the process of rule-following, one should then completely revert the priorities between the rules and the forms of life. In the same way as, in Hume's analysis of causation, one should not say that regularities manifest underlying causal powers, but rather that speaking of causal powers is a way of integrating the regularity within one's discourse, in Wittgenstein's analysis of rule-following, one should not say that regular behaviour manifests a real 'internal rail' called a rule, but rather that speaking of rule-following is a way of integrating the regular behaviour (and its more or less explicit normative ideal) within one's language game [81]. Such a reversal of priorities, if extended to science, has momentous consequences. It means renouncing the logical priority usually given to laws or to symmetries over the delegates of actuality called 'measurement outcomes'. And it pushes one to consider that these laws, or the propensities associated to these symmetries, are only a way of integrating the (deterministic or statistical, certified or *expected*) regularities of measurement outcomes within the project one ascribes to the experimental game of seeking and finding. This is the general version of the "Copernican revolution" of science which has already been documented in some readings of statistical physics and quantum mechanics (see §2 and §3 of this paper).

Finally, I must sketch briefly the third and most radical strategy tending to undermine the transcendental illusion. This strategy is that of the soteriological discourse of Indian thought. It goes beyond mere criticism of the hypostasis of both the heuristic principles of science and the regulative ideals of metaphysics; it goes beyond the therapy of the 'mental cramps' of those philosophers who look for substances underlying the substantives; it does not content itself with a philosophical cure of the philosophical temptation to reify the tacit guiding principles of everyday life and speech. It aims at drying up the very source of these wanderings, by moving to a level of awareness where the basic presuppositions which underly our action, our discourse, and even the way we *see* things, do not operate any longer. After all, one should not forget that endowing the regulative ideals of human investigations with a metaphysical significance, is a tendency which has its roots deep into the *natural ontological attitude* of the man-in-the-street. Whereas Kant and Wittgenstein only aimed at denouncing the *philosophical* consequences of the reification of substantives and theoretical entities used respectively in philosophical extrapolations and in scientific extrapolations of everyday speech, Indian soteriology had (and still has) the project of unrooting the *natural ontological attitude* of everyday life itself, by carefully identifying and defusing its existential motivations.

Before we examine the way this project is carried out, we have to direct our attention towards two aspects of the opposition between form and actuality which have been overlooked until now; two aspects which are really crucial if we are to understand the alien attitude of the Indian civilization with regard to form. The first aspect is the connection of form with the *future.* And the second aspect is what we could refer to as the *entanglement* of form and actuality. At first sight, actuality is restricted to the present whereas formal models allow one to master the future by means of their predictive contents. But things are not so simple; these two judgments have to be qualified in turn. On the one hand, pure actuality is not averse to an internal orientation, called intentionality, towards the future (see below for more details); yet this latent future of intentionality is likely to be more open than the enlisted future of predictive formalisms.

On the other hand, as N. Goodman emphasized [82], any formalized *projective* attempt is bound to have a basis in the present and the past. However, this basis is not necessarily restricted to present and past *facts*, as it would be the case in mere induction. According to Goodman, it rather extends to current and past successful *predictions.* A new projective hypothesis or formalism is not adopted if it only agrees with a finite set of past *facts*: it is accepted if it is more comprehensive than past *hypothesis*, and if it does not contradict the most entranched elements of the previous *overall projective network.* In other terms, a new projective formalism does not depend *anecdotally* on the past, but it depends *holistically* on it. A good illustration of this situation in physics is the way new theories take previous theories as their limiting case in a restricted domain of validity. It thus becomes clear that the relation between predictive formalisms and the future is quite ambivalent. They allow a reasonably reliable projection into the future, but they also tend surreptitiously to present the future as a more or less complete continuation of the past. Their very ideal of mastery of nature implies the belief that, some day, a 'Theory of everything' will enable us to behave with respect to the future with the same confidence and the same feeling of closure as with respect to the past [83]. The only two circumstances that leave this project in suspense is

(i) the current incompleteness of physical theories, which leaves room for further scientific revolutions, and
(ii) the element of irreducible indeterminism incorporated in these theo-

ries.

Let us now come to the problem of the entanglement between form and actuality. This issue is all the more important since it may retrospectively cast a doubt on the clearcut distinction we have accepted until now. At this point, we must take into account the thorough criticism which has been directed by the psychology and the philosophy of the twentieth century against the traditional Kantian divide between the pure 'matter' of sensation and the forms (of intuition and of thought). According to the Gestalt psychologists, to begin with, a perception does not split up into a purely passive sensorial input and an intellectual activity of interpretation; it so to speak carries its interpretation with it. There is not on the one side a pure present actuality devoid of any predictive element, and on the other side an intellectual projective form which takes the aspect of an explicitly stated set of hypothesis. Rather, as J. Bouveresse notices, "perception is the hypothesis" [84]. Perception incorporates tacit rules of anticipations which can be formalized retrospectively.

This point was repeatedly insisted upon by the phenomenological tradition. Husserl's analysis of what remains after the 'bracketing' of the *natural attitude* has taken place, involves what he calls a 'noema'. This 'noema' may be construed as a stable form which persists from one appearance to another, which incorporates an immanently intentional aspect in it, and which therefore operates as a generalization of the notion of meaning [85]. Now, despite its formal components, the noema is not separable from the whole act of perception. As Husserl explained, perception can be taken as a low-level epistemic operation, or an *implicit judgment* [86], even though the epistemic operations and judgements proper can only arise after a certain amount of elaboration of the ante-predicative layer of perception has taken place.

Later on, M. Merleau-Ponty amplified the anti-intellectualist stance of phenomenology. He especially emphasized that, unlike Kantian philosophy (especially in its neo-kantian reading), phenomenology deals with a formal and intentional component of perception which cannot be reduced to any operation of the understanding. According to Merleau-Ponty, "Each part (of experience) foretells more than its contents, and therefore this elementary perception is already loaded with *meaning*" [87]. Perception goes beyond the stage of a mere flux of sensations even before the intervention of the categorical forms of thought.

Wittgenstein himself undertook a remarkable 'grammatical' (and sometimes phenomenological) analysis of the formal aspect of actual perceptions, which he named 'seeing as'. The examples he gave range from the 'duck-rabbit' to the Necker cube. The drawing of a 'duck-rabbit' can (obviously) be perceived either *as* a duck or *as* a rabbit. As for the cube, it can be perceived under two three-dimensional orientations, which are equally compatible with the two- dimensional image on the paper; it can also be seen as the representation of many distinct objects. The key point of Wittgenstein's reflection is the major difference between *seeing* and *interpreting*. Even though we really see a certain illustration under one or another form, "(...) it is remarkable that we can equally use the term *interpretation* for describing what is immediately perceived!" [88] . We actually see something as such and such object, but, after careful retrospective analysis, we can also conceive this *view* as loaded with interpretation. Just in the same way as the fact that somebody's reliable behaviour can retrospectively be considered as a sign that this person is explicitly guided by a rule.

In view of this pre-verbal entanglement of actuality and form, as well as of present perceptions and intentionally aimed at future, the task of somebody who would like to defuse the *natural ontological attitude* appears very difficult, not to say hopeless. At any rate, disentanglement should involve an attempt at working below the level of verbal reasoning and argument. Now, this move from the verbal to the pre-verbal is exactly what most Indian philosophies would recommend. The universal method they use for that, to wit the *Yoga* [89], is several thousands years old. The yoga has some western analogs, ranging from psychological introspection to phenomenological reduction, not to mention mystical contemplation. But this very multiplicity of equivalents shows that no such analogy is perfect [90]. For instance, the yoga differs radically from its much criticized western counterpart (introspection), in so far as it is intrinsically non-dualistic. However, it has more interesting connections with phenomenological reduction (or 'bracketing'). Indeed, in phenomenological reduction, as in yoga, "(...) the only reason why [one] should bracket the belief in the existence of the world is to *see* it (...)" [91]. In phenomenological reduction, as in yoga, "(...) [one] must sink into the world instead of dominating it" [92]. But there are also two major differences between the yoga and the phenomenological reduction, bearing on their maturity and on their function. Firstly, the yoga has been much more carefully codified than the phenomenological reduction during its long life-span, and this allowed systematic teaching. Secondly, whereas the function of phenomenological reduction is primarily epistemic, yoga's is soteriological: it aims at freeing man from his/her worldly bondage.

Now, the fact that the methods recommended to disentangle form, intention, and actuality, *have* to be non-verbal while they are used, does not prevent one from expressing their essential features in retrospect. One must only remember that this expression does not provide a faithful description of what is at stake (this would be self-contradictory), but only a *guide* intended for other *practioners*. The Indian soteriologists as well as the western phenomenologists have made several attempts in this direction. A most remarkable point about these attempts is that Indian practioners of yoga and phenomenologists are in reasonable agreement with one another; and that moreover their views are both compatible, on the experiential side, with the objectifying descriptions of some specialists of cognitive science [93].

Let us begin with practitioners of yoga (for, after all, they have a historical priority). According to them, our bondage, to wit our being compelled to adopt the *natural ontological attitude* whereby we see things dualistically, (essentially) comes from desire, action, and grasping. Desire is motivated by our relating every appearance to our egocentric needs. Action arises from an understandable attempt at appeasing the desire. And grasping is the consequence of one's hope that it is possible to *freeze* the situation wherein the needs are satisfied (and the threats avoided). Accordingly, the yoga tends to dissolve the ego, to liberate the action from craving, and to release the urge of grasping:

(i) Promoting the immediate awareness of the cosmic unity (the 'That art thou [Tat tvam asi]' of the Upanishads [94]) is obviously a radical way of dissolving the individual ego.

(ii) The perfect immobility of the yoga practitioner is a first (though extrinsic and superficial) approach towards the solution of the problem of action. As Nâgârjuna writes, "The root of cyclic existence is action, Therefore the wise one does not act" [95]. However, the latter sentence is best understood not in the sense of complete and definitive withdrawal from active life, but in the sense of an attempt at freeing the acts from the chain of conditions. This idea is remarkably conveyed by M. Eliade: "the yoga recommends to live, but not to remain the instrument of life" [96].

(iii) Finally, since grasping is part of an attempt at casting out the effect of time, it has to be counterbalanced by a meditation on impermanence. This does not mean that a yogi (or yogini) does not care for the future; but he/she equanimously considers each present act as a seed which may or may not sprout, rather than as part of a heroic attempt at dominating time by reaching the immutable form of nature.

Once this process is completed, one reaches "The pacification of all objectification and the pacification of illusion (...)" [97]. The world is no longer *seen* as a collection of individual substances corresponding to the lexical substantives; rather, one "(...) *see(s)* things *as* they are - as merely (...) dependent, impermanent and non-substantial (...)" [98]. Unlike Kant, Indian (especially Buddhist) thinkers thus consider that the transcendental illusion *can* be overcome, and they give detailed and coherent instructions for that [99].

Further information about this process of emancipation from the very source of every metaphysical wandering can be found in contemporary philosophy and cognitive science. They also concern the three points which have already been documented, namely egocentration, action, and time.

To begin with, M. Merleau-Ponty repeatedly criticized the subjectivist tendency of transcendental philosophy. According to him, if I am to overcome the transcencental illusion, "(...) if I am to be ek-statically within the world and the things", then "nothing must retain me far away from them (...) not even this (philosophical) description of myself as a 'subject', as a 'mind' or as an 'ego' (...) which reintroduces in me a ghost of reality and arouses the belief that I am a *res cogitans* (...)" [100]. I must no longer represent myself as some separate entity facing Being, for my view on Being arises from within the midst of it [101]. This is an indispensible preliminary step in order to recognize that things and minds are all "(...) differences or extreme gaps of an unique something" [102]. And this obviously also prevents one from thinking that he/she can reach the position of a *Kosmotheoros* [103], a pure abstract detached observer seeing things from nowhere.

The second point is action. It is well accepted in contemporary philosophy that perception and intentionality cannot be separated from both the motivating forces of desire and the schemes of activity. In some early writings of Husserl's, one finds that these issues are completely intermingled: "The interest is not at rest, it is not bound to the image; it struggles to escape from it. Therefore, there arises an *intention*, a tensed interest, which tends to confront the matter. If nothing happens next, we feel tensed, dissatisfied; if it happens, we feel satisfied" [104]. Intentionality, and the 'noema' which is the formal basis of it, thus clearly have a partly pragmatic background which Husserl expressed in phenomenological terms.

This pragmatic background of the formal component of any perception (the *seeing-as*) has been suggested even more insistently by Wittgenstein, even though he finally reverted to a purely 'grammatical' analysis. As J. Bouveresse notices [105], seeing a drawing *as A* or *as B* might well depend on a different ordering of the eye-movements which come before recognition; it may also depend on the different activity schemes which are mobilized when something is perceived as one or as another possible objects of manipulation.

To summarize, Husserl and Wittgenstein both recognized a pragmatic component of intentionality or *seeing-as*. But they also both rejected the temptation of naturalizing it; they rather gave it the transcendental position of a phenomenological or grammatical pre-condition. This does not mean that the transcendentalist trend of thought dismisses any approach of the same issues by the methods of a natural science, but only that, if such an investigation were carried out, its outcomes should have to be related with the phenomenological / grammatical analysis in terms of parallelism [106] or supervenience, rather than in terms of reductionism or materialist foundationalism. This precaution being taken, it is very interesting, even from a phenomenological or Wittgensteinian standpoint, to notice that cognitive sciences have *also* recognized the intermingling of perception and action. A classical experiment [107] (commented by Varela et al. [108]) for instance shows that when kittens have been made completely passive, they lose the possibility of perceiving standard obstacles.

Along with these remarks, it becomes clearer than ever that disentangling actual perceptions from its formal-intentional component, to wit allowing one to *see* rather than to *see-as*, presupposes either suspension of action or lucid identification of the motor schemes involved in perception as soon as they operate. This is the price which would have to be paid if our life-long commitment to the *natural ontological attitude* were to be alleviated. In the same way as one has to step back from a scientific practice and from its regulative ideals in order to overcome the special variety of transcendental illusion which is associated to it, one would have to step back from *any* practice, or at least from any *commitment* to practices, in order to overcome the ultimate source of transcendental illusion. That this is possible is usually accepted without difficulty in the East. Whether this is desirable or not in our Western context, is obviously an open question, but the mere *possibility* of such a radical move should modify thoroughly, in the long term, our basic epistemological attitudes. This is one of the most important aspects of the *Renaissance* of the Western outlook [109], which could be prompted by our ever more insistant contact with the Eastern culture.

Finally, there is the question of time. As I mentioned in the introduction, science as a whole can be construed as the most advanced attempt of mankind for mastering the productiveness of/in time. Defining stable objects is the usual first step of this process. The second step consists in identifying the law-like behaviour of the changes in these objects, for, as M. Schlick explained: "(...) the permanent in an alteration is called its *law*" [110]. As for the third step, it is the project of elaborating an all-encompassing and hopefully immutable formalism (the 'final' theory 'of everything'), which nevertheless accounts for every 'apparently' fluent aspect of *all that is the case*. This is our civilization's most elaborate version of the urge to lock time up [111]. Of course, the attitude of scientists towards time is much more ambivalent than that in their everyday work. True, their project of formulating a theory of everything, as well as their attempts at encompassing every single transient feature of phenomena within the scope of their formalisms, manifests a neo-Parmenidean project. But as soon as a difficulty arises, implementation of the project is postponed until an indeterminate future. This openness of the scientific future might well be the way by which the blind spot of actuality manifests itself insistently in the midst of the all-embracing formalist design.

6. CONCLUSION

The purpose of the present paper was obviously not to prevent one from using the method of formalization in science, but rather to remind one repeatedly of its restricted *function*. This function is to elaborate a coherent, integrated, and universal system of *projection* (in Goodman's sense) for any activity of exploration of our environment. Coherence is provided by the precise (deductive and abductive) rules to which the formalism is subjected; integration prevails as a regulative ideal; and universality (namely validity irrespective of location, time, and individual) arises from the tendency of formalisms towards symmetry and abstraction. As long as these statements are borne in mind, nobody can loose sight of the fact that actuality is an indispensible *presupposition* of formalization, not a by-product of the entities postulated by a theory associated to that formalism; that formalisms are elaborated to *anticipate* particular actualities as efficiently and as universally as possible, not to *justify* the existence or the characteristics of any given actuality. It is only when one looses trace of very process of abstraction which led to the formalism that it becomes tempting to ascribe to the formalism the impossible task of accounting in retrospect for the existence and characteristics of actuality.

These remarks provide us with a useful insight into what should be expected from a *formal epistemology*. Instead of adding one more step to the process of formalization, thus favouring the process of forgetfulness of the restricted function of formalisms, a formal epistemology should promote the clarification of the origin and purpose of presently available predictive formalisms in physics and in other sciences. This can easily be done if one accepts, as I did in my paper "Formal epistemology, logic and grammar", that a formal epistemology is only meant to *show* or to *manifest* the coordinated structure of anticipation of physical theories, not to elaborate one more theory; exactly in the same way as logic and grammar are only meant to *show* or to *manifest* the tacit rules of use of (actual or possible) languages, not to add something to language.

REFERENCES

1. M. Bitbol and E. Ruhnau, eds., *Now, Time and Quantum Mechanics*

(Frontières, 1994).
2. G. W. F. Hegel, *The Phenomenology of Mind*, translation by J. B. Baillie (Library of Philosophy, 1931), Part A, Chap. I.
3. E. Ruhnau, "The now, the missing link between matter and mind," in M. Bitbol and E. Ruhnau, eds., *op. cit.* [1].
4. K. Popper, *Postscript to the Logic of Scientific Discovery* (Hutchinson, 1982), §26.
5. E. Ruhnau, "The now, the missing link between matter and mind," in M. Bitbol and E. Ruhnau, eds., *op. cit.* [1]. About the theme of the missing *now*, see also: J. Schneider, "The now, relativity theory and quantum mechanics"; S. Saunders, "Time and quantum mechanics," in M. Bitbol and E. Ruhnau, eds., *op. cit.* [1]. According to S. Saunders, for instance, *in both cases (quantum mechanics without state reduction and spacetime theory without 'time flow') something fundamental seems to be missing in physics*; M. Bitbol, "The missing now," *Contextos (University of Leon, Spain)* **VI** (11), 17-31 (1988).
6. B. C. Smith, *On the Origin of Objects* (MIT Press, 1996).
7. J. Hintikka and I. Kulas, *The Game of Language* (Reidel, 1983), p. 33. J. Hintikka, *Knowledge and the Known* (Reidel, 1974).
8. I. Kant, *Critique of Pure Reason*, new edition by V. Politis (Everyman's library, 1993), A32-B47.
9. E. Schrödinger, "The present situation of quantum mechanics," in J. A. Wheeler and W.H. Zurek, *Quantum Theory and Measurement* (Princeton University Press, 1983).
10. For example, E. T. Jaynes, "Gibbs versus Boltzmann entropies," in E. T. Jaynes, *Papers on Probability, Statistics, and Statistical Physics*, R. D. Rosenkrantz, ed. (Reidel, 1983).
11. I. Prigogine, *From Being to Becoming* (Freeman, 1980). I. Antoniou and I. Prigogine, "Intrinsic irreversibility and integrality of dynamics," *Physica* **192A**, 443 (1992). I. Prigogine, *Les lois du chaos* (Flammarion, 1994).
12. I. Stengers, *Cosmopolitiques 5. Au nom de la flèche du temps: le défi de Prigogine* (Editions La Découverte, 1997).
13. *Ibid.*, p. 150.
14. M. Mugur-Schächter, "From quantum mechanics to universal structures of conceptualization and feedback on quantum mechanics," *Found. Phys.* **23**, 37-122 (1993).
15. N. Bohr, "Chemistry and the theory of atomic constitution," *J. Chem. Soc.* 349-384 (1932), reprinted in N. Bohr, *Collected Works*, Vol. 6, J. Kalckar, ed. (North-Holland, 1985). Th. Görnitz, E. Ruhnau, and C. F. von Weizsäcker, "Temporal asymmetry as precondition of expe-

rience," *Int. J. Theor. Phys.* **31**, 37-46 (1992). M. Bitbol, "Prélude l'irréversibilité," *Sciences et avenir* (Numéro spécial, Les énigmes du temps), March 1994.
16. I. Stengers, *Cosmopolitiques 1. La guerre des sciences* (La Découverte, 1996), p. 107.
17. N. Bohr, *Collected Works, op. cit.* [15], p. 400.
18. *Ibid.*
19. Th. Görnitz, E. Ruhnau, and C.F. von Weizsäcker, *loc. cit.* [15].
20. H. D. Zeh, *The Physical Basis of the Direction of Time* (Springer, 1989). R. D. Sachs, *The Physics of Time-Reversal* (University of Chicago Press, 1987).
21. W. V. Quine, *The Roots of Reference* (Open Court, 1974), p. 2.
22. B. d'Espagnat, "Towards an empirical separable reality?," *Found. Phys.* **20**, 1147-1172 (1990).
23. G. C. Ghirardi, A. Rimini, and T. Weber, "Unified dynamics for microscopic and macroscopic systems," *Phys. Rev. D* **34**, 470-491 (1986).
24. J. S. Bell, *Speakable and Unspeakable in Quantum Mechanics* (Cambridge University Press, 1987), p. 202.
25. P. Mittelstaedt, *The Interpretation of Quantum Mechanics and the Measurement Process* (Cambridge University Press, 1997).
26. M. Dickson, "What is preferred about the preferred basis?," *Found. Phys.* **25**, 423-440 (1995).
27. R. Omnès, *The Interpretation of Quantum Mechanics* (Princeton University Press, 1994), p. 354.
28. J. P. Paz and W. H. Zurek, "Environment-induced decoherence, classicality, and consistency of quantum histories," *Phys. Rev. D* **48**, 2728-2737 (1993).
29. S. Saunders, "Decoherence, relative states, and evolutionary adaptation," *Found. Phys.* **23**, 1553-1585 (1993); "Time and quantum mechanics," in M. Bitbol and E. Ruhnau, *op. cit.* [1]. B. d'Espagnat, *loc. cit.* [22].
30. M. Gell-Mann and J. B. Hartle, "Classical equations for quantum systems," *Phys. Rev. D* **47**, 3345-3382 (1993).
31. A. Shimony, *Search for a Naturalistic World View, Vol. I* (Cambridge University Press, 1993); see also Gell-Mann's IGUS concept, in M. Gell-Mann and J. B. Hartle, *loc cit.* [30].
32. N. Bohr, *Essays 1958-1962 on Atomic Physics and Human Knowledge* (Ox Bow Press, 1987), p. 60.
33. D. Murdoch, *Niels Bohr's Philosophy of Physics* (Oxford University Press, 1987), p. 113.
34. M. Mugur-Schächter, "Mécanique quantique, réel, et sens," in M. Bit-

bol and S. Laugier, eds., *Physique et réalité, un débat avec Bernard d'Espagnat* (Frontières-Diderot, 1997), pp. 138 and 150. An English version of this paper was published in *Found. Phys.*
35. S. Saunders, "Time and quantum mechanics," in M. Bitbol and E. Ruhnau, *op. cit.* [1].
36. S. Saunders, *ibid.*
37. M. Bitbol, *Mécanique Quantique: Une Introduction Philosophique* (Flammarion, 1996); *Schrödinger's Philosophy of Quantum Mechanics* (Boston Studies in the Philosophy of Science, Kluwer Academic, 1996).
38. M. Richir, *La crise du sens et la phénoménologie* (Jérme Millon, 1990).
39. L. Wittgenstein, *Tractatus Logico-Philosophicus* (Routledge, 1974), Sec. 1: "The world is all that is the case."
40. H. Wang, *From Mathematics to Philosophy* (Routledge & Kegan Paul, 1974).
41. J. R. Lucas, "Minds, machines, and Gödel," in A. R. Anderson, ed., *Minds and Machines* (Prentice-Hall, 1964).
42. R. Penrose, *Shadows of the Mind* (Oxford University Press, 1994); "Why new physics is needed to understand the mind," in M. P. Murphy and L. A. J. O'Neill, eds., *What is Life?, The Next Fifty Years* (Cambridge University Press, 1997).
43. R. Penrose, "Why new physics is needed to understand the mind," *loc. cit.*
44. J. R. Lucas, "Minds, machines and Gödel," *Philosophy 1961*; reprinted in A. R. Anderson, ed., *op cit.* [41], p. 47.
45. H. Wang, *op. cit.* [40], p. 325.
46. P. Benacerraf, "God, the Devil, and Gödel," *Monist* **60**, 9-33 (1967).
47. J. C. Webb, *Mechanism, Mentalism and Metamathematics* (Reidel, 1980).
48. J. C. Webb, *ibid.*, p. 232.
49. J. Petitot, "Objectivité faible et philosophie transcendantale," in M. Bitbol and S. Laugier, eds., *op. cit.* [34].
50. I. Kant, *Inaugural Dissertation* (1770), §4, AK II 392.
51. I. Kant, *op. cit.* [8], A298-B354.
52. For a parallel between realism and morals, see R. Harré, *Varieties of Realism* (Blackwell, 1986) and M. Bitbol, *Schrödinger's Philosophy of Quantum Mechanics, op. cit.* [37].
53. H. Putnam, *Meaning and the Moral Science* (Routledge & Kegan Paul, 1978).
54. E. Cassirer, *Determinism and Indeterminism in Modern Physics* (Yale University Press, 1956)—for a cogent criticism of this position first

suggested by P. Jordan.

55. L. W. Beck, *A Commentary on Kant's Critique of Practical Reason* (University of Chicago Press, 1960).
56. P. Müllhausler and R. Harré, *Pronouns and People* (Blackwell, 1990).
57. J. Habermas, *Nachmetaphysisches Denken* (Suhrkamp, 1988).
58. G. H. Mead, "The mechanism of social consciousness," *Journal of Philosophy, Psychology and Scientific Methods* **IX**, 401-406 (1912); "The social self," *ibid.* **X**, 374-380 (1913); both papers are reprinted in G. H. Mead, *Selected Writings* (Bobbs-Merrill, 1964).
59. P. Müllhausler and R. Harré, *op. cit.* [56], p. 35.
60. L. Wittgenstein, *Philosophische Bemerkungen* (Blackwell, 1964), §68.
61. M. Mugur-Schächter, "Mécanique quantique, réalité et sens.," C. Schmitz, "Objectivité et temporalité," and the answers by B. d'Espagnat, in M. Bitbol and S. Laugier, eds., *op. cit.* [34].
62. On this point, see the long and thorough discussion of B. d'Espagnat, in M. Bitbol and S. Laugier, eds., *op. cit.* [34].
63. B. Williams, *Ethics and the Limits of Philosophy* (Harvard University Press, 1985).
64. H. Putnam, *Mathematics, Matter and Method, Vol. 1* (Cambridge University Press, 1975).
65. L Laudan, *Science and Values* (University of California Press, 1984), especially Chap. 5. B. Van Fraassen, *The Scientific Image* (Oxford University Press, 1980). For a reply from the realist side, see J. L. Aronson, R. Harré, and E. C. Way, *Realism Rescued* (Blackwell, 1993).
66. M. Bitbol, *Mécanique quantique, une introduction philosophique, op. cit.* [37]; "Some steps towards a transcendental deduction of quantum mechanics," *Philosophia Naturalis* **35**, 253-280 (1998).
67. This view of cognitive science was developed by F. Varela, E. Thompson, and E. Rosch, *The Embodied Mind* (MIT Press, 1993).
68. *Ibid.*
69. I. Kant, *Critique de la raison pure* (Flammarion, 1987); Preface by L. Ferry, p. xix.
70. T. Nagel, *The View from Nowhere* (Oxford University Press, 1986).
71. M. Lockwood, *Mind, Brain and the Quantum* (Blackwell, 1989), p. 159; "The grain problem," in H. M. Robinson, ed., *Objections to Physicalism* (Oxford University Press, 1991).
72. L. Wittgenstein, *Philosophical Investigations* (Blackwell, 1958), §§255 and 109.
73. L. Wittgenstein, *The Blue and Brown Books* (Blackwell, 1969), p. 1.
74. L. Wittgenstein, *On Certainty* (Blackwell, 1974), §476.
75. *Ibid.*, §7.

76. L. Wittgenstein, *op. cit.*, §110.
77. L. Wittgenstein, *op. cit.*, §248. See M. Bitbol, *Mécanique quantique, une introduction philosophique, op. cit.* [37].
78. L. Wittgenstein, *op. cit.* [72], §67.
79. H. Putnam, *Définitions* (L'éclat, 1992), interview with C. Bouchindhomme.
80. S. Kripke, *On Rules and Private Language* (Blackwell, 1982).
81. *Ibid.*, pp. 97-98.
82. N. Goodman, *Facts, Fiction, and Forecast* (Athlone, 1954).
83. C. Schmitz, "Objectivité et temporalité," in M. Bitbol and S. Laugier, eds., *op. cit.* [34].
84. J. Bouveresse, *Langage, perception et réalité, Vol. 1: Perception et jugement* (Chambon, 1995), p. 335.
85. H. L. Dreyfus, *Husserl, Intentionality and Cognitive Science* (MIT Press, 1982).
86. E. Husserl, *Erfahrung und Urteil* (Glaassen & Goverts, 1954), p. 62.
87. M. Merleau-Ponty, *Phenoménologie de la perception* (Gallimard, 1945), p. 9.
88. L. Wittgenstein, *Bemerkungen über die Philosophie der Psychologie* (Blackwell, 1980), §9.
89. M. Eliade, *Le yoga* (Payot, 1954). Patajali's *Yoga-Stras*, translated by F. Tola and C. Dragonetti (Banarsidass, 1987).
90. F. Varela, E. Thompson, and E. Rosch, *op. cit.* [67]—for a thorough discussion of this point.
91. M. Merleau-Ponty, *Le visible et l'invisible* (Gallimard, 1964), p. 61.
92. *Ibid.*
93. F. Varela, E. Thompson, and E. Rosch, *op. cit.* [67].
94. A. Huxley, *The Perennial Philosophy* (Triad Grafton Books, 1985), Chap. 1. E. Schrödinger, *My View of the World* (Ox Bow Press, 1983).
95. Nâgârjuna, Mlamadhyamakakârikâ, XXVI, 10, in J. Garfield, *The Fundamental Wisdom of the Middle Way* (Oxford University Press, 1995).
96. M. Eliade, *op. cit.* [89]: *Le yoga*, p. 67.
97. Nâgârjuna, Mlamadhyamakakârikâ, XXV, 24, in J. Garfield, *op. cit.* [95].
98. J. Garfield, *op. cit.* [95], p. 332.
99. T. R. V. Murti, *The Central Philosophy of Buddhism* (Allen & Unwin, 1955), p. 300.
100. M. Merleau-Ponty, *Le visible et l'invisible, op. cit.* [91], p. 77.
101. *Ibid.*, p. 154.
102. *Ibid.*, p. 117.

103. *Ibid.*, p. 152.
104. E. Husserl, *Aufsatze und Rezensionen, 1890-1910, Husserliana XXII,* B. Rang, ed. (Nijhoff, 1979), p. 411; French translation in E. Husserl, and K. Twardowski, *Sur les objets intentionnels* (Vrin, 1993), p. 262.
105. J. Bouveresse, *op. cit.* [84], p. 357.
106. About parallelism in the philosophy of mind and the philosophy of physics, see M. Bitbol, *Schrödinger's Philosophy of Physics, op. cit.* [37], Chap. 6.
107. R. Held and A. Hein, "Adaptation of disarranged hand-eye coordination contingent upon re-afferent stimulation," *Motor Skills* **8**, 87-90, (1958).
108. F. Varela, E. Thompson, and E. Rosch, *op. cit* [67], p. 237.
109. *Ibid.*, p. 51. R.-P. Droit, *L'oubli de l'Inde* (P.U.F., 1989).
110. M. Schlick, *Philosophical Papers* II, *1925-1936* (Reidel, 1979), p. 8.
111. C. Schmitz, "Objectivité et temporalité,"in M. Bitbol and S. Laugier, eds., *op. cit.* [34].

13

TO SUSPENDED INFORMAL TIME

Secret Note Penned by Michel Paty during a CeSEF Session on Time

> "That laughter and humour on thyself have the power to abolish of time the unyielding flight." — *Pseudo Rabelais-and-Montaigne.*

After endless discussions and arduous, though not uninteresting, debates about a formal construction of time, time in all its forms: physical, cosmic, biological, psychological, terrestrial and lunar, solar and planetary, social and structural, fractal and holistic, intrinsic or contingent, and others that might be experienced in this life or mentally conceived of in this or other lives, and parallel times (which in fact would be orthogonal because evidently in such times geometry would not be Euclidean), one participant in the Workshop on Formal Thought about the Absolute Relativity of Time, during the Session on the Relatively Absolute Character of the Instant, asked with a fresh ingenuousness that came unexpected in such a place at such an instant (about to conclude an austere afternoon of advanced abstract studies), albeit with the accents of unmistakable seriousness:

"What on earth is the use of all this?"

Someone answered spontaneously, without taking the time to think (which, too, might surprise under those circumstances):

"To kill time".

When at last the session came to its end, the studious assembly found itself tired but nevertheless quite satisfied with the results of its scholarly deliberations. In order to mitigate this admirable optimism, one of the members gave a reading of a philosophical tale he had just improvised for the occasion:

"They spent countless hours trying to construct time. During this time, time was passing. They died before their task was completed. This was a beautiful scandal: Time's irreversibility had irremissably struck before having been constructed."

14

THE CONSTRUCTED OBJECTIVITY OF MATHEMATICS AND THE COGNITIVE SUBJECT

Giuseppe Longo

Laboratoire d' Informatique
CNRS et École Normale Supérieure
45, Rue d'Ulm, 75005 Paris, France

Mathematics is engendered in conjunction with other forms of knowledge, physics in particular. It is a "genealogy of concepts" (Riemann), that stems from our active reconstruction of the world. Mathematics organizes space and time. It stabilizes notions and concepts as no other language, while isolating by them a few intelligible fragments of "reality" at the phenomenal level. Thus an epistemological analysis of mathematics is proposed, as a foundation that departs from and complements the logico-formal approaches: Mathematics is grounded in a formation of sense, of a congnitive and historical nature, which preceeds the explicit formulation of axioms and rules. The genesis of some conceptual invariants will be sketched (numbers, continua, infinity, proofs, etc.). From these, categories as structural invariants (objects) and "invariant preserving maps" (morphisms, functors) are derived, in a reflective equilibrium of theories that parallels our endeavour to gain knowledge of the physical world.

Key words: mathematics, cognitive foundation, epistemology.

> "The problems of Mathematics are not isolated problems in a vacuum; there pulses in them the life of ideas which realize themselves in concreto through out human endeavours in our historical existence, yet forming an indissoluble whole transcend any particular science."—*H. Weyl, 1949 [38].*

1. INTRODUCTION

This essay concerns the nature and the foundation of mathematical knowledge, broadly construed. The main idea is that mathematics is a human construction, but a very peculiar one, as it is grounded on forms of "invariance" and "conceptual stability" that single out the mathematical conceptualization from any other form of knowledge, and give unity to it. Yet, this very conceptualization is deeply rooted in our "acts of experience," as Weyl says, beginning with our presence in the world, first in space and time as living beings, up to the most complex attempts we make by language to give an account of it.

I will try to sketch the origin of some key steps in organizing perception and knowledge by "mathematical tools", as mathematics is one of the many practical and conceptual instruments by which we categorize, organise and "give a structure" to the world. It is conceived on the "interface" between us and the world, or, to put it in husserlian terminology, it is "designed" on that very "phenomenal veil" by which, simultaneously, the world presents itself to us and we give sense to it, while constituting our own "self".

The mathematical structures are literally "drawn" on that veil and, as no other form of knowledge, stabilize it conceptually: geometric images and spaces, or the linguistic/algebraic structures of mathematics, set conceptual "contours" to relevant parts of the enormous amount of information that arrives upon us. Yet, this drawing is not arbitrary, as it is grounded on key regularities of the world or that we "see" in the world. That is, on these regularities that we forcibly single out by "reading" them according to our own search or projection of similar patterns, as living beings: symmetries, physical and biological symmetries, or the connectivity and continuity of space and time, for example.

Intersubjectivity and history add up to the early cognitive processes; they modify our forms of "intuition", including mathematical intuition, which is far from being stable in history. Indeed, mathematical intuition is constructed in a complex historical process, which begins with our biological evolution: the analysis of "intuition", not as a "magic" or inspeakable form of knowledge, but as a relevant part of human cognition, is one of the aims of this ongoing project. A project which can be named a "cognitive foundation of mathematics", as opposed to, or more exactly, complementing the metamathematical analysis of foundations largely developed in this century. Indeed, the foundational analysis of mathematics cannot be only a mathematical challenge, as proposed by Frege and Hilbert's mathematical logic: Hilbert's metamathematics uses mathematical methods and, by this, it became part of mathematics, the very discipline whose methods or whose whole it was supposed to found. Mathematical logic gave us an essential analysis of (logic/syntactic) proof-principles, and more it is giving: yet, we also need to go further and evidentiate "what is behind" these linguistic principles, their meaning as rooted in our practices of life. Persisting *only* on the proof-theoretic, thus mathematical, analysis of mathematics, would leave us in a cognitive deadlock, actually in a philosophical or even conceptual vicious circle: one cannot found mathematical methods and tools by mathematical methods and tools. For example, no mathematical methods and tools can prove their own "consistency", which is the metamathematical way to assure meaning to a mathematical theory. Sufficiently expressive, finitary theories, such as arithmetic, have no *finitary* consistency proof; set theories which can represent a given infinity, need a larger one to be proven consistent. This is not a limitation to mathematics: in order to check the correctness of certain conceptual tools it is rather to be expected that one should "step back" from them and use different tools. Gödel's second incompleteness theorem proves exactly this: by Gödel's representation lemma, one can describe or encode the metatheory of arithmetic within arithmetic itself, which is thus part of the latter, and, then, prove that consistency is unprovable by that arithmetized metatheory. That is, the finitistic and mathematical metatheory of arithmetic is too weak to do the least job it was invented for, the proof of the consistency of arithmetic, since it is given by the same finitary tools as arithmetic itself.

And there begins the infinite regression of "relative consistency" results: if one wants to be sure that a given formal theory has a meaning (it is consistent), then one has to construct a stronger one and use it as metatheory (i.e., use induction over larger ordinal or, in set theory, assume the "existence" of larger cardinals). This is a non-well-founded chain or infinite regression of theories, which shows that the foundational relation of mathematics to mathematical logic is conceptually non well-founded, when it is exclusive. That is, when metamathematics is fully mathematized or the foundational analysis is reduced to be only mathematical, one ends into a conceptual vicious circle. This is our main motivation to go further and *step outside mathematics* and try to ground it into the networks of our forms of knowledge and of possibly pre-conceptual experiences, as part of an investigation on human cognition.

Clearly, also in the relation of cognition to mathematics there may be an (apparent) vicious circle. Jean Petitot in "The epistemology of physics versus a formalized epistemology", in this volume, raises the issues whether there may be a circularity in developing a cognitive approach to (the foundation of) mathematics, as suggested here, and, at the same time, a mathematical analysis of cognition, which he and many others are working out. I think that there is no danger of a conceptual circularity, in this case: indeed, a mathematical analysis of *some aspects* of human cognition is *possible*, exactly because mathematics is rooted in cognitive processes. The remarkable, yet not absolute, effectiveness of mathematics is due precisely to the fact that mathematics radicates in natural phenomena, from physical to biological ones, up to the "human endeavours towards knowledge, in our historical existence", as Weyl puts it. Mathematics, as a theory of conceptual invariants, grows out of the very same regularity of the world on which cognition is grounded: its effectiveness is due to its cognitive roots. Moreover, in the case of cognition, these two dual approaches are actually *needed*, as the challenge is enormous (mathematics is very expressive) and only a variety of enriching or converging view points may help us in some further scientific understanding, including the use of mathematical tools. There is no conceptual vicious circle, though, as mathematics cannot describe alone all cognitive and historical knowledge processes: these must be analyzed also by the other forms of knowledge, in the methodological pluralism mentioned below. There is no entirely mathematized metatheory of human knowledge: *this* would lead to a vicious circle, as mathematics is one and a specific form of knowledge. It is instead the network of methods, of mutual understandings, interpretations and descriptions, by different conceptual tools that gives meaning to each theoretical description, in the interactive reflection of theories and interpretations that form our human and scientific understanding of the world. Mathematics is grounded in this network and its foundation may be analysed by appealing to various (all?) forms of knowledge, while using it to clarify *some* aspects of it: there is no circularity in this, but a search for the unity of knowledge via interactions, conceptual bridges, mutual enrichments.

2. METHODOLOGICAL PLURALISM AND THE UNITY OF KNOWLEDGE

These "interacting" methods, cognition for mathematics and mathematics for cognition, are possible and needed, but not sufficient. I am a monist, as I do not believe in the distinction between body and soul, between concrete autonomous "objects" of reality and metaphysical ontologies, but I am not a *methodological* monist. That is, I do not believe that a unique method, the mathematical one for example, may provide the ultimate understanding of all phenomena, by a formal and complete reconstruction, or that "the essence of objectivity is mathematical". It may be so in physics, as mathematics has been largely designed over and with physical phenomena, in physical space and time; but this paradigm does not need to be transferable to other scientific analyses.

It should be clear that I do not doubt that, in the end, there is nothing else but "physical matter", whether we call it atoms, electrons or waves and strings or alike. However, I doubt that this (potential) reduction of "everything" to his material constituents implies that in other disciplines, beyond the one which proposed to us these notions, physics, must be reduced to the *same methodology*, the methodology of physics. Biological phenomena, say, do follow also "physical laws", but their biological behaviour cannot be *deduced* from the properties of physical matter: they lay at a different (further) phenomenal level. As suggested by [30] in a different, but relevant, context, "dependence does not imply reducibity". Indeed, there are *qualitative differences* between the phenomena that we analyze in physics and those of life, as well as between individual, biological life and human intelligence or historical organisations. The emergence of one from the other results in apparent discontinuities, yet to be understood, as if material components, when they "get together", give rise to new phenomena or, at least, are better described by a different scientific method (even within physics it is so: "enough" particles, when they cover a significant part of the universe, are better understood by relativistic mechanics than by quantum physics).

Mankind in its historical existence has been giving to itself a rich variety of forms of knowledge, exactly because the phenomenal world presents itself in different forms. It is too early, if ever possible, to pick up a method and propose it as absolute. Many of these forms of knowledge may or will be unified and reduced: the phenomenal descriptions of physics may be entirely reducible to (or handled by) mathematics for example, but, so far, that drawing of ours of the images and syntax of mathematics on the phenomenal veil, misses at least "colours", "nuances" and "intentions", which are at the core of life and human intelligence. There is no knowledge without intentionality for example (it even shows up in mathematical proofs, see below for references), without the guidance by feelings and passions, as "colours" and "nuances": we should not iterate Descartes' (and Kant's) error and introduce a primary break between rationality and the other forms of relation to the world we have. In particular, emotion and cognitive processes are not disjoint: intentionality sets the "direction" of analogies, say, or metaphors ("analogous to this", a "metaphor towards that"...). Thus, emotion is not just a possible stimulus to knowledge, but, as interlocked to intentionality, it contributes to constitute the content itself of analogies and metaphors, some key elements of human understanding and expression, even scientific one, as most scientific descriptions are "just" metaphors.

Our brain, indeed the simplest neural system, is first of all an "integrator", in the sense that it compares, integrates and synthesises very different perceptions and judgements. And so does our "historical brain", our human and collective form of intelligence. The objectivity and unity of knowledge is not given by a unique ultimate mathematization, but by the integration of a plurality of approaches, as a dialogue between different methodologies; it is given by setting bridges, by providing and relating common "supporting" points, here and there, as shared though spars, yet "communicating", foundations. The unity of scientific knowledge may be understood in terms of a "ring of disciplines", as beautifully described in [4]. Or, as already stressed here, knowledge is a network which supports itself by a reflective equilibrium of conjectures and theories, by the exchanges and enrichments between essentially different approaches; this is its unity. In this network, mathematics stems out by its strong two-ways relation to physics, but physical methodology does not completely cover biological analysis, for example: the peculiar internal unity of living beings as well as their unique relation to their ecosystem have little to do with the mathematical singling out and setting of conceptual unities in microphysics, say. Living beings impose to us their unity and their living relations. Any form of "découpage", as a key to the physical description, "kills" the systemic unity of life, which constitutes its *essence* and forces itself throughout the phenomenal veil. On this veil, in biology, we are no longer free to draw contours and single out (ever changing) unities with our mathematical tools, or not as free as in physics (yesterday atoms, later electrons or muons, today strings or cords, etc.): the phenomenal donation of living beings has a qualitatively different, pre-existing, form of unity.

As for comparing models, in physics the mathematical model is usually more rich than the phenomenon: in microphysics, say, a few sparks in a machine may lead to a complex theory of particles. In biology, the mathematical model is always poorer than the phenomenon, with its unbreakable living unity: it is only a "conceptual section" of this unity. Moreover, biology, as many little mathematized disciplines, deals with "examples" and "counterexamples" in a way which largely differs from the (physico)-mathematical practice: an example may constitute knowledge, it is not just an instance of a general "theorem"; a counterexample need not demolish a theory, but may suggest variations. Stability and invariance, repeatability of phenomena, as well as the absolute generality of a law are not at the core of life. Each living being is unique; more than elementary interactions, biology analizes individual totalities; contexts and ecosystems affect the repeatability of each experiment; more than stability and invariance, what really matters in biology is variation, non-isotropy, diversity; in biology, finalism contributes to explanation, on top of "physical" causality (living beings "want to" survive, to improve or maintain their metabolism: see [4] and [22] for references and further analogies and differences with physics). The successes of mathematics in physics should not leave us blind with respect to the peculiar phenomenology of life. The enterprise of applying mathematics to the sciences of life is particularly challenging, as it cannot be a simple "transfer" of techniques, but the methods should complement each other with no philosophical ultimate priority; on the contrary, we should try to make more explicit the "philosophical peculiarity" of the biological analysis, which may suggest novel mathematical tools, as it often happened in the historical relation between mathematics and physiscs. Thus, beyond using established tools from mathematical physics, say, as specific mathematical structures, one should also try to establish the interaction at the level of "conceptual invariants", as they are defined below. This are more "plastic", as I will try to argue, more subject to further mathematical specification, in particular in interaction with essentially different methodologies, as those of biology.

3. ON THE GENESIS OF CONCEPTUAL INVARIANTS

Invariance and stability are key components of mathematics. I would even dare to define mathematics, among our attempts to describe the world, by structuring it, as the locus of conceptual invariance and stability. We will try to analyze invariance and stability at the level of "concepts" (this section) and at the level of "mathematical structures" (§. 4). Husserl's philosophy, (see also [30] and more references in this volume), provides a guideline for our approach, as mathematics "transcendental objectivity" is the result of a path or genesis through "immanent cognitive and historical acts" [17].

3.1. Integers

The concept of integer number is perfectly stable: it is invariant w.r.t. notations and meaning as no other practical concept. It is (remarkably) stable through history. Indeed, an embryo of counting is surely pre-human.

There is large evidence today that the rat, the monkey and the human baby (at four and a half months!) distinguish between sets of 1 or 2 or even 3 objects, independently of the nature of the objects being enumerated [10]. That is, these various mammals exhibit similar reactions when they perceive two sounds, two flashes, or two dots on a screen. Pairs of objects or threes are recognised as such, even if they are in motion, distorted or continually changing: this shows that it is not a pattern that it is recognized. Evolution then appears to have constructed neurones which react to the number of events occurring in the environment, no matter what the nature of these events, providing there are a small number of them: two or three, rarely four (p. 37 of [10]), experiments conducted by R. Thompson; se also the references to K. Wynn's work.) For successful survival, it seems necessary to be able to recognise and compare at least a few samples of food, objects or useful markers, even when 'secondary' changes are affecting the perception of these; and this became part of our phylogenetic memory, as a common pre-conceptual experience.

Thus, in our life-world, we (the manual in us) constructed an early experience of "mathematical" invariance, a key concept in mathematics, to which history has given depth and complexity. The further experience of a variety of notations is part of the path leading to the explicit concept of "number", *as independent from reference and notations.* A far from obvious path: all early counting notations, as pointed out in [10], are identical up to 3 (I, II, III; also our Arabic 2 and 3 themselves are two or three connected horizontal lines, probably for writing convenience). Yet, the early sumerians numbers used to differ, beyond 3, according to the counted set (the 5 of 5 sheeps was written differently from that of 5 houses...); the late sumerians and the Egyptians made the fantastic step to get to a notation independent from meaning.

In short, I claim that the ancient appreciation of the independence of "counting" from examples and events, up to 3, as a shared praxis of many species, is *behind*, it is the primary ground on which lies the "foundation" of the concept of number, as an invariant: from the phylogenetic memory of this pre-conceptual invariant, to the practice of counting, to the difficult invention and experience of different possible notations, we got to the abstract concept, as what is common or stable. The historical construction wouldn't probably have been possible without this underlying appreciation of independence of elementary counting that we share with animals. This is its root and cognitive foundation: this is the early, but extremely deep link to the world, which makes counting so "objective" and so effective.

On these grounds, human language could start drawing on a phenomenal veil that we share with other animals and it gave to the underlying elementary invariance an intersubjective content. Husserl, in [17] stresses also the role of writings: oral communication is still missing the "persisting presence of" ideal objects", which last also in time... It misses being-in-perpetuity... This is the crucial role of written linguistic expression, of the expression that stabilizes, that of making possible communication without personal allocution, mediate or immediate, and of becoming, so to say, communication on a virtual mode. By this as well, human communication goes over a further step". Thus, writing adds a further level of objectivity to intersubjective communication: it further clarifies and stabilizes concepts, it increases the passiveness of the individual who accesses by this to human collective memory and it adds by this a further appreciation of the objectivity of our historical constructs.

In short, the invariant concept of number is primarily grounded on phylogenetic memory; it then acquires, by language and writing, a double form of stability. First, by symbols, it is detached form concrete experience; second, the very experience of many possible symbolic notations contributes to its further conceptual stability. It is then definitely given a "formal ontology", in the sense of Husserl. Yet, no platonistic metaphysics nor formal conventionalism is required to understand this human construction and its objective content, which entirely lie on the human praxis, throughout evolution and history.

3.2. Potential Infinity

Language and writing allow to conceive rigorously *indefinite iteration*, as an endless extension of the "go-to-the-next" operation or of a given notation. This is a possible route towards potential infinity. The ever increasing sequence of the natural numbers, a core mathematical structure, is thus the result of shared experiences by language and by writings, along generations, up to the symbolic extension of elementary counting, along potential infinity. This bold further step surely includes metaphysical appreciation of endless space and time, as never closing horizons.

As a result of this cognitive and historical experience, the structure of the potentially infinite collection of numbers is *"there"*, for all of us. We even put it back into space, into mental space: we *"see"* the discrete sequence of numbers, the so called "(integer) number line". It is well-known that, in western cultures, this mental line goes from left to right (don't you see it?), the opposite direction for Arabs and Iranians. Here is the influence of writings on the forms of stabilization of a mental construction. Yet, and again, one should not isolate in a vacuum the mathematical construction: the iteration towards infinity could only be dared and have a sense in conjunction to a metaphysical glance on the physical world, which include the sense of expectation, the future, the endless or iterated phenomena, which give us an appreciation of never ending sequences.

I said that we put the number line "back into (mental) space", as, according to many in "mathematical cognition" (see Dehaene's book, further references may be found in the journal with this name), the evolutionary counting-process is also analogical, as if represented in space, on a "logarithmic scale": highly precise at the lower end of the scale, but more and more crude when applied to larger quantities. According to numerous neuropsychological experiences, in humans and animals as well, it seems that large quantities, when only slightly differing, are operated upon in an approximate-analogical, but consistent fashion; in humans, they are also represented as appearing distant and ill-defined. These authors suggest then that representation of numbers, counting and comparing quantities are also a spatial, analogical process. Human language confers the precision of linguistically discrete expressions, and allows us to generalise from the precise handling of smaller numbers towards a similarly precise handling of larger quantities, by the arithmetic-linguistic operations. But the analogical counting-comparing process, in a mental space, appears again and again in a thousand different situations: in animals comparing quantities, in human everyday practice, in pathologies, in mathematical apprenticeship and in the "intuitive" work a mathematician does. We consciously and unconsciously "look" at numbers as a "well-ordered" structure, in space (and time, as Brouwer would stress), and we use it in approximate computations or size comparisons. The totally ordered, strictly increasing (well-ordered), sequence of numbers, is the solid *geometric* rock on which it is founded human arithmetic. Logical or formal induction (Peano-Dedekind-Frege) is the very late attempt to found it on purely logical (Frege) or formal language (Hilbert): it turned out to be incomplete. But we will briefly go back to this issue (see [24] for more).

3.3. Actual Infinity

The few hints given so far on the genesis of numbers concern only the *potentially* infinite sequence of integers (also and correctly called "natural"). The progressive conceptualization of the human notion of *actual* infinity is a further, long and complex story, largely embedded into metaphysical considerations. Weyl hints this, with reference to late Greek mathematics and oriental religions, [36]; in [34] one may find a remarkable historical reconstruction of how we gradually got to stabilize the idea of potential infinity, first, and later the difficult concept of actual infinity, a source of major disagreement and conflicts in history (see also [12]; a linguistic analysis of the metaphores implied, may be found in [19]). Projective geometry, a fall-out of renaissance pictorial perspective, played a crucial role in the conceptual specification of actual infinity: the convergence point of the perspective is "there", into infinity. It is no longer potential, as it is an *actual, visible* construction.

It is fair to say that with Cantor, on the grounds of the work of many, from Thomas Aquinas to Pascal, Newton, Leibniz, etc., we arrived to a robust concept of infinity, as Cantor even inserted it into operational contexts, he dared to work with it. Cantor invented ordinal and cardinal arithmetic and showed, beyond the experience of infinitesimal calculus, that we could have an infinity of infinities and operate on them. There is no better way to give solid ground to a mathematical idea and to turn it into a true conceptual invariant, than showing how it works in different operational contexts, by "manipulating" it as a operationally meaningful symbol.

Because, and this should be clear, the elements of Cantor's (and Veblen's) hierarchy, beginning with ω, the actual infinity of the natural numbers, and then $\omega + 1$, $\omega + 2$ ... $\omega + \omega = \omega \times 2, \omega \times 3, ... \omega \times \omega, ...$up to ω^{ω} and ... ω to the ω, ω times (named ϵ_0), and ϵ_1, ... ϵ_2, ... ϵ_{ϵ_0} and so on so forth, are not just symbols. They are not in the physical world either, as they have no meaning in it, nor they are a mere convention: *they synthesize a principle of construction, generalized iteration and limit; they are the symbolic, but meaningful result of an historical praxis, which turned them into a perfectly stable conceptual construction.* This conceptual construction is part of the genesis of our mathematical concept of infinity, as by this I intend both the route and the result of a mental and historical activity. It actually lead to a mathematical structure (see §.4).

3.4. The Continuum

In each key mathematical structure, pre-conceptual experiences are at the core of the conceptual construction, which is further specified by mathematical, "structured" invariants. A further crucial example is the "continuum". Since always its specification and use has been at the core of mathematics (see [23] for more discussions and references to part of the enormous literature on the topic).

We all "appreciate" the pre-conceptual continuum of gestures or movement in space and of the ongoing flux of phenomenal time. This is far from being uniquely determined, yet it is a robust experience on which further conceptual constructions are grounded. That is, we are able to *give meaning* to the subsequent concept by reference to the pre-conceptual experiences and actions. Or, the linguistic or geometric concept is meaningful for us, as rooted in our acts of life. These, on their hand, are better understood by the intersubjective specification of the concepts, by language and, then, writing, through history.

For a brief comparison of the continuum of space and that of analysis, consider now some of Euclid's definitions and "axioms". First ... what is a point? "A point is that which has no parts", definition 1. While "a line is a breathless length", definition 2. Observe now that Euclid's lines are not made out of points: "the *extremities* of lines are points", definition 3. By this, Euclid excludes the concept of "open" interval, which would miss exactly one point on each extremity. Thus, lines are parmenidean unities, flux, "compact, unidimensional" length ("breathless length"); a segment is a "rigid body" (like all figures of Greek geometry), ending, if finite, by points at extremities [11].

This is why, at their extremities or when they cross, unidimensional lines give rise to a point, as this has no parts, i.e., no "Cartesian dimension", in our terms (a point lies at the extremities of lines or is the *result* of two unidimensional lines which meet with no "local" parallelism). No more than this is needed for the continuum in the geometric constructions of Euclid's. In particular, and against the formalist fake reconstruction of history, the first theorem of the first book is perfectly proved, within his *mathematical* concept of *geometric* continuum: given a (finite) straight line, one can construct an equilateral triangle on top of it (proof: just construct the intersection point of the circles centred on the extremities of the line, and draw two straight lines from that point to each point of the extremities - this is an application of axiom 1). As said above, the existence of the intersection point is a *consequence* of Euclid's conception of the geometric continuum and his definition of point and line (see [23]; [21] [25] propose a cognitive analysis of the concept of mathematical line).

The modern use of the continuum in analysis, which uses infinitary limits processes, has been instead consistently found on Cantor-Dedekind "pointwise" construction. This is a possible and beautiful construction of the continuum, grounded on integer numbers not a necessary one as other approaches are possible, e.g., non-standard constructions (or the approach by Veronese, see the paper by R. Peiffer-Reuter in [32]). It is an historical abuse to assume Cantor's construction in order to analyse Euclid's geometry. Yet, this has been largely done during this century, by the prevailing formalist approach, which forced us to read back history in a biased fashion. Heath in his fundamental analysis of Euclid's books [15] opened the way, but also Leibniz, apparently, had raised the issue of a presumed flaw in Euclid's proof: one has to assume that the intersection point *exists* or *derive* it from a suitable construction (typically, one has to use the metaphysics of Leibniz's monads or, for Heath and followers, Cantor-Dedekind's real numbers!). A surprising way to impose backwards an analytic construction on top of Euclid's geometry of finite figures which does not require it, a geometry of structured lines as flux or parmenidean unities, of compact rigid bodies.

In other words, Euclid had in mind a different concept of continuum, different from the one specified by Cantor-Dedekind pointwise construction and specified it in different structures: a geometry of rigid figures and continuous and straight lines, like light rays, in his understanding (see [15]).

Clearly, also the analysis of time-variation or movement, by Newton and Leibniz, was based on the concept of "flux", yet understood in terms of limits or monads; the latter a remarkable concept, preliminary to the (non-) standard subsequent constructions and structures. Cantor and Dedekind proposed a precise, yet specific, mathematical structuring of this sort of analytic continuum, grounded on and used for limit operations, as a number-theoretic construction: consider the integer numbers, take the quotients (or rational numbers), collect all converging sequences (modulo identity of limits, defined *à la* Cauchy, say), call them real numbers. Then you have a totally ordered line, with no jumps nor lacunae: movement and its derivatives (i.e., speed and acceleration) are very well described or parametrized over this line. However, it is an "act of violence", as Weyl says, to assume the prefect coincidence of the analytic construction of the continuum with that of phenomenal space and time "... that is, the continuity given to us immediately by intuition (in the flow of time and in motion) has yet to be grasped mathematically" ([35]; see [23] for some reflections inspired by that classic).

It should be clear that there is no need of a transfinite ontology to use or to acknowledge the objectivity of any given specific construction of a mathematical continuum. Exactly as for the integer number line or for the hierarchy of infinities, *the objectivity of Cantor-Dedekind continuum, say, is in the construction itself. Its foundation is in the genesis of the conceptual invariant*, from the early cognitive grounds of the integer numbers to the mathematical construction of the rationals and, then, of the reals. No medieval ontology no "actual existence" of objects is involved in mathematics, but "just" the objectivity of mathematical constructions. The description of this objectivity is a difficult scientific challenge, by the need of a unified analysis of a large variety of often conceptually independent constructions.

The a posteriori, formal or logical investigation may help to distil some key proof-principles which unify various aspects of the mathematical deduction. Unfortunately these principles are incomplete even w.r.t. the simplest of these structures, the integer number line, let alone our geometric description of space. This is the result of (recent) incompleteness results, which evidentiate a "gap" between formal proof principles and various mathematical constructions, geometric ones for example (see [23] and [24]).

3.5. Proofs

The notion of "mathematical proof" has surely been evolving in history, yet "its" core structure is very old and based on a strong appreciation of invariance. Consider Pythagora's theorem. Egyptians had a long list of numbers in the ternary relation of the sides of right triangles: $3^2 + 4^2 = 5^2$, etc. Mathematics really begins when the Greek geometer makes a drawing of a right triangle on the sand, of a *specific* triangle, with inevitably some *given* length of the sides (or ratio of lengths). He then gives a proof of the theorem by projecting and comparing the squares on the sides and concludes: "observe that my proof *depends only* on the fact that this angle is right, not on the other properties of the triangle, such as the length I had to drew". The proof is *invariant* w.r.t. (the ratio of) lengths of the sides (which were drawn and, thus, specified in the construction!): this is the key remark that any mathematician has to do at the end of a proof. He has to single out what it depends on, as well as its "independence"; that is, the proof's exact level of generality. An "argument" turns a (universally quantified) statement into a *theorem* when its generality can be explicitly spelled out, or when one can single out w.r.t. to which assumptions it is an invariant. Sometimes a very hard task. Axioms and logical rules may help in this, but they are far from being sufficient. The point is that this key proof-theoretic invariance is more a *practical* invariant than a just a formal invariant; i.e., it is constructed in the conceptual, scientific praxis of mathematics, as part of human reasoning. In other words, this appreciation and specification of invariance is deeply rooted in the forms of invariance that are at the core of our general search for stability and generality in arguments (and in life: memory is the early cognitive ground for the invariance and the normative aspects of mathematics, see [25]).

The structures of mathematics brings conceptual invariance at its highest human level, by providing, often a posteriori, a re-construction of it as part of the proof, a proof which may well be given, as frequently in geometry, on a specific case (the drawing may be needed and it is always specific), post-factum generalized by an analysis of the proof itself. The use of (universally quantified) variables, in classical formal systems, sets on rigorous basis this analysis, in the algebraic settings, but it is not perfectly adequate in all mathematical frames. Yet, other systems of logic may give different or further levels of information, as it will be hinted in §4.4 in a comparison of classical and intuitionistic systems.

3.5.1. Methodological intermezzo: Possible stories vs. absolute objectivity

I am aware that, by this approach, I am violating a dogma of the prevailing foundational analysis in mathematics, well established since Frege's foundation of Mathematical Logic: I am referring to "history" and "psychology" (cognition, to be precise, as shared experience, through evolution and history, not just individual experience, as spelled out next). And I am not grounding mathematical knowledge into "universal (logical) laws", nor "absolute objectivity". Indeed, I believe, following the late Husserl, that there is no foundation without epistemological analysis, i.e., without an analysis of the "knowledge process". And that there is no epistemology without a *genetic analysis*, as analysis of a genesis, through history, provided that the meaning of history is meant in a broad sense [17]. This sense is given both as a (possibly pre-human and) human cognitive experience and as an ongoing intersubjective process, along generations.

In short, mathematical invariants sit on top of or are derived constructions from conceptual ones; these are the result of shared and manifolded experiences, "acts of experience" to be precise, as independent and meaningful extensions of our active presence in the world. Intersubjectivity grounds them in this plurality made possible by our communicating community of human beings, through history: we gain the human and historical level of conceptual invariance, by "specifying concepts" toghether, while comparing with the others' experience and, by this, by singling out the stable fragments. Mathematics must then be understood as an extension of a human praxis, the most objective of these communicable extensions, as the one with the highest degree of independence and invariance; it is the tip of the iceberg of human communication, the crystal clear tip, where ambiguities and contextual dependence are excluded as much as possible or as in no other form knowledge.

Of course, we can today work at this bold enterprise and resume the early intuitions on the "cognitive foundation" (in my terms) of mathematics, by Riemann, Helmotz, Mach, Poincaré, Enriques, and Weyl (see [7] for a broad survey and references), also because we are standing on the shoulders of giants like Frege and Hilbert, the founders of mathematical logic. After the enormous growth of XIX century's mathematics, often with little rigor, and the crisis of Euclidean geometry, it was absolutely needed to evidentiate core logical principles and to provide a precise notion of formal rigor (and even of what it means to give a "good definition", which was far from clear at the time): the advantages of Frege and Hilbert's work have been enourmous for the practice of mathematics, and for their fall outs as well. The main one is the analysis of deduction as computation, which lead to a precise notion of algorithm and computable function and originated computer science, the discipline of formal/purely-logical symbol pushers, the computers. Today, though, we can interact with other disciplines, such as biology or cognitive sciences, which have undergone an amazing and recent growth. Thus a phenomenal/cognitive analysis is not anymore a matter of introspection only, as from Riemann to Poincaré or Weyl, but it is becoming a scientific analysis of our forms of knowledge and their interactions. This interdisciplinary investigation does not aim at providing "absolute objectivity" and "unshakeable certainties", the quest of the founding fathers of mathematical logic: it aims at a scientific analysis and, thus, it can only propose "plausible stories" of the world or of our cognitive relation to it, like physics when it tells us plausible stories of the universe or of quanta. An analysis whose developments or very principles may be falsified or improved, discussed and modified, or, here and there, locally, verified, as usual for scientific knowledge, i.e., for our proposal to understand the world, by structuring it. In either case, in mathematics as well as in the other forms of knowledge, the key point, in our view, is that there are no "objets" (mathematical, physical, etc.) which precede the constituting of objectivity: the foundational analysis, in the various scientific realms, has to investigate this constitutive process.

In summary, the foundational analysis of mathematics has two complementary aspects. There is a *necessary* investigation of mathematical proof- principles, that tries to set on formally rigorous grounds mathematical theories, in particular by clarifying the rules by which, in each context, one may give "good definitions" and "formal proofs", by proving relative consistency results, etc. This is the relevant job of mathematical logic, proof theory in particular. But it is *not sufficient*, as mathematical logic itself has shown by the many incompleteness and independence results, a remarkable technical achievement (see [24] for a reflection based on recent "concrete" results). Then, in order to go further, one has to try to investigate the conceptual origin of the mathematical constructions and embed it in our human endeavour towards knowledge (see [37] as for an early hint towards this distinction between necessary and sufficient conditions in the foundational analysis). These investigations are just "plausible stories" in the typical sense of scientific descriptions mentioned above. This century physics has taught to us how to give up the Newtonian search for "universal laws" and "absolute universes" and develop an analysis of constituting principles of knowledge. The foundational analysis of mathematics has a lot to learn from the methods and epistemology of physics, quantum physics, in particular.

4. THE COGNITIVE SUBJECT: CONCEPTUAL CONSTRUCTIONS VS. ONTOLOGIES

But then, are we investigating by this an "independent reality", both in mathematics and in physics? Is this the reason why many observe that, for example, the "sequence of prime numbers has a more stable reality than material reality" ? (A quote from a leading mathematician; but I could also quote the "next-door mathematician" in my department, or Gödel: "numbers are at least as real as this table ..."). In this "realistic" perspective, often considered a form of naive platonism, the mathematical objects are "already there", independently of human definitions and conceptual constructions; they are even more stable or permanent than material ones (see [13] [14]; for the prevailing interpretation of current "naive platonism" in mathematics, see [5]; Introduction[1]).

[1]For a coherent and deep, not naive, platonistic understanding of the set-theoretic approach, and a parallel of Topos Theory to Aristotle and Leibniz views, see [3]. Another well-informed and interesting update of the platonistic view point may be found in [29].

I would agree that electrons or muons, let alone neural synapses, are far from "stable": they are our ongoing proposal to single out objects, in our attempt to organise "reality". A reality which is there, as it makes *friction* against our proposals to give *meaning* to it, a meaning which is in the organised interface between us, leaving and historical beings, and that "reality". Thus, electrons and muons as well are conceptual constructions, as theoretical unities, often very temporary ones, grounded on a few symptoms or sparks on a screen of a constructed measure instrument (thus conceived also on the grounds of a theory). In this, they resemble to a mathematical object (and their definition requires a lot of mathematics), except for the possible request to check their definition against a further friction on "reality", by more instruments. Not so differently, we *constructed* tables in (pre-)history at the same time as the concept of table: this individual table will be destroyed soon or late, and the very concept may become obsolete and forgotten, when tea cups will be held by antigravitational forces, in year 2435. As for synapses, not only our description changed over time, but these interconnections between neurones changed during evolution of species and keep evolving in phylogenesis or even ontogenesis.

Thus, there is no doubt, the concept of prime number is "more stable" than these fragments of reality or their related concepts. But, the concept of electron or muon, or even of synapses or of table, if *stabilized* by a mathematical definition or by a praxis which would only rely on the assumption of its perfect conceptual stability, then, this concept, may become as stable as the *concept* of prime number. Indeed, when using it just as formally defined, one would work in a branch of *mathematical* physics or biology (or "mathematical furniture theory", as for tables), such as rational mechanics, say, and not in (theoretical) physics or biology, the loci for (dynamical) conceptual descriptions. The point is that *mathematics is the paradigmatic language for this stabilizing praxis, while organising knowledge.* This is why its "constructed conceptual objects" are very stable, surely more than the (concept of) electron, synapses, table ... which follow the dynamic of (our knowledge of) physics, life or history, and are affected by their direct friction against the world.

And here comes the usual confusion proper to naive realism, in mathematics, a peculiar blend of idealism and empiricism: I call object this table, I appreciate its unity and independence from my "self", as it is invariant w.r.t. to perceiving it by seeing, touching, smelling ... . Enough sensorial invariants, in the common sense experience, usually guaranty that "there really is a table *there*", independently of me, the individual, psychological subject. This common sense remark is then transferred to mathematics: (the concept of) prime number is stable or invariant w.r.t. all reasonable "view points" and mathematical properties and theorems, it is then there, before and independently of any conceptual construction. Indeed, it is even more stable than "this shaky table" or uncertain electron, which can actually be broken or split into more fundamental pieces and particles and disappear, even from being considered a useful artefact or an existing unity or object. Of course! Mathematics works only at the conceptual level: it is the very language by which we try to structure the physical world, by singling out, naming and, possibly, formally defining objects; it has an autonomous development, which presents no friction, if not against itself or as interaction of its many structures. This language may be enriched also in reference to physics, by an indirect friction, as when proposing infinitesimal calculus to analize movement or building geometry on top of the notion of curbature of space, as in non-euclidean geometries.[2] But, once given, by definition, its concepts are as stable as Carol's "smile without the cat" or a frozen version of it, when the moving cat has gone: we work only on the smile, integrate, derivate, multiply it by itself ... in a non-arbitrary fashion, as its methods of conceptual construction and logical deduction are grounded in our cognitive relation to the world.

In summary, everyday realism refers to tables, electrons and positrons, as "objects", which exist independently of our being humans, who live and act in the world, while trying to organize and understand it; this very realism is then transferred into a naive platonism that confuses the objectivity of mathematics with pre-existing objects. In either case, the relative independence, w.r.t. *the psychological or individual subject*, of tables, particles and ... numbers, is considered as an absolute independence. This is another component of the ontological myth, in either realism: the lack of distinction between the psychological subject, with its individual history and being (I have no doubt that tables, but also electrons and numbers "exist" independently of me and will survive my death), and the *cognitive subject*. The later is the result of a phylogenetic and historical formation of being, it underlies the very life of each psychological subject, it is shaped by the living and historical community. Tables, electrons and numbers are not independent of the cognitive subject, as they are constructed by its own and ongoing constituting in an active presence in the world, simoultaneously to it, in a concurrent- interactive process. The cognitive subject ramifies into the psychological ones, each a peculiar historical instance, whose most stable and invariant part contains the cognitive subject. But the cognitive subject

[2]Both are dramatic changes in (or broadening of) paradigms. The first by the use of actual infinity to analyse finite movement, speed and acceleration. The second, by proposing a geometry of space, independently of the rigidity of bodies and the nature of light rays, in contrast to the Greek geometry of figures and rigid bodies (and straight light rays, to which Euclid explicitly refers as for defining straight lines, see [15].

is not independent of the individual ones, does not transcend them, as it is the shared, live and historical experience. By living and interacting, the individual egos form themselves, while shaping their own community and as part of it. Thus, the cognitive subject is immanent, as, at the same time, it constitutes and it is shaped by the communicating community of individual subjects.[3]

Within this philosophy, mathematics is analyzed here as the result of a cognitive praxis. A key aspect of the present approach is that it tries to relate mathematics to human cognition, both for the sake of mathematics and of cognitive sciences: mathematics is such a "conceptually simple" (even when technically difficult), deep and clear form of knowledge, that it may open the way to the analysis of more complex or involved forms of knowledge. Moreover, its unique generality, which transcends any specific form of knowledge, may help in singling out the "cognitive subject", as the shared part of our humanity, constructed in our phylogenetic and historical interaction with the world, at the various phenomenal levels: mathematics is grounded in core cognitive praxes, those which face the relevant regularities of the world and of life (or, at least, the regularities that matters for us, that we "see"). By the analysis of its foundation, mathematics may thus provide an early, simple and very general conceptual laboratory for human cognition.

5. STRUCTURAL INVARIANTS

Section 2 hints at those key *conceptual* invariants which are called "integer and real number", "potential and actual infinity", "mathematical proof". Many other mathematical notions and structures are derived from these: from effective computations to ... differential manifolds in geometry.

The idea here is that some "basic" *mathematical structures* are the specification of a conceptual invariant (the integer line, typically) while others are derived from operations on these structures. In a sense, say, there is a crucial difference between the *concept* of actual infinity and the *structure* of Cantor's ordinals or cardinals: the latter helped to specify the concept by operating on it, *within a specific construction*, yet actual infinity stands as a conceptual invariant (or as conceptual synthesis or integration of a variety of

[3]In the terminology of mathematical logic, I would call this ongoing biological and historical process-formation of the cognitive subject (or its description), as "impredicative": one cannot understand the elements without understanding the whole, which is in turn constituted by its elements (this kind of "circularities" are not vicious, but virtuous and are at the core of the dynamics of physics, life and cognition). The resulting (relatively) most invariant and stable properties, but not absolute, characterise the cognitive subject.

"acts of experience"). The concept of actual infinity, say, may be also specified by other mathematical structures: the point at infinity of projective geometry, for example.

Indeed, the order of infinities or projective points in geometry are mathematical structures. The first is "just" the indefinite extension of the successor operation and of limits; the other is the rigorous geometric handling of a proposal to describe pictorial perspective (L. B. Alberti and Piero della Francesca, the Italian renaissance painters, were the main creators both of the mathematics and of the artistic technique). The concept and the structures interact very fruitfully, as the first gets a specification by the latter, while these derive their "meaning" and external foundation in the historical constituting of the concept of (actual) infinity.

Similarly, the real numbers, as Cantor-Dedekind mathematical specification of the continuum, form also a derived structure (from integer and rational numbers); by this, they also acquire conceptual stability, as the result of a "structural invariance" (w.r.t. certain transformations, the continuous ones, see below). Yet, the phenomenal continua, of space and time, are far from giving uniquely determined mathematical invariants, in contrast to the integer numbers. As a matter of fact, one may provide non isomorphic structures for them, for example the continuum of non-standard analysis. Thus a concept does not need to receive always from mathematics a unique specification (as it may be done in the specific case of the integer numbers[4]). Moreover, the continuum of time is far from being described in a satisfactory way by the (reversible, pointwise) line of real numbers: an ongoing debate since Weyl's 1918 book (see [23]).

The key point, though, is that *the meaning of the intended mathematical construction, the structural invariant, for us human beings, relies in the reference to the underlying conceptual invariant*; or, more precisely, on the "knowledge process" which leads to the mathematical structure. In other words, the understanding, indeed the *foundation*, of the mathematical characterisation(s) is in the underlying pre-conceptual, first, and, then, conceptual experience(s) (numbers, continua, etc.), as progressively but interactively built on the interface between us and the world (they are "multilayered drawings" on the phenomenal veil, in the terminology already used: the mathematical structure adds the last, most stable and precise "touch"). Intersubjective exchange, of course, contributes essentially to this constituting of meaning: the sharing of the pre-conceptual and conceptual experiences, by language, writing and ... by doing mathematics, stabilizes the concept,

---

[4]Standard and non-standard models of the continuum are non-isomorphic, globally and locally. All models of Arithmetic, instead, have an initial segment which is order-isomorphic (is "identical") to the standard sequence of integer numbers.

allows its very explicitation or allows to single it out from an unfocused praxis.

The invariance and stability of a well established mathematical construction or tool, though, may give it some sort of rigidity, with respect to the underlying concept. As a matter of fact, the greater plasticity of "concepts" may help in proposing novel technical interactions between mathematics and other disciplines.

When Leibniz and Newton tackled the analysis of movement by limits and infinities, they did not use pre-existing mathematical tools and structures: they used the *concept* of actual infinity and, by this, they opened the way to its modern mathematical specification. When interacting with biologists, say, we need to learn from them their methodological tools, not just the technical ones. We should understand the way they handle "concepts", not just learn facts or data and deal with them with already given mathematical tools, mostly inherited from mathematical physics. This may lead to entirely novel mathematical structures or to a change in the discipline as dramatic as the one which followed the birth of infinitesimal calculus.

But, once certain mathematical structures are well specified, how further structural invariants are generated, in general?

Once we have "structures", we may work with the properties which are preserved under various sorts of transformations within or over the intended structure. *Structural invariance* may then be understood as an "internal" notion. It motivates the conceptual construction and, then, it is given *inside* specific and well-defined mathematical structures, within "categories" as pointed out next. And from categories one may obtain derived constructions, as we shall see, and invent new concepts.

In summary, mathematical or structural invariants specify conceptual invariants or are derived constructions from previously specified mathematical invariants. The specification does not need to be unique or "canonical", as already mentioned: the integer numbers are uniquely determined in existing constructions while the various standard and non-standard continua differently specify the concept of continuum or continuous variation. Yet, the mathematical experience helps to determine the pre-existing concept or may propose further ones. In all cases, I claim, the reference to the conceptual underlying invariants steps in the mathematical construction and in *proofs* (see below). By this, the so called "intuition" is permanently enriched by the mathematical work and it is far from stable: any pre-existing concept is affected by the mathematical structure, which specifies it and, by active work and experience, will contribute to a further conceptual construction. This mathematical "intuition or insight" ("... [mathematics] as knowledge or *insight*... whose organ is "seeing" in the widest sense... " [37]) is *the reference to meaning, as depending on underlying concepts or pre-given structures*; it is "seing" a structure as meaningful and alive, as specification of a meaningful concept. As for continua or the well-ordered sequence of numbers and even more so for all geometric constructions, these force themselves as "visions", since phenomenal space and time are the locus of the pre-conceptual constructions; we *see* them in the reconstructions of our dynamic memory, [25].

The mathematician's constructed intuition and insight are the reference to a network of meaningful conceptual constructions, often derived from previously obtained concepts: the key ones are in turn grounded on pre- conceptual experiences, largely embedded in phenomenal space and time ([25] discusses on the role of memory in this process).

5.1. Categories

Category Theory may greatly help to understand this essential aspect of mathematics, the construction of structures. Categories are defined as *structures* (or `objects`) and transformations that *preserve the structures* (or `morphisms`). Thus, each category has its own structural invariants. According to the given category, these may be topological invariants, say, such as the properties that are preserved by continuity, or geometric or algebraic invariants, which are preserved by the intended transformations (the metric or algebraic morphisms). Being a closed line, for example, is preserved by continuous transformations, in the sense of topology, i.e., in the category of topological spaces and continuous morphisms. One may make a more restrictive assumption and call invariants only those properties that are preserved not just by morphisms but by "isomorphisms" or "automorphisms" (transformations that preserve faithfully and fully the structure, and are within the intended object in the case of automorphisms). Distance, for example, is preserved by isometries, which are isomorphic embeddings of metric spaces.

Moreover, categories are inter-related by higher-order maps, called `functors`; functors transform objects into objects and morphisms into morphisms, in a consistent way. Then, functors have their own invariants, which are categorical constructions, i.e., entire collections of structures. Once more, one may just look at full and faithful functors and consider a more restricted class of invariants (or, even, at those functors which are isomorphisms in the category whose objects are categories and morphisms are functors).

This is not the end, as Category Theory was invented to describe (some "`natural`") `transformations` between functors. Then, these transformations, possibly as isomorphisms, preserve invariants at the level of categories, whose objects are functors.

Category Theory is full of examples of invariants constructed at these different orders. Of course, many (most) belong to a previous formation of sense, just specified and unified by the categorical approach. But new ones are derived by further mathematical constructions: duality, adjunction, or, more specifically, Sheaves, as categories of functors, Toposes etc. Toposes, in particular, provide relativized approaches to logic: the invariant properties, w.r.t. the intended transformations in a given topos, form a logical system; this system may be classical, intuitionistic or other, according to its peculiar way of "classifying" objects, [18].

One final remark: the role of structural unities in Category Theory reminds the search for unity and interconnections in the various approaches indebted to "Gestalt" (see [31]), in contrast to the unstructured approach typical of Set Theory.

5.2. Invariants in the Category of Sets

The category of sets (Set) is a particular category, with all functions as morphisms. The notion of `cardinality` (number of elements) is the invariant w.r.t. to isomorphisms in Set (the bijections). These are "generalized numbers", integer ones but also infinite ones: the latter may be characterized as the cardinalities of sets which are isomorphic to a proper subset (a remarkable mathematical specification of the concept of infinity!).

By this, we transformed into structural invariants much earlier conceptual ones and, at once, we extended the notion beyond finiteness (cardinalities as invariants w.r.t. bijections). However, what has been achieved it is not a *foundation* of the human notion of number, but a relevant *mathematical characterisation* of a concept and, most of all, by defining infinite cardinals by bijections with proper subsets, the way is opened to setting on formal grounds a generalization, towards larger and larger cardinalities. Yet, and exactly for describing infinities, the issue of consistency of the intended formal set-theoretic frames turns out to be crucial: formal theories need to be proved consistent, first. That is, the analysis of consistency becomes a necessary part of these characterization of infinity. While informative and mathematically deep, this analysis is essentially "non-well-founded", as hinted in the introduction and, by this, it is essentially insufficient or incomplete w. r. to its own "foundational" purposes (larger and larger infinities are required to prove relative consistency results). The analysis hinted here of the "cognitive-constitutive path" should complement it.

As already mentioned, one may also "experience" infinities in different conceptual contexts, as limits and iterations of limits (ordinals) or as geometric (projective) limits. These yield further and relevant *mathematical characterizations* of the *concept* of actual infinity, each presented in a different structural context or *category*. Functors relate them and unify the mathematical descriptions.

5.3. More Invariants

Some structural invariants may also help to define new abstract concepts and by this they sit on the bordering line with conceptual invariants, as in the case of categorical constructions. Yet another familiar example is given by the Recursive Functions, as the collection of number-theoretic functions that are computed by any of the systems for computations, so far formalized in Mathematical Logic (Herbrand-Gödel recursion, Turing Machines, Church's lambda-calculus, Kleene's equations, etc.). They are defined as the *extensional notion of function* that is as the invariant (class of functions) w.r.t. the various formalisms for computing.

Similarly, the three different geometries (Euclid, Bolyai-Lobacewskij, Riemann) may be formally described as the invariant properties w.r.t. different groups of transformations. Either construction may be carried on in suitable categories.

5.4. Proofs as Invariants

One of the examples above of conceptual invariant concerned *proofs* as invariants (the proof of Pythagora's theorem, §. 2.5). Again here, Proof- Theory, indeed Hilbert's metamathematics, looked upon proofs as objects of study, and, by this, it transformed proofs into mathematical "entities" (I do not dare to say structures, because, as I will say, this is exactly what classical formalism is still missing). Hilbert's proposal was a remarkable step, yet limited by the attempt to analyse proofs only as formal-linguistic (algebraic) entities, as sequences of symbols, independent of meaning and handled "mechanically". By this, for example, the classical (Tarski's) description and interpretation of the universal quantifier, "for all x...", as "for all instances ..." is inadequate to grasp the nature of the generality of "for all right triangle ...", say, in the proof of a theorem of geometry as in §. 2.5. That treatment of universal quantification is eminently linguistic/algebraic: in algebra (often in analysis) one proves a result for an arbitrary element of the intended structure, by the explicit formal/algebraic manipulation of a variable, as a "generic" instance. The universally quantified variable stands for an arbitrary element and there is no need to use any of its specific properties during the proof and, then, prove that the proof is invariant w.r.t. them (like one has to do, instead, in Pithagora's proof, with the specific length of the sides, or their ratio, which must be drawn).

Of course, similarly as in that theorem, also in algebra or in analysis, one has to check that no other properties of the intended variable are used, i.e., that only its "type" is used in the proof, but this is easy then, as no explicit use of its individual properties has, in general, been made. When there is no (possible) reduction to an algebraic treatment, the geometric interpretation of "For all ..." is qualitatively different from the linguistic/algebraic one, which guided Frege's logical approach, Hilbert's formalisms and their tarskian semantics. With reference to our example (Pythagora's theorem), one cannot formally manipulate a variable, subsequently interpreted (instantiated) by any, generic, right triangle: there is no such a proof.

A better insight is given by the intuitionistic and the categorical (Lawvere-Grothendiek) description/interpretations of quantification. These approaches depart from Tarski's and are much more insightful, yet still different from the one hinted here for geometry

In short, in Intuitionistic Theories, both implications and universal quantifications are understood as *functions* or, more precisely, as *effective transformations*, i.e., computations with a well defined structure of function, see [33]; [9]. In a perfect correspondence to intuitionistic systems (by the extended "Curry-Howard isomorphism": "Types - as - Propositions - as - Objects of Categories") Category Theory interprets proofs as morphisms: the nature of proofs as invariants is then understood by the morphisms which structure the intended categories. By this, universal quantification becomes a (fibred) product, where fibration is a deep way to understand "variations", or as indexed product (in the second order case). This interpretation departs from Tarski's and it is much more insightful, yet still different from the one hinted here for geometry (see the Lawvere-Grothendiek interpretation of quantification [18] [20] and, for the second order case, [16, 2, 27].)

5.4.1. Proofs as prototypes

Following then the intuitionistic approach, let's try to introduce some basic ideas for a type-theoretic analysis of "universal" proofs, which stresses the role of "invariance". Given a mathematical theory which allows universal quantification, i.e., sentences such as $\forall x.P(x)$, how does one prove these kind of propositions? If the range of quantification is infinite, even uncountable, there is no way to explore and check each individual case. For the analysis of proofs, the understanding of $\forall x.P(x)$ as "for all x..." (the tarskian semantics) has little interest, or is even misleading (for the second order, impredicative case, in particular, see [8] nd [26]). Indeed, the practice of mathematics is usually the following: check what is the "intended range of quantification", i.e., over which set, domain or structure the universally quantified x is meant to be interpreted (the set of real numbers, for example, or any other countable or uncountable structure). Then prove $P(a)$, where a is an arbitrary or `generic` element of that domain ("take a to be a real", typically), that is, give a proof of $P(a)$ where the only property of a, used in the proof, is that a belongs to the intended domain. In Type Theory we would say: only the type of a is used in the proof. By this, the proof of

$P(a)$ is a `prototype` or paradigm or pattern for the proof of $P(b)$, for any other b in that domain ([26] presents a detailed type-theoretic approach, for second order theories). Thus, one has a proof of $\forall x.P(x)$, i.e., a proof that $P(x)$ holds everywhere in the intended domain of interpretation of x. In Category Theory, a prototype proof should be understood as a fiber in a fibred product (or a morphism, in an indexed product, as for the second order case): an issue of further investigation.

Consider now the special case when the (intended) domain of interpretation is the set of natural numbers or any countable well-ordering. How does this technique relate to induction? The implicit "regularity" of a prototype proof - all proofs are given by the same reasons relatively to a given formal frame - may be not present in induction. Once proved $P(0)$, which may be "ad hoc", one proves, in an inductive argument, the implication $P(n)$ $P(n+1)$, *for all n*. This is the proof that, *soon* (in general, at the first level of the application of the rule) *or late* (in case of nested induction), *must be* prototype in a generic n. Note also that this proof is based only on the assumption of $P(n)$ not on its proof (as for derivable vs. admissible rules in logic) So, formally, $P(n)$ and $P(n+1)$ may be true for different raisons or their individual proofs may follow different patterns and yet, the proof of $P(n) \Rightarrow P(n+1)$ must be prototype (in case of nested induction, this will show up after a finite number of nesting).

In summary, intuitionistic systems of Types and Category Theory allow a finer analysis of (universal) quantification in proofs, by giving a mathematical structure to the conceptual underlying invariant. More work should be done in order to understand universal quantification in proofs of Geometry, as we lack a proof theory of this discipline.

5.4.2. The geometry of proofs

Going back to Geometry, a *dual* link may be established. G. Gentzen first, in the 1930s, but more broadly J.Y. Girard, in these days, gave to proofs also a geometric structure. Some rules, for Girard, are there just because of symmetries or dualities; deductions are carried on the ground of nets, as spatial connections of formulae by lines, designed along the derivations; nodes of these lines are inspected, as three dimensional properties, in order to go to the next "deductive" step. This may be unrelated to the logical meaning of formulae; in a sense, Girard's Geometry of Interactions let geometry come back "through the window" into proofs. Proofs are investigated in a structured space, a novel mathematical construction, given on top of deductions, well beyond the formalist perspective and a turning point for Proof Theory. Once more, mathematics (geometry) is giving some structure to the independent "game of symbols" by which the formalists planned to found mathematics, independently of concepts and disregarding space (and time).

6. CONCLUSION: A TWO-WAYS FOUNDATIONAL RELATION

It should be clear that the interaction between conceptual and structural invariants is a two ways interaction. First, we may turn into mathematical or structural invariants some conceptual ones. Moreover, purely formal treatments may be proposed for the mathematical structures: symbols are then detached from meaning and manipulated according to mechanical rules. A further step, not a starting point. Both these "clarifications" are a crucial part of the mathematical activity, in particular as they may provide the technical tools for considering (structural) specifications and (formal) generalizations of the very notion, in various mathematical settings. The first, at least, is a necessary step, in order to set on rigorous grounds some general, often ambiguous, practical and conceptual experiences (e.g., the continuum, the infinite); but also full (axiomatic) formalization may contribute to this.

Yet, this activity is insufficient to "found", in the epistemological sense we mean here, the originating or underlying concept, which sets the "meaning" of the mathematical construction. As already said here, *this meaning or sense is in the cognitive and historical path that lead us to the abstract notion, as a mathematical structure.*

Thus, on one hand, a mathematical structure, with its properties, possibly in its complete formalization, may be considered to "found" a pre-conceptual or conceptual experience, only in the limited sense that it specifies it (and possibly gives an account of its formal derivability from least axiomatic frames). On the other hand, the foundation of the mathematical construction (and of its formalization, whenever made or possible) lies on its "meaning", as reference to a pre-existing or ongoing formation of sense. This underlying meaning may be possibly "pulled back", up to the level of axioms, in case of an axiomatic treatment, but even then it is grounded in our practices of life and conceptual constructions which give sense to the axiomatic choice.

Clearly, structural or formal invariants may in turn suggest new concepts, by operations or by conceptual contaminations: further infinite cardinalities, in Sets, or the many "universal notions" or the derived constructions obtained by functors, duality or adjunctions in Category Theory. Moreover, the practice of mathematics may change the very "intuition", and, thus, the nature and extent of the underlying concepts: the search for invariance and conceptual stability of (pre-)conceptual experiences is at the core of the mathematical work.

Category Theory is very effective in setting the "right level" of invariance which is being used: (iso-)morphisms, functors, natural transformations... w.r.t. the intended categories. This is the relevant foundational role it plays, which accommodates the novel constructions of Mathematics, as a growing, open-ended human activity. New categories may be proposed and the unity of mathematical knowledge is recomposed by relating them by functors and transformations, at the right (intended) level of invariance. This part of the foundational analysis brings in the sense of "relativized", yet not "relativistic", constructions so typical of modern science, physics in particular. As in modern physics, Category Theory does not search for "absolute and ultimate" constructions: proposals are made of new categories (similarly as for theories, in physics), in order to make intelligible other parts of mathematics (of the physical universe), or to single out brand new constructions (objects of investigation); unity is brought back by looking for links and mutual explanations, as functors and natural transformations. This practice is enriching and it departs from the *Newtonian absolute* of Set Theory, with its fixed universe of reference, a bunch of axioms, sort of "universal laws of thought", into which all possible mathematical constructions (present and future ones!) already exist or can be embedded or derived.

The interaction with other disciplines may require a further digging into the "knowledge process" which leads to the mathematical invariants. We may need to go back to conceptual experiences and specify them into novel mathematical constructions. This was done in the fruitful relation between mathematics and physics; it may require a further change in our mathematical tools when interacting with biology. Or, perhaps, we must be ready to modify even the perfect stability of the mathematical structures and allow the "ambiguities" and individualization (or context dependence) which are so strong in biological phenomena: dynamics may need to get not only into the structures and theorems of the mathematics of "dynamical systems", typically, but into the underlying concepts as well, following the distinction proposed here between "concepts" and "structures". Probably then, mathematics would become a new discipline or at least a broader understanding of it may be required, grounded on new conceptual experiences. A change comparable to the one which lead us from the Greek geometry of figures and the algebra of Arabs to infinitesimal analysis and to the geometry of spaces as Riemannian manifolds. In these cases, the phenomenal description radically changed, beginning with the very concepts of space and time involved; the intelligibility of movement, first, then of space and time themselves was enriched by (and lead to) entirely novel mathematical tools. But this is how the growth of (mathematical) knowledge goes: surely not by derivations from fixed sets of axioms, but as a mutually enriching interplay between us and the world, on continually changing and restructured phenomenal plans.

REFERENCES

1. Preliminary or revised versions of Longo's papers are downloadable at www.dmi.ens.fr/users/longo.
2. A. Asperti and G. Longo, *Categories, Types and Structures* (M.I.T. Press, 1991).
3. A. Badiou, "Platon et/ou Aristote/Leibniz: Théorie des ensembles et théorie des Topos sous l'oeil du philosophe," in [28].
4. F. Bailly, *Revue Internationale de Systémique* **5** (3) (1991), a special issue: *L'anneau des Disciplines.*
5. P. Benacerraf and H. Putnam, eds., *Philosophy of Mathematics: Selected Readings* (Prentice-Hall, 1964).
6. A. Berthoz, *Le Sens du Mouvement* (English translation by Odile Jacob: Harvard University Press, 1997).
7. L. Boi, *Le problème mathématique de l'espace* (Springer, 1995).
8. R. Carnap, "The logicist foundation of mathematics," 1931; reprinted in [5].
9. D. van Dalen and A. Troelstra, *Constructivism in Mathematics, Vols. I and II* (North-Holland, 1988).
10. S. Dehaene, *The Number Sense* (Oxford University Press, 1998).
11. D. Fowler, *The Mathematics of Plato's Academy: A New Reconstruction* (Oxford University Press, 1999).
12. G. Gardies, *Pascal entre Eudoxe et Cantor* (Vrin, 1884).
13. K. Gödel, "Russell's mathematical logic," in *The Philosophy of Bertrand Russell*, P. Schilpp, ed. (Open Court, 1944); reprinted in [5].
14. K. Gödel, "What is Cantor's continuum problem?," *Am. Math. Monthly* **54** (1947); reprinted in [5].
15. T. L. Heath, *The Thirteen Books of Euclid's Elements* (Cambridge University Press, 1908).
16. M. Hyland and A. Pitts, "The theory of constructions: Categorical semantics and topos theoretic models," *Categories in C.S. and Logic* (AMS Notes, Boulder, 1987).
17. E. Husserl, *L'origine de la Géométrie* (French translation, PUF, 1962).
18. P. Johnstone, *Topos Theory* (Academic, 1977).
19. G. Lakoff and R. Nunez, *Where Mathematics Comes From: How the Embodied Mathematics Creates Mathematics* (Basic Books, 2000).
20. J. Lambek and P. J. Scott, *Introduction to Higher Order Categorical Logic* (Cambridge University Press, 1986).
21. G. Longo, "Géométrie, mouvement, espace: Cognition et mathématiques," *Intellectica* **2** (25) (1997).

22. G. Longo, "Mathematics and the biological phenomena," in *Proceedings, International Symposium on Foundations in Mathematics and Biology: Problems, Prospects, Interactions* (Pontifical Lateran University, Vatican City, November, 1998), to appear.
23. G. Longo, "The mathematical continuum, from intuition to logic," in *Naturalizing Phenomenology: Issues in Comtemporary Phenomenology and Cognitive Sciences*, J. Petitot et al., eds. (Stanford University Press, 1999).
24. G. Longo, "Mathematical intelligence, infinity and machines: Beyond the Gödelitis," *J. Consciousness Studies* (special issue on cognition) **6**, 11-12 (1999).
25. G. Longo, "Mémoire et objectivité en mathématiques," in *Proceedings, Le réel en Mathématiques* (Cérisy, September, 1999), to appear.
26. G. Longo, "Prototype proofs in type theory," in *Mathematical Logic Quaterly* (formerly: *Z. Math. Logik Grundl. Math.*) **46** (3) (2000).
27. G. Longo and E. Moggi, "Constructive natural deduction and its omega-set interpretation," *Math. Struc. Comp. Sci.* **1** (2) (1991).
28. M. Panza and J.-M. Salanskis, *L'objectivité Mathématique* (Masson, 1995).
29. M. Piazza, *Intorno ai Numeri*, to appear.
30. J. Petitot, "Pour un platonisme transcendantal," in [28].
31. V. Rosenthal and Y.-M. Visetti, "Sens et temps de la Gestalt," preprint, LTM-ENS, Paris, 1999.
32. J.-M. Salanskis and H. Sinaceur, *Le Labyrinthe du Continu* (Springer, 1992).
33. A. S. Troelstra, *Metamathematical Investigation of Intuistionistic Logic and Mathematics* (LNM **344**) (Springer, 1973).
34. P. Zellini, *Breve Storia dell'Infinito*, 1980.
35. H. Weyl, *Das Kontinuum*, 1918.
36. H. Weyl, *Die heutige Erkenntnislage in der Mathematik* (Symposium, Berlin, 1926); French translation by Largeault in *Le continu* (Vrin, 1994).
37. H. Weyl, "Comments on Hilbert's second lecture on the foundations of mathematics," 1927, in J. van Heijenoort, *From Frege to Gödel*, 1967.
38. H. Weyl, *Philosophy of Mathematics and Natural Science* (Princeton University Press, 1949).
39. H. Weyl, "Axiomatic versus constructive procedures in mathematics," T. Tonietti, ed., *The Mathematical Intelligence* **7** (4) (Springer, 1985).

15

ON COMPLEXITY

Vincent Schächter

Hybrigenics
3-5 impasse Reille
75014 Paris, France

Under the generic label "complexity", this chapter includes 3 related contributions of different natures. The first one (C1) is a synthetic exposition by Vincent Schächter of James P. Crutchfield's attempt [1] at a general computational representation of the notion of complexity; the second one (C2) is an account of a CeSEF-session devoted to a general discussion of this attempt; the third contribution (C3) consists of a brief note by Vincent Schächter, on the construction of complexity measures, within the framework of the method of relativized conceptualization exposed in the second part of this volume. Considered globally, the sequence indicated above should convey an illustration of the way in which, accordingly to the aim announced in the general introduction, we try to extract from modern scientific representations, methodological essences to be progressively incorporated in the researched formalized epistemology. In the following, Crutchfield is denoted by Cr.

Key words: complexity measures, models of computation, discrete representations, abstract machines.

C1. COMPLEXITY, EMERGENCE, INNOVATION: A FORMALIZATION ATTEMPT BY J. R. CRUTCHFIELD: "HIERARCHICAL ϵ-MACHINE RECONSTRUCTION". SOME IMPORTANT ISSUES

The question that Crutchfield addresses is the following: how is it possible to identify in a given system the appearance of the so-called "emergence" phenomenon?

Prerequisites include a definition of the notion of "emergence", and a formal framework inside which that definition takes its meaning. In the article [1], Crutchfield tries to construct such a framework. He attempts to capture—through a series of successive abstractions—what constitutes, in his view, the essence of the process of modeling of a phenomenon by an observer. Indeed, as the notion of "novelty" is relative to an observer and to his frame of reference, it is necessary to include into the explanatory framework, a model of the observer and in particular of how he may detect that "novelty".

Given a physical process and an observer: "novelty" is inferred by the observer by comparing the information perceived about the process with an *internal model* constructed from past perceptions.

Crutchfield defines the category of models to which the observer will try to fit his observations as the set of abstract machines of the computational hierarchy. More specifically, the model that an observer will build from a process—perceived through some kind of sensory apparatus as a flux of symbols—*is defined* as the simplest abstract machine capable of simulating—that is, of reproducing—the process with the limited resources available to the observer (i.e., the machine cannot be too "large") and within a fixed error range. In this framework, one can define formally the complexity of a process as well as the conditions in which a new representation of the process by the observer will emerge. Crutchfield's approach draws its inspiration from recent work aimed at modeling the dynamics of so-called "complex system" via the computational capabilities that can be supported by that dynamics.[1]

C1.1. The General Framework

Crutchfield considers dynamical systems, represented as prototypical universes involving by definition an "environment" in which internal constituents of the system, called agents, interact with one another and with the environment. Each agent attempts to survive, by constantly increasing his capacity to predict the behavior of its surroundings: an agent is a stochastic dynamical system that tries to build and maintain a maximally predictive internal model of its environment. In this context, agents are anthropomorphic in nature: no distinction is made between, e.g., a [(human agent)-(human agent)] interaction, or a [(human agent)-(environment)] interaction, or an interaction between two "non-conscious" components of the system. Crutchfield adopts in this a "computational view of nature", by which finality is abstracted away. Agents are endowed with sensory apparatuses, i.e., filters

[1]This is sometimes summarized by the following ambiguous formulation: the physical system "computes".

through which they receive information from their environment. Consider an agent observing a process, i.e., a part of its environment that it is trying to model, for example another agent:

Reduction of a process to a stream of symbols

> The process can be viewed as a dynamical system, possibly with a stochastic component. A measurement accuracy constant ϵ is set. The state-space s—of dimension M—is partitioned in cells c_i of volume ϵ^M, yielding k cells ($k \approx \epsilon^{-m}$) that are labeled by k symbols s_n. A measurement is carried every τ time units (τ can be set to 1). The sequence of symbols is obtained by indexing measurement results with dates of measurement. The process is thus abstracted as a stream of symbols.

The environment is thus perceived as a stream of symbols

$$\{s_0, \dots, s_i, \dots\}$$

from a finite alphabet, resulting from measures of the sensory apparatus indexed by discrete time. The agent relies on a predetermined set of resources that define and constrain the type of models that it may construct. The modeling problem can thus be reexpressed as follows: how should an agent proceed to build the best model—i.e., the most accurately predictive model—of a stream of symbols, given its limited set of resources such as memory or computational power. The *complexity* of the process producing the stream will be identified with the complexity of the agent's model of the process. In this context, measuring the *complexity* of the observed phenomenon will be seen both as a subgoal of the modeling task, and as a way to guide that task: the agent will monitor the complexity of its internal model and force a change in representation if that complexity threatens to overcome its available resources, the so-called *emergence* phenomenon. Now, defining and measuring complexity, identifying emergence, presuppose definitions of what is called pattern, order, etc. The "objectivity" of such definitions is hindered by the "subjective" features that mark any distinction between data and noise as well as any distinction between pertinent events or features and accidental ones. This brings about a "modelling dilemma", just as in the case of descriptional relativities if no method of representation and control is specified.

It is in order to escape the "modelling dilemma" that Cr wants to build a "computational mechanics" able to take us, "beyond statistics, to pattern", to structure. In a sense, this is a quest for an absolute structure, a structure in-itself: the whole question of realism is there, skin-deep. It is at this point that Cr's first fundamental idea comes in: to use computational concepts, designed to help define and measure the complexity of a structure for processing information, as referred to the definition and measure of the structure "contained" in natural physical evolutions.

First, the "universe of possible models" must be specified, as well as the nature of the agent's resources: together, these will constitute the reference relative to which the observer will measure the complexity or "novelty" of a process.

C1.2. The Computational Hierarchy

An "abstract machine" is a formal mechanism characterized by a set of internal states and capable of producing strings of symbols through the application of a finite set of elementary transition rules from abstract state to abstract state. The language associated to an abstract machine M is the set of all words (strings) that M can produce. Abstract machines are traditionally distributed in classes according to their state space, their set of transition rules, and their use of time and memory space. Within a given class, an infinite number of different machines can be specified. Roughly speaking, describing a specific machine amounts to giving a "program" in addition to the "generic" mechanism typical of the class. A class A of abstract machines is more expressive than a class B if, for every language L recognizable by a machine in B, there exists a machine in A that recognizes L. The set of classes partially ordered by this relation is referred to as the "computational hierarchy". Turing machines are situated at the top of this hierarchy, while finite-state automata are situated at the bottom.

We can now specify the nature of the models that an agent constructs from its environment in Crutchfield's approach: *an internal model is, by definition, a specific abstract machine.* What remains to be defined is the set of criteria for choosing a "good" model: how well the model fits the perception of the environment, and how complex it is.

C1.3. Complexity Measures

Kolmogorov complexity

One fairly natural definition of the complexity of a string s is: the minimal amount of information necessary to effectively reconstruct s, i.e., to reconstruct s following a sequence of unambiguous elementary operations given in a certain formalism. It has been shown that such a quantity is actually independent (when s grows arbitrarily long, and if one neglects a constant factor) from the specific mechanism chosen to carry these operations.[2] This

[2]Intuitively, this is related to the fact that there is an "upper bound" to the expressive-

legitimates the following definition:

The Kolmogorov complexity of a string s is the size of the smallest program which, given as input to a Universal Turing Machine (UTM), will produce s as its output.

$$K(s) = \|M_{min}(s|UTM)\|,$$

where $M_{min}(x|V)$ stands for the minimal representation of object x using vocabulary V, in the present case that of programs for UTMs.

For instance, a string exhibiting simple regularities—e.g., a periodic string—will be describable by a short program. The more complex the "underlying structure" of the string, the longer the minimal program for string reconstruction. At the limit, a string for which no deterministic law can be identified, for example a random string, will have a complexity close to its length, as the minimal program will just explicitely encode the string.[3]

In the case of strings s^L (L being the length of the string) produced by an information source (in the information-theoretic sense) of entropy h_μ, we have

$$\frac{K(s^L)}{L} \underset{L \to +\infty}{\longrightarrow} h_\mu.$$

One can thus say informally (and somewhat abusively) that K(x) measures the "randomness" of the object x.

Statistical complexity

Strings resulting from sequences of coin flip results cannot be "compacted" using the above definition. Such strings, however, obey a very simple law, albeit one that is statistical in nature. It is however possible to capture the *probabilistic* structure of the string in the following sense: the structure of a string s is described perfectly if one knows how to generate a string s' which, while not identical on a symbol-per-symbol basis, is identical according to the definition below:

> **(R)** For every pair (a,b) of substrings of s, the probability that an occurrence of b follows an occurrence of a is identical in s and s'

---

ness of imaginable computational mechanisms, and that any Universal Turing Machine can compute all that we intuitively think of as being covered by the concept of "algorithm", and thus simulate any other computational mechanism (the Church-Turing thesis).

[3]Actually, one of Kolmogorov's main motivations for introducing this complexity measure was to attempt at a formal definition of the notion of "random string". That attempt failed.

This definition takes on meaning in the case of infinite strings of which only a finite "sample" part is known: the comparison criteria between two such infinite strings can only be probabilistic. With the weaker equivalence relationship (R), all infinite sequences of coin flips are equivalent: it is sufficient to provide their probabilistic law (the description of the information source, in information-theoretic parlance) to completely characterize these sequences, thus making them *primitives* of the description language.

A complexity measure corresponding to this statistical view must obey the "boundary condition" that the complexity of an ideal random string is null, in addition to the condition that the complexity of a string vanishes as its "regularity" increases. Formally, Crutchfield introduces the class of Bernoulli-Turing Machines (BTM): Turing machines with an additional "stochastic" instruction, i.e., which randomly produce a 0 or a 1. The *statistical complexity* of the string can then be defined as:

$$C_\mu(x) = \|M_{min}(x|BTM)\|$$

If s^L represents the first L symbols of x, the relationship between the two complexity measures is given by:

$$K(s^L) \approx C_\mu(s^L) + h_\mu L,$$

where h_μ is the entropy of the information source.

When string length increases to infinity, Kolmogorov complexity can be seen as a measure of the "stochasticity" of the string, and grows like its (information-theoretic) entropy, while statistical complexity measures the quantity of information necessary to predict the "deterministic part" of the string, stochastic aspects excluded.

$C_\mu(x)$ can also be interpreted as the minimal amount of historical information necessary to optimally predict bits of x within an h_μ error range: in other words, as the size of the minimal model yielding an optimal prediction. As the agent attempts to model infinite strings, the equivalence defined above can be used to compare the observed stream of symbols to the string produced by an internal model. Likewise, the statistical complexity measure defined above can be used to measure the size of these internal models.

C1.4. Building Models

How can the structure "hidden" within a stream of symbols be discovered and captured inside a finite model ? First, the fundamental hypothesis underlying such an endeavor is that this structure reflects more fundamental laws that govern the dynamics of the system perceived through the symbol stream. The aim being to simulate/predict that stream, the focus is on "causal structure", i.e., which "causes" produce which "effects". Specifically, the agent's aim is to identify the minimal set of "hidden states" $(S_0, S_1, \ldots, S_n) = \mathbf{S}$ such that knowledge of the state S_i of the system at a given point in time will be sufficient to (optimally) predict its evolution. In this context, "optimally" means that any historical information on the system in addition to knowledge of its state S_i would be redundant, i.e., would not improve the prediction.

Again, the only information available to the agent on the phenomenon to be modeled is the symbol stream. The search for "cause" implies "effect" regularities and "effects" can only be specific strings of symbols. To every instant t, one tries to associate the *causal state* S_i of the system at that instant. In the context of discrete time series, a causal state is defined to be the set of subsequences that render the future conditionally independent of the past, i.e., as an equivalence class of future predictability. Times t and t' are thus identified as corresponding to the same state as follows:

$$(\sim)\ t \sim t'\ if\ and\ only\ if\ Pr(\overrightarrow{s} \mid \overleftarrow{s_t}) = Pr(\overrightarrow{s} \mid \overleftarrow{s_{t'}}),$$

where $Pr(\overrightarrow{s} \mid \overleftarrow{s_t})$ is the probability distribution on semi-infinite symbol sequences $\overrightarrow{s}$ beginning at time t (i.e., possible futures), given that sequence $\overleftarrow{s_t}$ has been observed before time t. It is worth emphasizing that the whole point of modeling is that there are less states than instants.

The equivalence relation R defined earlier has now been recovered, as two strings are equivalent with respect to R if and only if they can be identified at every time point using $\sim$. Thus, the agent does capture the finest structure possible within the limits of its perceptive powers. The set $\mathbf{S}$ of causal states S_i can thus be understood as the most compact way to characterize elements of the R-equivalence class of string s. When $\mathbf{S}$ is determined, the function $T : \mathbf{S} \to \mathbf{S}$ defining the transition rule from state to state is also known. The pair ($\mathbf{S}$, T) is called an ϵ-machine by Crutchfield. Note that the transition rule may be probabilistic, in which case T can be described as a transition matrix.

This leads to the relationship between ϵ-machine and the statistical complexity measure: C_μ measures the "size" of the reconstructed machine. Specifically, if p_S is the asymptotic probability distribution of causal states (which is also the left eigenvector of matrix T), we have:

$$C_\mu = H(p_S)$$

where H is the average of self-information $-log_2(Pr(S_i))$ over $Pr(S_i)$.

Practically speaking, ϵ-machine reconstruction is based on a *finite* sample s_f of the input data stream. One option is to compute the frequencies of the various strings s^N of length N following a specific string, also of length N: this yields a frequency matrix of size at most $(2^N)^2$. Identical rows, corresponding to strings that have the same "past-conditioned future distribution" and therefore finish in the same causal state, can then be identified. A minimal finite automaton M can thus be constructed that reproduces the stream with maximum accuracy given the authorized "depth of vision" N towards the past and the future. Whenever the set **S** remains identical as N increases, one may claim that the "structure" of the stream has been captured.

In principle, this approach can be generalized from the reconstruction of a finite state automaton from a stream to the reconstruction of representation (an abstract machine) at a given level of the computational hierarchy from a representation at a lower level. The idea is again to regroup states within equivalence classes, the equivalence relation capturing those characteristics that are identified at the respective level of abstraction. From a practical standpoint, there doesn"t seem to exist at the moment a general and effective method to progress from one level of the hierarchy to the next. The "right" abstraction has to be chosen from the outset, and we are not aware of any existing method to automate that choice. It is finally possible to describe the behavior of the modeling agent, and to introduce the notion of "emergence".

Hierarchical ϵ-machine reconstruction

The agent is equipped with a set of tools—abtract machines from the computational hierarchy—with the help of which it attempts to reconstruct a stream of symbols as close as possible to the stream observed through its sensory apparatus. The agent is also endowed with a finite memory. It focuses on a given class of machines (e.g., finite-state automata at the beginning of the process), decides on an "observation depth" N, and then reconstructs the corresponding ϵ-machine. The model can then be improved within the same class by:

- adjusting the value of N: if the reconstructed ϵ-machine is stable above a certain threshold N_0, then the structure of the stream has been extracted, and the process has been described up to error level ϵ;
- adjusting the value of ϵ, i.e., increasing the accuracy of measurements.

If the procedure stabilizes on a given machine, the agent keeps that machine as an "exact" model of the observed process. If, however, the size of the reconstructed model increases as $\epsilon \to 0$ or $N \to +\infty$ (practically speaking, if the size of the model threatens to grow beyond the finite information storage capacity of the agent), the agent is forced to "innovate". It jumps to the next level of the computational hierarchy and seeks a representation of the stream at that level, trying to identify equivalence-classes in lower-level abstract-machine descriptions. The procedure iterates until stability is achieved.

The resulting ϵ-machine is thus a *minimal model at the least computationally powerful level yielding a finite description of the observed process.*

Crutchfield calls *emergence of a new representation* the transition, driven by resource limitations, to a higher level in the computational hierarchy.

C1.5. Conclusion

Cr considers evolving, dynamical systems, each one represented by a "prototypical universe" involving by definition (containing inside itself) an "environment" as well as "agents" (i.e., internal constituents of the system which interact with one another and with the environment). Each agent is supposed to try to survive and to optimize his survival, by constantly increasing his capacity to predict as well as his "functionalities", on both of which he tends to "capitalize". An evolving system of this sort undergoes complexification (of its structure) by emergence of order *or of disorder.* The aim of the observer is to *measure* the complexity of the system in each stage of its evolution—which might require "innovation" from the part of the observer—and also to identify *what* "emergences" (in the environment, in the interactions between the agents and the environment, and in the interactions between agents) have brought forth modifications in the previous complexity of the system.

It then is decided to suppress the initially fundamental distinction between the external human observer and the agents (human *or not*) and to speak only of "agents" in a generic sense. This suppression entails that any agent, whatever his nature, no matter whether a conscious entity *or not*, handles "information" and "calculates" on it. This amounts to designating in a henceforth *uniformly anthropomorphic way*, indistinctly, either a [(human agent)-(human agent)] interaction, or a [(human agent)-(environment)] interaction, or an interaction between two non conscious components of the system: this is the "computational view of nature" criticized in C2, by which finality is abstracted away.

Now, defining and measuring complexity, identifying emergence, presuppose definitions of what is called *pattern, order*, etc. The "objectivity" of such definitions is hindered by the "subjective" features that mark any distinction between *data* and *noise* as well as any distinction between *pertinent* events or features and *accidental* ones. This generates the "modelling dilemma" (just *descriptional relativities*; cf. Mugur-Schächter in this volume). In any case—according to Cr—there "are" (in some absolute sense!) both, statistical features, i.e., noise, and pattern, i.e., order. And the criteria for distinguishing them from one another are lacking. *It is in order to escape the "modelling dilemma" that Cr wants to build a "computational mechanic" able to take us, "beyond statistics, to pattern"*, to structure: a quest of an absolute structure, a structure in-itself, not only a structured intrinsic model in the sense of D19.2. The whole question of realism is there, skin-deep. It is at this point that Cr's first fundamental idea comes in:

> In order to define and to measure the structure "contained" in *natural physical evolutions*, to adapt recent concepts and procedures from the theory of computation, which *there* have permitted to define and measure the complexity of a structure for *processing information.*

In the theory of computation the object to be studied consists of a language L containing words written with the help of some alphabet of signs. The problem to be solved is to recognize whether yes or not a given word belongs to a given language. What has to be measured is the complexity of `the recognition-algorithm`, and this complexity depends on the machine made use of. A measure of this sort, for instance, is the "Kolmogorov-Chaitin algorithmic complexity", also called "deterministic complexity", defined as follows.

$[K(s^L)]_{MTU} = ||M_{min}(s^L)/UTM|| =$ [the number of bits from the shortest program which, if performed by a *deterministic* Universal Turing Machine UTM, produces the sequence of signs s^L of length L]

It can be shown that

$$Lim.L \to \infty[K(s^L)]_{MTU}/L = h_\mu,$$

where h_μ designates the informational entropy of the utilized information-source. This expresses that the mean length per sign in the program, possesses a `non-null` *lower bound* "information-source" that emits each sign σ_I with some definite *probability* $p(\sigma_I)$, the probability law $\{p(\sigma_I), i = 1, 2 \ldots n\}$ being an essential feature of the definition of that source. So, the source being probabilistic, in a long sequence of signs, the number of occurrences of a given sign cannot be predicted, notwithstanding that both the program and the functioning of the machine are *deterministic*. In other terms, the Kolmogorov-Chaitin algorithmic complexity does not concern the "deterministic bits", i.e., the individual messages intentionally sent with the help of the considered information-source, *it concerns the activity of the source as a whole* (cf. note 3); and the informational entropy h_μ of the source measures "the rate of error *in the* `predictability` of the successive signs", which is non null. (For chaotic systems depending on continuous parameters which are re-expressed in terms of discrete symbols with e-"granulation", one obtains $Lim.(L \to \infty, \epsilon \to 0)[K(s^L)]_{MTU}/L \cong h_\mu L)$.

Cr adapts this approach to the problem of defining and measuring "the degree of structure contained in a *natural* process". In order to achieve this, he first endows himself with a "computational hierarchy" (see in 1) and then makes a fundamental remark, namely that, besides "pure" randomness, there exists also "probabilistic randomness" which, on the probabilistic level of observation, is legal, ruled, expressible by probabilistic *laws*. To these Cr assigns explicitly a definite place in the "computational hierarchy": he introduces the concept of *probabilistic computation*, with respect to which the Turing-Bernoulli Machines TBM play a role equivalent to that played by the deterministic Universal Turing Machines UTM with respect to *deterministic computation*: a Turing-Bernoulli Machine TBM includes by definition a (random) information-source, but on the other hand it also admits of programs of *probabilistic* recognition. On this basis Cr defines a *statistical complexity* $[C_\mu(s^L)]_{TBM}$ (where C reads Cr) as follows.

$[C_\mu(s^L)]_{TBM} = ||M_{min}(s^L)/TBM|| =$ [the number of bits from the shortest program which, if performed by a Turing-Bernoulli Machine TBM, produces the sequence of signs s^L of length L] $\cong [K(s^L)]_{MTU} - h_\mu L$

So the Cr complexity measure $[C_\mu(s^L)]_{TBM}$ estimates the minimal quantity of information (of bits) necessary in order to make *optimal* probabilistic predictions concerning the signs from a sequence of signs s^L emitted by an information-source, i.e., in conditions in which there exists a non suppressible lower bound h_μ of the rate of error in such predictions. Then, making use of the two concepts introduced by him, of a computational hierarchy and of statistical complexity $[C_\mu(s^L)]_{TBM}$, Cr defines his method for "modelling by ϵ-machine reconstruction" (cf. the contribution C1).

C2. INTRODUCTION

The discussion reproduced below consists of two parts. The part C2.1 is an exchange of opinions. The part C2.2 is a summary of the conclusions that can be drawn from C2.1.

C2.1. Exchange of Opinions

Francis Bailly. The concept of "emergence" is related to a change of the considered level of organization. When this occurs "fulguration" takes place, that is, breaking of certain awaited symmetries which were specific of the preceding spacetime scale. *Cr's ϵ-machines reconstruction presents analogies with the renormalization procedures.* I would say that Cr's method is a sort of attempt at *a "legalization" of data-fluxes without the use of* conceptual, intellectual principles, or in other terms, in the absence of any condition of coherence with the context. But is this really possible?

Furthermore, in the usual modelizations, semantics involves an arrow which is absent in Cr's approach: what is lacking is a representation of the finalistic features, of a theory of finality.

Mioara Mugur-Schächter. I am particularly interested in the remark concerning an analogy with the methods of renormalization. According to me, we are here in presence of a remarkable manifestation of the principle of separation PS in the sense defined in my contribution to Part II of this volume. Indeed I hold that the renormalization procedures from modern Physics compensate for a violation of the principle of separation when a passage from an initial level of conceptualization to another level has occurred *surreptitiously*, i.e., without a corresponding explicit change of the involved epistemic referential.

Hervé Barreau. It seems quite improbable, and naïve, to suppose that computers might ever work like humans: they will always lack a dimension where are inscribed the aims and the decision.

Michel Bitbol. By the anthropomorphic way in which the "agents" are characterized in it, Cr's approach reminds of the thinkers from the preceding centuries who were introducing everywhere demons and homunculi.

Paul Bourgine. Cr's approach suggests that it would be interesting to specify clearly the definitions accepted for the concepts of deduction, induction, eduction, abduction, and the relations between these concepts. The insects, reptiles, etc. manifest a sort of primary consciousness in that they seem to simulate beforehand in their "minds" what should happen, what their action should encounter, which triggers their action; and if what they awaited is not confirmed, this expresses—for them—a sort of emergence, which brings forth a defensive attitude or action.

Giuseppe Longo. The methods of formal computation have been generayed by logic, not by physics. So probably Cr's computational modelization is not the most appropriate for natural phenomena. *How could one estimate the value, the degree of appropriation of a way of modelizing?* What we should understand in order to get nearer to a possibility of estimating the value of a computational method, is how mathematics emerged.

There exists an abstract complexity, independent of any machine, and a concrete complexity which depends upon a machine. Both can be (variously) "measured". But they cannot be identified to one another.

Michel Paty. Cr, in sum, seems to presuppose, or would like to reach the conclusion that *a natural physical phenomenon somehow, by an appropriate dynamics, can generate its own representation.* But such an assumption or project seems to me immensely nave. It involves a false naturalism, an adulterated "objectivity". It reminds of induction where one would like to reconstruct the whole by small steps. But this impoverishes the whole. In this sense Cr's approach involves a sort of reductionism. A human modelization goes on in a radically different way. One begins by imagining something material, a designatum, a whole of a semantic nature. But then, while the analytic understanding of this thing progresses, the initially imagined semantic whole becomes more and more syntactical. For instance, initially the ether has been conceived by Maxwell as an invisible material medium which supports the fields of forces. But later it got identified to the electromagnetic fields, and finally to the equations which express these fields : it dissolved in syntax. *Without occurrence of some reduction.*

Jean Petitot. In Cr's approach the semantic gets lost, or at least it remains hidden. Compared to the approach practised in mathematical physics, one finds oneself in presence of another way of representing, namely by the "computational price" of the reproduction of a given data-flux, by a given computing-machine. So there comes into being a *bi-modality* of modelization, a computational modality where any semantic is occulted, and a modality of the type of that practised in physics which integrates explicitly the semantic features, in the generated formal structures. Then *the question is to understand what* `relations` *exist between these two distinct modalities.* For instance, is it possible to reconstruct by Cr's method the movement of a planet? One has the feeling that the spatial structure is irremediably lost, while in physics it is essential. And in this context ons should remember Putnam's argument: "any program can be implemented in a dynamical system ; it suffices to dissect the system in a convenient way". I add the following remarks concerning "emergence". Cr Speaks of "intrinsic emergence". he seems to conceive it to be "objective". So he supposes the possibility of a sort of objectivity of representation, a sort of naturalization of representation. But what could that mean, exactly? According to Cr Nature itself computes. He seems to practise a sort of computational realism, a tendency to "legalize" not only the physical reality itself any more, but also the calculus. But what could mean a "computation" that would be "intrinsic" to the physical processes? *A man-made machine which reconstructs a data-*

flux is just a model of a human-like process of treatment of information. So there must exist a connection between this sort of model, and the model involved in physics, but which connection? For instance, does there exist a hierarchy of the physical models? And if there does, in what relation to Cr's computational hierarchy does it stay?

According to the nowadays cognitivists *space is constitutive of objectivity.* The data registered in the brain bear certain marks of objectivity, of possibility of intersubjective consensus, and only space permits to reconstruct explicitly this objectivity by starting from the registered marks. When an "emergence" occurs, according to the physical models there always is, in a substrate, breaking of a symmetry presupposed by the observer, and this always involves some *spatial* structure (more or less explicitly and more or less combined with other structural elements, temporal or abstract). Then what are the connections between the breaking of symmetries (emergence) in the sense of physics, and computational emergence in Cr's sense? In other terms, and in short, what is the computational status of space?

Finally, concerning Longo's remark, I ask: why would certain languages be "material", in contradistinction to other metalanguages that are abstract? *What* is said with help of a language is never material. The theory of algorithm is different from the theory of the device which accomplishes the algorithm, and from its functionality. Semantics involves a non-material finality, an aim-arrow of the algorithms. But for the process of embryogenesis itself there is no distinction between material and abstract. We should examine what happens in mathematics: there one finds at the same time semantics, syntax, and algorithms.

Robert Vallée. In connection with Paty's remark on Cr's approach: Naïveté is often very fertile. It would be an error to suppress it. It would be appropriate to compose a "praise of naiveté".

Vincent Schächter. What meaning could be assigned to the assertion that "nature itself computes"? What does "intrinsic emergence" mean? Could "computational view" be understood to mean a universal view, efficacious with respect to any "law" whatever the nature of the concerned facts?

For Cr to "model" means to be able to reproduce. *Semantic is eliminated from the process of modelling.* This is a revolution in the concept of modelling. *Correlatively, Cr's definition of complexity and of its measure concern specifically* this *sort of process of modelling, by direct reproduction of fluxes of data, semantic being court-circuited.*

Mioara Mugur-Schächter. In Cr's approach the object to be studied consists of a flux of data obtained by measurements performed upon an "environment". This environment is characterized by parameters which in general are continuous, but because any measurement introduces finite units, the flux of data generated by the measurements is an "ϵ-approximate" discrete sequence of signs. Keeping Cr's notation such a discrete sequence of signs is denoted s^L if its length is L. This is what has to be studied. Let furthermore M_ϵ denote the considered ϵ-machine, a Turing- Bernoulli machine TBM that incorporates an *information-source*; and let $R(M_\epsilon)$ be a *reconstruction-operator* representing an operation of reconstruction of s^L by M_ϵ. Now, when closely examined, the whole essence of Cr's method of "modelling by ϵ-machine reconstruction" appears to be expressible synthetically by the equation

$$R(M_\epsilon)s^L = C_\mu(s^L)s^L,$$

where $C_\mu(s^L)$ is a real number, namely Cr's "statistical" complexity measure of the sequence s^L as defined in C1:

$$[C_\mu(s^L)]_{TBM} = ||M_{min}(s^L)/TBM|| \cong [K(s^L)]_{MTU} - h_\mu L.$$

(This, in fact, is a *probabilistic* complexity measure, not a "statistical" one, since the statistics to which Cr refers is quite essentially supposed to be probabilistically stable, i.e., to consist of relative frequencies of realization of the signs from s^L which, at the limit of big numbers, obey a limiting *convergence law*). Now, the equation $R(M_\epsilon)s^L = C_\mu(s^L)s^L$ is an equation for the calculation of the eigen-sequences s^L and the corresponding eigen-values $C_\mu(s^L)$ of the reconstruction-operator $R(M_\epsilon)$. But in *any* equation

$$R(EE_J) = V_J(EE_J)$$

for calculating eigen-elements EE_J and the corresponding eigen-values V_J of an operator R, this operator is a "re-construction" operator. Indeed the operator from $R(EE_J) = V_J\ (EE_J)$ is *such* that when it works upon an eigen-element EE_J it re-constructs, it re-produces this same eigen-element, *and* multiplied with a numerical constant V_J which measures a characteristic of the eigen-element EE_J, by a number from the "spectrum" of R, which, as a whole, is part of the definition of R (an operator R of this kind can be considered to be a somewhat cryptic but quite pertinent mathematical representation of an aspect-view).

Now, given on the one hand the definition of the complexity measure $C_\mu(s^L)$, and on the other hand the general significance of an equation for eigen-elements and eigen-values of an operator R, the equation $R(M_\epsilon)s^L = C_\mu(s^L)s^L$ can be read as follows. When the reconstruction-operator $R(M_\epsilon)$ involved by the ϵ-machine M_ϵ works upon the sequence of data s^L, it reproduces this sequence s^L by spending a quantity of resources and work which, by definition, is measured by the "statistical" complexity $C_\mu(s^L)$ of s^L. So:

> *The statistical complexity* "of s^L " *is measured by the quantity of work spent* by the ϵ-machine M_ϵ *to re-construct* s^L.

Expressed in this way, the essential relativity of Cr's method, to a "machine-reconstruction-view" $R(M_\epsilon)$ is obvious, like also the relativity to the considered object-entity s^L. These two basic descriptional relativities are there. If they are removed the whole approach dissolves. Of course, like in practically any nowadays approach, the relativity to the operation of *generation* of the considered object-entity remains entirely unexpressed. But this third fundamental descriptional relativity *also* is there, quite essentially: who, what produces the sequence s^L? Nature? The machine M_ϵ alone? Neither nature nor a Turing-Bernoulli machine M_ϵ could ever *devise measurements*, nor can they establish by themselves what mark imprinted by an "environment" upon the machine, has to be interpreted in terms of *what* result of measurement. For this, *semantic and coding have to be brought in*, by man, purposefully.

Throughout Cr's work one can feel an aching need of explicating the involved descriptional relativities: a systematic translation of the whole work, in relativized terms, would be an illuminating exercise. To form an intuitive background, one should read E. Morin's works [2].

The fact that Cr's approach can be synthetically expressed in terms of an equation for eigen-elements and corresponding eigen-values of a reconstruction operator, while on the other hand it deals with sequences of signs produced by an *information*-source, is very interesting from the point of view of a formalized epistemology. Indeed *both* the quantum theory of measurements and the information theory, deal with transmissions of "information", i.e., with sequences of signs that involve only probabilistically-stable observable features. But the mathematical representations of these two theories are very different from one another. A unification of these representations should be possible and it would increase the efficiency of the theory of information, while it would throw light upon the stubborn quantum mechanical pseudo-problem of hidden parameters ; furthermore, together with the quantum mechanical formalism and the informational one, such a unification would be a guide for a mathematical representation of the unified *general* logico-probabilistic structure drawn from quantum mechanics (cf. Chap. 7 in this volume): *Cr's complexity measure—duly relativized—can measure the computational complexity of any branch-probability-measure from a probability tree.* This shows that the basic object-entity from a probability tree of a basic epistemic referential plays the role of an information source, while the involved basic view, with its branch-views, is equivalent to branch-channels of information.

C2.2. Summary of the Discussion

Mioara Mugur-Schächter. I shall try to summarize the conclusions that can be drawn from our exchange of opinions.

On the one hand:

Cr's approach does not produce a representation that exhausts the questions of "complexity and of "emergence such as they are spontaneously perceived. This is so—mainly—for the following two reasons.

* *It loses explicit semantic significance.* In the first place because it occults the spatial structures. But also, more generally, because it does not incorporate certain *principles of hierarchical coherence* which in the natural human conceptualization do constantly act. These, in the natural processes of conceptualization, connect the absorbed semantic contents, to the syntactic structure drawn from these, and to the algorithms associated with these structures. In this way there appears a conceptual whole—a "conception", a "theory"—involving three explicit levels, a semantic level, a syntactic level, and an algorithmic level. These levels are essentially different from one another, but (more or less explicitly) they communicate by a sort of mutual convertibility. While in Cr's approach the basic semantic level remains implicit and it seems to be assumed not even to exist.

* *Cr's approach looses also "final significance"*, in the following sense. The "reconstruction-operations" of Cr's machines is devoid of also another—quite fundamental—element, namely of *an explicit arrow of aims*, of an explicit finality. This is correlative to the absence of an explicit semantic. While in the human cognitive actions the arrow of the aims plays such a basic and constant role: it imprints upon what we call "natural laws", the marks of the human pragmatic thinking, which orients the curiosities (interesting-uninteresting), the feeling of "understanding", the impression of "coherence", and even that of beauty. The human conceptual wholes, the views, conceptions, theories, are crystallized under the pressure of strong fluxes of finality which act throughout the process of elaboration and then remain incorporated in the result by a sort of geometrization which, in a coded form, renders them perennial. But in Cr's approach these fluxes of finality are occulted away.

The concepts of complexity, of innovation, of emergence—in the sense of Cr—are flawed by a certain lack of generality: in the contexts of usual language, and even in the exact sciences, the just enumerated words possess also—mainly, I would say—*other* acceptations than those assigned to them by Cr's definitions, they involve relations to the human mind. An emergence is a breaking of a symmetry *expected by the conceptor-observer*, etc. Of course, any formalized representation introduces a certain impoverishing precision. But in this case the impoverishment is very important indeed.

On the other hand:

Cr introduces in the concept of modelization a certain revolution of which the content is worth while being thoroughly explicated. Indeed, even if Cr's concepts of complexity, innovation and emergence are suppressive, they suggest questions and formal analogies which, if developed, could come out to be fertile.

Furthermore—considered globally—Cr's approach raises at least one precise question, endowed with utmost importance from both a point of view of principle and a technical point of view:

What are the relations between natural human conceptualization and a computational modelization in the sense of Crutchfield?

A thorough answer to this question could act as a guide for reaching a genuine definition of the mind-machine relations as they now stand; and may be for transcending them, since possibly, what is clearly defined can be transcended.

C3. COMPLEXITY MEASURES VIEWED THROUGH THE METHOD FOR RELATIVIZED CONCEPTUALIZATION

"Unfortunately, " complexity" has been used without qualification by so many authors, both scientific and non-scientific, that the term has almost been stripped of its meaning. Given this state of affairs, it is even more imperative to state clearly why one is defining a measure of complexity and what it is intended to capture."
—J.P. Crutchfield, in *Measures of Statistical Complexity: Why ?*

The aim of this short note is not to provide an overview nor even a classification of complexity measures, but rather to clearly identify the degrees of freedom and corresponding relativities that exist in the process of constructing such a measure, and to position them within the framework of the method for relativized conceptualization (MRC) described in Chap. 7 of this book [3].

We show that MRC can be used as a high-level conceptual guide for the comparison and design of complexity measures, one that is complementary and almost orthogonal to the existing approaches based on classifications of the mathematical techniques involved.

In the following, we will try to sketch the first steps in the construction of the two essential notions of complexity of a view and complexity of a relative description. Note that we will manipulate only finite abstract entities, as imposed by method MRC.

C3.1. Complexity of a View

The content of an aspect-view V_i can be denoted as a set of values $V_i = a_{1_i}, ..., a_{p_i}$. A fictive value $\perp$ is added, symbolizing inexistence relatively to the chosen aspect i, and the completed set is denoted $V_i^* = V_i \cup \perp$. The application of the view containing the aspect V_i to an object-entity $œ_G$ produced by a generator of object-entity G, yields either an element of V_i, or $\perp$ if the situation is one of relative inexistence ; in either case, the result is an element of V_i^*. The main purpose of this notation for relative inexistence is to be able to reason solely on the number of values, without having to check explicitely for relative existence. A view V consisting of n aspect-views will involve the corresponding sets $V_1^*, ..., V_n^*$. We can now define the complexity of the view V as the product of the cardinalities of all the sets Vi^* from V:

$$C_V = \prod_{i=1}^{n} (p_i + 1),$$

which is the number of different potential qualifications permitted by V, i.e., of all the different possible combinations of aspect-view values offered by V.

Indeed, this definition captures the essence of interest: it is not a function of the 'nature' of the view, but rather of its structure, which is reduced here to the number and cardinalities of its aspect-view components. The product was chosen (somewhat arbitrarily) as the 'aggregating function' because it ensures that the complexity of a view resulting from the union of two initially given views, is the product of the complexities of these.

C3.2. Formal Structure of Qualifications and Descriptions Result

We explore the construction of an hypothetical complexity measure assignable to qualifications and to descriptions by following the definitions of MRC. Consider an epistemic referential (G,V). Let us first focus on a qualification—or "instantaneous description"—produced by only one realization of the succession [G.V] defined as the application of the view V to the object-entity $œ_G$ produced by the generator G.

A space frame must be introduced if G is a physical entity (while for a non-physical entity, the spatial aspect may be omitted or is altogether irrelevant). Let S be a spatial grid, formally, a subset of Z^n—e.g. Z^3 for 3D space, more or of a higher dimensionality if the focus is on some variety of phase space, etc.

A qualification can then be seen formally as a function

$$q : G \to \prod V_i^*$$

that maps every point in the spatial grid to the aspect-values from V that are realized at this point. In other words, q is a pattern on the grid S.

A description is then defined as the result of N repetitions of the achievement of a qualification q by repetitions of the same succession [G.V], the clock being reset at each iteration. In other words, individual qualifications that constitute a description are carried out under identical conditions, "time included". Such a description can be dubbed "individual" or "statistical", depending on whether the results of its component qualifications are identical or not. In the latter case, the description is probabilistic if these results satisfy certain convergence criteria [3]. The resulting description can therefore be seen as a set of N qualification patterns.

One step further, an *evolution* of a given object-entity can be defined as a sequence of (similar) descriptions of this object-entity operated at points in time $t_1 < t_2 < t_3 ... < t_m$ placed at different and increasing time-distances from the generation of this object-entity. In other words, the clock here is not reset after each descriptional action, and one is interested in the temporal evolution of the resulting description. The situation is similar to that of a description, only with the additional "distinguished" dimension of time added to the spatial grid. Changes from one description to another thus take on a dynamic meaning.

C3.3. The Universe of Potential Complexity Measures

Without losing any generality, the choice of a complexity measure for a qualification, a description or a sequence of descriptions indexed by time can thus be construed respectively as the choice of a complexity measure for a (p+1)-colored pattern on a spatial grid S, for a set of N such patterns, and for an indexed sequence of pattern sets. The universe of possibilities for any one of these choices, obviously, is huge.

To convey a sense of the size of the landscape of potential complexity measures, we need only take a look at the variety of approaches that have been proposed to measure the "complexity" of dynamical systems: efforts to describe the degree of unpredictability of dynamical systems built on the first general measure of the uncertainty associated with the behavior of a probabilistic process—the Shannon entropy of the underlying distribution—to yield measures familiar to theoretical physicists such as the metric entropy; Lyapunov exponents and fractal dimensions, that quantify the degree of deterministic chaotic behavior in a system, etc. While these measures focus on the randomness of a system, they hardly concern themselves with the *structure* of the system, a wide and fairly vague but intuitive notion that points at the relationships between the components of the system. From a different scientific front, computation theory, came the idea of complexity measures defined as the quantity of information necessary for a certain type of abstract computational device to reproduce the object (typically, a string) whose complexity is being measured. The corresponding family of *deterministic complexity measures* [5,8,11] captures deterministic structure (i.e., correlation between system components) relatively to very general and abstract classes of structural templates: the abstract machine classes of the of the computational hierarchy [8,10]. They do not discount for randomness, however, in the sense that an ideal "perfectly random" system would show maximally high complexity. *Statistical complexity measures* were introduced to answer the intuitive need for measures which capture regularities "above and beyond randomness", i.e., such that both a perfectly random process and a perfectly ordered process—relatively to specified notions of randomness and order—would exhibit minimal complexity [4,8,9]. Among these statistical complexity measures is the Bernoulli-Turing machine variety of the ϵ-machine reconstruction scheme described in [1].

As the latter category of measures subsumes conceptually the first two categories, and emphasizes the "computable" nature of a complexity measure, it appears as a good candidate framework to structure at least part of the universe of potential complexity measures [8].

C3.4. Complexity Measures for Qualifications and Descriptions

We can now return to our main topic, the choice/construction of complexity measures for qualifications and descriptions within the framework of the MRC, and the degrees of freedom that exist for that choice.

To fit within the framework of MRC, a complexity measure must be effectively realizable, i.e., computable. Moreove, the measure must be dependant on the qualification or description that is being evaluated, that is, it must take this—a pattern or set of patterns—as input to a computation that shall yield the complexity score. That computation can range from the very simple (e.g., just counting "defined" aspect values on the grid) to the arbitrarily complex.

In the case of Crutchfield's "ϵ-machine" complexity measure, and supposing that the pattern has been linearized in a given order to yield an input stream, the complexity function results in the size of the smallest machine in the lowest class of the computational hierarchy that will reproduce the (linearized) pattern. The more general idea here is to use a measure of algorithmic complexity of an algorithm that reconstructs the spatial pattern, as "the complexity of the pattern", building on the fact that a complexity measure can be more or less complex in its definition, and *quantifiably* so.

In summary, a vast landscape of technical ideas, only sketched above, can be exploited to construct the complexity "function" or algorithm. The choice of a given function will correspond to a choice of balance between emphasis on "randomness" and emphasis on a certain notion of structure, itself *relative*.

Some of the intuitive requirements that are often associated with complexity measures can be expressed as boundary conditions. For example, as mentioned above, the complexity of constant patterns should be minimal, while the complexity of "random" patterns (according to a given definition of randomness) should be either maximal or minimal, depending on one's point of view.

Given a definition for the complexity of an individual pattern (a qualification), one could try to build the complexity measure for a probabilistic description (N patterns) in a compositional way, i.e., by combining the complexities of the individual patterns. Such a compositional measure would be very uninformative, however, as the detailed information on individual patterns cannot be directly used in the measure (probabilities are just countings), which will therefore not allow comparison between patterns...Using a different approach, if one starts from an individual pattern and extends the pattern set, building the complexity measure along the way, one probably wants "additional" complexity produced by multiple instances of the same pattern to be minimal, to increase as "differences between identifiable structures in the patterns" can be specified—again, a highly relative notion—and to be low again when there are no identifiable similarities between patterns. Moving from a pattern to a set of patterns thus adds a whole new dimension to the space of potential complexity measures, by forcing explicit choices on how to handle similarities and differences between patterns.

Finally, in the case of an evolution, the aim is to measure the complexity of a "movement" from pattern to pattern. Intuitively, the focus here could be on "small differences" reflecting the evolution of structural features that can be identified as "being the same" in several patterns, and thus ascribed an identity. Again, the complexity measure can be designed as an algorithmic complexity measure on a space-*time* grid reconstruction algorithm.

C3.5. Complexity Measures for Object-Entities?

An object-entity either is a precedingly achieved description, or it is a basic physical object-entity in the sense of MRC. The first case falls inside the categories discussed above. As for the second one, inside the method it follows that a basic physical object-entity is generated strictly non-qualified, so speaking of its complexity is devoid of significance (this, inside the method now, brings back to our initial remark concerning the illusory character of the concept of "structure- in-itself"). One of the major confusions that often appears in research concerning complexity is dissolved by recognizing that a basic physical object- entity possesses complexity only as involved in some description, and thus only relatively to some view.

C3.6. Conclusion

Many approaches to the question of complexity are flawed by gross false absolutes. Crutchfield's approach, on the contrary, is characterized by strong relativizing tendencies, but these are left more or less implicit. When immersed inside the method of relativized conceptualization, all the relativities involved by the complexity problem become explicit and stay under control. This opens up a field of research of explicitly relativized concepts of complexity and of the corresponding relativized complexity measures.

REFERENCES

1. J. P. Crutchfield, "The calculi of emergence: Computation, dynamics, and induction," *Physica D* **75**, 11-54 (1994).
2. E. Morin, *La Méthode, Vols. 1-4* (Le Seuil, 1977-1991).
3. M. Mugur-Schächter, "Quantum mechanics versus a method of relativized conceptualization," in this book.
4. C. H. Bennett, "How to define complexity in physics, and why," in *Complexity, Entropy and the Physics of Information*, W. H. Zurek, ed. (Addison-Wesley, Redwood City, CA, 1990), pp. 137-148.
5. G. J. Chaitin, "On the length of programs for computing finite binary sequences," *J. Assoc. Comp. Machinery* **13**, 547-569 (1966).
6. G. J. Chaitin, *Algorithmic Information Theory* (University Press, Cambridge, 1987).

7. J. P. Crutchfield and C. R. Shalizi, "Thermodynamic depth of causal states: Objective complexity via minimal representations," *Phys. Rev. E* **59**, 275-283 (1999).
8. D. P. Feldman and J. P. Crutchfield, "Measures of statistical complexity: Why?," *Phys. Lett. A* **238**, 244-252 (1998).
9. P. Grassberger, "Information and complexity measures in dynamical systems," in *Information Dynamics*, H. Atmanspacher and H. Scheingraber, eds. (Plenum, New York, 1991), pp. 15-33.
10. J. E. Hopcroft, and J. D. Ullman, *Introduction to Automata Theory, Languages, and Computation* (Addison-Wesley, Reading, MA, 1979).
11. M. Li and P. M. B. Vitanyi, *An Introduction to Kolmogorov Complexity and Its Applications* (Springer, New York, 1993).

BIOGRAPHICAL NOTES

Evelyne Andreewsky is Directeur de Recherche at the French Medical Research Institute INSERM. Formerly a computer scientist, she turned to Cognitive Science: Multidisciplinary Systematic Approaches to Cognition and Language. At INSERM she directed a research group on Cognitive Neuropsychology and Artificial Intelligence, working within the framework of System Science around a fundamental question: What does it mean to understand language?

Francis Bailly is a physicist at the Laboratory of Solid State Physics (LPSC-UVSQ), National Center for Scientific Research (CNRS), France.

Hervé Barreau is a philosopher and a member of the Académie Internationale de Philosophie des Sciences. He is a former director of research at CNRS-Strasbourg, where he established a research team in Epistemology and Philosophy of Science. He authored the books *Aristote et l'analyse du savoir* (1972), *L'épistemologie* (1990), and *Le temps* (1996).

Élie Bernard-Weil, after his specialization in Neurology and Endocrinology, was appointed Professor at the Collège Hospitalier Pitié-Salpêtrière. His interest and engagement in therapies of a new kind using bipolar strategies led him to integrate these methods into his subsequent biomathematical research (leading to a Ph.D.) and into epistemological considerations around System Science, which approach caused him to propose a translevel and transdisciplinary model for Biology and Human Sciences.

Michel Bitbol, born in 1954, was trained in Paris in Medicine, Physics, and Philosophy. He pursued research in biophysics from 1978 to 1990. Then, from 1990 on, he began a new career as a philosopher of modern physics. One of his basic concerns has been to think about quantum mechanics by taking contextuality as a principle rather than as a problem. He is presently CNRS Directeur de Recherche at CREA in Paris.

Giuseppe Longo, a former Professor of Mathematical Logic and Computer Science at the University of Pisa, is a CNRS Directeur de Recherche at the École Normale Supérieure in Paris. He is Editor-in-Chief of *Mathematical Structures in Computer Science*, a leading scientific journal of Cambridge University Press. Having worked on various applications of Mathematical Logic to Computer Science, he is currently interested in the mathematics of Cognition and the Epistemology of Mathematics; see his website *http://www.di.ens.fr/users/longo*

Mioara Mugur-Schächter is a physicist. While Professor of Theoretical Physics at the University of Reims, France, she founded and directed the Laboratory of Quantum Mechanics and Structures of Information. Her research is devoted to the foundations of quantum mechanics, the theory of probabilities, and the theory of information. She is the President of the Centre pour la Synthèse d'une Épistémologie Formalisée (Paris).

Michel Paty is Directeur de Recherche at the Centre National de la Recherche Scientifique (CNRS). Formerly a quantum particle physicist (Ph.D. on neutrino physics, 1965), he turned to Philosophy, Epistemology, and History of Science (Ph.D. on d'Alembert's philosophy of knowledge, 1977). He co-founded and directed for years a research group in this field at the CNRS and Université Paris 7-Denis Diderot. His main works are about contemporary physics and the enlightenment period; they include: *La matière dérobée* (1988), *Einstein philosophe* (1993), *Einstein, les quanta et le réel* (forthcoming), *La physique au xxè siècle* (2003), *L'intelligibilité du monde quantique* (forthcoming), and *Matière et concepts* (forthcoming).

Jean Petitot is Professor at the École des Hautes Études en Sciences Sociales (Paris) and Director of the CREA (École Polytechnique). His publications include *Morphogenèse du Sens* (1985), *Les Catastrophes de la parole: de Roman Jakobson à René Thom* (1985), *Physique du Sens* (1992), and more than 200 papers dealing with Epistemology and Cognitive Science. His e-mail address is: petitot@poly.polytechnique.fr

Vincent Schächter is currently Director of Bioinformatics Research at Hybrigenics, a Paris-based biotechnology company. He also co-founded and manages the international BioPathways Consortium. His research focuses on modeling, reconstruction, and analysis of cellular pathways through a variety of formal methods. Prior to joining Hybrigenics in 1998, his research at the Computer Science Department of the École Normale Supérieure focused on programming language theory, linear logic, and constraint-solving algorithms. He also served in the Scientific Department of the Embassy of France in Washington, D.C., analyzing and forecasting trends in the information technology industry. He is a graduate of École Normale Supérieure and holds a Ph.D. in Theoretical Computer Science from the Université de Paris.

Robert Vallée, born in 1922 in Poitiers, France, is an alumnus of École Polytechnique. He earned a Dr. ès Sciences (Mathematics) at the Université de Paris. In the interval 1961-1987 he was a Lecturer at École Polytechnique and occupied Professorships at the Universities of Besançon, Paris-Nord, and Paris I. Since 1987, he has been Director-General of the World Organisation of Systems and Cybernetics. He serves on the boards of international societies and journals and authored 150 articles as well as the monograph *Cognition et Système: Essai d'Epistémo-Praxéologie* (1995). He is a honorary fellow of WOSC, a Norbert Wiener Memorial Gold Medalist, Vice-President of the Cybernetics Academy Odobleja, and a recipient of a Dr. Honoris Causa from the University of Petrosani.

Alwyn van der Merwe, a graduate of the Universities of Amsterdam and Bern, is Professor of Physics at the University of Denver and Editor of *Foundations of Physics* as well as Founder and Editor of *Foundations of Physics Letters* and the international book series *Fundamental Theories of Physics.* He has coauthored or (co)edited numerous books covering a wide range of topics in Theoretical Physics and Philosophy of Science.

AUTHOR AND SUBJECT INDEX

9 781402 011207

GPSR Compliance

The European Union's (EU) General Product Safety Regulation (GPSR) is a set of rules that requires consumer products to be safe and our obligations to ensure this.

If you have any concerns about our products, you can contact us on ProductSafety@springernature.com

In case Publisher is established outside the EU, the EU authorized representative is:

Springer Nature Customer Service Center GmbH
Europaplatz 3
69115 Heidelberg, Germany

Batch number: 10584518

Printed by Printforce, the Netherlands